$3.90

REVIEWING SEQUENTIAL MATHEMATICS

COURSE I

Marilyn Occhiogrosso

Director of Mathematics
Amsco School Publications, Inc.

Former Assistant Principal, Mathematics
Erasmus Hall High School, New York City

Dedicated to serving our nation's youth

When ordering this book, please specify *either* **R 288 P** *or* REVIEWING SEQUENTIAL MATHEMATICS: COURSE I

AMSCO SCHOOL PUBLICATIONS, INC.
315 Hudson Street New York, N.Y. 10013

ISBN 0-87720-255-9

PRINTED IN THE UNITED STATES OF AMERICA

PREFACE

The traditional study of mathematics in American high schools has been through an in-depth concentration on individual branches of the subject. Hoping to give students a more meaningful appreciation of the nature of mathematics, educators are reexamining this traditional approach. Some advocate a sequential approach in which topics are integrated and certain unifying themes are introduced. In the sequential curriculum, the first two years of high school mathematics consist of a simultaneous study of algebra and geometry. Trigonometry is studied in the third year, along with algebra and geometry. Logic is a major unifying theme. Topics from probability, combinatorics, and statistics are also included.

The aim of *Reviewing Sequential Mathematics: Course I* is to provide a second look for teachers and students at the materials of Course I. The book can serve the following useful purposes:

- as preparation for a formal examination toward the end of Course I.
- as a study guide and a supplement to whatever text is used during Course I.
- as a review of Course I throughout Course II.
- as an aid to students transferring from a traditional program into the sequential course of study.

The elements of this book include:

- crisp summaries of principles and procedures highlighted in tables, in diagrams, and in handy lists.
- detailed model examples, with step-by-step solutions that are easy to follow.
- abundant batteries of exercises at the end of each section within a chapter and also at the end of each chapter. These exercises include questions drawn from actual Regents examinations dating back to the inception of this new course of study. In addition, cumulative summary exercises cover the topics of the first four chapters and then the first eight chapters. A final collection of recent Regents examinations covers the topics of all thirteen chapters.
- a complete, cross-referenced index.
- a separate comprehensive answer key.

The mathematics presented in this book:

- begins with a review of essentials studied before Course I.
- continues with all the materials required by the New York State syllabus, grouped for understanding of related materials.
- includes additional materials to help develop algebraic skills needed for future study.

Drawing on her many years of experience in teaching high school students and training teachers, the author has presented the materials in a format and language that should be easy for the reader to understand. By anticipating trouble spots wherever possible, she has tried to smooth the way for the student.

The author wishes to thank Lance Modell, West Hempstead High School, New York; Marion Morganstein, Southside Senior High School, Rockville Centre, New York; and Susan Smith, East Meadow High School, New York for reading the manuscript and offering valuable suggestions. She also wishes to thank her colleagues at Amsco School Publications who gave of themselves in the effort to produce a book that upholds and enhances the Amsco tradition of excellence.

—MARILYN OCCHIOGROSSO

CONTENTS

CHAPTER 1 Logic

CHAPTER 2 Introduction to Algebra

CHAPTER 3 Monomials and Polynomials

CHAPTER 7 Algebraic Procedures

CHAPTER 8 Further Aspects of Algebra

CHAPTER 9 Measurement of Geometric Figures

CHAPTER 10 Probability

CHAPTER 11 Statistics

CHAPTER 12 The Rectangular Coordinate System

CHAPTER 13 Systems of Equations and Inequalities

CHAPTER 1

Logic

1.1 INTRODUCTION

Symbolic Forms of Single Statements and Their Truth Values

The basic statements of logic, each representing a single thought, are written as p, q, r, etc. Each statement is assigned a ***truth value:*** either T (true) or F (false).

The ***negation*** of a statement, $\sim p$ (read "not p"), always has a truth value that is the opposite of the truth value of the original statement. The negation of the two truth values of p is shown in tabular form in the accompanying diagram.

p	$\sim p$
T	F
F	T

To understand the meaning of the negation of a verbal statement, it is easiest to think of the original statement as beginning with the condition "It is true that . . ." Then, the negation can begin with the statement "It is *not* true that . . ." In this way, we can eliminate incorrect negations resulting from inserting the word "not" into various other places in the original word sentence.

Examples

Write negations of each of the following true sentences.

1. Today's weather is sunny.

2. Alice is not going to the play.

3. $x > 2$.

4. y is an odd integer.

Solutions:

The negation of a statement must have the opposite truth value. Since each of the original statements is given as true, any reasonable verbal sentence that describes the given circumstance and is false is a correct negation. Sometimes, we are not sure about the truth value of a sentence that simply includes the word "not." Such a sentence cannot be offered as the correct negation of the original statement.

1. Original Statement: Today's weather is sunny. (Given that this statement is true.)

   Correct Negation: Today's weather is not sunny. (This statement must be false.)

   Incorrect Negation: Not today's (yesterday's, tomorrow's, last week's, etc.) weather is sunny. (The truth values of these statements are not certain.)

   Answer: It is not true that today's weather is sunny.
   Today's weather is not sunny.

2. Original Statement: Alice is not going to the play. (Given as true.)

   Correct Negation: Alice is going to the play. (Must be false.)

   Incorrect Negation: Not Alice (Sam, Jane, etc.) is not going to the play. (Not certain.)

   Incorrect Negation: Not Alice is going to the play. (Not certain.)

   Answer: It is not true that Alice is not going to the play.
   Alice is going to the play.

3. Original Statement: $x > 2$. (Given as true.)

   Correct Negation: $x \ngtr 2$. (Must be false.)
   If $x \ngtr 2$, then either $x < 2$ or $x = 2$. These two statements may be written as one: $x \leq 2$.

   Answer: It is not true that $x > 2$.
   $x \ngtr 2$.
   $x \leq 2$.

4. Original Statement: y is an odd integer. (Given as true.)

   Correct Negation: y is not an odd integer. (Must be false.)

   Correct Negation: y is an integer that is not odd. (Must be false.)

Correct
Negation: y is an even integer. (Must be false.)

Answer: It is not true that y is an odd integer.
y is not an odd integer.
y is an integer that is not odd.
y is an even integer.

For two single statements, p and q, there are four possible combinations of truth values, as shown in the accompanying diagram.

p	q
T	T
T	F
F	T
F	F

Connectives and Their Truth Values

Two statements can be connected in various ways. The ***connectives*** determine the truth values of the resulting combinations. The connectives, along with their related words, symbols, and truth values, are as follows:

Connective	Related Words	Symbol	Truth Value
Conjunction	p AND q.	$p \wedge q$	True only when both p and q are true.
Disjunction	p OR q.	$p \vee q$	False only when both p and q are false.
Conditional or ***Implication***	IF p, THEN q. or p IMPLIES q.	$p \rightarrow q$	False only when p is true and q is false.
Biconditional	IF p, THEN q AND IF q, THEN p. or p IMPLIES q AND q IMPLIES p. or p IF AND ONLY IF q. (*Note 1:* We use "iff" to mean "if and only if.") (*Note 2:* A biconditional is a conjunction of two conditionals.)	$p \leftrightarrow q$	True only when p and q have the same truth values (that is, p and q are both true or both false).

SUMMARY OF TRUTH VALUES FOR THE CONNECTIVES

		Conjunction	Disjunction	Conditional	Biconditional
p	q	$p \wedge q$	$p \vee q$	$p \rightarrow q$	$p \leftrightarrow q$
T	T	T	T	T	T
T	F	F	T	F	F
F	T	F	T	T	F
F	F	F	F	T	T

Examples

1. Which should be the heading for column 3?
(1) $\sim p$
(2) $\sim q$
(3) $p \wedge q$
(4) $p \vee q$

Col. 1	Col. 2	Col. 3
p	q	
T	T	F
T	F	F
F	T	T
F	F	T

Solution:

The truth values in column 3 are the opposite of those in column 1. Since column 1 represents the truth values for p, column 3 must represent the truth values for the negation of p, $\sim p$.

Answer: (1)

2. If $p \wedge q$ is true, then:
(1) p is true and q is false.
(2) p is false and q is true.
(3) p and q are both true.
(4) p and q are both false.

Solution:

The conjunction, $p \wedge q$, of two statements, (p, q), is true only when both p and q are true.

Answer: (3)

3. If $p \vee q$ is false, then:
(1) p and q must have the same truth values and that value must be "true."
(2) p and q must have the same truth values and that value must be "false."
(3) p and q must have different truth values.
(4) no judgment can be made about the truth values of p and q.

Solution:

The disjunction, $p \vee q$, of two statements, (p, q), is false only when both p and q are false.

Answer: (2)

4. If p represents "$x + 3 \neq 7$" and q represents "$2x = 8$," then which statement is true when $x = 4$?
(1) p (2) $\sim q$ (3) $p \wedge q$ (4) $p \vee q$

Solution:

When $x = 4$, $x + 3 \neq 7$ is a false statement. Thus, p is false, and choice (1) is eliminated.

When $x = 4$, $2x = 8$ is a true statement. Thus, q is true. Since q is true, $\sim q$ is false. Therefore, choice (2) is eliminated.

In order for choice (3) to be true, both p and q have to be true. Since this is not the case, choice (3) is eliminated.

It is true that $p \vee q$ is true when p is false and q is true. The only time $p \vee q$ is false is when both p and q are false.

Answer: (4)

EXERCISES

In 1–5, write a negation for each statement.

1. I am hungry.
2. Lisa can do today's math assignment.
3. x is false.
4. $x = 3$.
5. $x < 6$.

In 6–9, which should be the heading for column 3?

6. (1) $p \vee q$ (3) $p \rightarrow q$
(2) $p \wedge q$ (4) $p \leftrightarrow q$

Col. 1	Col. 2	Col. 3
p	q	
T	T	T
T	F	F
F	T	F
F	F	F

7. (1) $p \wedge q$ (3) $p \rightarrow q$
(2) $p \vee q$ (4) $p \leftrightarrow q$

Col. 1	Col. 2	Col. 3
p	q	
T	T	T
T	F	T
F	T	T
F	F	F

8. (1) $p \wedge q$ (3) $p \to q$
(2) $p \vee q$ (4) $q \to p$

Col. 1	Col. 2	Col. 3
p	q	
T	T	T
T	F	F
F	T	T
F	F	T

9. (1) $p \to q$ (3) $p \leftrightarrow q$
(2) $q \to p$ (4) $q \leftrightarrow p$

Col. 1	Col. 2	Col. 3
p	q	
T	T	T
T	F	T
F	T	F
F	F	T

10. If $p \vee q$ is false, then:
(1) p is true and q is false. (3) both p and q are true.
(2) p is false and q is true. (4) both p and q are false.

11. If $p \wedge q$ is true, then:
(1) p is true and q is false. (3) p and q are both true.
(2) p and q are both false. (4) p is false and q is true.

12. When $p \to q$ is false, then:
(1) p is true and q is false. (3) p and q are both true.
(2) p is false and q is true. (4) p and q are both false.

13. If $p \leftrightarrow q$ is true, then:
(1) p and q must have the same truth values.
(2) p and q must have different truth values.
(3) the truth values of p and q are not important.
(4) no judgment can be made about the truth values of p and q.

14. If $p \leftrightarrow q$ is true, which statement must also be true?
(1) $p \wedge q$ (2) $p \vee q$ (3) $p \wedge \sim q$ (4) $p \to q$

15. When $p \vee q$ is false, which statement is also false?
(1) $p \to q$ (2) $p \leftrightarrow q$ (3) $p \wedge q$ (4) $\sim(p \wedge q)$

16. When $p \wedge q$ is true, which statement is also true?
(1) $\sim p \wedge \sim q$ (2) $\sim p \vee \sim q$ (3) $p \to \sim q$ (4) $\sim p \to \sim q$

17. When $p \to q$ is false, which statement is true?
(1) $p \wedge q$ (2) $q \wedge p$ (3) $p \vee q$ (4) $q \leftrightarrow p$

18. If p represents "x is divisible by 2" and q represents "x is divisible by 3," then which statement is true when $x = 14$?
(1) $p \wedge q$ (2) $p \vee q$ (3) $\sim p$ (4) q

19. Which whole number, when substituted for x, makes the following sentence true?

$$(3x > 6) \wedge (x < 4)$$

20. Let p represent "$x + 2 = 7$" and let q represent "$x - 1 = 4$." For what value of x is $p \wedge q$ true?

21. Let p represent "x is odd." Let q represent "$x > 8$." When x is 6, which statement is true? (1) p (2) q (3) $p \to q$ (4) $p \wedge q$

1.2 TRANSLATING VERBAL STATEMENTS INTO SYMBOLIC FORM AND MAKING JUDGMENTS ABOUT THEIR TRUTH VALUES

The principles of logic can be used to judge the truth values of verbal statements if these statements are first translated into symbols. To accomplish the translations, remember:

not	means	THE OPPOSITE OF:	$\sim p$
and	means	CONJUNCTION:	$p \wedge q$
or	means	DISJUNCTION:	$p \vee q$
if–then or *only if*	means	CONDITIONAL:	$p \rightarrow q$
if and only if	means	BICONDITIONAL:	$p \leftrightarrow q$

◆ To determine the truth value of the sentence in question, remember:

1. Conjunction is true only when both p and q are true.
2. Disjunction is false only when both p and q are false.
3. Conditional is false only when p is true and q is false.
4. Biconditional is true whenever p and q have the same truth values.

Examples

1. If p represents "The cat is cute" and q represents "The cat is clever," write in symbolic form:

   a. The cat is cute and the cat is not clever.

   b. The cat is clever or the cat is cute.

   c. If the cat is clever, then the cat is cute.

   d. If the cat is not cute, then the cat is clever.

   e. The cat is cute if and only if the cat is clever.

Solution: *Answer:*

a. $\underbrace{\text{The cat is cute}}_{p}\ \underbrace{\text{and}}_{\wedge}\ \underbrace{\text{the cat is not clever.}}_{\text{(not)}\ \sim q}$ $\qquad p \wedge \sim q$

b. $\underbrace{\text{The cat is clever}}_{p}\ \underbrace{\text{or}}_{\vee}\ \underbrace{\text{the cat is cute.}}_{q}$ $\qquad p \vee q$

c. If $\underbrace{\text{the cat is clever,}}_{q}$ then $\underbrace{\text{the cat is cute.}}_{p}$ $\quad \underset{\text{(If . . . then)}}{\rightarrow}$ $\qquad q \rightarrow p$

d. If $\underbrace{\text{the cat is not cute,}}_{\text{(not)}\ \sim p}$ then $\underbrace{\text{the cat is clever.}}_{q}$ $\quad \underset{\text{(If . . . then)}}{\rightarrow}$ $\qquad \sim p \rightarrow q$

e. $\underbrace{\text{The cat is cute}}_{p}\ \underbrace{\text{if and only if}}_{\leftrightarrow}\ \underbrace{\text{the cat is clever.}}_{q}$ $\qquad p \leftrightarrow q$

2. Let p represent "$x + 3 = 7$" and let q represent "$x - 1 = 3$."

a. For what value of x is $p \wedge q$ true?

b. Which of the following is true when $x = 5$?
(1) $p \wedge q$ (2) $p \vee q$ (3) $\sim p \wedge q$ (4) $\sim p \wedge \sim q$

Solution:

a. The conjunction $p \wedge q$ is true only when both p and q are true. p is true when $x = 4$ ($4 + 3 = 7$ is true); q is also true when $x = 4$ ($4 - 1 = 3$ is true).

Answer: $x = 4$

b. $x = 5$ makes both p ($5 + 3 = 7$ is false) and q ($5 - 1 = 3$ is false) false. Thus $\sim p$ ($5 + 3 \neq 7$ is true) and $\sim q$ ($5 - 1 \neq 3$ is true) are both true. The conjunction of two true statements is true.

Answer: (4)

EXERCISES

1. Let p represent "You are correct" and let q represent "I agree." Using p and q, write in symbolic form: "You are correct and I agree."
2. If p represents "I have studied" and q represents "I will pass this test," write in symbolic form, using p and q: "I have studied, or I will not pass this test."
3. If p represents "He is tall" and q represents "He is handsome," write in symbolic form, using p and q: "He is not tall and he is handsome."
4. Let p represent "It is summer" and q represent "I go swimming." Using p and q, write in symbolic form: "If I go swimming, then it is summer."
5. Let p represent "I have studied" and q represent "I will pass this test." Write in symbolic form, using p and q: "If I have not studied, then I will not pass this test."
6. If p represents "Today is Monday" and q represents "I am tired," write in symbolic form, using p and q: "Today is Monday and I am not tired."
7. p represents "It is cold" and q represents "I will go skiing." Using p and q, write in symbolic form: "If it is cold, then I will not go skiing."
8. If p represents "Two sides of a triangle are congruent" and q represents "Two angles of a triangle are congruent," write in symbolic form, using p and q: "Two sides of a triangle are congruent if and only if two angles of a triangle are congruent."
9. If p represents "Today is Monday" and q represents "Tomorrow is Wednesday," write in symbolic form, using p and q: "If today is Monday, then tomorrow is not Wednesday."

10. If p represents "It is sunny" and q represents "It is warm," then which is a correct translation of $p \wedge \sim q$?
   (1) It is sunny and it is warm.
   (2) If it is sunny, then it is not warm.
   (3) It is sunny and it is not warm.
   (4) It is sunny or it is not warm.
11. If p represents "I do try" and q represents "I do succeed," then which is a correct translation of $\sim p \wedge q$?
   (1) I do try and I do succeed.
   (2) I do not try and I do succeed.
   (3) If I do not try, then I do succeed.
   (4) If I do not try, then I do not succeed.
12. If p represents "x is an even integer" and q represents "x^2 is an even integer," then which is a correct translation of $p \rightarrow \sim q$?
   (1) If x is an even integer, then x^2 is an odd integer.
   (2) If x is an odd integer, then x^2 is an even integer.
   (3) If x is an even integer, then x^2 is an even integer.
   (4) If x is an odd integer, then x^2 is an odd integer.
13. If p represents "Three sides of a triangle are congruent" and q represents "Three angles of a triangle are congruent," then which is a correct translation of $\sim p \leftrightarrow \sim q$?
   (1) Three sides of a triangle are not congruent if and only if three angles of a triangle are congruent.
   (2) Three sides of a triangle are not congruent if three angles of a triangle are congruent.
   (3) Three sides of a triangle are not congruent if three angles of a triangle are not congruent.
   (4) Three sides of a triangle are not congruent if and only if three angles of a triangle are not congruent.
14. If p represents "Today is Saturday" and q represents "Tomorrow is Sunday," then which is a correct translation of $q \rightarrow \sim p$?
   (1) If tomorrow is not Sunday, then today is not Saturday.
   (2) If tomorrow is not Sunday, then today is Saturday.
   (3) If tomorrow is Sunday, then today is Saturday.
   (4) If tomorrow is Sunday, then today is not Saturday.

1.3 TRUTH TABLES; TAUTOLOGY; LOGICAL EQUIVALENCE

The truth values of compound statements involving combinations of connectives can be determined through the use of ***truth tables.*** We construct a truth table by breaking down a compound statement into its connective components and using these components as the headings of the columns of the table. The first two columns of the truth table will always be the four possible truth combinations of p and q. The remaining columns are completed by inserting the truth values associated with the four connectives ($\wedge$, $\vee$, $\rightarrow$, and $\leftrightarrow$).

Consider the compound statement $(p \wedge q) \vee (p \rightarrow q)$. This statement is essentially a *disjunction:*

$(p \wedge q)$	$\vee$	$(p \rightarrow q)$
left side of the disjunction	disjunction	right side of the disjunction

Each side of the disjunction contains a connective. A truth table will allow us to read the truth values for the left and right sides of the compound statement so that we can make a judgment about the truth values of the entire compound statement.

The headings for an appropriate truth table would be:

(1)	(2)	(3)	(4)	(5)
p	q	$p \wedge q$	$p \rightarrow q$	$(p \wedge q) \vee (p \rightarrow q)$

Columns 1 and 2 are the starting place. They will contain the four possible truth values of the simple statements p and q. Then, we will enter the truth values for the basic connectives $\wedge$ and $\rightarrow$ in columns 3 and 4. Once we know the values for columns 3 and 4, we can enter the values for the disjunction in column 5.

The truth table is completed as follows:

Columns 1 and 2: Insert the four truth combinations for p and q.

Column 3: The conjunction $p \wedge q$ is true only when p and q are both true.

Column 4: The implication $p \rightarrow q$ is false only when p is true and q is false.

Column 5: The disjunction "column 3 $\vee$ column 4" is false only when both a value in column 3 and the corresponding value in column 4 are false.

(1) p	(2) q	(3) $p \wedge q$	(4) $p \rightarrow q$	(5) $(p \wedge q) \vee (p \rightarrow q)$
T	*T*	*T*	*T*	*T*
T	*F*	*F*	*F*	*F*
F	*T*	*F*	*T*	*T*
F	*F*	*F*	*T*	*T*

A compound statement that is *always true,* regardless of the truth values of its components, is called a ***tautology.***

Consider the compound statement $(p \wedge q) \rightarrow p$ and its truth table:

p	q	$p \wedge q$	$(p \wedge q) \rightarrow p$
T	T	T	T
T	F	F	T
F	T	F	T
F	F	F	T

Since the statement $(p \wedge q) \rightarrow p$ is always true, it is a tautology.

Two compound statements that *always have the same truth values* are called ***logically equivalent statements.*** Consider the following truth table:

(1) p	(2) q	(3) $\sim p$	(4) $\sim p \rightarrow q$	(5) $p \vee q$	(6) $(\sim p \rightarrow q) \leftrightarrow (p \vee q)$
T	T	F	T	T	T
T	F	F	T	T	T
F	T	T	T	T	T
F	F	T	F	F	T

From the table, we see that columns 4 and 5 have exactly the same truth values. Thus, the statement $\sim p \rightarrow q$ is *logically equivalent* to the statement $p \vee q$.

Column 6 shows the two logically equivalent statements connected as a *biconditional.* Since a biconditional is true whenever its left and right sides have the same truth values, the *biconditional of two logically equivalent statements is a tautology.*

Examples

1. a. Copy and complete the truth table for the statement:

$$[(p \vee q) \wedge (\sim p)] \rightarrow q$$

p	q	$p \vee q$	$\sim p$	$(p \vee q) \wedge (\sim p)$	$[(p \vee q) \wedge (\sim p)] \rightarrow q$

b. Determine if the statement is a tautology.

Solution:

a. Columns 1 and 2: Insert the four truth combinations for p and q.
Column 3: The disjunction $p \vee q$ is false only when p and q are both false.
Column 4: Insert the opposite values of those in column 1.
Column 5: The conjunction of columns 3 and 4 is true only when a value in column 3 and the corresponding value in column 4 are both true.
Column 6: The conditional "column 5 → column 2" is false only when a value in column 5 is true and the corresponding value in column 2 is false. There is no such case here.

(1)	(2)	(3)	(4)	(5)	(6)
p	q	$p \vee q$	$\sim p$	$(p \vee q) \wedge (\sim p)$	$[(p \vee q) \wedge (\sim p)] \rightarrow q$
T	T	T	F	F	T
T	F	T	F	F	T
F	T	T	T	T	T
F	F	F	T	F	T

b. Since the given statement is always true, it is a tautology.

2. a. Copy and complete the truth table for the statement:

$$\sim(p \rightarrow q) \rightarrow (p \wedge \sim q)$$

p	q	$p \rightarrow q$	$\sim(p \rightarrow q)$	$\sim q$	$p \wedge \sim q$	$\sim(p \rightarrow q) \rightarrow (p \wedge \sim q)$

b. Determine if the statement is a tautology.

c. Let p represent "I am hungry." Let q represent "I eat." Which sentence is logically equivalent to $\sim(p \rightarrow q)$?
(1) I am not hungry or I do not eat.
(2) I am hungry and I do not eat.
(3) If I am not hungry, then I do not eat.
(4) If I do not eat, then I am not hungry.

Solution:

a. Columns 1 and 2: Insert the four truth combinations for p and q.
Column 3: The conditional $p \rightarrow q$ is false only when p is true and q is false.
Column 4: Insert the opposite values of those in column 3.
Column 5: Insert the opposite values of those in column 2.

Column 6: The conjunction of columns 1 and 5 is true only when a value in column 1 and the corresponding value in column 5 are both true.

Column 7: The conditional "column 4 → column 6" is false only when a value in column 4 is true and the corresponding value in column 6 is false. There is no such case here.

(1) p	(2) q	(3) $p \to q$	(4) $\sim(p \to q)$	(5) $\sim q$	(6) $p \wedge \sim q$	(7) $\sim(p \to q) \to (p \wedge \sim q)$
T	T	T	F	F	F	T
T	F	F	T	T	T	T
F	T	T	F	F	F	T
F	F	T	F	T	F	T

b. Since the given statement is always true, it is a tautology.

c. Translate each of the given verbal sentences into symbolic form:
(1) $\sim p \vee \sim q$ (2) $p \wedge \sim q$ (3) $\sim p \to \sim q$ (4) $\sim q \to \sim p$

From the truth table, we can see that column 4, $\sim(p \to q)$, has the same truth values as (is logically equivalent to) column 6, $p \wedge \sim q$.

Answer: (2)

3. a. Copy and complete the truth table for the statement:

$$[(p \to q) \wedge \sim p] \to \sim q$$

p	q	$p \to q$	$\sim p$	$(p \to q) \wedge \sim p$	$\sim q$	$[(p \to q) \wedge \sim p] \to \sim q$

b. Determine if the statement is a tautology.

Solution:

a. Columns 1 and 2: Insert the four truth combinations for p and q.

Column 3: The conditional $p \to q$ is false only when p is true and q is false.

Column 4: Insert the opposite values of those in column 1.

Column 5: The conjunction of columns 3 and 4 is true only when a value in column 3 and the corresponding value in column 4 are both true.

Column 6: Insert the opposite values of those in column 2.

Column 7: The conditional "column 5 → column 6" is false only when a value in column 5 is true and the corresponding value in column 6 is false.

(1) p	(2) q	(3) $p \to q$	(4) $\sim p$	(5) $(p \to q) \wedge \sim p$	(6) $\sim q$	(7) $[(p \to q) \wedge \sim p] \to \sim q$
T	T	T	F	F	F	T
T	F	F	F	F	T	T
F	T	T	T	T	F	F
F	F	T	T	T	T	T

b. Since the given statement is *not always true,* it is *not* a tautology.

EXERCISES

1. a. Copy and complete the truth table for the statement:

$$[p \wedge (p \to q)] \to q$$

p	q	$p \to q$	$p \wedge (p \to q)$	$[p \wedge (p \to q)] \to q$

b. Is $[p \wedge (p \to q)] \to q$ a tautology?

2. a. Copy and complete the truth table for the statement:

$$(p \to q) \leftrightarrow (q \vee \sim p)$$

p	q	$p \to q$	$\sim p$	$q \vee \sim p$	$(p \to q) \leftrightarrow (q \vee \sim p)$

b. Is $(p \to q) \leftrightarrow (q \vee \sim p)$ a tautology?

c. In $(p \to q) \leftrightarrow (q \vee \sim p)$, let p represent "We pollute the water" and let q represent "The fish will die." Which statement is logically equivalent to "If we pollute the water, then the fish will die"?
(1) The fish will die or we do not pollute the water.
(2) We pollute the water and the fish will die.
(3) If we do not pollute the water, then the fish will not die.
(4) If the fish die, then we pollute the water.

3. a. Copy and complete the truth table for the statement:

$$(\sim p \to q) \leftrightarrow (p \vee q)$$

p	q	$\sim p$	$\sim p \to q$	$p \vee q$	$(\sim p \to q) \leftrightarrow (p \vee q)$

b. Is $(\sim p \to q) \leftrightarrow (p \vee q)$ a tautology?

c. Let p represent "I do my homework." Let q represent "I get into trouble." Which sentence is equivalent to $(p \lor q)$?
(1) If I do not do my homework, I will get into trouble.
(2) I do my homework or I do not get into trouble.
(3) If I do my homework, I get into trouble.
(4) I do my homework and I get into trouble.

4. a. Copy and complete the truth table for the statement:

$$[(\sim p) \land (p \lor q)] \to q$$

p	q	$\sim p$	$p \lor q$	$(\sim p) \land (p \lor q)$	$[(\sim p) \land (p \lor q)] \to q$

b. Is $[(\sim p) \land (p \lor q)] \to q$ a tautology?

5. a. Copy and complete the truth table for the statement:

$$\sim[p \land (\sim p \land q)]$$

p	q	$\sim p$	$\sim p \land q$	$p \land (\sim p \land q)$	$\sim[p \land (\sim p \land q)]$

b. Is $\sim[p \land (\sim p \land q)]$ a tautology?

6. a. Copy and complete the truth table for the statement:

$$[\sim p \land (p \to q)] \to \sim q$$

p	q	$\sim p$	$p \to q$	$\sim p \land (p \to q)$	$\sim q$	$[\sim p \land (p \to q)] \to \sim q$

b. Is $[\sim p \land (p \to q)] \to \sim q$ a tautology? Justify your answer.

7. a. Copy and complete the truth table for the statement:

$$(q \to \sim p) \leftrightarrow \sim(p \land q)$$

p	q	$\sim p$	$q \to \sim p$	$p \land q$	$\sim(p \land q)$	$(q \to \sim p) \leftrightarrow \sim(p \land q)$

b. Is $(q \to \sim p) \leftrightarrow \sim(p \land q)$ a tautology? Justify your answer.

c. Let p represent "I laugh." Let q represent "The joke is funny." Which statement is logically equivalent to $\sim(p \land q)$?
(1) If the joke is funny, I laugh.
(2) If the joke is funny, I do not laugh.
(3) If I laugh, the joke is funny.
(4) If I do not laugh, the joke is not funny.

8. a. Copy and complete the truth table for the statement:

$$(p \to q) \leftrightarrow (\sim p \to \sim q)$$

p	q	$p \to q$	$\sim p$	$\sim q$	$\sim p \to \sim q$	$(p \to q) \leftrightarrow (\sim p \to \sim q)$

b. Is the statement $(p \to q) \leftrightarrow (\sim p \to \sim q)$ a tautology?

9. a. Copy and complete the truth table for the statement:

$$[(p \to q) \wedge \sim q] \leftrightarrow \sim p$$

p	q	$p \to q$	$\sim q$	$(p \to q) \wedge \sim q$	$\sim p$	$[(p \to q) \wedge \sim q] \leftrightarrow \sim p$
T	T					
T	F					
F	T					
F	F					

b. Is $[(p \to q) \wedge \sim q] \leftrightarrow \sim p$ a tautology?

10. a. Copy and complete the truth table for the statement:

$$\sim(p \to q) \leftrightarrow (p \wedge \sim q)$$

p	q	$p \to q$	$\sim(p \to q)$	$\sim q$	$p \wedge \sim q$	$\sim(p \to q) \leftrightarrow (p \wedge \sim q)$
T	T					
T	F					
F	T					
F	F					

b. Is $\sim(p \to q) \leftrightarrow (p \wedge \sim q)$ a tautology?

c. Let p represent "I am tired." Let q represent "I sleep." Which statement is logically equivalent to $\sim(p \to q)$?
(1) I am not tired and I sleep.
(2) I am not tired and I do not sleep.
(3) If I am not tired, then I do not sleep.
(4) I am tired and I do not sleep.

11. a. Construct a truth table for the statement $[(p \vee q) \wedge (\sim p)] \to q$.
b. Is the statement in part **a** a tautology?

12. a. Construct a truth table for the statement $(p \wedge q) \to \sim(p \to q)$.
b. Is the statement in part **a** a tautology?

13. a. Construct a truth table for the statement $[p \wedge \sim(p \vee q)] \to q$.
b. Is the statement in part **a** a tautology?

14. **a.** Construct a truth table for the statement $[\sim q \wedge (p \rightarrow q)] \rightarrow \sim p$.
b. Is the statement in part **a** a tautology?

15. **a.** Construct a truth table for the statement $[(p \rightarrow q) \vee \sim q] \rightarrow (p \wedge q)$.
b. Is the statement in part **a** a tautology?
c. Which of the following statements is logically equivalent to the statement in part **a**?
(1) $p \wedge q$ (2) $p \vee q$ (3) $p \rightarrow q$ (4) $p \leftrightarrow q$

1.4 FORMING THE CONVERSE, INVERSE, AND CONTRAPOSITIVE OF A CONDITIONAL STATEMENT AND DETERMINING THE TRUTH VALUES

The following table shows how to form the *converse, inverse,* and *contrapositive* of a conditional statement.

Name	What to Do	"If–Then" Form	Symbols
Conditional	given	if p, then q	$p \rightarrow q$
Converse	interchange p and q	if q, then p	$q \rightarrow p$
Inverse	negate p and negate q	if not p, then not q	$\sim p \rightarrow \sim q$
Contrapositive	interchange and negate	if not q, then not p	$\sim q \rightarrow \sim p$

Note: The contrapositive is the converse of the inverse. The contrapositive is also the inverse of the converse.

If the given conditional statement is true, only the contrapositive will always be true. The contrapositive is logically equivalent to the given conditional.

If the given conditional statement is true, the converse and the inverse may be true or false. Neither the converse nor the inverse is logically equivalent to the given conditional. The converse and the inverse are logically equivalent to each other.

When working with verbal sentences, first translate them into symbolic form. Choose the thought to be represented by p and the thought to be represented by q. Adjust words as necessary to make grammatical sense, but be careful not to interchange thoughts.

Example

Given the following *true* conditional statements, write the converse, inverse, and contrapositive. Determine the truth value of each statement.

a. If David is a resident of New York City, then he is a resident of New York State.
b. If Joan is a vegetarian, then she does not eat meat.

Solution:

a. Let p = is a resident of New York City.
Let q = is a resident of New York State.

Conditional (if p, then q)	If David is a resident of New York City, then he is a resident of New York State.	(True)
Converse (if q, then p)	If David is a resident of New York State, then he is a resident of New York City.	(False)
Inverse (if not p, then not q)	If David is not a resident of New York City, then he is not a resident of New York State.	(False)
Contrapositive (if not q, then not p)	If David is not a resident of New York State, then he is not a resident of New York City.	(True)

b. Let p = is a vegetarian.
Let q = does not eat meat.
(*Note:* Since q is itself "negative," $\sim q$ will be "positive.")

Conditional (if p, then q)	If Joan is a vegetarian, then she does not eat meat.	(True)
Converse (if q, then p)	If Joan does not eat meat, then she is a vegetarian.	(True)
Inverse (if not p, then not q)	If Joan is not a vegetarian, she *does* (the negation of *does not*) eat meat.	(True)
Contrapositive (if not q, then not p)	If Joan *does* (the negation of *does not*) eat meat, then she is not a vegetarian.	(True)

If the given verbal statement is not in the "if-then" form, first write it in this form by using all the words up to but not including the verb as the "if" part and the verb plus the words following it as the "then" part. Insert words as needed to complete a grammatically correct sentence. Sometimes words have to be moved from one side of the verb to the other to help the sentence make sense, but do not change the basic conditions. (*Note:* The "if" part is called the ***hypothesis, antecedent,*** or ***premise.*** The "then" part is called the ***conclusion*** or ***consequent.***)

Examples

1. Write the following statements in the "if-then" form:

   a. An even integer is exactly divisible by 2.

   b. All right angles are congruent.

   c. The base angles of an isosceles triangle are congruent.

   d. An integer is a prime if it has no factors other than itself and 1.

   e. Two lines are perpendicular when they intersect at right angles.

Solution:

a. $\underbrace{\text{An even integer}}_{\text{if}}\ \underbrace{\overset{\text{(verb)}}{\text{is}}\ \text{exactly divisible by 2}}_{\text{then}}.$

Answer: If an integer is even, then it is exactly divisible by 2.

b. $\underbrace{\text{All right angles}}_{\text{if}}\ \underbrace{\overset{\text{(verb)}}{\text{are}}\ \text{congruent}}_{\text{then}}.$

Answer: If two angles are right angles, then they are congruent.

c. $\underbrace{\text{The base angles of an isosceles triangle}}_{\text{if}}\ \underbrace{\overset{\text{(verb)}}{\text{are}}\ \text{congruent}}_{\text{then}}.$

Answer: If a triangle is isosceles, then its base angles are congruent.

d. $\underbrace{\text{An integer is a prime}}_{\text{then}}\ \underbrace{\text{if it has no factors other than itself and 1}}_{\text{if}}.$

Answer: If an integer has no factors other than itself and 1, then it is a prime.

e. $\underbrace{\text{Two lines are perpendicular}}_{\text{then}}\ \underbrace{\text{when they intersect at right angles}}_{\text{if}}.$

Answer: If two lines intersect at right angles, then they are perpendicular.

2. What is the converse of $p \rightarrow \sim q$?

(1) $\sim q \rightarrow p$ (2) $\sim q \rightarrow \sim p$ (3) $q \rightarrow \sim p$ (4) $p \leftrightarrow \sim q$

Solution:

The converse of a statement interchanges the hypothesis (if) and the conclusion (then).

Statement: $p \rightarrow \sim q$

Converse: $\sim q \rightarrow p$

Answer: (1)

3. What is the inverse of $\sim p \rightarrow q$?

(1) $p \rightarrow q$ (2) $p \rightarrow \sim q$ (3) $\sim p \rightarrow q$ (4) $\sim p \rightarrow \sim q$

Solution:

The inverse of a statement negates the hypothesis and negates the conclusion.

Statement: $\sim p \rightarrow q$
Inverse: $\sim(\sim p) \rightarrow \sim q$
or $p \rightarrow \sim q$

Answer: (2)

EXERCISES

1. Which is the converse of the statement "If John is a poet, then he is poor"?
(1) If John is not a poet, then he is poor.
(2) If John is poor, then he is a poet.
(3) If John is not poor, then he is not a poet.
(4) John is a poet and he is poor.

2. What is the inverse of the statement "If n is an odd integer, then $n + 2$ is an odd integer"?
(1) If n is an odd integer, then $n + 2$ is not an odd integer.
(2) If n is not an odd integer, then $n + 2$ is not an odd integer.
(3) If $n + 2$ is an odd integer, then n is an odd integer.
(4) If $n + 2$ is not an odd integer, then n is not an odd integer.

3. What is the contrapositive of the statement "If Rachel is 18, then she may vote"?
(1) If Rachel is not 18, then she may not vote.
(2) If Rachel is 18, then she may not vote.
(3) If Rachel may vote, then she is 18.
(4) If Rachel may not vote, then she is not 18.

4. What is the converse of $\sim p \rightarrow q$?
(1) $p \rightarrow q$ (2) $p \rightarrow \sim q$ (3) $\sim q \rightarrow p$ (4) $q \rightarrow \sim p$

5. Which is the inverse of $\sim p \rightarrow q$?
(1) $q \rightarrow \sim p$ (2) $p \rightarrow \sim q$ (3) $\sim q \rightarrow p$ (4) $p \rightarrow q$

6. What is the contrapositive of the statement $p \rightarrow q$?
(1) $p \leftrightarrow q$ (2) $q \rightarrow p$ (3) $\sim p \rightarrow \sim q$ (4) $\sim q \rightarrow \sim p$

7. Which is the converse of "If it rains today, I'll stay home"?
(1) If I stay home, then it will rain today.
(2) If it rains today, then I'll stay home.
(3) If I don't stay home, then it won't rain today.
(4) If it doesn't rain today, then I won't stay home.

8. What is the inverse of the statement "If I study, then I will pass"?
(1) If I pass, then I did study.
(2) If I do not pass, then I did not study.
(3) If I do not study, then I will not pass.
(4) If I study, then I will not pass.

9. Which is the contrapositive of the statement "If an integer is a multiple of 2, then it is even"?
(1) If an integer is not even, then it is not a multiple of 2.
(2) If an integer is even, then it is a multiple of 2.
(3) If an integer is not even, then it is a multiple of 2.
(4) If an integer is even, then it is not a multiple of 2.

10. Which is the inverse of $\sim p \rightarrow \sim q$?
(1) $p \rightarrow \sim q$ (2) $q \rightarrow \sim p$ (3) $p \rightarrow q$ (4) $\sim q \rightarrow \sim p$

11. What is the contrapositive of $\sim p \rightarrow \sim q$?
(1) $p \rightarrow q$ (2) $q \rightarrow p$ (3) $p \rightarrow \sim q$ (4) $\sim p \rightarrow q$

12. Which is the contrapositive of $\sim p \rightarrow q$?
(1) $p \rightarrow \sim q$ (2) $p \rightarrow q$ (3) $\sim q \rightarrow \sim p$ (4) $\sim q \rightarrow p$

13. Which is the contrapositive of $p \rightarrow \sim q$?
(1) $q \rightarrow \sim p$ (2) $q \rightarrow p$ (3) $\sim q \rightarrow p$ (4) $\sim q \rightarrow \sim p$

14. Which is the converse of $\sim p \rightarrow \sim q$?
(1) $q \rightarrow p$ (2) $q \rightarrow \sim p$ (3) $\sim q \rightarrow p$ (4) $\sim q \rightarrow \sim p$

15. What is the inverse of $\sim q \rightarrow \sim p$?
(1) $\sim p \rightarrow \sim q$ (2) $p \rightarrow q$ (3) $q \rightarrow p$ (4) $\sim q \rightarrow p$

The *contrapositive* of a conditional statement is the only statement that *must be true* whenever the conditional statement is true.

Example

Given the true statement: "If it is August in Texas, it is likely to be sunny," which of the following statements must be true?

(1) If it is not likely to be sunny, it is August in Texas.
(2) If it is not likely to be sunny, it is not August in Texas.
(3) If it is not August in Texas, it is not likely to be sunny.
(4) If it is August in Texas, it is not likely to be sunny.

Solution:

Given: If it is August in Texas, it is likely to be sunny.

p = is August in Texas; q = is likely to be sunny

Form the contrapositive (if not q, then not p).

If it is not likely to be sunny, then it is not August in Texas.

Answer: (2)

EXERCISES

1. If a conditional statement is true, which related conditional is also always true?
(1) converse (2) inverse (3) contrapositive (4) biconditional

2. Given the true statement: "If a figure is a square, then it is a rectangle." Which sentence is also true?
(1) If a figure is a rectangle, then it is a square.
(2) If a figure is not a square, then it is not a rectangle.
(3) If a figure is a square, then it is not a rectangle.
(4) If a figure is not a rectangle, then it is not a square.

3. This sentence is true: "If it is raining, then the ground gets wet." Which sentence must also be true?
(1) If the ground gets wet, then it is raining.
(2) If it is not raining, then the ground does not get wet.
(3) If the ground gets wet, then it is not raining.
(4) If the ground does not get wet, then it is not raining.

4. Given the true statement: "If a figure is a triangle, then it is a polygon." Which statement must also be true?
(1) If a figure is a polygon, then it is a triangle.
(2) If a figure is not a triangle, then it is not a polygon.
(3) If a figure is not a polygon, then it is not a triangle.
(4) If a figure is not a triangle, then it is a polygon.

5. Given the true statement: "If two angles are right angles, then they are congruent." Which statement must also be true?
(1) If two angles are not right angles, then they are not congruent.
(2) If two angles are congruent, then they are right angles.
(3) If two angles are not right angles, then they are congruent.
(4) If two angles are not congruent, then they are not right angles.

6. Given the true statement: "If a triangle is equilateral, then it is isosceles." Which statement must also be true?
(1) If a triangle is not equilateral, then it is not isosceles.
(2) If a triangle is not equilateral, then it is isosceles.
(3) If a triangle is not isosceles, then it is not equilateral.
(4) If a triangle is isosceles, then it is equilateral.

7. Which has the same truth value as the statement "If Bill pitched, then we won the game"?
(1) If we won the game, then Bill pitched.
(2) If we did not win the game, then Bill did not pitch.
(3) If Bill pitched, then we did not win the game.
(4) If Bill did not pitch, then we did not win the game.

8. Which statement has the same truth value as the statement "If x is true, then y is true"?
(1) If x is false, then y is true.
(2) If x is false, then y is false.
(3) If y is false, then x is true.
(4) If y is false, then x is false.

9. Which statement has the same truth value as the statement "If x is correct, y is incorrect"?
   (1) If y is incorrect, x is incorrect. (3) If y is correct, x is incorrect.
   (2) If y is incorrect, x is correct. (4) If y is correct, x is correct.
10. Which statement has the same truth value as the statement "If x is real, then y is imaginary"?
   (1) If y is imaginary, then x is real.
   (2) If y is real, then x is imaginary.
   (3) If y is imaginary, then x is imaginary.
   (4) If y is real, then x is real.
11. Which statement has the same truth value as the statement "If x is irrational, then y is rational"?
   (1) If y is irrational, then x is rational.
   (2) If y is irrational, then x is irrational.
   (3) If y is rational, then x is rational.
   (4) If y is rational, then x is irrational.
12. Given the true statement: "Right angles are equal in measure." Which statement must also be true?
   (1) If two angles are right angles, they are not equal in measure.
   (2) If two angles are not right angles, they are equal in measure.
   (3) If two angles are not equal in measure, they are not right angles.
   (4) If two angles are not equal in measure, they are right angles.

1.5 SUMMARY EXERCISES

In 1–6, fill in the word "true" or the word "false."

1. The conjunction of two statements p and q is ____ only when both p and q are ____.
2. The disjunction of two statements p and q is ____ only when both p and q are ____.
3. The conditional of two statements p and q is ____ only when p is ____ and q is ____.
4. The biconditional of two statements p and q is ____ whenever p and q have the same truth values.
5. It is ____ to say that a statement and its negation always have the same truth values.
6. A tautology is always ____.

In 7–16, fill in the symbol $\wedge$, $\vee$, $\rightarrow$, $\leftrightarrow$, or $\sim$.

7. ____ means "or."
8. ____ means "if and only if."
9. ____ means "the opposite of."
10. ____ means "and."
11. ____ means "if-then."
12. p ____ q is false only when p and q are both false.
13. p ____ q is true only when p and q have the same truth values.

14. _____ p always has the opposite truth value of p.
15. p _____ q is true only when p and q are both true.
16. p _____ q is false only when p is true and q is false.

In 17–21, fill in the words "conditional," "biconditional," "conjunction," "disjunction," or "negation."

17. The _____ of a statement always has the opposite truth value.
18. The _____ of two statements is true only when both statements are true.
19. The _____ of two statements is false only when the first statement is true and the second statement is false.
20. The _____ of two statements is false only when both statements are false.
21. The _____ of two statements is true only when both statements have the same truth value.

In 22–26, fill in the words "converse," "inverse," or "contrapositive."

22. The _____ of a true conditional statement must be true.
23. The _____ of a conditional statement interchanges the "if" and the "then."
24. The _____ of a conditional statement negates the "if" and the "then."
25. The _____ of a conditional statement is the converse of the _____.
26. Neither the _____ nor the _____ of a true conditional statement must be true.

In 27–31, p represents "x is an integer" and q represents "x is even." Write each of the verbal statements in 27–31 in symbolic form, using p and q.

27. x is an integer and x is odd.
28. x is odd if and only if x is an integer.
29. If x is not odd, then x is not an integer.
30. If x is an integer, then x is even.
31. x is even or x is not an integer.

In 32–36, p represents "I smile" and q represents "I am happy." Match each of the verbal statements in 32–36 with a symbolic equivalent chosen from **a–h**.

32. I smile if and only if I am happy.
33. I am not happy and I do not smile.
34. I smile and I am not happy.
35. If I am not happy, then I do not smile.
36. If I am not happy, then I smile.

a. $p \wedge \sim q$
b. $\sim q \rightarrow p$
c. $p \leftrightarrow \sim q$
d. $\sim p \vee q$
e. $p \leftrightarrow q$
f. $\sim q \rightarrow \sim p$
g. $\sim q \vee \sim p$
h. $\sim q \wedge \sim p$

In 37–41, use the given truth table. Supply the missing column headings by matching the column numbers in 37–41 with the proper headings chosen from **a–f**.

Col. 1	Col. 2	37. Col. 3	38. Col. 4	39. Col. 5	40. Col. 6	41. Col. 7
p	q					
T	T	T	T	T	T	F
T	F	F	F	F	T	T
F	T	F	T	F	T	F
F	F	F	T	T	F	T

a. $p \wedge q$
b. $p \vee q$
c. $p \leftrightarrow q$
d. $p \rightarrow q$
e. $\sim p$
f. $\sim q$

42. If $p \rightarrow q$ is false, which statement is true?
(1) $p \wedge q$ (2) $p \vee q$ (3) $p \leftrightarrow q$ (4) $\sim p \leftrightarrow \sim q$

43. If $p \vee q$ is false, which statement is true?
(1) $\sim p \rightarrow q$ (2) $p \leftrightarrow q$ (3) $p \wedge q$ (4) $p \wedge \sim q$

44. If $p \leftrightarrow q$ is true and $p \wedge q$ is false, what is the truth value of $p \vee q$?

45. Which is the inverse of $\sim q \rightarrow p$?
(1) $p \rightarrow \sim q$ (2) $p \rightarrow q$ (3) $\sim q \rightarrow \sim p$ (4) $q \rightarrow \sim p$

46. Which is the contrapositive of $q \rightarrow \sim p$?
(1) $\sim p \rightarrow q$ (2) $\sim p \rightarrow \sim q$ (3) $p \rightarrow q$ (4) $p \rightarrow \sim q$

In 47–49, given the statement: "If x is good, then y is bad." Match each of the verbal sentences in 47–49 with a name chosen from **a–e**.

47. If x is bad, then y is good.
48. If y is bad, then x is good.
49. If y is good, then x is bad.

a. conjunction
b. disjunction
c. converse
d. inverse
e. contrapositive

50. Given the true statement: "July 4th is a national holiday." Which statement must also be true?
(1) If today is a national holiday, today is July 4th.
(2) If today is not a national holiday, today is not July 4th.
(3) If today is not July 4th, today is not a national holiday.
(4) If today is July 4th, today is not a national holiday.

51. Given the true statement: "If $x > 3$, then $x > 1$." Which statement must also be true?
(1) If $x \leq 3$, then $x \leq 1$.
(2) If $x > 3$, then $x \leq 1$.
(3) If $x \leq 1$, then $x > 3$.
(4) If $x \leq 1$, then $x \leq 3$.

52. Given the true statement: "An equilateral triangle is isosceles." Which statement must also be true?
(1) If a triangle is isosceles, then it is equilateral.
(2) If a triangle is not isosceles, then it is not equilateral.
(3) If a triangle is not isosceles, then it is equilateral.
(4) If a triangle is isosceles, then it is not equilateral.

53. a. Copy and complete the truth table for the statement:

$$[(p \to q) \wedge \sim q] \to \sim p$$

p	q	$p \to q$	$\sim q$	$(p \to q) \wedge \sim q$	$\sim p$	$[(p \to q) \wedge \sim q] \to \sim p$

b. Is $[(p \to q) \wedge \sim q] \to \sim p$ a tautology? Justify your answer.

54. a. Copy and complete the truth table for the statement:

$$(p \to q) \leftrightarrow (\sim q \to \sim p)$$

p	q	$p \to q$	$\sim q$	$\sim p$	$\sim q \to \sim p$	$(p \to q) \leftrightarrow (\sim q \to \sim p)$

b. Is $(p \to q) \leftrightarrow (\sim q \to \sim p)$ a tautology?
c. Which statement is true?
(1) A conditional and its converse always have the same truth value.
(2) A conditional and its inverse always have the same truth value.
(3) A conditional and its contrapositive do not always have the same truth value.
(4) A conditional and its contrapositive always have the same truth value.

55. a. Copy and complete the truth table for the statement:

$$[(p \wedge q) \vee \sim q] \to \sim p$$

p	q	$p \wedge q$	$\sim q$	$(p \wedge q) \vee \sim q$	$\sim p$	$[(p \wedge q) \vee \sim q] \to \sim p$

b. Is $[(p \wedge q) \vee \sim q] \to \sim p$ a tautology?
c. Which of the following symbols could be used to replace the ? in the statement $\{[(p \wedge q) \vee \sim q] \to p\}$? $\sim p$ so that the resulting statement is a tautology?
(1) $\to$ (2) $\leftrightarrow$ (3) $\wedge$ (4) $\vee$

CHAPTER 2

Introduction to Algebra

2.1 BASIC SET THEORY

A ***set*** is a collection of items (***elements***) that are generally grouped together because they are "of the same kind." For example, it would make sense to group the following items into two sets: skirts, oranges, shoes, apples, pears, neckties, coats. One set {oranges, apples, pears} has as its elements fruits to be eaten. The second set {skirts, shoes, neckties, coats} has as its elements clothes to be worn.

A set may be written so that its elements are ***listed, described*** in words, or ***notated*** in symbols. Following is an example in which the same set is listed, described in words, and notated in symbols:

Listing: $\{2, 4, 6, 8, . . . \}$

Word Description: The set of positive integers that are divisible by 2.

Set Notation:

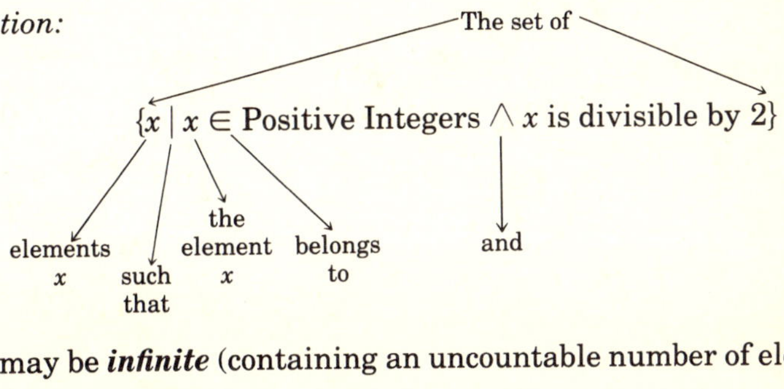

A set may be ***infinite*** (containing an uncountable number of elements), such as:

$\{\text{positive integers}\} = \{1, 2, 3, . . .\}$

or

$\{x \mid x \in \text{Integers} \land x \text{ is divisible by } 3\} = \{. . ., -9, -6, -3, 0, 3, 6, 9, . . .\}$

A set may be ***finite*** (containing a specific number of elements), such as:

{states of the U.S.A.} (Fifty elements)

or

{letters in the English alphabet} (Twenty-six elements)

Two sets are ***equivalent*** if they have the same number of elements. For example, {1, 2, 3, 4} and {a, b, c, d} are equivalent sets since each contains four elements.

Two sets may be combined to make one listing. The combination that consists of all of the elements appearing in each set, with no element appearing more than once, is called the ***union,*** $\cup$, of two sets. The combination that consists of only the elements appearing in both sets is called the ***intersection,*** $\cap$, of two sets.

Consider the two sets $A = \{1, 2, 3, 4, 5, 6\}$ and $B = \{3, 5, 7, 9, 11\}$.

The union of A and B is $A \cup B = \{1, 2, 3, 4, 5, 6, 7, 9, 11\}$.

The intersection of A and B is $A \cap B = \{3, 5\}$.

A set may be *empty* (contain no elements). The symbols $\varnothing$ and { } are used to represent the ***empty set*** or ***null set.*** (*Note:* Do not confuse the number 0 or a set containing only the element 0, {0}, with the empty set. When used to represent the null set, the symbol $\varnothing$ is written without braces.) A part of a set is called a ***subset.***

Consider the set $A = \{1, 2, 3\}$.

The ***proper subsets*** of A (containing fewer than all of the elements of A) are: {1}, {1, 2}, {1, 3}, {2}, {2, 3}, {3}, { } or $\varnothing$. (*Note:* $\varnothing$ is a subset of every set.)

The ***improper subset*** is the set itself: {1, 2, 3}.

In a specific problem, the elements of a set usually all belong to some given set or ***universe.*** For example, in arithmetic or algebra the universe is usually a set of numbers. In geometry, the universe may be a set of points or a set of polygons. Thus, a set may be thought of as a subset of its universe.

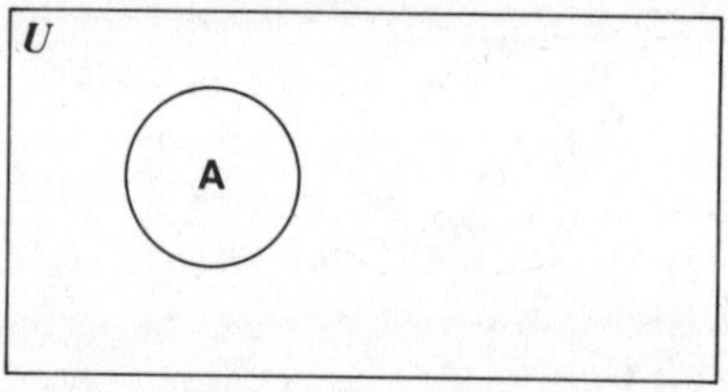

The diagram shows that set A is a subset of universe U. The points outside of the circular region representing A represent the elements of U that *do not belong to* A. The set of elements that do not belong to A is called the ***complement** of* A (in symbols, $\sim A$ or A' or $\overline{A}$).

Consider a universe $U = \{1, 2, 3, 4, 5, 6\}$ and set $A = \{1, 2, 5\}$. Then the complement of A is $\sim A = \{3, 4, 6\}$.

The concepts of sets and those of logic are closely related. The following table summarizes some of these relations:

Sets	Logic
Complement of a set $\sim A$ or A' or $\overline{A}$	Negation of a statement $\sim p$
Union of two sets $A \cup B$	Disjunction of two statements $p \vee q$
Intersection of two sets $A \cap B$	Conjunction of two statements $p \wedge q$

Recall that an ***open sentence*** contains a variable (an element subject to change). An open sentence cannot be judged true or false. If the variable is replaced by a specific element, the sentence "closes" and may then be judged for its truth value.

Consider the following examples:

1. x is a state of the U.S.A. (Open sentence)
   Africa is a state of the U.S.A. (False)
   Alabama is a state of the U.S.A. (True)

   Any one of the fifty states might have been used to replace x to make the sentence true. Thus, there are fifty ***solutions*** to the given open sentence.

2. $x + 2 = 11$ (Open sentence)
   $3 + 2 = 11$ (False)
   $9 + 2 = 11$ (True)

   Any number might have been chosen as a replacement for x, but only the choice 9 makes the sentence true. Thus, {9} is the ***solution set*** for the given open sentence.

When a specific set of elements offers choices for the replacement of the variable, that set is called the ***replacement set*** or ***domain of the variable.***

Example

Using the domain {Arthur, Africa, Afghanistan, Alabama, Asia, airplane}, find the solution set of the open sentence "x is a continent."

Answer: {Africa, Asia}

EXERCISES

1. The set $\{a, b, c, d\}$ is equivalent to which set?
(1) $\{d\}$ (2) $\{a, b, c\}$ (3) {5, 9, 13} (4) {1, 2, 3, 4}

2. Which of the following sets is *not* equivalent to the set $A = \{2, 4, 6\}$?
(1) {2, 4} (2) {1, 3, 5} (3) {6, 4, 2} (4) $\{\frac{1}{2}, \frac{1}{4}, \frac{1}{6}\}$

3. Which of the following sets is equivalent to the null set?
(1) {triangles having two equal sides}
(2) {triangles having two equal angles}
(3) {triangles having three equal sides}
(4) {triangles having only two sides}

4. List the elements of the set $\{x \mid x \in \text{Positive Integers} \wedge x < 5\}$.

5. Which is an infinite set?
   (1) $\{x \mid x$ is a United States Senator$\}$
   (2) $\{x \mid x$ is a human being who was born in 1970$\}$
   (3) $\{x \mid x$ is an odd integer$\}$
   (4) $\{x \mid x$ is a positive integer less than one million$\}$
6. Which is an example of an infinite set?
   (1) the citizens of the United States
   (2) positive integers less than 100
   (3) the even integers
   (4) all schools in New York State

In 7–9, A is the set of odd integers between 0 and 6. B is the set of integers greater than 0 and less than 6.

7. List the members of the set that is the intersection of sets A and B.
8. List the members of the set that is the union of sets A and B.
9. Which of the following statements is true?
   (1) $A \cap B = A$ (3) $A \cup B = A$
   (2) $A \cap B = B$ (4) $A \cup B = B$

In 10–13, the universe is {names of the days of the week}. A = {names of the days that do not begin with "S"}. B = {names of the days that begin with "T"}.

10. List the elements in the complement of A.
11. List the elements in the complement of B.
12. List the elements in the intersection of A and B.
13. List the elements in the intersection of the complement of A and the complement of B.

14. List *all* the subsets of $\{a, b, c, d\}$.
15. Using the domain $\{1, 2, 3\}$, find the solution set for the open sentence "x is less than 0."

2.2 CLASSIFICATION OF NUMBERS

Over the centuries, numbers have been "invented" as the need to do more with them arose. In this course, we will work with numbers in the following ways:

Numbers can be used to count; to tell direction.

Two numbers can be added or subtracted; multiplied or divided.

One number can be raised to a power; have a root extracted.

Numbers have been grouped into sets, and names have been assigned to these sets describing the most important aspect of each type of number.

In this course, we will study the sets of numbers shown in the following diagram:

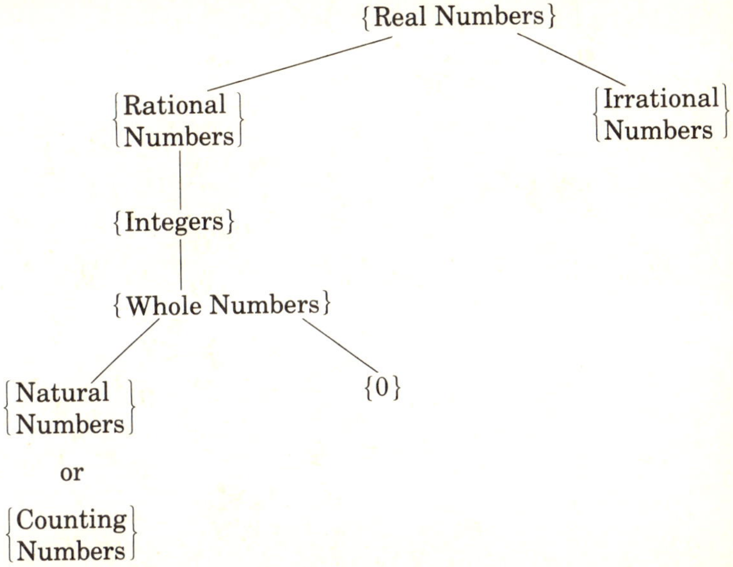

This diagram shows that ***Real Numbers*** is the most encompassing term; that is, each of the sets shown is a subset of the set of Real Numbers. (*Note:* If there are "real" numbers, then there must be numbers that are "not real." See Chapter 7.)

The set of ***Counting Numbers*** or ***Natural Numbers*** is the simplest set of numbers. These numbers have no sign and do not include 0.

$$\{\text{Counting Numbers}\} = \{1, 2, 3, 4, \ldots\}$$

When the set of Counting Numbers is combined with the number 0, the new set is called the set of ***Whole Numbers.*** These numbers have no sign.

$$\{\text{Whole Numbers}\} = \{0, 1, 2, 3, 4, \ldots\}$$

The set of ***Integers*** is the first set of numbers to have signs. Integers are whole numbers with signs; 0 is included.

$$\{\text{Integers}\} = \{\ldots, -3, -2, -1, 0, 1, 2, 3, \ldots\}$$

(*Note: **Even numbers*** are integers that are exactly divisible by 2. ***Odd numbers*** are integers that have a remainder when divided by 2.)

Rational Numbers are numbers that can be expressed as a ***ratio*** *of two integers* (excluding the case where the denominator is 0). Examples of rational numbers are: $\frac{1}{2}$ or .5, $-.7$ or $-\frac{7}{10}$, $\frac{1}{3}$ or .3333 . . ., $\sqrt{36}$ or 6, $-\sqrt{121}$ or -11, $-\frac{1}{8}$ or $-.125$, $\frac{5}{3}$ or 1.6666 . . ., .131313 . . . or $\frac{13}{99}$, 0 or $\frac{0}{1}$. (*Note:* It is not possible to list the set of rational numbers, but they are described by the rule "ratio of two integers.")

Irrational Numbers are *not rational;* that is, irrational numbers *cannot* be expressed as the ratio of two integers. Examples of irrational numbers are: $\sqrt{2}$, $-\sqrt{3}$, π, $\sqrt{37}$, $-\sqrt{122}$. (*Note:* Operations with irrational numbers will be reviewed in Chapter 8.)

◆ In general, the following are rational numbers:

1. *integers* (since each integer can be expressed as the ratio of itself and 1. For example, $6 = \frac{6}{1}$.)
2. *fractions* (since they are already expressed as a ratio of two integers. For example, $\frac{2}{3}$, $\frac{7}{4}$.)
3. *decimals that end* (numbers of this nature can always be converted to fractions that are, in turn, rational numbers. For example, $.5 = \frac{1}{2}$, $.25 = \frac{1}{4}$, $.125 = \frac{1}{8}$.)
4. *decimals that repeat* (numbers of this nature can always be converted to fractions that are, in turn, rational numbers. For example $.333\ldots = \frac{1}{3}$, $.666\ldots = \frac{2}{3}$, $.111\ldots = \frac{1}{9}$, $.232323\ldots = \frac{23}{99}$, $.135135135\ldots = \frac{135}{999}$.)
5. *the square root of perfect squares* (For example, $\sqrt{25} = 5$, $\sqrt{100} = 10$, $\sqrt{.81} = .9$, $\sqrt{\frac{1}{4}} = \frac{1}{2}$.)
6. *the cube root of perfect cubes* (For example, $\sqrt[3]{8} = 2$, $\sqrt[3]{27} = 9$.)
7. *the* n*th root of perfect* n*th powers* (For example, $\sqrt[4]{16}$, $\sqrt[5]{32}$, $\sqrt[6]{64}$.)

(*Note:* Square roots, perfect squares, etc. are reviewed in Chapter 8.)

◆ In general, the following are irrational numbers:

1. *square roots of nonperfect squares* (For example, $\sqrt{3}$, $\sqrt{39}$, $\sqrt{50}$.)
2. *cube roots of nonperfect cubes* (For example, $\sqrt[3]{9}$, $\sqrt[3]{28}$, $\sqrt[3]{65}$.)
3. n*th roots of nonperfect* n*th powers* (For example, $\sqrt[5]{25}$, $\sqrt[6]{37}$.)
4. *the number* π
5. *decimals that do not end or repeat* (For example, 1.7320508 . . .)

EXERCISES

1. Which is an example of an infinite set?
(1) $\{x \mid x$ is a positive even number less than $7\}$
(2) $\{x \mid x$ is a whole number less than $7\}$
(3) $\{x \mid x$ is a natural number less than $7\}$
(4) $\{x \mid x$ is a positive rational number less than $7\}$

2. Which is equivalent to the empty set?
(1) the set of all even natural numbers greater than 12
(2) the set of integers that are multiples of 3
(3) the set of all natural numbers greater than 9 and less than 10
(4) the set of integers that are divisible by 3

3. The set of rational numbers is a subset of the set of
(1) integers (3) real numbers
(2) irrational numbers (4) whole numbers

4. If a and b are natural numbers, which expression *must* represent a natural number?
(1) $a - b$ (2) $a + b$ (3) $\frac{a}{b}$ (4) $b \div a$

5. Which is a rational number?
(1) $\sqrt{6}$ (2) $\sqrt{16}$ (3) $\sqrt{18}$ (4) $\sqrt{24}$

6. Which represents an irrational number?
(1) $-\frac{4}{5}$ (2) π (3) $\sqrt{9}$ (4) 0

7. Which represents an irrational number?
(1) 0 (2) $\sqrt{2}$ (3) $\sqrt{9}$ (4) $-\frac{5}{13}$

8. Which is an irrational number?
(1) 0 (2) $\frac{1}{2}$ (3) $\sqrt{3}$ (4) $\sqrt{4}$

9. Which of the following is an irrational number?
(1) $\frac{1}{2}$ (2) $\sqrt{5}$ (3) $\sqrt{9}$ (4) 0

10. For *each* incomplete sentence in **a** through **e** below, write the *numeral* preceding the name of the set, *chosen from the list below,* that best completes that sentence.

Names of Sets

(1) Positive integers
(2) Rational numbers
(3) Irrational numbers

a. The set that contains all the natural numbers and does *not* contain the member $\frac{1}{2}$ is the set of
b. The set that contains the member $\sqrt{2}$ is the set of
c. The set that has a smallest member is the set of
d. The set that contains the member $\frac{1}{2}$ is the set of
e. The set that contains the member π is the set of

11. Which of the following is an irrational number?
(1) .375 (2) .375375375 . . . (3) .375124637142 . . . (4) 37.5

12. The set of integers is *not* a subset of the set of
(1) real numbers (3) irrational numbers
(2) rational numbers (4) integers

2.3 SIGNED NUMBERS

Signed numbers are positive (+) or negative (−) or zero (0). Zero is neither positive nor negative. (*Note:* Sometimes, ***nonnegative numbers*** are mentioned. This means the positives and zero.)

Signed numbers are arranged on a ***number line*** in the following *order:*

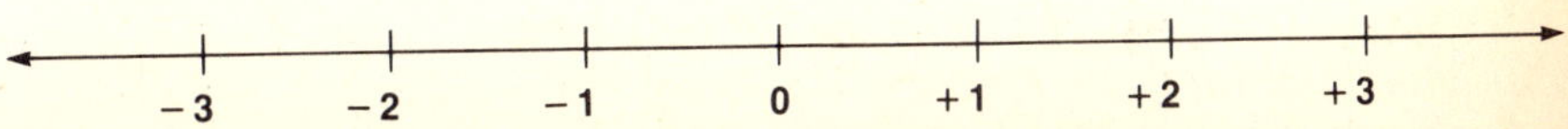

(*Note:* See Chapter 4, Section 4.4, for a discussion of a number line.)

Two signed numbers may have either ***like signs*** (both positive or both negative) or ***unlike signs*** (one positive and the other negative).

Absolute Value

The ***absolute value*** (symbol $| \;|$) of a nonzero number is always positive. The absolute value of zero is zero: $|0| = 0$.

One way to describe the absolute value of a number is as a ***distance.*** For example, the point representing $+3$ is a distance of 3 units from zero on a number line. The point representing -3 is also a distance of 3 units from zero. Thus, the absolute value of both $+3$ and -3 is 3. In symbols, we write:

$$|+3| = 3 \qquad |-3| = 3 \qquad |+3| = |-3| = 3$$

Another way to describe the absolute value of a nonzero number is to say that for every pair of opposite numbers, the greater number of the pair is the absolute value of both numbers. For example, $+3$ and -3 represent a pair of opposite numbers. Since $+3 > -3$, $+3$ is the absolute value of both numbers. That is, $|+3| = |-3| = +3$.

A third way to describe the absolute value of a nonzero number is to say that:

1. The absolute value of a positive number is that number. (For example, $|+3| = +3$.)
2. The absolute value of a negative number is the negative of the number. (For example, $|-3| = -(-3) = +3$.)

Examples

1. Evaluate $|-4| - |7|$.

Solution:

$|-4| - |7|$
$= 4 - 7$
$= -3$ *Ans.*

2. Evaluate $|7 - 4|$.

Solution:

$|7 - 4|$
$= |3|$
$= 3$ *Ans.*

3. Evaluate $|5 - 8|$.

Solution:

$|5 - 8|$
$= |-3|$
$= 3$ *Ans.*

Operations With Signed Numbers

Two signed numbers can be added, subtracted, multiplied, or divided according to the summary on the following page.

Operation	Rule	Example
Addition of Signed Numbers	**Like Signs:** Keep the sign, and add.	**a.** $\begin{array}{r} {}^{+}3 \\ + {}^{+}2 \\ \hline {}^{+}5 \end{array}$ **b.** $\begin{array}{r} {}^{-}3 \\ + {}^{-}2 \\ \hline {}^{-}5 \end{array}$ or **a.** $({}^{+}3) + ({}^{+}2) = {}^{+}5$ **b.** $({}^{-}3) + ({}^{-}2) = {}^{-}5$
	Unlike Signs: Use the sign of the number with the larger *absolute value*, and subtract.	**a.** $\begin{array}{r} {}^{+}3 \\ + {}^{-}2 \\ \hline {}^{+}1 \end{array}$ **b.** $\begin{array}{r} {}^{-}3 \\ + {}^{+}2 \\ \hline {}^{-}1 \end{array}$ or **a.** $({}^{+}3) + ({}^{-}2) = {}^{+}1$ **b.** $({}^{-}3) + ({}^{+}2) = {}^{-}1$
Subtraction of Signed Numbers	Change the sign of the ***subtrahend*** (bottom number), and follow the rules for addition. or Change the problem to addition by changing the sign of the second number.	**a.** $\begin{array}{r} {}^{+}3 \\ - \overset{-}{\oplus}2 \\ \hline {}^{+}1 \end{array}$ **b.** $\begin{array}{r} {}^{-}3 \\ - \overset{+}{\ominus}2 \\ \hline {}^{-}1 \end{array}$ **c.** $\begin{array}{r} {}^{+}3 \\ - \overset{+}{\ominus}2 \\ \hline {}^{+}5 \end{array}$ **d.** $\begin{array}{r} {}^{-}3 \\ - \overset{-}{\oplus}2 \\ \hline {}^{-}5 \end{array}$ or **a.** $({}^{+}3) - ({}^{+}2) = ({}^{+}3) + ({}^{-}2) = {}^{+}1$ **b.** $({}^{-}3) - ({}^{-}2) = ({}^{-}3) + ({}^{+}2) = {}^{-}1$ **c.** $({}^{+}3) - ({}^{-}2) = ({}^{+}3) + ({}^{+}2) = {}^{+}5$ **d.** $({}^{-}3) - ({}^{+}2) = ({}^{-}3) + ({}^{-}2) = {}^{-}5$
Multiplication and Division of Signed Numbers	**Like Signs:** Answer is positive.	**a.** $({}^{+}10) \times ({}^{+}2) = {}^{+}20$ **c.** $({}^{+}10) \div ({}^{+}2) = {}^{+}5$ **b.** $({}^{-}10) \times ({}^{-}2) = {}^{+}20$ **d.** $({}^{-}10) \div ({}^{-}2) = {}^{+}5$
	Unlike Signs: Answer is negative.	**a.** $({}^{+}10) \times ({}^{-}2) = {}^{-}20$ **c.** $({}^{+}10) \div ({}^{-}2) = {}^{-}5$ **b.** $({}^{-}10) \times ({}^{+}2) = {}^{-}20$ **d.** $({}^{-}10) \div ({}^{+}2) = {}^{-}5$

EXERCISES

In 1–10, perform the indicated operations.

1. $(-2) + (+1) + (-4)$
2. $(-7) - (+2)$
3. $(+8) - (-3)$
4. $(-6) \times (+3)$
5. $(-8) \times (-4)$
6. $(-2) \times (-7) + (+3) \times (+4)$
7. $(-3) \times (+5) + (-2) \times (-6)$
8. $(+9) \times (-1) - (-8) \times (-1)$
9. $(-12) \times (-3)$
10. $(+9) \times (+4) \div (-3) \times (+4)$

In 11–16, find the value of each expression.

11. $|-8| - |2|$ **12.** $|7| + |-7|$ **13.** $|-6| - |2|$ **14.** $|7| - |-4|$

15. The value of $|-3 - 8|$ is
(1) 5 (2) -5 (3) 11 (4) -11

16. The value of $|3| + |-4| - |-2|$ is
(1) 1 (2) 9 (3) -3 (4) 5

17. Which statement is true?
(1) $|-5| < |+5|$
(2) $|+6| = |-6|$
(3) $|4| > |-4|$
(4) $|-3| - 2 = |-5|$

In 18 and 19, tell whether the sentence is true or false.

18. $\dfrac{|-16|}{|-4|} = -|+4|$ **19.** $|6| \cdot |-2| - \dfrac{|-24|}{|2|} = 0$

2.4 IDENTITIES AND INVERSES

Zero and ***one*** are special numbers for the following reasons:

1. When zero is added to any number, the number remains unchanged. Thus, zero is called the ***additive identity*** (that is, zero "preserves the identity" of the number to which it is added; adding zero does not change the number). More formally, this information is written:

   For every number, a, $a + 0 = a$ and $0 + a = a$.

2. When zero is subtracted from any number, the number remains unchanged.
3. When any number is multiplied by zero, the product (answer) is zero.
4. Zero cannot be used as a divisor. Division by zero is *undefined*. For example, the fraction $\dfrac{6}{x-2}$ is not defined when $x = 2$ since the denominator would then be zero.
5. When any number is multiplied by one, the number remains unchanged. Thus, one is called the ***multiplicative identity*** (that is, one

"preserves the identity" of the number it multiplies; multiplying by one does not change the number). More formally, this information is written:

For every number, a, $a \cdot 1 = a$ and $1 \cdot a = a$.

6. When any number is divided by one, the number remains unchanged.

Every signed number has an opposite in sign. That is, every positive number has an opposite that is negative; every negative number has an opposite that is positive. Zero is neither positive nor negative; zero is its own opposite.

When a signed number and its opposite are added, the result is zero. For example, $(+3) + (-3) = 0$; $(-2) + (+2) = 0$. The opposite of a signed number is called its ***additive inverse.*** More formally, this information is written:

For every number, a, there is an additive inverse, $-a$, such that $(a) + (-a) = 0$.

The additive inverse of a number is its opposite with respect to addition. A number also has an opposite with respect to multiplication. The opposite of a number with respect to multiplication is called its ***multiplicative inverse*** or ***reciprocal.***

The reciprocal of $\frac{3}{2}$ is $\frac{2}{3}$. Also, 5 and $\frac{1}{5}$ are reciprocals; $-\frac{1}{7}$ and -7 are reciprocals.

When a number and its reciprocal (multiplicative inverse) are multiplied, the result is one. (Examples: $\frac{3}{2} \cdot \frac{2}{3} = 1$; $5 \cdot \frac{1}{5} = 1$; $-\frac{1}{7} \cdot -7 = 1$.) **Zero does not have a reciprocal since $\frac{1}{0}$ is undefined.**

More formally, this information is written:

For every number, a, there is a multiplicative inverse (reciprocal), $\frac{1}{a}$, such that $a \cdot \frac{1}{a} = 1$. $a \neq 0$.

EXERCISES

1. For what value of x is $\frac{3}{x}$ undefined?
2. For what value of x is the expression $\frac{8}{x-7}$ undefined?
3. The expression $\frac{4}{(x-3)}$ is undefined for what value of x?
4. For what value of w is $\frac{7}{8-w}$ undefined?
5. What value of x will make $\frac{3}{x+2}$ undefined?
6. What number is the additive inverse of 9?
7. The additive inverse of 7 is (1) 1 (2) 0 (3) -7 (4) $\frac{1}{7}$

8. The additive inverse of -3 is
(1) $-\frac{1}{3}$ (2) $\frac{1}{3}$ (3) 3 (4) 0

9. What is the reciprocal of $\frac{1}{a}$? $(a \neq 0)$
(1) $-\frac{1}{a}$ (2) $\frac{1}{a^2}$ (3) $-a$ (4) a

10. The reciprocal of $\frac{1}{5}$ is (1) $-\frac{1}{5}$ (2) $\frac{1}{25}$ (3) -5 (4) 5

11. Which number does *not* have a reciprocal?
(1) 1 (2) -2 (3) $\frac{1}{3}$ (4) 0

12. The multiplicative inverse of $-\frac{1}{3}$ is
(1) $\frac{1}{3}$ (2) -3 (3) 3 (4) $.33\frac{1}{3}$

13. What is the multiplicative inverse of 7?

14. Which statement is true for the set of whole numbers (0, 1, 2, 3, etc.)?
(1) The multiplicative identity element is 0.
(2) The additive identity element is 1.
(3) The multiplicative identity element is 1.
(4) The multiplicative identity element is -1.

15. Which of the following numbers has an additive inverse and a multiplicative inverse equal to each other?
(1) 0 (2) 1 (3) -1 (4) there is no such number

2.5 THE FUNDAMENTAL LAWS

Addition and multiplication have special properties in that two numbers may be added or multiplied "forwards" or "backwards" and still give the same results. For example, $2 + 3 = 5$ and $3 + 2 = 5$; $2 \cdot 3 = 6$ and $3 \cdot 2 = 6$. More formally, this information:

is written	and is called
$a + b = b + a$	***The Commutative Property of Addition***
$a \cdot b = b \cdot a$	***The Commutative Property of Multiplication***

The commutative property is not true for subtraction or division. For example, $2 - 3 = -1$ is not equal to $3 - 2 = 1$. Also, $10 \div 2 = 5$ is not equal to $2 \div 10 = \frac{1}{5}$.

No matter how rapidly you may add (or multiply) a list of numbers, you are really working on them two at a time. Three or more numbers may be added (or multiplied) in any group of two and still yield the same result. For example, $2 + 3 + 4$ may be grouped as $(2 + 3) + 4$, giving the result $5 + 4$, or 9. Or, $2 + 3 + 4$ may be grouped as $2 + (3 + 4)$, giving the result $2 + 7$, or 9. Similarly, different groupings in multiplication yield equal results. More formally, this information:

is written	and is called
$(a + b) + c = a + (b + c)$	***The Associative Property of Addition***
$(a \cdot b) \cdot c = a \cdot (b \cdot c)$	***The Associative Property of Multiplication***

The associative property is not true for subtraction or division. The following examples show why not:

$$\begin{aligned}(2-3)-4 &\stackrel{?}{=} 2-(3-4)\\ -1-4 &\stackrel{?}{=} 2-(-1)\\ -5 &\stackrel{?}{=} 2+1\\ -5 &= 3 \quad \text{False.}\end{aligned}\qquad\qquad \begin{aligned}(10\div 2)\div 5 &\stackrel{?}{=} 10\div(2\div 5)\\ 5\div 5 &\stackrel{?}{=} 10\div \tfrac{2}{5}\\ 1 &\stackrel{?}{=} 10\times\tfrac{5}{2}\\ 1 &= 25 \quad \text{False.}\end{aligned}$$

When multiplication is combined with addition (or subtraction), sometimes the order in which the operations are carried out has no effect on the answer.

For example, consider:

$$\begin{aligned}2\cdot(3+4) &\stackrel{?}{=} 2\cdot 3+2\cdot 4\\ 2\cdot 7 &\stackrel{?}{=} 6+8\\ 14 &= 14\checkmark\end{aligned}\qquad\qquad \begin{aligned}2\cdot(3-4) &\stackrel{?}{=} 2\cdot 3-2\cdot 4\\ 2\cdot(-1) &\stackrel{?}{=} 6-8\\ -2 &= -2\checkmark\end{aligned}$$

More formally, this information:

is written	and is called
$a\cdot(b+c)=a\cdot b+a\cdot c$	***The Distributive Property of Multiplication Over Addition***
$a\cdot(b-c)=a\cdot b-a\cdot c$	***The Distributive Property of Multiplication Over Subtraction***

Sometimes, the order in which the operations are carried out is important. For example,

$$\begin{aligned}&3+4\times 5\\ =\ &3+20\\ =\ &23\end{aligned}$$

(*Note:* See Chapter 3, Section 3.2, for more about the order of operations.)

EXERCISES

1. Which is an illustration of the commutative property for multiplication?
(1) $a(b+c)=ab+ac$
(2) $a(bc)=(ab)c$
(3) $a+b=b+a$
(4) $ab=ba$

2. The equation $x+4=4+x$ is an illustration of the
(1) associative property of addition
(2) commutative property of addition
(3) associative property of multiplication
(4) commutative property of multiplication

3. Which statement illustrates the associative property for multiplication?
(1) $2(\frac{1}{2})=1$
(2) $2(1)=2$
(3) $2(3+4)=2(3)+2(4)$
(4) $2(3\cdot 4)=(2\cdot 3)4$

4. The equation $6(x + y) = 6x + 6y$ is an example of the
(1) associative law of multiplication
(2) associative law of addition
(3) commutative law of addition
(4) distributive law

5. Which is an illustration of the distributive property?
(1) $a(b + c) = ab + ac$
(2) $ab = ba$
(3) $a + (b + c) = (a + b) + c$
(4) $a + b = b + a$

6. Which is an illustration of the distributive property?
(1) $(12 \cdot 25)(4) = (12)(25 \cdot 4)$
(2) $7n + 2 - 3n = 7n - 3n + 2$
(3) $(17)(b^2)(5) = (17)(5)(b^2)$
(4) $x(x - 2) = x^2 - 2x$

7. For *each* statement in **a** through **e** write the *number* of the property of the real number system, *chosen from the list below,* that justifies that statement.

Properties

(1) Additive inverse property
(2) Multiplicative identity property
(3) Commutative property of addition
(4) Commutative property of multiplication
(5) Associative property of addition
(6) Associative property of multiplication
(7) Distributive property of multiplication over addition

a. $7 + (3 + 2) = (7 + 3) + 2$
b. $(-5)(1) = -5$
c. $3(x + 2) = 3x + 6$
d. $4 + (-4) = 0$
e. $7(8) = 8(7)$

8. Write the *number* of the property of the real number system, *chosen from the list below,* that justifies *each* of the statements **a** through **e.**

Properties

(1) Additive identity property
(2) Multiplicative inverse property
(3) Distributive property of multiplication over addition
(4) Associative property of addition
(5) Associative property of multiplication
(6) Commutative property of addition
(7) Commutative property of multiplication

a. $2(\frac{1}{2}) = 1$
b. $2(1 + 3) = 2(1) + 2(3)$
c. $2 + 1 = 1 + 2$
d. $(1 + 2) + 3 = 1 + (2 + 3)$
e. $2 + 0 = 2$

2.6 OPERATIONS WITH FRACTIONS AND DECIMALS

We have said that the set of rational numbers contains the following types of numbers:

integers
fractions
decimals that end
decimals that repeat
the square root of perfect squares
the cube root of perfect cubes, etc.

Rational numbers are also signed numbers. At the beginning of this chapter, we reviewed the rules of signs as applied to the operations of addition, subtraction, multiplication, and division. These rules must be followed when dealing with rational numbers. Let us now review the operations with fractions.

Fractions can be reduced, added, subtracted, multiplied, or divided according to the following summary:

Operation	Rule	Example
Reducing Fractions	Cancel common factors. An improper fraction can be written as a mixed number.	$\frac{7}{14} = \frac{\cancel{7} \times 1}{\cancel{7} \times 2} = \frac{1}{2}$ $\frac{5}{3} = 1\frac{2}{3}$
Addition or Subtraction of Fractions	**Like denominators:** Keep the denominator, and add or subtract the numerators.	**a.** $\frac{3}{4} + \frac{2}{4} = \frac{5}{4}$ or $1\frac{1}{4}$ **b.** $\frac{7}{8} - \frac{1}{8} = \frac{6}{8}$ or $\frac{3}{4}$
	Unlike denominators: Obtain the lowest common denominator (L.C.D.), and rewrite each fraction using the L.C.D. Now the denominators are alike and the numerators may be added or subtracted.	**a.** $\frac{1}{4} + \frac{3}{8}$ L.C.D. of 4 and 8 is 8. $= \frac{2}{8} + \frac{3}{8}$ $\frac{1}{4} \cdot \frac{2}{2} = \frac{2}{8}$ $= \frac{5}{8}$ Add numerators. **b.** $\frac{2}{3} - \frac{1}{2}$ L.C.D. of 3 and 2 is 6. $= \frac{4}{6} - \frac{3}{6}$ $\frac{2}{3} \cdot \frac{2}{2} = \frac{4}{6}, \frac{1}{2} \cdot \frac{3}{3} = \frac{3}{6}$ $= \frac{1}{6}$ Subtract numerators.

Operation	Rule	Example
Multiplication of Fractions	Multiply numerators; multiply denominators. Cancel first when possible.	**a.** $\frac{2}{3} \times \frac{1}{5} = \frac{2 \times 1}{3 \times 5} = \frac{2}{15}$ **b.** $\frac{2}{\cancel{3}_1} \times \frac{\cancel{6}^2}{7} = \frac{2 \times 2}{1 \times 7} = \frac{4}{7}$
Division of Fractions	Change to multiplication by inverting the second fraction, or multiply by the reciprocal.	$\frac{3}{4} \div \frac{1}{8}$ $= \frac{3}{4} \times \frac{8}{1}$ $= \frac{3}{\cancel{4}_1} \times \frac{\cancel{8}^2}{1}$ $= \frac{3 \times 2}{1 \times 1} = \frac{6}{1} = 6$

Examples

1. Add: $\left(+\frac{2}{5}\right) + \left(-\frac{1}{3}\right)$

Solution:

The L.C.D. of 5 and 3 is 15. Rewrite $+\frac{2}{5}$ as $+\frac{2 \cdot 3}{5 \cdot 3} = +\frac{6}{15}$. Rewrite $-\frac{1}{3}$ as $-\frac{1 \cdot 5}{3 \cdot 5} = -\frac{5}{15}$. Apply the rule of signs for addition of numbers with unlike signs: keep the sign of the number with the larger absolute value and subtract.

$$\left(+\frac{2}{5}\right) + \left(-\frac{1}{3}\right) = \left(+\frac{6}{15}\right) + \left(-\frac{5}{15}\right) = +\frac{1}{15} \quad \textit{Ans.}$$

2. Multiply: $-\frac{3}{10} \cdot -\frac{2}{9}$

Solution:

Cancel common factors. Apply the rule of signs for multiplication of numbers with like signs: the answer is positive.

$$-\frac{\cancel{3}^1}{\cancel{10}_5} \cdot -\frac{\cancel{2}^1}{\cancel{9}_3} = +\frac{1}{15} \quad \textit{Ans.}$$

When two numbers are compared by division, thus forming a fraction, the numbers are said to be in a certain ***ratio.*** For example, the fraction $\frac{1}{3}$ is the ratio of 1 to 3, also written as 1:3.

An infinite number of fractions can be reduced to the fraction $\frac{1}{3}$, the ratio $\frac{1}{3}$, or 1:3. Examples of fractions that reduce to the ratio $\frac{1}{3}$ are $\frac{3}{9}, \frac{4}{12}, \frac{5}{15}, \frac{20}{60}, \frac{100}{300}, \frac{x}{3x}$.

(*Note:* Ratio will be discussed at length in Chapter 6. Before that discussion, however, we will at times use the word *ratio*. The explanation just offered will be sufficient for understanding how to use ratio at those times.)

◆ Two fractions are compared to each other as follows:

1. If they have the *same numerator*, the fraction with the larger denominator is the smaller: $\frac{5}{8}, \frac{5}{12}$ (smaller: $\frac{5}{12}$)

2. If they have the *same denominator*, the fraction with the larger numerator is the larger: $\frac{3}{10}, \frac{7}{10}$ (larger: $\frac{7}{10}$)

3. If neither numerators nor denominators are the same, use the L.C.D. to make the denominators the same:
$$\frac{3}{5}, \frac{8}{15}$$
$$\frac{3 \cdot 3}{3 \cdot 5}, \frac{8}{15}$$
$$\frac{9}{15}, \frac{8}{15}$$
(larger: $\frac{9}{15}$)

Decimals can be added, subtracted, multiplied, or divided much like whole numbers. The important difference comes in dealing with the decimal point. The rules are summarized as follows:

Operation	Rule	Example
Addition or Subtraction of Decimals	Align the decimal points and "carry" over the decimal point of the answer.	Add: .2 + 1.3 + 2.76 = 4.26 Subtract: 2.91 − 1.23 = 1.68
Multiplication of Decimals	The number of decimal places in the product is equal to the sum of the number of decimal places in the multiplicand and the multiplier.	**a.** 7.32 (2 decimal places) × .4 (1 decimal place) 2.928 (3 decimal places) **b.** 9.14 (2 decimal places) × .23 (2 decimal places) 2742 1828 2.1022 (4 decimal places)

Operation	Rule	Example
Division of Decimals	$\text{whole number}\overline{)\text{decimal}}$ Mark the decimal point of the dividend in the quotient.	$\begin{array}{r} .9 \\ 3\overline{)2.7} \end{array}$
	$\text{decimal}\overline{)\text{whole number}}$ Remove the decimal point from the *divisor* by moving the required number of places in both the divisor and dividend. Add zeros as needed. Mark the new decimal place in the quotient.	$.3\overline{)27}$ $.3.\overline{)27.0.}$ $\begin{array}{r} 90. \\ 3\overline{)270.} \end{array}$
	$\text{decimal}\overline{)\text{decimal}}$ Remove the decimal point from the *divisor* by moving the required number of places in both the divisor and the dividend. Mark the new decimal place in the quotient. Add zeros as needed.	$.3\overline{)2.73}$ $.3.\overline{)2.7.3}$ $\begin{array}{r} 9.1 \\ 3\overline{)27.3} \end{array}$

Examples

1. Add $(-.6) + (-.41)$.

Solution:

Align the decimal points. Apply the rule of signs for addition of numbers with like signs: keep the sign and add.

$$\begin{array}{r} -.6 \\ +\ -.41 \\ \hline -1.01 \end{array} \quad \textit{Ans.}$$

2. Multiply $(-.3)(-.11)$.

Solution:

The number of decimal points in the answer will be three. Apply the rule of signs for multiplication of numbers with like signs: the answer is positive.

$$\begin{array}{r} -.11 \\ \times\ -.3 \\ \hline +.033 \end{array} \quad \textit{Ans.}$$

A fraction can be changed to a decimal by dividing the denominator of the fraction into the numerator of the fraction. Mark a decimal point in the dividend and add as many zeros to the dividend as are needed.

Examples

1. Write $\frac{1}{2}$ as a decimal.

Solution:

Divide the denominator 2 into the numerator 1.

$$\frac{1}{2} = 2\overline{)1.0}\quad \text{quotient } .5$$

One zero is needed.

Answer: $\frac{1}{2} = .5$

2. Write $\frac{3}{8}$ as a decimal.

Solution:

Divide the denominator 8 into the numerator 3.

$$\begin{array}{r} .375 \\ \frac{3}{8} = 8\overline{)3.000} \\ \underline{2\,4} \\ 60 \\ \underline{56} \\ 40 \\ \underline{40} \\ 0 \end{array}$$

Three zeros are needed.

Answer: $\frac{3}{8} = .375$

3. Write $\frac{2}{3}$ as a decimal.

Solution:

Divide the denominator 3 into the numerator 2.

$$\begin{array}{r} .666 \\ \frac{2}{3} = 3\overline{)2.000} \\ \underline{1\,8} \\ 20 \\ \underline{18} \\ 20 \\ \underline{18} \\ 2 \end{array}$$

This division will never end.

Answer: $\frac{2}{3} = .666\ldots$

A decimal can be changed to a fraction. This is done by remembering how to read the decimal places:

.	___	___	___	___	. . .
	tenths	hundredths	thousandths	ten-thousandths	

Read the decimal by its proper name (for example, .9 is read "nine tenths," not "point 9"), and write what you have read (.9 = $\frac{9}{10}$). Similarly, .64 = sixty-four hundredths = $\frac{64}{100}$. This fraction can be reduced:

$$\frac{64}{100} = \frac{\cancel{4} \times 16}{\cancel{4} \times 25} = \frac{16}{25}$$

EXERCISES

In 1–3, write the fraction in simplest form.

1. $\frac{30}{60}$ **2.** $\frac{75}{100}$ **3.** $\frac{54}{81}$

In 4–6, add. Express the answers in simplest form.

4. $(+\frac{2}{3}) + (+\frac{1}{3})$ **5.** $(-\frac{1}{5}) + (-\frac{2}{5})$ **6.** $(+\frac{3}{8}) + (-\frac{1}{16})$

In 7–9, subtract. Express the answers in simplest form.

7. $(+\frac{5}{8}) - (+\frac{3}{16})$ **8.** $(+\frac{7}{12}) - (-\frac{5}{6})$ **9.** $(-\frac{1}{9}) - (+\frac{7}{18})$

In 10–12, multiply. Express the answers in simplest form.

10. $(-\frac{1}{3})(-\frac{2}{3})$ **11.** $(-\frac{3}{10})(+\frac{2}{9})$ **12.** $(+\frac{1}{18})(-\frac{9}{10})$

In 13–15, divide. Express the answers in simplest form.

13. $(-\frac{27}{15}) \div (-\frac{9}{20})$ **14.** $(+\frac{8}{11}) \div (+2)$ **15.** $\dfrac{-\frac{1}{5}}{-\frac{1}{10}}$

16. Which fraction has the greatest value?
(1) $\frac{1}{2}$ (2) $\frac{5}{8}$ (3) $\frac{3}{5}$ (4) $\frac{3}{4}$

17. Which fraction has the smallest value?
(1) $\frac{3}{5}$ (2) $\frac{3}{7}$ (3) $-\frac{2}{7}$ (4) $-\frac{4}{7}$

18. Which fraction lies between $\frac{2}{3}$ and $\frac{4}{5}$?
(1) $\frac{5}{6}$ (2) $\frac{17}{20}$ (3) $\frac{7}{10}$ (4) $\frac{13}{15}$

19. Write .72 as a fraction in simplest form.

20. Write .375 as a fraction in simplest form.

21. Write $\frac{3}{10}$ as a decimal. **22.** Write $\frac{7}{20}$ as a decimal.

23. Write $\frac{5}{8}$ as a decimal.

24. Which decimal has the greatest value?
(1) .01 (2) .1 (3) .101 (4) .11

25. Which decimal lies between .64 and .65?
(1) .604 (2) .640 (3) .644 (4) .654

In 26–34, perform the indicated operation.

26. $(-2.1) + (+3.26)$ **29.** $(+8.1) - (+11.9)$ **32.** $.7\overline{)2.8}$

27. $(+1) + (-.64)$ **30.** $(-4.8)(+6.3)$ **33.** $.7\overline{)28}$

28. $(-7.4) - (+3.8)$ **31.** $(+3.4)(-4.1)(-1.1)(0)$ **34.** $6\overline{)12.18}$

2.7 WRITING AND ROUNDING OFF NUMBERS

A ***numeral*** is the name or symbol for a number. In the ***decimal system,*** a numeral is composed of the digits 0, 1, 2, 3, 4, 5, 6, 7, 8, 9 arranged in ***places*** that have the following names:

hundred-millions	ten-millions	millions	hundred-thousands	ten-thousands	thousands	hundreds	tens	ones	.	tenths	hundredths	thousandths	ten-thousandths

The numerical value of each place is read from its name. For example, the thousands place has a value of 1,000 and the tenths place has a value of $\frac{1}{10}$.

(*Note:* The places are actually arranged as ***consecutive integral powers*** of 10:

$$\underline{}\;\underline{}\;\underline{}\;\underline{}\;\bullet\;\underline{}\;\underline{}\;\underline{}$$
$$10^3\;10^2\;10^1\;10^0\quad 10^{-1}\;10^{-2}\;10^{-3}$$

Thus, as we move left from the decimal point, each place is 10 times the previous place; as we move right from the decimal point, each place is $\frac{1}{10}$ of the previous place.)

For example, the numeral 643 means $(6 \times 100) + (4 \times 10) + (3 \times 1)$ or $600 + 40 + 3$.

When we read the symbols for a number, we should be sure to read the place names. The decimal point is read as "and." The number 0 is a place-holder. For example:

32.7	Thirty-two and seven tenths.
4,176.01	Four thousand, one hundred seventy-six and one hundredth.

Sometimes it is more convenient to work with an ***approximation*** of an exact number. For example, it is easier to work with the number 500,000 than it is to work with the number 501,742. In the example, the number 500,000 is an approximate value of the number 501,742 when the number 501,742 has been *rounded off to the nearest hundred-thousand.* That is, 501,742 is closer to 500,000 than it is to any other hundred-thousand.

A number may be ***rounded off*** to any of its places. For example, consider the number 374.16.

374.16 is approximated as:

400 when rounded off to the *nearest hundred*
370 when rounded off to the *nearest ten*
374 when rounded off to the *nearest integer* (one)
374.2 when rounded off to the *nearest tenth*

◆ To round off a number to any of its places:

1. Identify the digit in the place to be rounded off.
2. Examine the digit in the place immediately to the right of the digit in the place to be rounded off.
   a. If the "examination digit" is less than 5, leave the digit to be rounded off as is.
   b. If the "examination digit" is exactly 5 or is more than 5, add 1 to the digit to be rounded off.
3. Replace all the digits to the right of the rounded-off digit by zeros.

Examples

Round off 9,738.562 to the nearest:

1. hundred **2.** integer **3.** tenth

Solutions:

1. 9,⑦38.562

The hundreds digit is to be rounded off.

This digit is less than 5: leave the circled digit as is. Replace all the digits to the right of the circled digit by zeros. 9,700.00 or 9,700 *Ans.*

2. 9, 73⑧.562

The ones digit is to be rounded off.

This digit is exactly 5: add 1 to the circled digit. Replace all the digits to the right of the circled digit by zeros. 9,739.000 or 9,739 *Ans.*

3. 9,738.⑤62

The tenths digit is to be rounded off.

This digit is more than 5: add 1 to the circled digit. Replace all the digits to the right of the circled digit by zeros. 9,738.600 or 9,738.6 *Ans.*

EXERCISES

In 1–3, write the number in symbols.

1. One thousand, two hundred fourteen and eight tenths.
2. Six hundred five and three hundredths.
3. Three thousand seven and twenty-three thousandths.

In 4 and 5, write the number in words.

4. 101,742.3 **5.** 20,007.05

In 6–10, round off as indicated.

6. 768.14 to the nearest ten

7. 347.58 to the nearest integer

8. 4,821.01 to the nearest thousand

9. 539.7 to the nearest integer

10. 709.5 to the nearest integer

2.8 SUMMARY EXERCISES

In 1–5, the universe = {positive integers less than 10}.
If A = {positive odd integers less than 5} and
B = {positive integers through 5}, list the elements in

1. $A \cap B$ **2.** $A \cup B$ **3.** $\sim A$ **4.** $\sim B$ **5.** $\sim A \cap \sim B$

6. Which represents a rational number?
(1) π (2) $\sqrt{2}$ (3) .121212 . . . (4) $\sqrt{3}$

7. Which represents an irrational number?
(1) $\frac{1}{2}$ (2) $\sqrt{2}$ (3) .125 (4) 12.5

8. Which of the following is *not* a subset of the {Rational Numbers}?
(1) {Integers} (2) {Natural Numbers} (3) {Irrational Numbers} (4) {Whole Numbers}

9. Which of the following statements is false?
(1) $|5 - 2| = |-3|$ (2) $|5 - 2| = |5| - |2|$ (3) $|5 - 2| = |2 - 5|$ (4) $|5 - 2| = |2| - |5|$

10. Which of the following statements is true?
(1) $\dfrac{|-20|}{-4} = |-5|$ (2) $\dfrac{-|20|}{-4} = |-5|$ (3) $\dfrac{|-20|}{|-4|} = -5$ (4) $\dfrac{-|20|}{|-4|} = 5$

11. For what value of x is $\dfrac{7}{x - 2}$ undefined?

12. The additive inverse of 10 is (1) 0 (2) 1 (3) -10 (4) $\frac{1}{10}$

13. The reciprocal of $-\frac{1}{2}$ is (1) $+\frac{1}{2}$ (2) -2 (3) $+2$ (4) 1

For each statement in 14 through 18, write the *number* of the property of the real number system, *chosen from the list below,* that justifies the statement.

14. $(3 + 4) + 5 = 3 + (4 + 5)$

15. $3 + 0 = 3$

16. $4 \cdot \frac{1}{4} = 1$

17. $2(3 + 4) = 2 \cdot 3 + 2 \cdot 4$

18. $7 \cdot 2 = 2 \cdot 7$

Properties

(1) Additive inverse property
(2) Multiplicative inverse property
(3) Additive identity property
(4) Multiplicative identity property
(5) Associative property of addition
(6) Commutative property of multiplication
(7) Distributive property of multiplication over addition

19. Which fraction has the greatest value?

(1) $\frac{3}{11}$ (2) $\frac{4}{11}$ (3) $\frac{4}{13}$ (4) $-\frac{5}{12}$

20. Which fraction is equivalent to $-2\frac{1}{4}$?

(1) $\frac{-9}{4}$ (2) $\frac{-7}{4}$ (3) $\frac{-7}{-4}$ (4) $\frac{-9}{-4}$

21. Which fraction lies between $\frac{4}{15}$ and $\frac{2}{5}$?

(1) $\frac{1}{2}$ (2) $\frac{1}{3}$ (3) $\frac{1}{20}$ (4) $\frac{1}{30}$

In 22–25, perform the indicated operation. Express the answer in simplest form.

22. $(-\frac{1}{4}) + (-\frac{3}{8})$

23. $(-\frac{5}{27}) - (-\frac{2}{9})$

24. $(-\frac{1}{3})(-\frac{6}{5})(-\frac{15}{20})$

25. $(+\frac{3}{8}) \div (-\frac{9}{16})$

26. Write .28 as a fraction in simplest form.

27. Write $\frac{1}{20}$ as a decimal.

28. Write $\frac{3}{8}$ as a decimal.

29. Which decimal has the greatest value?

(1) .2 (2) .02 (3) .22 (4) .022

30. Which decimal lies between .73 and .74?

(1) .730 (2) .740 (3) .731 (4) .741

In 31–36, perform the indicated operation.

31. $(-3.9) + (-9.3)$

32. $(+1.2) - (+1.2)$

33. $(+3.14)(+\frac{1}{2})$

34. $(-6.1)(-8.2)$

35. $(.4) \div (\frac{1}{2})$

36. $(1.2) \div (.2)$

In 37 and 38, write the number in symbols.

37. Two hundred four and one tenth.

38. Three thousand three and three thousandths.

39. Write the number 3,042.02 in words.

In 40–43, round off as indicated.

40. 76.57 to the nearest integer

41. 41.98 to the nearest tenth

42. 732.6 to the nearest hundred

43. 5,643.1 to the nearest thousand

CHAPTER 3

Monomials and Polynomials

3.1 DEFINITIONS AND OPERATIONS

Definitions

An algebraic ***term*** is composed of three parts:

$$\underset{\textit{coefficient}\leftarrow}{3}\,\underset{\rightarrow\textit{base}}{x}^{2\rightarrow\textit{exponent}}$$

The coefficient may be 1, in which case it is not written (for example: x^2 means $1x^2$).

The exponent may be 1, in which case it is not written (for example: $3x$ means $3x^1$). (*Note:* "Exponent" will be discussed at length in the next section of this chapter.)

The coefficient may appear without a base and exponent (for example: 3). This kind of term is called a ***constant term,*** or a ***constant.***

In a term, the coefficient is attached to the base by multiplication or division only. That is, $3x^2$ is a term and $\frac{3}{x^2}$ is a term; but $3 + x^2$ and $3 - x^2$ are not examples of one term.

In a term, there may be more than one letter base with a corresponding exponent. For example: $3xy$, $3x^2yz^3$, and $\frac{3xy^2}{z^2}$ are terms.

If two or more terms have the *same letter base and the same exponent,* they are called ***like terms.*** Only like terms may be ***combined*** (added or subtracted) to form fewer terms. The combining of like terms is done by adding or subtracting the coefficients; the like base and exponent are carried along.

For example:

1. $3x^2 + 4x^2 = 7x^2$.
2. $3x^2 + 4y^2$ cannot be combined.
3. $3x^2 - 4x^3$ cannot be combined.
4. $3xy + 4xy = 7xy$.
5. $3x^2y + 4x^2y = 7x^2y$.
6. $3x^2y + 4xy^2$ cannot be combined.

Some terms are also called ***monomials.*** Excluded from the classification "monomial" are those terms in which the letters appear in the denominator. Thus, the term $3x^2$ is a monomial, but the term $\frac{3}{x^2}$ is not called a monomial. (*Note:* An explanation for this classification will be given later in this chapter.) The term $\frac{x^2}{3}$ is a monomial.

The sum or difference of two monomials that cannot be combined is called a ***binomial.*** For example, $3x^2 + 4y^2$ is a binomial.

Similarly, the sums or differences of three monomials that cannot be combined is called a ***trinomial.*** For example, $3x + y - 2z$ is a trinomial.

In general, the sums or differences of a series of monomials is called a ***polynomial.***

Combining Polynomials by Addition or Subtraction

We have seen that only like monomials can be combined by addition or subtraction. Similarly, when adding or subtracting two or more polynomials, we can combine only the like terms.

The addition of polynomials can be shown vertically or horizontally.

Examples

1. Add $(3x + 4y - z)$ and $(2x - 3y + 6z)$.

Solution:

$$\begin{array}{r} 3x + 4y - z \\ + \; 2x - 3y + 6z \\ \hline 5x + y \; + 5z \end{array} \quad \textit{Ans.}$$

or

$$\begin{aligned} &(3x + 4y - z) + (2x - 3y + 6z) \\ &= (3x + 2x) + (4y - 3y) + (-z + 6z) \\ &= 5x + y + 5z \quad \textit{Ans.} \end{aligned}$$

2. Find the sum of $(4x^2 + 6x - 3yz)$ and $(-2x^2 - 8yz)$.

Solution:

$$\begin{array}{r} 4x^2 + 6x - 3yz \\ + \; -2x^2 \qquad - 8yz \\ \hline 2x^2 + 6x - 11yz \end{array} \quad \textit{Ans.}$$

or

$$\begin{aligned} &(4x^2 + 6x - 3yz) + (-2x^2 - 8yz) \\ &= (4x^2 - 2x^2) + (6x) + (-3yz - 8yz) \\ &= 2x^2 + 6x - 11yz \quad \textit{Ans.} \end{aligned}$$

3. Find the sum of $(2x + 4y - z)$, $(-3y + z)$, and $(3x - y - 3z)$.

Solution:

$$\begin{array}{r} 2x + 4y - z \\ - 3y + z \\ + \; 3x - y \; - 3z \\ \hline 5x + 0 \cdot y - 3z \\ = 5x - 3z \quad \textit{Ans.} \end{array}$$

or

$$\begin{aligned} &(2x + 4y - z) + (-3y + z) + (3x - y - 3z) \\ &= (2x + 3x) + (4y - 3y - y) + (-z + z - 3z) \\ &= 5x + 0 \cdot y - 3z \\ &= 5x - 3z \quad \textit{Ans.} \end{aligned}$$

The subtraction of polynomials can also be shown vertically or horizontally. Remember that algebraic subtraction requires a change of sign, since we are working with signed numbers. In the vertical subtraction, we show this change by changing every sign in the *subtrahend* (the bottom polynomial). In the horizontal subtraction, we show this change by changing the problem to addition. To do this, we change the subtraction sign to addition and we change the sign of *every term* of the second polynomial.

Examples

1. From $(3x + 4y - 2z)$ subtract $(2x - 3y - 6z)$.

Solution:

$$\begin{array}{r} 3x + 4y - 2z \\ -\ \overset{-}{\oplus}2x \overset{+}{\ominus} 3y \overset{+}{\ominus} 6z \\ \hline x + 7y + 4z \quad \textit{Ans.} \end{array}$$

or

$$\begin{aligned} &(3x + 4y - 2z) - (2x - 3y - 6z) \\ &= (3x + 4y - 2z) + (-2x + 3y + 6z) \\ &= (3x - 2x) + (4y + 3y) + (-2z + 6z) \\ &= x + 7y + 4z \quad \textit{Ans.} \end{aligned}$$

Note that it is a good idea to circle the old sign and write the new sign above it.

2. Subtract $(3x - y)$ from $(3x + y)$.

Solution:

$$\begin{array}{lr} \text{From} & 3x + y \\ \text{Subtract} & \overset{-}{\oplus}3x \overset{+}{\ominus} y \\ \hline & 0 \cdot x + 2y \\ & = 2y \quad \textit{Ans.} \end{array}$$

or

From $(3x + y)$ subtract $(3x - y)$.

$$\begin{aligned} &= (3x + y) - (3x - y) \\ &= (3x + y) + (-3x + y) \\ &= (3x - 3x) + (y + y) \\ &= 0 \cdot x + 2y \\ &= 2y \quad \textit{Ans.} \end{aligned}$$

Note that the word "from" signals the top or first polynomial (the *minuend*) and "subtract" signals the subtrahend.

3. From the sum of $(x + 2y)$ and $(2x - y)$, subtract $(3x - y)$.

Solution:

This problem requires an addition followed by a subtraction:

$$\begin{array}{r} x + 2y \\ +\ \ 2x - y \\ \hline 3x + y \\ -\ \overset{-}{\oplus}3x \overset{+}{\ominus} y \\ \hline 0 \cdot x + 2y \\ = 2y \quad \textit{Ans.} \end{array}$$

or

$$\begin{aligned} &(x + 2y) + (2x - y) - (3x - y) \\ &= (x + 2y) + (2x - y) + (-3x + y) \\ &= (x + 2x - 3x) + (2y - y + y) \\ &= 0 \cdot x + 2y \\ &= 2y \quad \textit{Ans.} \end{aligned}$$

EXERCISES

1. Express as a binomial: $9x - 5y - 4x - y$
2. Find the sum of $5x^2 - 3x + 2$ and $3x^2 - 4x - 1$.
3. Find the sum of $4a + 2b - c$ and $3a - 5b - 2c$.
4. Find the sum of $2x^2 + 3x - 1$ and $3x^2 - 2x + 4$.
5. Find the sum of $3x^2 + 2x - 4$ and $2x^2 - 3x + 1$.
6. Express the sum of $3n^2 - n - 2$ and $n^2 + 2n - 1$ as a trinomial.

7. From $5x^2 - 3x + 2$ subtract $3x^2 - 4x - 1$.
8. Subtract $4m - h$ from $4m + h$.
9. The sum of two binomials is $4x^2 + 7x$. One of the binomials is $3x^2 + 2x$. What is the other binomial?
10. From the sum of $5x - y$ and $-3x + 2y$, subtract $-2x + 2y$.
11. Which must be added to $2x - 4$ to produce a sum of 0?
    (1) 0 (2) $x + 2$ (3) $2x + 4$ (4) $-2x + 4$
12. From the sum of $-6x + y$ and $2x - y$, subtract $-3x + y$.

Multiplication

◆ Any two terms can be multiplied. To perform the multiplication, we proceed as follows:

1. Multiply the numerical coefficients and write that product as a single number. Remember the laws of signs for multiplication.
2. Multiply the like letter bases by adding their exponents.
3. Multiply the unlike letter bases by simply showing a product.

Examples

1. Multiply $(3x^4)(-2x^2)$.

Solution:

$$(3x^4)(-2x^2) = (3)(-2) \cdot x^{4+2} = -6x^6 \quad Ans.$$

2. Multiply $(-2x)(-3x^3)$.

Solution:

$$\begin{aligned}(-2x)(-3x^3) = (-2)(-3) \cdot x^{1+3} &= +6x^4 \\ &= 6x^4 \quad Ans.\end{aligned}$$

3. Multiply $(-2x^2y)(4xy^2z)$.

Solution:

$$\begin{aligned}(-2x^2y)(4xy^2z) &= (-2)(4) \cdot x^{2+1} \cdot y^{1+2} \cdot z \\ &= -8x^3y^3z \quad Ans.\end{aligned}$$

4. Multiply $(-3xy^2)^2$.

Solution:

$$\begin{aligned}(-3xy^2)^2 &= (-3xy^2)(-3xy^2) \\ &= (-3)(-3) \cdot x^{1+1} \cdot y^{2+2} \\ &= +9x^2y^4 \\ &= 9x^2y^4 \quad Ans.\end{aligned}$$

A polynomial can be multiplied by a monomial. To do this multiplication, we apply the distributive property. That is, we multiply each term of the polynomial by the monomial, keeping the rules for the multiplication of two terms in mind.

Examples

1. Multiply $(3x^3 + 2x^2 - 7x - 3)$ by $4x^2$.

Solution:

$$\begin{aligned} 4x^2(3x^3 + 2x^2 - 7x - 3) &= 4x^2(3x^3) + 4x^2(2x^2) + 4x^2(-7x) + 4x^2(-3) \\ &= (4)(3) \cdot x^{2+3} + (4)(2) \cdot x^{2+2} + (4)(-7) \cdot x^{2+1} \\ &\quad + (4)(-3) \cdot x^2 \\ &= 12x^5 + 8x^4 - 28x^3 - 12x^2 \quad \textit{Ans.} \end{aligned}$$

2. Multiply $3x^2y(2x^2 + 3y^2)$.

Solution:

$$\begin{aligned} 3x^2y(2x^2 + 3y^2) &= 3x^2y(2x^2) + 3x^2y(3y^2) \\ &= (3)(2) \cdot x^{2+2}y + (3)(3)\, x^2 \cdot y^{1+2} \\ &= 6x^4y + 9x^2y^3 \quad \textit{Ans.} \end{aligned}$$

A polynomial can be multiplied by another polynomial. To do this multiplication, we multiply each term of the first polynomial by each term of the second polynomial, and combine like terms.

(*Note:* We will use a vertical arrangement for this type of multiplication. An alternate form is shown for the case of a "binomial times a binomial.")

Examples

1. Multiply $(2x - 3)$ by $(-3x + 2)$.

Solution 1:

$$\begin{array}{r} 2x - 3 \\ -3x + 2 \\ \hline \end{array}$$

$-6x^2 + 9x$ Multiply each term of $2x - 3$ by $-3x$.

That is, $$\begin{aligned} -3x(2x - 3) &= -3x(2x) + (-3x)(-3) \\ &= (-3)(2) \cdot x^{1+1} + (-3)(-3) \cdot x \\ &= -6x^2 + 9x \end{aligned}$$

$+ 4x - 6$ Multiply each term of $2x - 3$ by $+2$.

That is, $$\begin{aligned} +2(2x - 3) &= (+2)(2x) + (+2)(-3) \\ &= (+2)(2) \cdot x + (+2)(-3) \\ &= 4x - 6 \end{aligned}$$

$-6x^2 + 13x - 6$ Align the like terms and combine.

Answer: $-6x^2 + 13x - 6$

Solution 2: This method is sometimes called the FOIL method of multiplying two binomials.

First, Inner, Last, Outer

$$(2x - 3)(-3x + 2) = \underbrace{(2x)(-3x)}_{\text{First}} + \underbrace{(2x)(+2)}_{\text{Outer}} + \underbrace{(-3)(-3x)}_{\text{Inner}} + \underbrace{(-3)(+2)}_{\text{Last}}$$

$$= -6x^2 + \underbrace{4x + 9x}_{\substack{\text{combine}\\ \text{Outer and Inner}}} - 6$$

$$= -6x^2 + 13x - 6 \quad Ans.$$

A shorter way of showing this FOIL method is:

$$\overset{+9x}{}$$
$$(2x - 3)(-3x + 2) = -6x^2 + 13x - 6 \quad Ans.$$
$$\underset{+4x}{}$$

Here we do the "First" multiplication and the "Last" multiplication mentally, and we write their results in the first and last places of the answer. We do the "Inner" multiplication and the "Outer" multiplication mentally, and we write their results above and below the problem. Then, we mentally combine the results of the "Inner" and "Outer" multiplications, and we write that result in the middle place of the answer. Note that the results of the "Inner" and "Outer" multiplications cannot always be combined. For example: $(x + y)(x + z)$.

2. Multiply $(2x + y)(3x - y)$.

Solution:

$$\begin{array}{r} 2x + y \\ \underline{3x - y } \\ 6x^2 + 3xy \\ \underline{- 2xy - y^2} \\ 6x^2 + xy - y^2 \end{array} \; Ans.$$

or

$$\overset{+3xy}{}$$
$$(2x + y)(3x - y) = 6x^2 + xy - y^2 \quad Ans.$$
$$\underset{-2xy}{}$$

3. $(3x - 4)(3x + 4)$

Solution:

$$\begin{array}{r} 3x - 4 \\ \underline{3x + 4 } \\ 9x^2 - 12x \\ \underline{+ 12x - 16} \\ 9x^2 + 0 \cdot x - 16 \end{array} = 9x^2 - 16 \quad Ans.$$

or

$$\overset{-12x}{}$$
$$(3x - 4)(3x + 4) = 9x^2 - 16 \quad Ans.$$
$$\underset{+12x}{}$$

(*Note:* When the two binomials are identical except for opposite second signs, the combination of the inner and outer multiplications will always be zero.)

The procedure for multiplying a polynomial by a polynomial can be extended to include any number of terms. Our work is limited to the product of two binomials.

EXERCISES

1. The product of $7y^3$ and $4y^5$ is
   (1) $11y^8$ (2) $11y^{15}$ (3) $28y^8$ (4) $28y^{15}$
2. The product of $6x^3$ and $5x^4$ is
   (1) $11x^{12}$ (2) $11x^7$ (3) $30x^{12}$ (4) $30x^7$
3. The product of $3x^5$ and $2x^4$ is
   (1) $5x^9$ (2) $5x^{20}$ (3) $6x^9$ (4) $6x^{20}$
4. What is the product $(-3x^2)(4x^3)$?
   (1) $-12x^5$ (2) $-12x^6$ (3) $7x^5$ (4) $12x^5$
5. When $4x^3y^3$ is multiplied by $8xy^2$, the product is
   (1) $12x^3y^6$ (2) $12x^4y^6$ (3) $32x^3y^6$ (4) $32x^4y^5$
6. When $4x^3y^5$ is multiplied by $3xy^2$ the product is
   (1) $7x^3y^7$ (2) $7x^4y^{10}$ (3) $12x^4y^7$ (4) $12x^3y^{10}$
7. When $2x^2y$ is multiplied by $3x^3y^4$, the product is
   (1) $5x^5y^5$ (2) $6x^5y^5$ (3) $5x^6y^4$ (4) $6x^6y^4$
8. The product $(3x^2y^2)(4xy^3)$ is
   (1) $12x^3y^5$ (2) $12x^2y^6$ (3) $7x^3y^5$ (4) $7x^2y^6$
9. When $5x^4y^3$ is multiplied by $2x^2y^3$, the product is
   (1) $10x^8y^9$ (2) $10x^6y^6$ (3) $7x^8y^9$ (4) $7x^6y^6$

In 10–15, express each product as a trinomial.

10. $(x - 5)(x + 3)$
11. $(x + 2)(x - 5)$
12. $(x + 5)(x + 7)$
13. $(2x - 3)(x + 4)$
14. $(5x + 1)(x - 2)$
15. $(3x - 2)(2x + 3)$

In 16–20, express each product as a binomial.

16. $(a - b)(a + b)$
17. $(x - 3)(x + 3)$
18. $(x + 5)(x - 5)$
19. $(x + .2)(x - .2)$
20. $(x + \frac{1}{2})(x - \frac{1}{2})$

Division

◆ Any two terms can be divided. To perform the division, we proceed as follows:

1. Divide the numerical coefficients, and write the quotient in simplest form. Remember the laws of signs for division.
2. Divide like letter bases by subtracting their exponents; attach this result to the coefficient by multiplication.

(*Note:* The discussion of division is limited to that of two terms. Division involving polynomials is left for future study.)

Examples

1. Divide $20x^5$ by $4x^3$.

Solution:

$(20x^5) \div (4x^3) = (20 \div 4) \cdot x^{5-3} = 5x^2$ *Ans.*

2. Divide $14x^4y^3$ by $2x^3y$.

Solution:

$(14x^4y^3) \div (2x^3y) = (14 \div 2) \cdot x^{4-3} \cdot y^{3-1} = 7xy^2$ *Ans.*

An alternate form for showing the division is to write the problem as a fraction and to ***cancel.*** Cancellation here means to *factor* the numerical coefficients and to *divide* by the common factors, and then to do the same for the letter bases with their exponents. See how this is done for the preceding examples:

1. $(20x^5) \div (4x^3) = \dfrac{20x^5}{4x^3} = \dfrac{5 \cdot \cancel{4} \cdot \cancel{x} \cdot \cancel{x} \cdot \cancel{x} \cdot x \cdot x}{1 \cdot \cancel{4} \cdot \cancel{x} \cdot \cancel{x} \cdot \cancel{x}} = 5x^2$ *Ans.*

2. $(14x^4y^3) \div (2x^3y) = \dfrac{14x^4y^3}{2x^3y} = \dfrac{7 \cdot \cancel{2} \cdot \cancel{x} \cdot \cancel{x} \cdot \cancel{x} \cdot x \cdot \cancel{y} \cdot y \cdot y}{1 \cdot \cancel{2} \cdot \cancel{x} \cdot \cancel{x} \cdot \cancel{x} \cdot \cancel{y}} = 7xy^2$ *Ans.*

EXERCISES

1. Divide $18x^{12}$ by $3x^3$ and express the result as a *single term.*

2. When $-15x^6$ is divided by $-5x^3$, the quotient is

(1) $3x^2$ (2) $-3x^2$ (3) $3x^3$ (4) $-3x^3$

3. The quotient $\dfrac{12x^3y^6}{4xy^2}$ is equivalent to

(1) $3x^2y^3$ (2) $3x^2y^4$ (3) $3x^3y^4$ (4) $3x^3y^3$

4. The expression $\dfrac{-12x^6y^2}{3x^2y}$ is equivalent to

(1) $9x^3y^2$ (2) $4x^4y$ (3) $-4x^3y$ (4) $-4x^4y$

5. The quotient $\dfrac{8x^4y^3}{2xy^3}$ is equivalent to

(1) $6x^3$ (2) $4x^3$ (3) $4x^4$ (4) $4x^3y$

In 6–8, divide and express the result as a single term.

6. $\dfrac{-24x^2y}{-2xy}$ **7.** $\dfrac{21a^2b^2}{-7a^2b^2}$ **8.** $\dfrac{-56xyz^2}{8xyz}$

3.2 EXPONENTS

The Laws of Exponents and the Order of Operations

From the two examples on page 58 showing cancellation, we also see the meaning of the word "exponent." An *exponent tells how many times a base is used as a factor.*

The base may be a letter: x^3 means $x \cdot x \cdot x$

The base may be a number: 6^3 means $6 \cdot 6 \cdot 6$, which is equivalent to 216

The base may be a combination of a number and letter(s):

$(6x)^3$ means $6x \cdot 6x \cdot 6x$, which is equivalent to $216x^3$
$(6xy)^3$ means $6xy \cdot 6xy \cdot 6xy$, which is equivalent to $216x^3y^3$
$(x^2)^3$ means $x^2 \cdot x^2 \cdot x^2$, which is equivalent to x^6

Combining the preceding explanation with the rules previously given for multiplication and division of terms, we may now state the following laws for exponents:

1. $x^a \cdot x^b = x^{a+b}$ ***The Law of Exponents Relating to Multiplication***
2. $x^a \div x^b = x^{a-b}$ ***The Law of Exponents Relating to Division***
3. $(x^a)^b = x^{a.b}$ ***The Law of Exponents Relating to Powers***
   Note that x^a is called a *power* of x.

◆ Let us summarize the rules for exponents:

1. When multiplying with like bases, we add the exponents. For example:

$$x^2 \cdot x^3 = x^{2+3} = x^5$$
$$2^2 \cdot 2^3 = 2^{2+3} = 2^5 \text{ or } 32$$
$$4x^2 \cdot 2x^3 = 8x^{2+3} = 8x^5$$

2. When dividing with like bases, we subtract the exponents. For example:

$$x^7 \div x^2 = x^{7-2} = x^5$$
$$\text{or} \quad \frac{x^7}{x^2} = x^{7-2} = x^5$$
$$\text{or} \quad \frac{x^7}{x^2} = \frac{\not{x} \cdot \not{x} \cdot x \cdot x \cdot x \cdot x \cdot x}{\not{x} \cdot \not{x}} = x^5$$

3. When a power (a base with an exponent) has an exponent, the result is the same as if the exponents were multiplied. For example:

$$(x^3)^2 = x^3 \cdot x^3 = x^{3+3} = x^6$$

4. Only like terms can be combined by addition or subtraction. In these cases, the exponents are just carried along. For example:

$$3x^2 + 4x^2 = 7x^2$$
but $3x^2 + 4x^3$ cannot be combined

Following are numerical examples applying the meaning of exponents and the laws telling how to operate with exponents.

Examples

1. Evaluate (find the numerical value of) 3^4.

Solution:

$$3^4 = 3 \cdot 3 \cdot 3 \cdot 3$$
$$= 81 \quad \textit{Ans.}$$

2. Evaluate x^2y^3 when $x = 3$ and $y = 2$.

Solution:

$$x^2 \cdot y^3$$
$$= 3^2 \cdot 2^3 = 3 \cdot 3 \cdot 2 \cdot 2 \cdot 2 = 9 \cdot 8 = 72 \quad \textit{Ans.}$$

3. Evaluate $(xy)^2$ when $x = 2$ and $y = 4$.

Solution:

$$(xy)^2$$
$$= (2 \cdot 4)^2 = 8^2 = 8 \cdot 8 = 64 \quad \textit{Ans.}$$

Note that the work within the parentheses was done first.

4. Evaluate $x^2y - 4xy^2$ when $x = 3$ and $y = 2$.

Solution:

$$x^2y - 4xy^2$$
$$= 3^2 \cdot 2 - 4 \cdot 3 \cdot 2^2$$
$$= 3 \cdot 3 \cdot 2 - 12 \cdot 2 \cdot 2$$
$$= 9 \cdot 2 - 12 \cdot 4$$
$$= 18 - 48$$
$$= -30 \quad \textit{Ans.}$$

Note that the work with the exponents was done first. The work with multiplication was done next. Finally, the work with subtraction (or addition) was done.

When doing calculations, we follow a certain order in carrying out the operations.

◆ In general, the ***order of operations*** is:

1. Do all work within parentheses first.
2. Do all other work with exponents next.
3. Complete multiplications and divisions next, working from left to right.
4. Do additions and subtractions last, working from left to right.

Examples

1. Evaluate $4(x^2 + y^3)^2$ when $x = 2$ and $y = -1$.

Solution:

$$\begin{aligned} &4(x^2 + y^3)^2 \\ &= 4[2^2 + (-1)^3]^2 \\ &= 4[2 \cdot 2 + (-1)(-1)(-1)]^2 \\ &= 4(4 - 1)^2 \\ &= 4(3)^2 \\ &= 4 \cdot 3 \cdot 3 = 4 \cdot 9 = 36 \quad \textit{Ans.} \end{aligned}$$

2. Evaluate $100 \div 25 \times 2 - 5 \times 3$.

Solution:

$$\begin{aligned} &100 \div 25 \times 2 - 5 \times 3 \\ &= 4 \times 2 - 5 \times 3 \\ &= 8 - 15 \\ &= -7 \quad \textit{Ans.} \end{aligned}$$

EXERCISES

1. What is the value of $3x^3$ if $x = 2$?
2. Find the value of xy^2 if $x = 3$ and $y = -2$.
3. Find the value of the expression $2xy^2$ when $x = -2$ and $y = 3$.
4. If $x = -3$ and $y = 2$, find the value of $(xy)^2$.
5. If $x = \frac{1}{2}$, find the value of $(4x)^3$.
6. Find the value of the expression $2x^2y$ when $x = -1$ and $y = 3$.
7. If $x = 3$ and $y = -1$, find the value of $2x + y^2$.
8. Find the numerical value of the expression $c + xy$ when $c = 3, x = 4$, and $y = -5$.
9. If $y = ax^2$, what is the numerical value of y when $a = 2$ and $x = 3$?
(1) 12 (2) 18 (3) 25 (4) 36
10. If $b = -2$ and $c = 3$, find the value of $b^2 + c$.
11. If $x = 2$ and $y = \frac{1}{2}$, find the value of $x^2 - 4y$.
12. Find the value of $6b^2 - 4a^2$ when $b = 2$ and $a = 1$.
13. What is the value of the expression $4x + 2$ when $x = 3$?
14. The product of $5x$ and $(3x)^2$ is
(1) $30x^2$ (2) $15x^3$ (3) $45x^3$ (4) $225x^4$
15. What is the product of $(2x)^3$ and $(3x)^2$?

The Zero Exponent

The laws of exponents lead to situations in which zero and negative numbers do arise as exponents. (*Note:* Fractional numbers as exponents can also arise under certain circumstances. These are left for future study.)

Suppose we wish to divide x^5 by x^5. One way to do this division is to write the problem as a fraction and to cancel common factors. That is:

$$x^5 \div x^5 = \frac{x^5}{x^5} = \frac{\overset{1}{\cancel{x} \cdot \cancel{x} \cdot \cancel{x} \cdot \cancel{x} \cdot \cancel{x}}}{\underset{1}{\cancel{x} \cdot \cancel{x} \cdot \cancel{x} \cdot \cancel{x} \cdot \cancel{x}}} = 1 \quad Ans.$$

Another way to do the same division is to follow the law of exponents relating to division. Since the bases are alike, the law applies. We keep the base and subtract the exponents. That is:

$$x^5 \div x^5 = x^{5-5} = x^0 \quad Ans.$$

What has happened is that we did one problem in two different ways and came up with what appear to be two different answers, x^0 and 1. It is important to keep mathematics *consistent,* that is, to make sure that one rule does not contradict another rule or that no matter what method is used, the answer to one problem is always the same. We must therefore say that the two answers, x^0 and 1, are equivalent. Thus, we define what we mean by a zero exponent:

$$x^0 = 1 \text{ where } x \text{ is a real number} \neq 0$$

◆ In calculations, replace whatever base has been raised to the zero power by the number 1.

For example:

1. $3x^0$ means $3(1) = 3$
2. $(3x)^0$ means 1
3. $3(2x^0)$ means $3(2 \cdot 1) = 3(2) = 6$
4. $3(2x)^0$ means $3(1) = 3$
5. $3^0 \cdot x$ means $1 \cdot x = x$

Negative Exponents

Suppose we wish to divide x^5 by x^7. One way to do this division is to write the problem as a fraction and to cancel common factors. That is:

$$x^5 \div x^7 = \frac{x^5}{x^7} = \frac{\overset{1}{\cancel{x} \cdot \cancel{x} \cdot \cancel{x} \cdot \cancel{x} \cdot \cancel{x}}}{\cancel{x} \cdot \cancel{x} \cdot \cancel{x} \cdot \cancel{x} \cdot \cancel{x} \cdot x \cdot x} = \frac{1}{x^2} \quad Ans.$$

Another way to do the same division is to follow the law of exponents relating to division. Since the bases are alike, the law applies. We keep the base and subtract the exponents. That is:

$$x^5 \div x^7 = x^{5-7} = x^{-2} \quad Ans.$$

Again, for the sake of consistency, the two answers must be equivalent. That is:

$$x^{-2} = \frac{1}{x^2}$$

◆ In general, the definition of a negative exponent is:

$$x^{-n} = \frac{1}{x^n} \quad \text{where } x \text{ is a real number} \neq 0$$

We limit our study to the case where n is an integer.

For example:

1. 5^{-1} means $\frac{1}{5^1} = \frac{1}{5}$
2. 5^{-2} means $\frac{1}{5^2} = \frac{1}{5 \cdot 5} = \frac{1}{25}$
3. 2^{-4} means $\frac{1}{2^4} = \frac{1}{2 \cdot 2 \cdot 2 \cdot 2} = \frac{1}{16}$
4. $3 \cdot 4^{-2}$ means $3 \cdot \frac{1}{4^2} = 3 \cdot \frac{1}{4 \cdot 4} = 3 \cdot \frac{1}{16} = \frac{3}{16}$
5. $\frac{2}{3^{-3}}$ means $\frac{2}{\frac{1}{3^3}} = \frac{2}{\frac{1}{3 \cdot 3 \cdot 3}} = \frac{2}{\frac{1}{27}} = \frac{2}{1} \cdot \frac{27}{1} = 54$

Note that another way of thinking of negative exponents in a calculation is to *replace the base having the negative exponent by* THE RECIPROCAL *of the base having the same absolute value positive exponent.* See the preceding examples again.

EXERCISES

1. What is the value of $5x^0$?
(1) 0 (2) 1 (3) 5 (4) $\frac{1}{5}$

2. What is the value of $4(2x)^0$?
(1) 0 (2) 4 (3) 8 (4) $\frac{1}{8}$

3. The value of 3^{-2} is
(1) -9 (2) -6 (3) $-\frac{1}{9}$ (4) $\frac{1}{9}$

4. The value of 4^{-3} is
(1) -12 (2) $\frac{1}{12}$ (3) $\frac{1}{64}$ (4) 64

5. The value of $2 \cdot 5^{-2}$ is
(1) $\frac{2}{25}$ (2) $\frac{1}{50}$ (3) 50 (4) -20

6. The value of $\frac{2}{4^{-2}}$ is
(1) $\frac{1}{4}$ (2) 4 (3) 32 (4) 64

3.3 ORDER AND DEGREE OF A POLYNOMIAL

At the beginning of this chapter, we said that terms in which the letters appear in the denominator are not called monomials. Thus, the term $\frac{3}{x^2}$ is not called a monomial. By relating what we have just learned about negative exponents to the concept of the order and degree of a polynomial, we can see why *only some single terms are classified as monomials.*

When we discuss a polynomial in one variable, we usually arrange the polynomial in *order*. We can arrange the polynomial in ***descending order*** (the highest exponent of the variable down to the lowest exponent of the variable, as in $5x^3 + 7x^2 + 8x - 4$). We can also arrange the polynomial in ***ascending order*** (the lowest exponent of the variable up to the highest exponent of the variable, as in $-4 + 8x + 7x^2 + 5x^3$). When a polynomial in one variable has been arranged in either descending or ascending order, we speak of the ***degree of the polynomial,*** which is the highest exponent of the variable. The degree of the polynomial $5x^3 + 7x^2 + 8x - 4$ is 3. The degree of the polynomial $7 - x + 5x^2$ is 2. (*Note:* It is not necessary to first arrange the polynomial in some order. The degree of the polynomial is always the highest exponent.)

If we were to allow terms such as $\frac{3}{x^2}$ to be classified as monomials, the exponent of that term, -2 $\left(\frac{3}{x^2}\right.$ is equivalent to $\left.3 \cdot x^{-2}\right)$, would add an unnecessary complication to the idea of the degree of a polynomial. (Similarly, single terms that lead to fractional exponents are not classified as monomials.) The degree of a polynomial becomes important in future studies. We will have occasion in this course to speak of the degree of an equation.

At this time, you should be able to order a polynomial and tell its degree.

Examples

1. Arrange the following polynomial in *descending* order and tell its degree: $5x + 7x^3 - 2 + 8x^2$

   Solution:

   Descending order means the highest exponent of the variable down to the lowest exponent of the variable. Thus, $7x^3 + 8x^2 + 5x - 2$ is in descending order. *Ans.*

   The degree of the polynomial means the value of the highest exponent of the variable. The degree of the polynomial is 3. *Ans.*

2. Arrange the following polynomial in *ascending* order and tell its degree: $7x^4 - 3x^2 + 8 + 6x$

   Solution:

   Ascending order means the lowest exponent of the variable up to the highest exponent of the variable. Thus, $8 + 6x - 3x^2 + 7x^4$ is in ascending order and the degree of the polynomial is 4. *Ans.*

Note that a constant can be thought of as a polynomial of degree zero. For example, the constant 3 is equivalent to $3x^0$, since $x^0 = 1$.

When a polynomial has two variables, we usually arrange it if we can so that one variable is in descending order, while the other variable is in ascending order, or conversely. For example:

$x^3 + 2x^2y - 3xy^2 + y^3$ is in descending order of x and ascending order of y

$7y^3 + xy^2 - x^2y + 8x^3$ is in ascending order of x and descending order of y

EXERCISES

1. Arrange the polynomial $3x^2 - 7 + 4x$ in ascending order.
2. Arrange the polynomial $6 - x^3 + 3x$ in descending order.
3. What is the degree of the polynomial $7x^2 - 8x + 4$?
4. What is the degree of the polynomial $5x - 3$?
5. The degree of a polynomial that is a constant is
   (1) 0 (2) 1 (3) that constant (4) not defined
6. Arrange the polynomial $x^4y - 3x^2y^3 + x^3y^2$ in descending order of x and ascending order of y.
7. Arrange the polynomial $x^2 + y^2 - 2xy$ in ascending order of x and descending order of y.

3.4 SUMMARY EXERCISES

1. Which of the following terms is a monomial?
   (1) $2x^2$ (2) $\frac{2}{x^2}$ (3) $2 + x^2$ (4) $2 - x^2$
2. Which of the following is a pair of like monomials?
   (1) $8x$ and $9x^2$ (3) $8x$ and $9y$
   (2) $8x^2y$ and $9x^2y$ (4) $8y$ and $9y^2$
3. Express as a binomial: $7x - 3y + 2x + 8y$
4. Express the sum of $x^2 + 2x - 1$ and $x^2 - x + 1$ as a binomial.
5. Express the sum of $4a^2 - 2a + 3$ and $a^2 - a - 4$ as a trinomial.
6. From $9x^2 - x + 1$ subtract $2x^2 + x - 3$.
7. Subtract $-6x^2 + 3x + 2$ from $8x^2 - 4x - 3$.
8. The sum of two binomials is $8x + 3$. One of the binomials is $2x - 1$. What is the other binomial?
9. Which of the following must be added to $3x - 7$ to produce a sum of 0?
   (1) 0 (2) $3x + 7$ (3) $-3x + 7$ (4) $-3x - 7$
10. From the sum of $a + 2b$ and $2a - b$, subtract $a - b$.
11. The product of $3x^2$ and $4x^3$ is
    (1) $7x^6$ (2) $7x^5$ (3) $12x^6$ (4) $12x^5$
12. When $2x^3y$ is multiplied by $-3x^2y^4$, the product is
    (1) $-6x^6y^4$ (2) $-6x^5y^4$ (3) $-6x^5y^5$ (4) $-6x^6y^5$

In 13–17, multiply each pair of binomials.

13. $(x + 6)(x + 4)$ **15.** $(x + y)(x - y)$ **17.** $(5x + 2)(5x - 2)$
14. $(a - 2)(a + 3)$ **16.** $(2x - 3)(4x - 1)$

18. The quotient of $27x^6$ and $3x^2$ is
(1) $24x^3$ (2) $9x^3$ (3) $24x^4$ (4) $9x^4$

19. The quotient $\dfrac{15x^2y^5}{3x^2y}$ is equivalent to
(1) $5y^4$ (2) $5xy^4$ (3) $5y^5$ (4) $5xy^5$

20. Divide and express the result as a single term:

$$\frac{24x^3y^2z}{-3xyz}$$

In 21–24, find the value of n that makes each statement true.

21. $x^2 \cdot x^3 = x^n$ **22.** $3^n = 9$ **23.** $2^n \cdot 2^4 = 2^6$ **24.** $y^n \div y^2 = y^5$

In 25–30, find the numerical value of the expression. Use $x = 2, y = 3, z = -1$.

25. $(xy)^2$ **27.** $x^3 - z$ **29.** $x + yz$
26. x^2z^3 **28.** $4x^2yz$ **30.** $xz^2 - y$

31. What is the product of $(3x)^2$ and $(2x)^3$?

32. What is the value of $7x^0$?
(1) 0 (2) 1 (3) 7 (4) $\frac{1}{7}$

33. What is the value of $5(3x)^0$?
(1) 0 (2) 5 (3) 15 (4) $\frac{1}{15}$

34. The value of 4^{-1} is
(1) -4 (2) $\frac{1}{4}$ (3) $-\frac{1}{4}$ (4) 0

35. The value of 5^{-3} is
(1) -15 (2) $\frac{1}{125}$ (3) 125 (4) -125

36. The value of $3 \cdot 4^{-2}$ is
(1) $\frac{3}{16}$ (2) -24 (3) -48 (4) $-\frac{3}{16}$

In 37–39, use the polynomial $7x^2 - 1 + 4x^3 - 3x$.

37. Arrange the polynomial in ascending order.

38. Arrange the polynomial in descending order.

39. What is the degree of the polynomial?

40. Which of the following is the degree of the polynomial 8?
(1) 8 (2) 1 (3) 0 (4) undefined

CHAPTER 4

Open Sentences in One Variable

4.1 SOLVING EQUATIONS

An ***equation*** is an open sentence containing two ***members*** (sides) separated by an equals sign, =. The left member of the equation consists of all the terms to the left of the equals sign, and the right member of the equation consists of all the terms to the right of the equals sign. The terms of the equation consist of numbers, or ***constants,*** and letters, or ***variables.*** We are concerned with *solving the equation.* That is, we wish to find the value(s) of the variable, or the ***solution set,*** that will make the open sentence true.

In order to solve an equation, we work toward one major goal: *to get the variable alone on one side of the equals sign.* When the variable is alone, it is easy to read the solution set, since all that will remain will be, for example, $x = 3$. In this case, {3} is the solution set.

◆ We will be able to get the variable alone if we carefully "undo" every operation that ties the variable to the other terms of the equation. In this "undoing" process, we must remember two things:

1. Whatever we do, we must do to both members of the equation so that we do not upset the equality.
2. Every operation has an opposite (inverse) operation that will "undo" it. Since addition and subtraction are opposites (inverse operations), each will "undo" the other. Since multiplication and division are opposites (inverse operations), each will "undo" the other.

Let us consider a series of examples in which we will solve for x, that is, obtain the solution set for the open sentence containing x.

Examples

Solve for x:

1. $x + 3 = 11$

Solution:

Since x is tied to 3 by addition and since subtraction undoes addition, subtract 3 from both members of the equation.

$$\begin{array}{rcr} x + 3 &=& 11 \\ -3 &=& -3 \\ \hline x &=& 8 \end{array}$$

To determine if $x = 8$ is correct (if it is the solution set of the open sentence), we must *check*. We check by substituting $x = 8$ into the *original* equation, and we work the arithmetic on each side of the equals sign. *We do not repeat the method of solution.*

Check:

$$x + 3 = 11$$
$$8 + 3 \stackrel{?}{=} 11$$
$$11 = 11 \checkmark$$

Answer: $x = 8$ or $\{8\}$

2. $x - .4 = -6$

Solution:

Since x is tied to .4 by subtraction and since addition undoes subtraction, add .4 to both members of the equation.

$$\begin{array}{rcl} x - .4 &=& -6 \\ + .4 &=& +.4 \\ \hline x &=& -5.6 \end{array}$$

Answer: $x = -5.6$ or $\{-5.6\}$

Check:

$$x - .4 = -6$$
$$-5.6 + .4 \stackrel{?}{=} -6$$
$$-6 = -6 \checkmark$$

3. $3x = 21$

Solution:

Since x is tied to 3 by multiplication and since division undoes multiplication, divide both members of the equation by 3.

$$\frac{3x}{3} = \frac{21}{3}$$
$$x = 7$$

Answer: $x = 7$ or $\{7\}$

Check:

$$3x = 21$$
$$3(7) \stackrel{?}{=} 21$$
$$21 = 21 \checkmark$$

4. $\frac{x}{.4} = 7$

Solution:

Since x is tied to .4 by division and since multiplication undoes division, multiply both members of the equation by .4.

$$.4\left(\frac{x}{.4}\right) = 7(.4)$$
$$x = 2.8$$

Answer: $x = 2.8$ or $\{2.8\}$

Check:

$$\frac{x}{.4} = 7$$
$$\frac{2.8}{.4} \stackrel{?}{=} 7$$
$$\frac{28}{4} \stackrel{?}{=} 7$$
$$7 = 7 \checkmark$$

5. $2x + 3 = 11$

Solution:

Here x is tied to two numbers: x is tied to the 2 by multiplication and to the 3 by addition. *Undo the addition first.*

$$\begin{array}{rl} 2x + 3 = 11 & \\ -3 = -3 & \text{Subtract 3.} \\ \hline \frac{2x}{2} = \frac{8}{2} & \text{Divide by 2.} \\ x = 4 & \end{array}$$

Answer: $x = 4$ or $\{4\}$

Check:

$$\begin{aligned} 2x + 3 &= 11 \\ 2(4) + 3 &\stackrel{?}{=} 11 \\ 8 + 3 &\stackrel{?}{=} 11 \\ 11 &= 11\checkmark \end{aligned}$$

6. $3x - .2 = 13$

Solution:

Here x is tied to two numbers: x is tied to the 3 by multiplication and to the .2 by subtraction. *Undo the subtraction first.*

$$\begin{array}{rl} 3x - .2 = 13 & \\ +.2 = +.2 & \text{Add .2.} \\ \hline \frac{3x}{3} = \frac{13.2}{3} & \text{Divide by 3.} \\ x = 4.4 & \end{array}$$

Answer: $x = 4.4$ or $\{4.4\}$

Check:

$$\begin{aligned} 3x - .2 &= 13 \\ 3(4.4) - .2 &\stackrel{?}{=} 13 \\ 13.2 - .2 &\stackrel{?}{=} 13 \\ 13 &= 13\checkmark \end{aligned}$$

7. $12x + 3 = 8x - 5$

Solution:

There are x terms on both sides of the equation; there are also number terms on both sides of the equation. *Get the* x *terms on one side and the number terms on the other side.*

$$\begin{array}{rl} 12x + 3 = 8x - 5 & \\ -8x \quad = -8x & \text{Subtract } 8x. \\ \hline 4x + 3 = -5 & \\ -3 = -3 & \text{Subtract 3.} \\ \hline \frac{4x}{4} = \frac{-8}{4} & \text{Divide by 4.} \\ x = -2 & \end{array}$$

Check:

$$\begin{aligned} 12x + 3 &= 8x - 5 \\ 12(-2) + 3 &\stackrel{?}{=} 8(-2) - 5 \\ -24 + 3 &\stackrel{?}{=} -16 - 5 \\ -21 &= -21\checkmark \end{aligned}$$

Answer: $x = -2$ or $\{-2\}$

8. $2x - 3 + 4x = 3x + 12$

Solution:

Collect like terms first. That is, combine the x terms (or the number terms) that are on the *same side* of the equation.

$$\begin{array}{rcll} 2x - 3 + 4x & = & 3x + 12 & \\ 6x - 3 & = & 3x + 12 & \text{Collect like terms.} \\ -3x & = & -3x & \text{Subtract } 3x. \\ \hline 3x - 3 & = & 12 & \\ +3 & = & +3 & \text{Add 3.} \\ \hline \frac{3x}{3} & = & \frac{15}{3} & \text{Divide by 3.} \\ x & = & 5 & \end{array}$$

Check:

$$\begin{array}{rcl} 2x - 3 + 4x & = & 3x + 12 \\ 2(5) - 3 + 4(5) & \stackrel{?}{=} & 3(5) + 12 \\ 10 - 3 + 20 & \stackrel{?}{=} & 15 + 12 \\ 7 + 20 & \stackrel{?}{=} & 27 \\ 27 & = & 27 \checkmark \end{array}$$

Answer: $x = 5$ or $\{5\}$

9. $5 - 7x + 3 = 8x - 11 + 4x$

Solution:

$$\begin{array}{rcll} 5 - 7x + 3 & = & 8x - 11 + 4x & \\ 8 - 7x & = & 12x - 11 & \text{Collect like terms.} \\ -12x & = & -12x & \text{Subtract } 12x. \\ \hline 8 - 19x & = & -11 & \\ -8 & = & -8 & \text{Subtract 8.} \\ \hline \frac{-19x}{-19} & = & \frac{-19}{-19} & \text{Divide by } -19. \\ x & = & 1 & \end{array}$$

Note that the x terms could have been brought to the right side and the number terms to the left. That is:

$$\begin{array}{rcll} 8 - 7x & = & 12x - 11 & \\ +7x & = & +7x & \text{Add } 7x. \\ \hline 8 & = & 19x - 11 & \\ +11 & = & +11 & \text{Add 11.} \\ \hline \frac{19}{19} & = & \frac{19x}{19} & \text{Divide by 19.} \\ 1 & = & x \text{ or } x = 1 & \end{array}$$

Check:

$$\begin{array}{rcl} 5 - 7x + 3 & = & 8x - 11 + 4x \\ 5 - 7(1) + 3 & \stackrel{?}{=} & 8(1) - 11 + 4(1) \\ 5 - 7 + 3 & \stackrel{?}{=} & 8 - 11 + 4 \\ -2 + 3 & \stackrel{?}{=} & -3 + 4 \\ 1 & = & 1 \checkmark \end{array}$$

This type of choice exists in every solution.

Answer: $x = 1$ or $\{1\}$

10. $4(2x - 1) = 20$

Solution:

$$4(2x - 1) = 20$$
$$8x - 4 = 20 \quad \text{Remove parentheses.}$$
$$+4 = +4 \quad \text{Add 4.}$$
$$\frac{8x}{8} = \frac{24}{8} \quad \text{Divide by 8.}$$
$$x = 3 \text{ or } \{3\} \quad \textit{Ans.}$$

Check:

$$4(2x - 1) = 20$$
$$4(2 \cdot 3 - 1) \stackrel{?}{=} 20$$
$$4(6 - 1) \stackrel{?}{=} 20$$
$$4(5) \stackrel{?}{=} 20$$
$$20 = 20\checkmark$$

The previous examples show that the solution set for an open sentence in one variable can be obtained by a series of steps. Just where to begin depends upon how complicated the sentence is.

◆ In general, the procedure for obtaining the solution set for an open sentence in one variable is:

1. Remove parentheses by multiplication (the distributive property).
2. Collect like terms in each member of the equation. Use addition or subtraction as indicated.
3. Get the terms containing the variable on one side and the number terms on the other side. Using the operation opposite to the one indicated (the inverse of the operation) will move a term from one side of the equation to the other.
4. Separate the variable from its coefficient. Use division or multiplication as indicated.
5. Check in the original equation by separately doing the arithmetic on each side of the equals sign.

When several terms of an equation contain decimals, it is easiest to clear decimals by multiplying both members of the equation by the power of 10 needed.

Example

Solve for x: $.3x + .04 = .64$

Solution:

Multiply each term by 100, since the greatest number of decimal places is two.

$$.3x + .04 = .64$$
$$30x + 4 = 64 \quad \text{Multiply by 100.}$$
$$-4 = -4 \quad \text{Subtract 4.}$$
$$\frac{30x}{30} = \frac{60}{30} \quad \text{Divide by 30.}$$
$$x = 2 \text{ or } \{2\} \quad \textit{Ans.}$$

Check:

$$.3x + .04 = .64$$
$$.3(2) + .04 \stackrel{?}{=} .64$$
$$.6 + .04 \stackrel{?}{=} .64$$
$$.60 + .04 \stackrel{?}{=} .64$$
$$.64 = .64\checkmark$$

When more than one fraction appears in an equation, it is easiest to clear fractions by multiplying both members of the equation by the L.C.D.

Example

Solve for x: $\frac{x}{2} - \frac{x}{3} = 4$.

Solution:

$$\text{L.C.D.} = 6$$
$$6\left(\frac{x}{2} - \frac{x}{3}\right) = 4(6)$$
$$3x - 2x = 24$$
$$x = 24$$

Answer: $x = 24$ or $\{24\}$

Check:

$$\frac{x}{2} - \frac{x}{3} = 4$$
$$\frac{24}{2} - \frac{24}{3} \stackrel{?}{=} 4$$
$$12 - 8 \stackrel{?}{=} 4$$
$$4 = 4\checkmark$$

(*Note:* Equations with fractions will be considered in Chapter 7.)

EXERCISES

In 1–25, solve and check the equation.

1. $\frac{x}{3} = 12$
2. $4x - 3 = 41$
3. $3x - 2 = 16$
4. $3x - 5 = 16$
5. $\frac{1}{3}c + 2 = 4$
6. $\frac{y}{3} + 2 = 5$
7. $3a + 0.2 = 5$
8. $x - 0.4 = 1.6$
9. $0.4x + 2 = 12$
10. $0.2x + 6 = 80$
11. $0.2x + 0.3 = 8.1$
12. $.03x = 1.5$
13. $4x = x + 21$
14. $3x - 2 = x + 8$
15. $3z + 9 = 1 - z$
16. $4x - 8 = 2x + 4$
17. $3(2 + c) = 12$
18. $3(2x - 1) = 21$
19. $8x = 2(x + 15)$
20. $3(x + 4) - x = 18$
21. $\frac{n}{2} + \frac{n}{4} = 6$
22. $\frac{a}{2} + \frac{a}{6} = 2$
23. $11x = 40 - (-7x - 4)$
24. $\frac{1}{2}(8y - 6) = 25$
25. $5z - 4 = 3(z + 2)$

4.2 LITERAL EQUATIONS

Sometimes an equation contains letters instead of numbers. Such an equation is called a ***literal equation.*** In some literal equations, there is still really only one variable. For example, $ax + b = c$ is much like $2x + 3 = 11$ in that x is considered the only variable and a, b, and c represent constants.

Once the idea of using more than one letter in an equation arises, it turns out that you can solve a literal equation for any letter you choose. In a problem, the instructions will tell you which letter is to be solved for "in terms of" the other letters. Of course, you cannot expect a numerical value as an answer since you are dealing only with letters. (*Note:* Formulas are literal equations.)

◆ The idea is to think of the letter you are asked to solve for as THE variable and to think of all the other letters as constants. Then you can apply the general goal of solving an equation: get the letter you are asked to solve for to be alone on one side of the equals sign.

Examples

1. Solve for x in terms of a, b, and c: $ax + b = c$

Solution:

We wish to get x alone.

$$\begin{array}{rcll} ax + \cancel{b} &=& c & \\ -\cancel{b} &=& -b & \text{Subtract } b. \\ \hline \dfrac{\cancel{a}x}{\cancel{a}} &=& \dfrac{c-b}{a} & \text{Divide by } a. \\ x &=& \dfrac{c-b}{a} & \end{array}$$

Note that we have treated a, b, and c as constants; and we undid each operation that tied x to them just as we did when dealing with actual numbers.

Check:

$$\begin{aligned} ax + b &= c \\ \cancel{a}\left(\frac{c-b}{\cancel{a}}\right) + b &\stackrel{?}{=} c \\ c - b + b &\stackrel{?}{=} c \\ c + 0 &\stackrel{?}{=} c \\ c &= c\checkmark \end{aligned}$$

Answer: $x = \dfrac{c-b}{a}$

2. Solve for R in terms of D and T: $R \cdot T = D$

Solution:

We wish to get R alone.

$$\begin{aligned} \frac{R \cdot \cancel{T}}{\cancel{T}} &= \frac{D}{T} \quad \text{Divide by } T. \\ R &= \frac{D}{T} \end{aligned}$$

Check:

$$\begin{aligned} R \cdot T &= D \\ \frac{D}{\cancel{T}} \cdot \cancel{T} &\stackrel{?}{=} D \\ D &= D\checkmark \end{aligned}$$

Answer: $R = \dfrac{D}{T}$

Note that the literal equation $R \cdot T = D$ is a formula for distance:

Rate · Time = Distance

3. Solve for r in terms of π and C: $C = 2\pi r$

Solution:

We wish to get r alone.

$$\frac{C}{2\pi} = \frac{2\pi r}{2\pi} \quad \text{Divide by } 2\pi.$$

$$\frac{C}{2\pi} = r \quad \text{or} \quad r = \frac{C}{2\pi}$$

Check:

$$C = 2\pi r$$
$$C \stackrel{?}{=} 2\pi\left(\frac{C}{2\pi}\right)$$
$$C = C\checkmark$$

Answer: $r = \dfrac{C}{2\pi}$

Note that the literal equation $C = 2\pi r$ is a formula for finding the circumference of a circle.

4. Solve for y in terms of a, b, and c: $ay - by = c$

Solution:

$ay - by = c$ Apply the reverse of the distributive property

$(a - b)y = c$ (factor y).

$\dfrac{(a - b)y}{(a - b)} = \dfrac{c}{a - b}$ Divide by the coefficient of y.

$y = \dfrac{c}{a - b}$

Check:

$$ay - by = c$$

$a\left(\dfrac{c}{a - b}\right) - b\left(\dfrac{c}{a - b}\right) \stackrel{?}{=} c$ Work with the left member only.

$\dfrac{ac}{a - b} - \dfrac{bc}{a - b} \stackrel{?}{=} c$ Multiply to remove parentheses.

$\dfrac{ac - bc}{a - b} \stackrel{?}{=} c$ Combine fractions.

$\dfrac{c(a - b)}{a - b} \stackrel{?}{=} c$ Factor c in the numerator.

$c = c\checkmark$ Cancel the common factor $(a - b)$.

Answer: $y = \dfrac{c}{a - b}$

(*Note:* The solution of this literal equation required the reverse of the distributive property or *factoring*. Factoring is discussed at length in Chapter 7.)

5. Solve for W in terms of P and L: $P = 2(W + L)$

Solution:

$$P = 2(W + L)$$

$$P = 2W + 2L \quad \text{Remove parentheses by multiplication.}$$

$$-2L = \quad -2L \quad \text{Subtract } 2L.$$

$$\frac{P - 2L}{2} = \frac{2W}{2} \quad \text{Divide by 2.}$$

$$\frac{P - 2L}{2} = W$$

Check:

$$P = 2(W + L)$$

$$P \stackrel{?}{=} 2\left(\frac{P - 2L}{2} + L\right) \quad \text{Work with the right member only.}$$

$$P \stackrel{?}{=} 2 \cdot \frac{P - 2L}{2} + 2L \quad \text{Multiply to remove parentheses.}$$

$$P \stackrel{?}{=} P - 2L + 2L \quad \text{Cancel the common factor of 2.}$$

$$P \stackrel{?}{=} P - 0 \quad \text{Combine like terms.}$$

$$P = P\checkmark$$

Answer: $W = \dfrac{P - 2L}{2}$

Note that the literal equation $P = 2(W + L)$ is a formula for finding the perimeter of a rectangle.

EXERCISES

1. Solve for x in terms of a, b, and c: $ax - b = c$
2. Solve for r in terms of S, π, and h: $S = 2\pi rh$
3. Given the formula $V = LWH$, solve for H in terms of V, W, and L.
4. Solve for h in terms of A and b: $A = \dfrac{bh}{2}$
5. Solve for B in terms of V and h: $V = \frac{1}{3}Bh$
6. Solve for x in terms of a, b, and c: $ax + bx = c$
7. Solve for t in terms of A, p, and r: $A = p + prt$
8. Solve for y in terms of p and x: $p = 2(x + y)$
9. Solve for y in terms of c, d, and h: $dy - c = h$
10. Solve for x in terms of a, b, and c: $a + bx = c$
11. Solve for x in terms of c and d: $3c + 2x = d$
12. Solve for x in terms of b, m, and y: $y = mx + b$

4.3 SOLVING INEQUALITIES

An ***inequality*** is an open sentence containing two members (sides) separated by a symbol telling that the members are not equal. The symbols of inequality are $>$, read "greater than," and $<$, read "less than." The symbol $>$ may be combined with $=$ to make $\geq$, read "greater than or equal to." Also, $<$ may be combined with $=$ to make $\leq$, read "less than or equal to."

As in the solution of an equation (equality), our major goal in the solution of an inequality ***(inequation)*** is to get the variable alone. When we undo each of the operations in an equation that ties the variable to the other terms of the equation, we are careful always to operate on both members of the equation so that the equality is maintained. In those instances, we are using the properties of equalities that allow us to maintain the equality. That is, when we:

add equals to equals,
subtract equals from equals, or
multiply or divide equals by equals,

the results are always equal.

The following numerical examples show that not all the properties of equality apply to inequalities.

$$\begin{array}{rcl} 14 & > & 8 \\ +\ 2 & = & +\ 2 \\ \hline 16 & > & 10\surd \end{array} \qquad \begin{array}{rcl} 14 & > & 8 \\ -\ 2 & = & -\ 2 \\ \hline 12 & > & 6\surd \end{array}$$

The examples above tell us that when equals are added to unequals or subtracted from unequals, the order of the inequality is maintained.

$$\begin{array}{rcl} 14 & > & 8 \\ \times\ (2) & = & \times\ (2) \\ \hline 28 & > & 16\ \surd \end{array} \qquad \begin{array}{rcl} 14 & > & 8 \\ \times\ (-2) & = & \times\ (-2) \\ \hline -28 & > & -16 \end{array} \text{ false; } -28 < -16$$

$$\begin{array}{rcl} 14 & > & 8 \\ \div\ (2) & = & \div\ (2) \\ \hline 7 & > & 4\ \surd \end{array} \qquad \begin{array}{rcl} 14 & > & 8 \\ \div\ (-2) & = & \div\ (-2) \\ \hline -7 & > & -4 \end{array} \text{ false; } -7 < -4$$

These examples tell us that when we multiply or divide unequals by positive equals, the order of the inequality is maintained. However, when we multiply or divide unequals by negative equals, the order of the inequality is reversed.

Thus, we can say that, in general:

◆ The solution set for an inequality is obtained in much the same way as for an equality. A major difference is that if we are multiplying or dividing by a negative number, we must reverse the order of the inequality sign.

Examples

1. Solve for x: $x + 2 > 4$

Solution:

$$\begin{array}{rcl} x + 2 &>& 4 \\ -2 &=& -2 \\ \hline x &>& 2 \end{array}$$ Subtract 2.

Check:

Choose any number > 2, for example, 3. Substitute $x = 3$ into the original inequality:

$$\begin{array}{r} x + 2 > 4 \\ 3 + 2 \overset{?}{>} 4 \\ 5 > 4 \checkmark \end{array}$$

Answer: The original inequality *is equivalent to* $x > 2$, or:

$\{x \mid x > 2\}$ is the solution set.

(The set of — elements x — such that — x is greater than 2)

2. Find the solution set of the open sentence: $x - 3 \leq 5$

Solution:

$$\begin{array}{rcl} x - 3 &\leq& 5 \\ +3 &=& +3 \\ \hline x &\leq& 8 \end{array}$$ Add 3.

Check:

Choose any number ≤ 8, for example, 5. Substitute $x = 5$ into the original inequality.

$$\begin{array}{r} x - 3 \leq 5 \\ 5 - 3 \overset{?}{\leq} 5 \\ 2 \leq 5 \checkmark \end{array}$$

Answer: The original inequality is equivalent to $x \leq 8$, or $\{x \mid x \leq 8\}$ is the solution set.

3. If the replacement set for x is $\{5, 6, 7, 8\}$, find the solution set for $2x - 3 \geq 11$.

Solution:

$$\begin{array}{rcl} 2x - 3 &\geq& 11 \\ +3 &=& +3 \\ \hline \frac{2x}{2} &\geq& \frac{14}{2} \\ x &\geq& 7 \end{array}$$

Add 3.

Divide by 2.

The only members of the replacement set that satisfy the conditions are 7 and 8. *Answer:* $\{7, 8\}$

Alternate Solution:

Substitute each of the elements of the replacement set into the original inequality. Those elements that satisfy are the elements of the solution set.

4. Find the solution set of the open sentence: $x - 2 > 3x + 4$

Solution:

$$\begin{array}{rcll}
x - 2 & > & 3x + 4 & \\
-3x & = & -3x & \text{Subtract } 3x. \\
\hline
-2x - 2 & > & 4 & \\
+2 & = & +2 & \text{Add 2.} \\
\hline
\frac{-2x}{-2} & > & \frac{6}{-2} & \text{Divide by } -2. \\
x & < & -3 & \text{Reverse the order of the inequality.}
\end{array}$$

Check:

Choose any number < -3, for example, -5.
Substitute $x = -5$ into the original inequality:

$$\begin{aligned}
x - 2 &> 3x + 4 \\
(-5) - 2 &\overset{?}{>} 3(-5) + 4 \\
-7 &\overset{?}{>} -15 + 4 \\
-7 &> -11 \checkmark
\end{aligned}$$

Answer: The original inequality is equivalent to $x < -3$, or $\{x | x < -3\}$ is the solution set.

5. Solve for x: $2(x - 3) \leq 3x + 7$

Solution:

$$\begin{array}{rcll}
2(x - 3) & \leq & 3x + 7 & \\
2x - 6 & \leq & 3x + 7 & \text{Distributive property.} \\
-3x & & -3x & \text{Subtract } 3x. \\
\hline
-x - 6 & \leq & 7 & \\
+6 & = & +6 & \text{Add 6.} \\
\hline
\frac{-x}{-1} & \leq & \frac{13}{-1} & \text{Divide by } -1. \\
x & \geq & -13 & \text{Reverse the order of the inequality.}
\end{array}$$

Check:

Choose any number ≥ -13, for example, -12.
Substitute $x = -12$ into the original inequality:

$$\begin{aligned}
2(x - 3) &\leq 3x + 7 \\
2(-12 - 3) &\overset{?}{\leq} 3(-12) + 7 \\
2(-15) &\overset{?}{\leq} -36 + 7 \\
-30 &\leq -29 \checkmark
\end{aligned}$$

Answer: The original inequality is equivalent to $x \geq -13$, or $\{x \mid x \geq -13\}$ is the solution set.

The other major difference between the solution of an equation and the solution of an inequality is in the nature of the solution set. The solution set for an equation is usually finite. That is, there is a specific number of elements in the solution set. In our work so far, the solution set for an equation contained exactly one element, meaning that there was only one value of the variable that made the open sentence true.

(*Note:* There may be more than one element that makes an equation true. For example, $\{+2, -2\}$ is the solution set for the equation $x^2 = 4$. The solution set for an equation may be empty, that is, contain no elements. For example, $\emptyset$, the empty set, is the solution set for the open sentence $x + 1 = x + 2$. The solution set for an equation may be infinite. For example, the entire set of real numbers is the solution set for the open sentence $x + 1 = x + 1$.)

The solution set for an inequality is usually infinite, as was the case in all the examples previously done in this section. (*Note:* We will leave the exceptions for future study.)

EXERCISES

1. Solve for x: $7x < 4x + 18$
2. The inequality $2x > x + 3$ is equivalent to
   (1) $x > 1$ (2) $x > \frac{3}{2}$ (3) $x = 3$ (4) $x > 3$
3. The inequality $2x + 5 > x + 3$ is equivalent to
   (1) $x > -2$ (2) $x > 2$ (3) $x > 8$ (4) $x > 4$
4. If the replacement set for x is $\{-2, -1, 0, 1, 2\}$, find the solution set of $2x \leq 1$.
5. If the replacement set is $\{0, 1, 2, 3\}$, find the solution set of $3x + 1 \geq 10$.
6. If the replacement set for x is $\{0, 1, 2, 3, 4\}$, the solution set of $3(x - 1) \leq x + 3$ is
   (1) $\{1, 2, 3\}$ (2) $\{0, 1, 2\}$ (3) $\{0, 1, 2, 3\}$ (4) $\{3, 4\}$
7. The inequality $-2x < 8$ is equivalent to
   (1) $x < -4$ (2) $x > -4$ (3) $x < 4$ (4) $x > 4$
8. Solve for x: $2(x + 3) < x - 4$
9. Solve for x: $5x + 1 \geq 2x + 16$
10. Solve for x: $3 - 2x \leq 11$

4.4 GRAPHING A SOLUTION SET ON A NUMBER LINE

A solution set for an equation or for an inequality can be pictured as a graph on a number line. A number line is infinitely long and contains an infinite number of points. Each of the real numbers is matched with a point on the number line. We show a plan for the matching by writing 0, a few positive integers, and a few negative integers on the number line. We show that the line is infinitely long by including arrowheads. We must remember, however, that *the number line contains an infinite number of points and that there is an infinite number of points between any two points.*

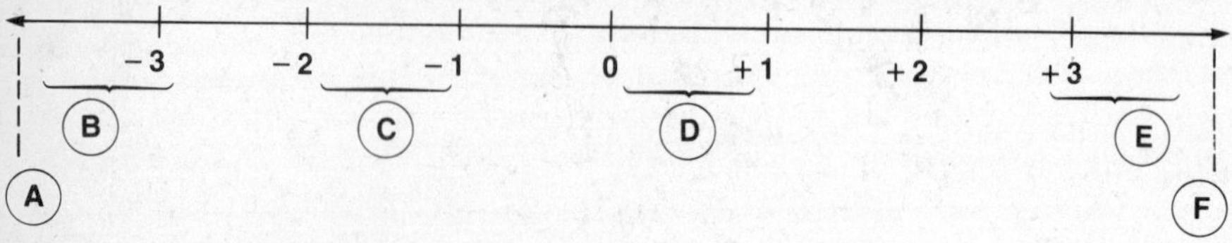

A. The left arrowhead indicates that the number line extends infinitely in the negative direction.

B. The numbers continue in the negative direction: −4, −5, . . . , and all the numbers between.

C. This interval contains an infinite number of points. Some of these points are: $-\frac{7}{6}$, $-\frac{4}{3}$, $-\frac{3}{2}$, $-\frac{5}{3}$, −1.4, $-\sqrt{3}$.

D. This interval contains an infinite number of points. Some of these points are: $\frac{1}{1000}$, $\frac{1}{200}$, $\frac{1}{146}$, $\frac{1}{2}$, .6, .71, $\sqrt{.5}$, $\sqrt{\frac{1}{3}}$.

E. The numbers continue in the positive direction: +4, +5, . . . , and all the numbers between.

F. The right arrowhead indicates that the number line extends infinitely in the positive direction.

Now let us see how to graph a solution set on the number line.

Suppose the solution set contains just one element, 3. Then the graph of the solution set {3} will show just one point. The graph of {3} is shown below.

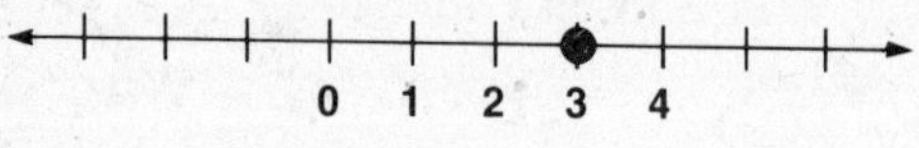

If the solution set contains four elements, −3, 0, 1, 7, then the graph of the solution set {−3, 0, 1, 7} will show four points. The graph of {−3, 0, 1, 7} is shown below.

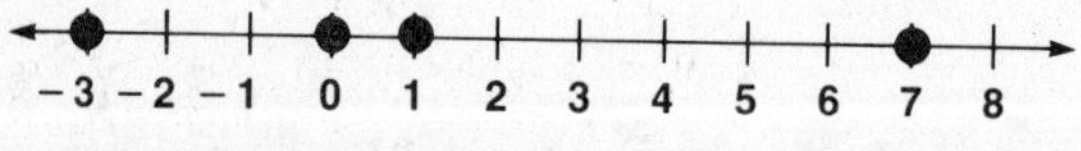

◆ In general, if a solution set is finite, the graph consists of as many points as there are elements in the solution set.

If a solution set is infinite, the graph must show that all the points between numbers are represented. If the solution set continues infinitely, the graph must also show that fact. See how various infinite solution sets are represented by the following graphs.

1. The graph of $\{x \mid x > 2\}$ is shown below. Note that at $x = 2$, we show an "open hole" to represent the fact that $x = 2$ is *not included* in the solution set.

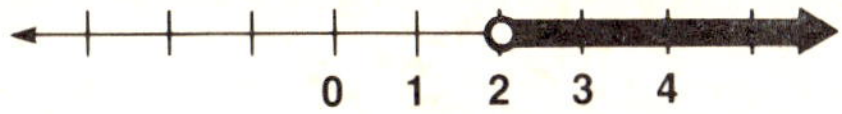

2. In the graph of $\{x \mid x \leq 1\}$, shown below, note the "closed hole" at $x = 1$ to indicate that $x = 1$ *is included* in the solution set.

3. The graph of $\{x \mid x \geq -3\}$ is shown below.

4. Study the graph of $\{x \mid -2 \leq x \leq 3\}$, shown below.

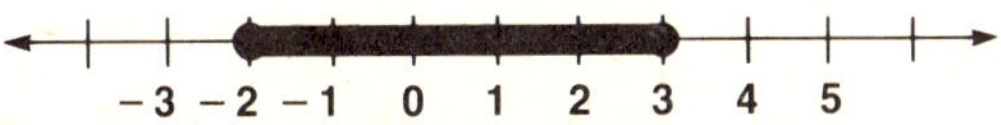

The following ideas are expressed by the graph:

a. The solution set extends only between -2 and $+3$.
b. Both -2 and $+3$ are included in the solution set.
c. All of the numbers between -2 and $+3$ are included in the solution set.
d. The solution set is infinite.

The solution set $\{x \mid -2 \leq x \leq 3\}$ can also be represented algebraically as $-2 \leq x \leq 3$. This solution set is read "-2 is less than or equal to x and x is less than or equal to 3." Another way to read this solution set is "x is *between* -2 and 3, *inclusive*."

5. The graph of $1 \leq x < 3$ or $\{x \mid 1 \leq x < 3\}$ is shown below. This solution set is read "1 is less than or equal to x and x is less than 3" or "x is *between* 1 and 3, including 1 but not including 3 (or, *excluding* 3)."

6. The graph of $-4 < x < 0$ or $\{x \mid -4 < x < 0\}$ is shown below. This solution set is read "-4 is less than x and x is less than 0" or "x is *between* -4 and 0, *exclusive*."

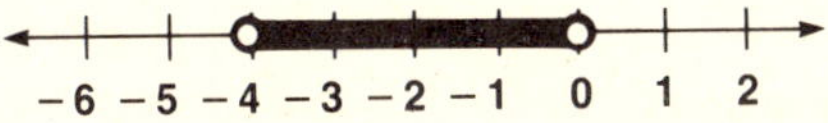

7. The graph of $2 < x \leq 5$ or $\{x \mid 2 < x \leq 5\}$ is shown below. This solution set is read "2 is less than x and x is less than or equal to 5" or "x is *between* 2 and 5, excluding 2 and including 5."

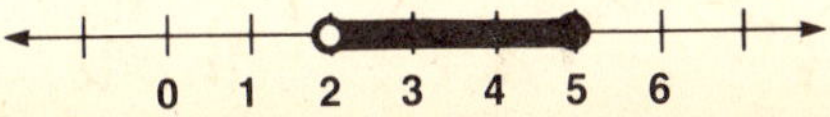

An open sentence of inequality that involves two signs of inequality is equivalent to a combination of two single sentences. From our work in logic, we know that a combination of sentences can be:

1. a conjunction, "and," $\wedge$
2. a disjunction, "or," $\vee$

(*Note:* There are other combinations of two sentences, but they are not applicable here.)

The examples we have examined so far are of the conjunction variety. That is:

$$-2 \le x \le 3 \text{ is equivalent to } (-2 \le x) \wedge (x \le 3)$$
$$1 \le x < 3 \text{ is equivalent to } (1 \le x) \wedge (x < 3)$$
$$-4 < x < 0 \text{ is equivalent to } (-4 < x) \wedge (x < 0)$$
$$2 < x \le 5 \text{ is equivalent to } (2 < x) \wedge (x \le 5)$$

We have seen how these graphs are pictured on the number line. Now let us picture the graph of a disjunction. Study the following examples:

1. The graph of $x \le 5$ is shown here. Note that $x \le 5$ means $(x < 5) \vee (x = 5)$.

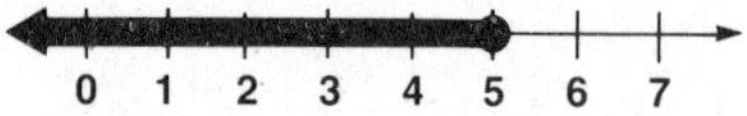

2. The graph of $(x < 2) \vee (x \ge 5)$ is shown here. (*Note:* There is no more compact way to write this disjunction.)

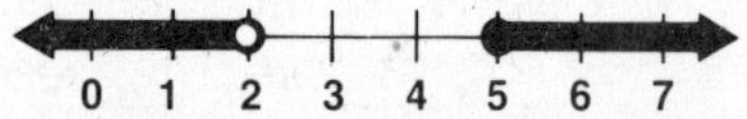

3. The graph of $(x \le -1) \vee (x \ge 0)$ is shown here.

4. The graph of $(x \le 2) \vee (x > 5)$ is shown here.

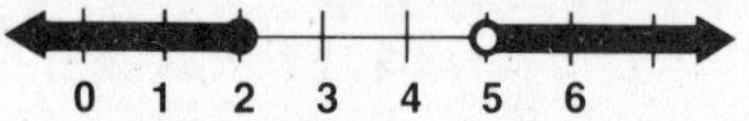

We have seen how to graph a solution set on a number line. Before we leave this topic, let us look at a few examples of reading a solution set.

Examples

1. A member of the solution set of $-2 \le x < 3$ is
(1) -3 (2) 3 (3) 4 (4) -2

Solution:

The inequality is read "x is between -2 and 3, including -2 but excluding 3." Examination of the choices eliminates (1), (2), and (3).

Answer: (4)

2. If x is an integer, the solution set of the inequality $-3 \leq x < 0$ is
(1) $\{-3, -2, -1, 0\}$
(2) $\{-2, -1, 0\}$
(3) $\{-3, -2, -1\}$
(4) $\{-3, -2\}$

Solution:

The inequality $-3 \leq x < 0$ is read "x is between -3 and 0, including -3 but excluding 0." Since we are given a replacement set of only integers, we need not concern ourselves with all the numbers between integers.

Consider $-3, -2, -1, 0$
include (-3)   exclude (0)

Thus, $\{-3, -2, -1\}$ is the solution set.

Answer: (3)

3. If x is an integer, find a value of x that will make the following expression true: $\frac{1}{4} < \frac{x}{8} < \frac{3}{4}$ (1) 1   (2) 2   (3) 3   (4) 7

Solution:

The given inequality is read "x-eighths is between one-fourth and three-fourths, exclusive." In order to determine the value(s) of x, it is easiest to express all the fractions in terms of the same denominator. That is:

$$\frac{1}{4} < \frac{x}{8} < \frac{3}{4}$$

$$\text{is equivalent to } \frac{2}{8} < \frac{x}{8} < \frac{6}{8}$$

Now, it is clear that x must be an integer between 2 and 6, exclusive. That is, x can only be 3, 4, 5. Knowing this fact eliminates choices (1), (2), and (4).

Answer: (3)

EXERCISES

1. Which solution set is represented by the accompanying graph?
(1) $\{x \mid x \geq -1\}$
(2) $\{x \mid x \leq -1\}$
(3) $\{x \mid x < -1\}$
(4) $\{x \mid x > -1\}$

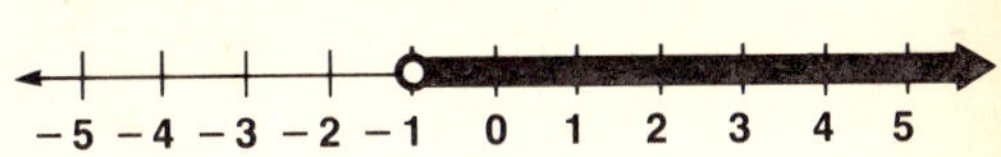

2. A graph of $2x \leq 6$ is

(1) 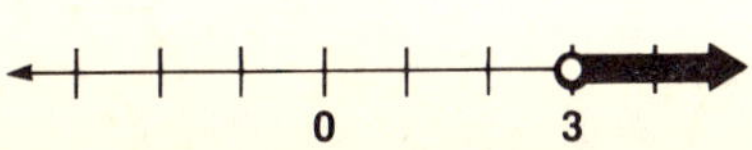

(3) 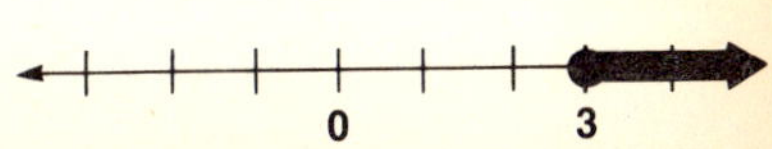

(2) 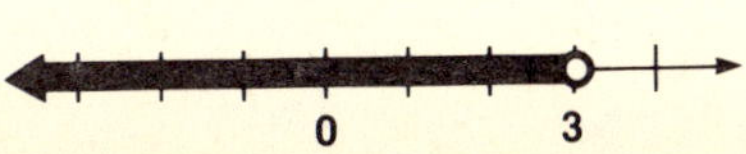

(4) 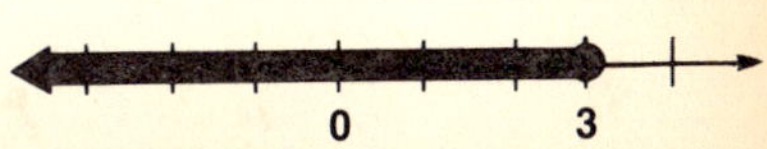

3. Which graph represents the solution set of the inequality $3x > 6$?

(1) 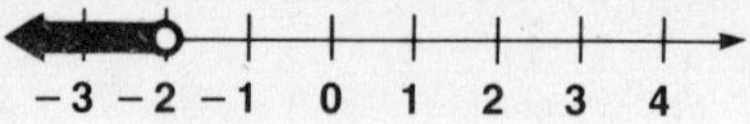

(2) 

(3) 

(4) 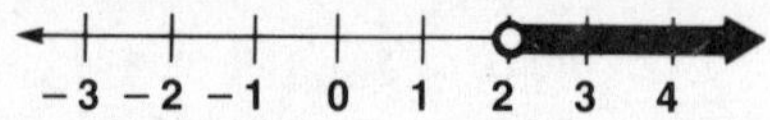

4. Which graph represents the solution set of $3x > 9$?

(1) 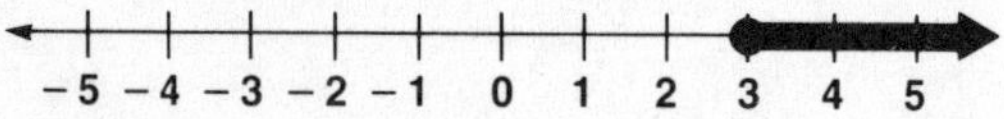

(2) 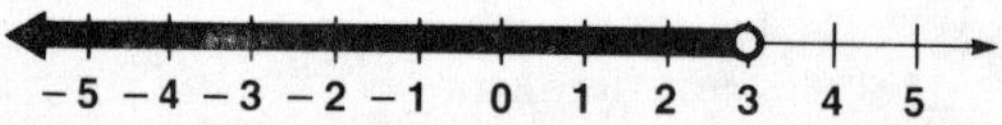

(3) 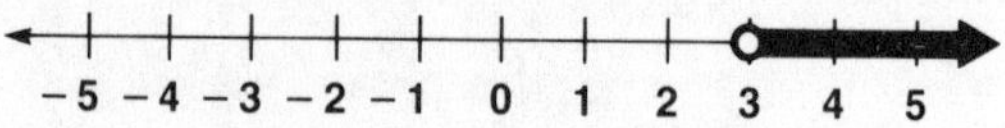

(4) 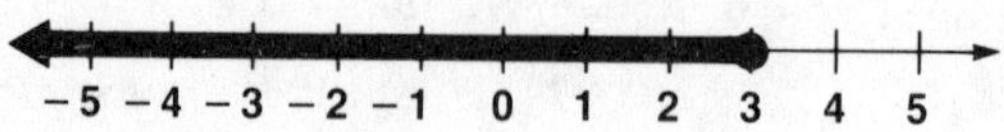

5. Which graph represents the solution set of $2x \leq 10$?

(1) 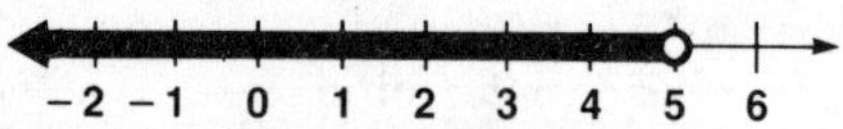

(2) 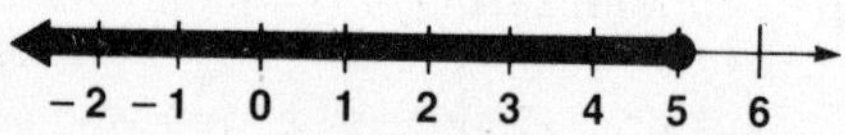

(3) 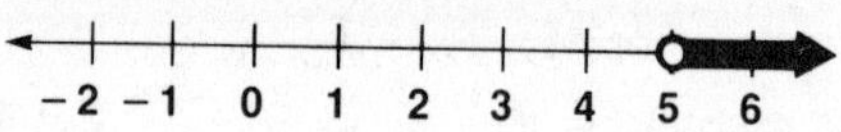

(4) 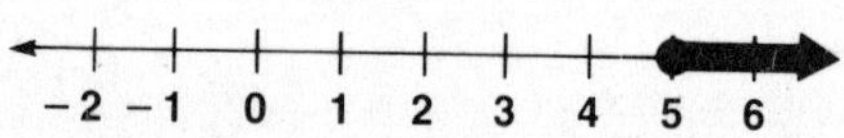

6. Which inequality is represented by the accompanying graph?

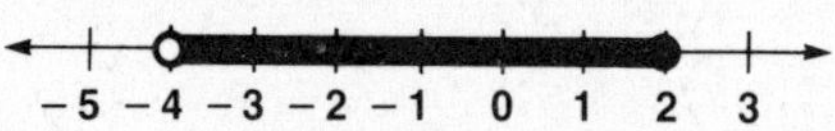

(1) $2 \leq x < -4$
(2) $-4 \leq x \leq 2$
(3) $-4 < x \leq 2$
(4) $-4 \leq x < 2$

7. Which inequality is represented by the accompanying graph?

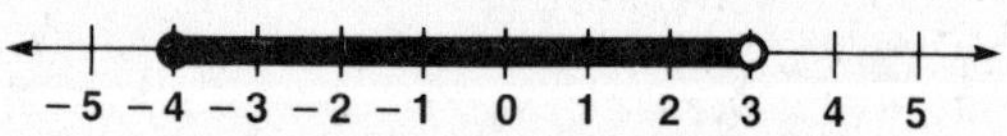

(1) $3 \leq x < -4$
(2) $-4 \leq x \leq 3$
(3) $-4 < x \leq 3$
(4) $-4 \leq x < 3$

8. Which inequality is represented by the accompanying graph?

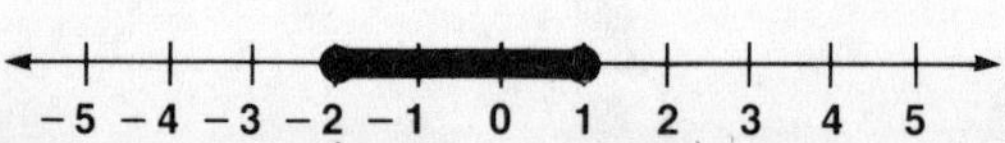

(1) $\{x \mid -2 < x < 1\}$
(2) $\{x \mid -2 \leq x < 1\}$
(3) $\{x \mid -2 < x \leq 1\}$
(4) $\{x \mid -2 \leq x \leq 1\}$

9. The accompanying number line shows the solution set of which inequality?

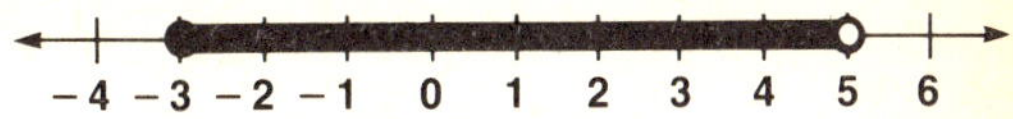

(1) $-3 < x < 5$
(2) $-3 < x \leq 5$
(3) $-3 \leq x < 5$
(4) $-3 \leq x \leq 5$

10. Which inequality is represented by the graph?

(1) $-2 < x \leq 3$
(2) $-2 \leq x < 3$
(3) $3 < x \leq -2$
(4) $3 \leq x < -2$

11. Which expression is shown by the accompanying graph?

(1) $1 \leq x < -4$
(2) $-4 \leq x \leq 1$
(3) $-4 < x \leq 1$
(4) $-4 \leq x < 1$

12. Which set is shown on the accompanying graph?

(1) $\{x \mid -2 < x < 4\}$
(2) $\{x \mid -2 \leq x < 4\}$
(3) $\{x \mid -2 < x \leq 4\}$
(4) $\{x \mid -2 \leq x \leq 4\}$

13. Which graph shows the solution to $(x < 3) \vee (x \geq 5)$?

(1)
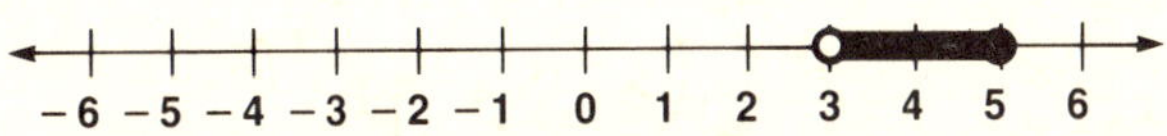

(2)
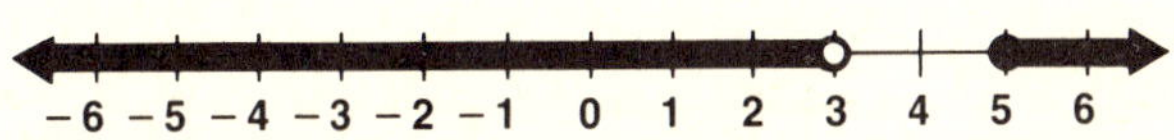

(3)
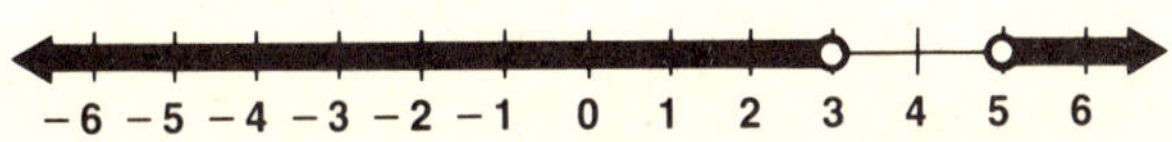

(4)
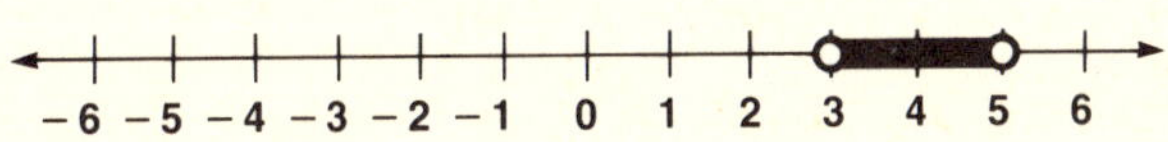

14. Which graph shows the solution to $(x \leq -2) \vee (x \geq 0)$?

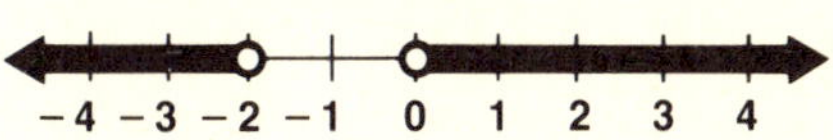

(1)

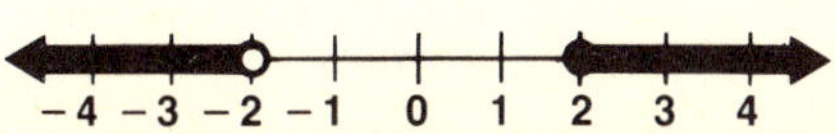

(2)

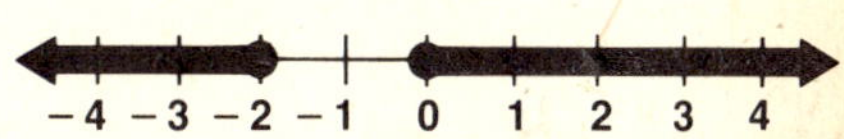

(3)

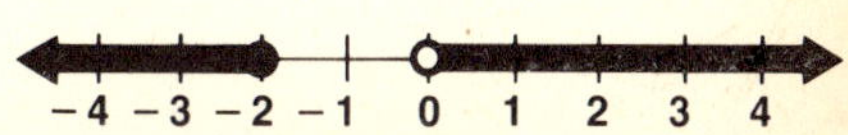

(4)

15. Which graph shows the solution to $(x \leq 3) \vee (x > 6)$?

(1) 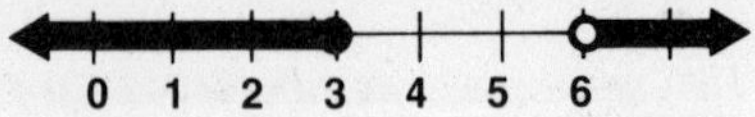

(2) 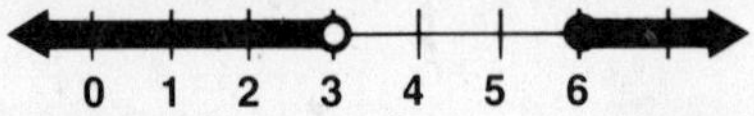

(3) 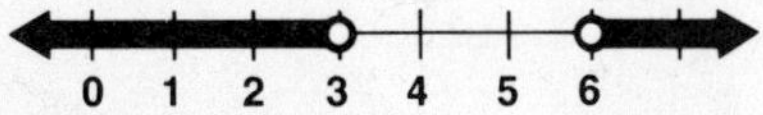

(4) 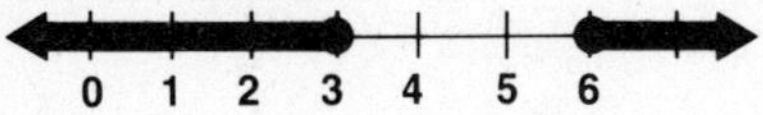

16. Which graph shows the solution to $(x < -1) \vee (x \geq 3)$?

(1) 

(2) 

(3) 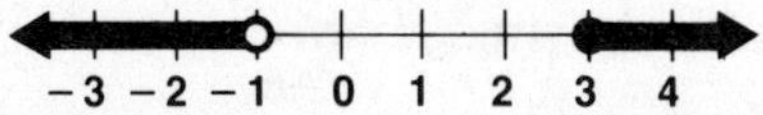

(4) 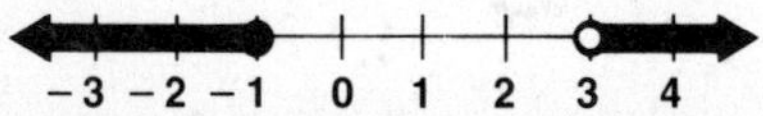

17. What are the numbers in the solution set of $4 \leq x < 7$ if x is an integer?
(1) 5, 6 (2) 5, 6, 7 (3) 4, 5, 6 (4) 4, 5, 6, 7

18. One member of the solution set of $-5 < x \leq -1$ is
(1) -1 (2) -5 (3) 3 (4) -6

19. If the replacement set for x is $\{-3, -1, 0, 1, 3\}$, write the members of the solution set for $3x \leq 0$.

20. If the replacement set for x is $\{-4, -3, -2, -1, 0, 1, 2, 3, 4\}$, write the members of the solution set for $-2x > 4$.

21. If the replacement set for x is $\{-3, -2, -1, 0, 1, 2, 3\}$, write the members of the solution set for $-x \leq -2$.

22. If x is a whole number, find the value of x that will make the following expression true: $\frac{1}{3} < \frac{x}{6} < \frac{2}{3}$

4.5 PROBLEM SOLVING

Translating From Words Into Symbols

Open sentences can be used to solve a word problem if the problem is first translated from words into algebraic symbols. To help with the translation, the table on the following page tells which of the symbols $+$, $-$, $\times$, $\div$, and $=$ can be used to replace certain word phrases.

$+$	$-$	$\times$	$\div$	$=$
the sum of (the sum of a and b is $a + b$)	*the difference of; the difference between* (the difference of [between] a and b is $a - b$)	*the product of* (the product of a and b is $a \times b$)	*the quotient of* (the quotient of a and b is $a \div b$)	*is equal to*
more than (6 more than x is $x + 6$)	*less than* (6 less than x is $x - 6$)	*times* (a times b is $a \times b$)	*divided by* (a divided by b is $a \div b$)	*equals*
increased by (x increased by 6 is $x + 6$)	*decreased by* (x decreased by 6 is $x - 6$)	*twice* means *multiply by 2*	*half of* means *divide by 2*	*the result is*
exceeds means *is more than*	*diminished by* means *decreased by*	*double* means *multiply by 2*	*one-third of* means *divide by 3*	*is*

◆ In general, the procedure for solving a word problem is:

1. Read the problem carefully and decide what is the one basic value to be found (the unknown).
2. Represent the unknown by a letter. At the beginning of the solution, write the words that letter stands for. (If more than one value is to be found, represent all the unknown values in terms of the one basic unknown.)
3. Translate each word phrase into algebraic symbols, using the previous table as a guide, so that an algebraic open sentence results.
4. Find the solution set for the open sentence.
5. Check each element of the solution set *in the words of the problem.*
6. Write the answer to the problem in the form of a word sentence.

Number Problems

Several types of word problems will be considered here and throughout this book. The first type of word problem, *Number Problems,* comes in a variety of forms. One simple form requires no preliminary discussion.

Examples

1. If three times a number is increased by 4, the result is the same as when five times the number is decreased by 20. What is the number?

Solution:

1. Decide what is to be found.	1. one number
2. Represent the unknown value by a letter, telling what the letter represents.	2. Let x = the number.
3. Translate the word phrase into symbols and write an equation:	3. $3x + 4 = 5x - 20$

If three times	a number	is increased by	4,	the result is the same as
$3 \cdot$	(x)	$+$	4	$=$

when five times	the number	is decreased by	20.
$5 \cdot$	(x)	$-$	20

4. Solve the equation:	4. $3x + 4 = 5x - 20$
Subtract $3x$.	$4 = 2x - 20$
Add 20.	$24 = 2x$
Divide by 2.	$12 = x$
5. Check in the words of the problem.	5. If three times a number, $3 \cdot 12$, is increased by 4, $36 + 4$, the result, *40,* is the same as when five times the number, $5 \cdot 12$, is decreased by 20, $60 - 20$ or *40.*✓
6. Write the answer in the form of a word sentence.	6. *Answer:* The number is 12.

2. The larger of two numbers is twice the smaller. If the smaller number is increased by 4, the result is 6 less than the larger number. Find both numbers.

Solution:

$$\text{Let } x = \text{the smaller number.}$$
$$\text{Then } 2x = \text{the larger number.}$$

If the smaller number is increased by 4, the result is

$x \qquad +4 \qquad =$

(6 less than) the larger number.

$2x \qquad -6$

$$x + 4 = 2x - 6$$

Subtract x. $\quad 4 = x - 6$

Add 6. $\quad 10 = x \quad$ smaller number

$20 = 2x \quad$ larger number

Check: The larger number, *20,* is twice the smaller number, 10.√
If the smaller number, 10, is increased by 4, 10 + 4, the result, *14,* is 6 less than the larger number, 20 − 6 or *14*.√

Answer: The smaller number is 10 and the larger number is 20.

Some number problems involve the word *ratio*. In Chapter 2, ratio was defined as a fraction. The fractions $\frac{6}{10}, \frac{9}{15}, \frac{15}{25}, \frac{24}{40}, \frac{3x}{5x}$, etc. are equivalent fractions in that they all reduce to the ratio $\frac{3}{5}$ or 3:5. In each case, if the numerator and denominator have the same factor removed, the fraction reduces to the fraction $\frac{3}{5}$.

For example, in the fraction $\frac{6}{10}$, if the same factor of 2 is removed from the numerator and from the denominator, the ratio $\frac{3}{5}$ is easily seen:

$$\frac{6}{10} = \frac{3 \cdot \cancel{2}}{5 \cdot \cancel{2}} = \frac{3}{5}$$

To solve number problems involving ratio, we will use the converse of this idea. That is, if two unknown numbers are to be in the ratio 3:5, we do not know what common factor is to appear in both the numerator and the denominator. It could be 2 as in $\frac{6}{10} = \frac{3 \cdot 2}{5 \cdot 2} = \frac{3}{5}$; it could be 4 as in

$\frac{12}{20} = \frac{3 \cdot 4}{5 \cdot 4} = \frac{3}{5}$; it could be any number x as in $\frac{3x}{5x} = \frac{3}{5}$. Thus, if two unknown numbers are to be in a particular ratio, we introduce an unknown ***ratio factor*** as we set up the problem.

Example

Two numbers are in the ratio 4:7. If the sum of the numbers is 33, find the numbers.

Solution:

Let x = the common ratio factor.
Then $4x$ and $7x$ = the numbers whose ratio is 4:7.

$$\underbrace{\text{The sum of the numbers}}_{4x + 7x} \ \underbrace{\text{is}}_{=} \ \underbrace{33.}_{33}$$

$$\begin{aligned} 11x &= 33 \\ x &= 3 \\ 4x &= 12 \\ 7x &= 21 \end{aligned}$$

Check: Are 12 and 21 in the ratio 4:7?

$$\frac{12}{21} = \frac{4 \cdot \cancel{3}}{7 \cdot \cancel{3}} = \frac{4}{7} \checkmark$$

Is 33 the sum of 12 and 21?

$$\begin{aligned} 12 + 21 &\stackrel{?}{=} 33 \\ 33 &= 33 \checkmark \end{aligned}$$

Answer: The numbers are 12 and 21.

EXERCISES

1. If 16 is 4 more than $3a$, find a.
2. If three times a number is subtracted from seven times the same number, the result is 12. What is the number?
3. If 7 is subtracted from four times a certain number, the difference is 25. What is the number?
4. If three times a certain number is decreased by 7, the result is 20. What is the number?
5. If 19 is subtracted from three times a certain number, the difference is 110. What is the number?
6. The product of two numbers is p. If one number is represented by n, then the other number can be represented by

   (1) $p - n$ (2) pn (3) $\frac{n}{p}$ (4) $\frac{p}{n}$

7. The difference between two numbers is 9. If four times the larger number is ten times the smaller, what are the numbers?
8. One number is 4 more than another number. If four times the smaller number is decreased by twice the larger number, the result is 12. Find *both* numbers.
9. The ratio of two numbers is 1:5 and their sum is 24. Find the *smaller* number.
10. Two numbers are in the ratio 4:5 and their sum is 54. Find the *smaller* number.
11. Two numbers are in the ratio of 5 to 13. If the sum of the two numbers is 36, what is the *larger* number?
12. Two numbers are in the ratio of 5:1 and their difference is 28. What is the *smaller* number?
13. The sum of three numbers is 22. The second is 3 more than the first and the third is 1 less than twice the first. Find the numbers.
14. The sum of four numbers is 48. The second is 3 more than the first, the third is 1 more than twice the first, and the fourth is 1 more than the third. Find the numbers.

Consecutive Integer Problems

A special type of number problem deals with ***consecutive integers.*** The word "consecutive" means "immediately following"; "integer," of course, means a whole number with a sign (zero is included). Notice in the set of integers {. . ., −4, −3, −2, −1, 0, 1, 2, 3, 4, . . .} that, for example, 2 immediately follows 1, 3 immediately follows 2, and 4 immediately follows 3. In general, *two integers are consecutive if the second is one more than the first.*

Using symbols, we can express the idea of consecutive integers as:

Let x = the first of the consecutive integers
$x + 1$ = the second of the consecutive integers
$x + 2$ = the third of the consecutive integers, etc.

Recall that *even numbers* are exactly divisible by 2 (for example, 2, 4, 10, 0, −4, −6, etc. are even numbers), and *odd numbers* are not exactly divisible by 2 (for example, 1, 3, 9, 13, −5, −9, etc. are odd numbers). Two ***consecutive even integers*** are two even integers such that the second even integer immediately follows the first even integer.

Examples of sets of consecutive even integers are:

{2, 4}, {2, 4, 6}, {2, 4, 6, 8}, {14, 16, 18}, {−6, −4, −2}, etc.

Similarly, examples of sets of ***consecutive odd integers*** are:

{1, 3}, {1, 3, 5}, {1, 3, 5, 7}, {13, 15, 17}, {−7, −5, −3}, etc.

The set of integers is arranged as follows:

$$\{\ldots, \text{even}, \text{odd}, \text{even}, \text{odd}, \text{even}, \text{odd}, \text{even}, \text{odd}, \text{even}, \ldots\}$$
$$\downarrow \quad \downarrow \quad \downarrow \quad \downarrow \quad \downarrow \quad \downarrow \quad \downarrow \quad \downarrow \quad \downarrow$$
$$\{\ldots, -4, -3, -2, -1, 0, 1, 2, 3, 4, \ldots\}$$

Thus, each even integer is two more than the even integer just before it, and each odd integer is two more than the odd integer just before it. The symbolic representation for consecutive even integers and for consecutive odd integers is, therefore, the same. The difference comes in what the symbols represent. Using symbols, we can represent consecutive even integers and consecutive odd integers in the following manner:

Let x = the first of the consecutive even integers
$x + 2$ = the second of the consecutive even integers
$x + 4$ = the third of the consecutive even integers.

Let x = the first of the consecutive odd integers
$x + 2$ = the second of the consecutive odd integers
$x + 4$ = the third of the consecutive odd integers.

The following examples show how to use the representations for consecutive integers, consecutive even integers, and consecutive odd integers, and the general procedure for solving word problems.

Examples

1. The sum of three consecutive integers is 36. Find those integers.

Solution:

1. Decide what is to be found. — 1. Three consecutive integers.

2. Represent the unknown values in terms of one letter, telling what the letters represent. — 2. Let x = the first of the consecutive integers
$x + 1$ = the second of the consecutive integers
$x + 2$ = the third of the consecutive integers.

3. Translate the word phrases into symbols and write an equation: — 3. $(x) + (x + 1) + (x + 2) = 36$

$$\underbrace{\text{The sum of}}_{+}\ \underbrace{\text{three consecutive integers}}_{x,\ x+1,\ x+2}\ \underbrace{\text{is}}_{=}\ \underbrace{36.}_{36}$$

4. Solve the equation.	4. $x + x + 1 + x + 2 = 36$ $3x + 3 = 36$ $3x = 33$ $x = 11$ $x + 1 = 11 + 1 = 12$ $x + 2 = 11 + 2 = 13$
5. Check in the words of the problem.	5. 11, 12, 13 are three consecutive integers.√ The sum of the three consecutive integers, 11 + 12 + 13, is 36.√
6. Write the answer in the form of a word sentence.	6. *Answer:* The three consecutive integers are 11, 12, 13.

2. Find three consecutive even integers such that 8 more than the third is twice the first.

Solution:

Let x = the first of the consecutive even integers
$x + 2$ = the second of the consecutive even integers
$x + 4$ = the third of the consecutive even integers.

8 more than the third is twice the first

$$(x + 4) + 8 = 2 \cdot (x)$$
$$x + 4 + 8 = 2x$$
$$x + 12 = 2x$$
$$12 = x \text{ or } x = 12$$
$$x + 2 = 12 + 2 = 14$$
$$x + 4 = 12 + 4 = 16$$

Check: 12, 14, 16 are three consecutive even integers.√
8 more than the third, 16 + 8 or *24*, is twice the first, 2(12) or *24*.√

Answer: The three consecutive even integers are 12, 14, 16.

3. Find three consecutive odd integers such that twice the sum of the first and second is 5 more than three times the third.

Solution:

Let x = the first of the consecutive odd integers
$x + 2$ = the second of the consecutive odd integers
$x + 4$ = the third of the consecutive odd integers.

twice the sum of the first and second is

$$2 \cdot [(x) + (x + 2)] =$$

5 more than three times the third

$$3 \cdot (x + 4) + 5$$

$2(x + x + 2) = 3(x + 4) + 5$	
$2(2x + 2) = 3(x + 4) + 5$	Combine like terms.
$2(2x) + 2(2) = 3(x) + 3(4) + 5$	Distributive property.
$4x + 4 = 3x + 12 + 5$	Multiply to remove parentheses.
$4x + 4 = 3x + 17$	Combine like terms.
$x + 4 = 17$	Subtract $3x$.
$x = 13$	Subtract 4.
$x + 2 = 13 + 2 = 15$	
$x + 4 = 13 + 4 = 17$	

Check: 13, 15, 17 are three consecutive odd integers.✓
Twice the sum of the first and second, 2(13 + 15) or 2(28) or *56*, is 5 more than three times the third, 3(17) + 5 or 51 + 5 or *56*.✓

Answer: The three consecutive odd integers are 13, 15, 17.

EXERCISES

In 1–10, only an algebraic solution will be accepted.

1. If n represents an even integer, express the next larger even integer in terms of n.
2. If n represents an odd integer, express the next larger odd integer in terms of n.
3. If $n + 7$ is an even number, the next larger even number is
(1) $n + 5$ (2) $n + 9$ (3) $10n + 7$ (4) $2n + 14$
4. If the sum of two consecutive integers is 13, find the smaller integer.
5. The sum of two consecutive odd integers is 40. Find the value of the smaller integer.
6. Find three consecutive even integers such that twice the largest is 2 less than three times the smallest.
7. Find three consecutive odd integers such that the sum of the first and third is 37 more than the second. Check.
8. Find three positive consecutive odd integers such that the largest decreased by twice the second is equal to 10 less than the smallest.
9. Find three consecutive odd integers such that twice the sum of the second and third is 43 more than three times the first.
10. Find four consecutive odd integers such that the sum of the first three exceeds twice the fourth by 5.

Problems With Inequalities

Some problems can be solved by using an open sentence of inequality.

Examples

1. Let x represent the smaller of two integers whose sum is greater than 49. The larger integer is six times the smaller. Find the smallest possible value of x.

Solution:

Let x = the smaller integer.
Then $6x$ = the larger integer.

The sum of the two integers is greater than 49.

$$\underbrace{x + 6x}_{} \quad > \quad 49$$

$7x > 49$ Combine like terms.
$x > 7$ Divide by 7.

The solution set, $x > 7$, of the open sentence has an infinite number of elements. That is, all numbers greater than 7 are elements of the solution set. However, the problem requires only one number as an answer: the smallest integer greater than 7. Thus, 8 is the required number.

Check: If $x = 8$, the numbers are 8 and $6 \cdot 8$ or 48.
The larger integer, *48,* is six times the smaller integer, 8.√
The sum of the two integers, 8 + 48 or *56, is greater than 49.* $(56 > 49)$√

Answer: The smallest possible integral value of x is 8.

2. The sum of two consecutive integers is greater than 21. If x represents the first of these integers, what is the smallest possible value of x?

Solution:

Let x = the first of the consecutive integers
$x + 1$ = the second of the consecutive integers.

The sum of two consecutive integers is greater than 21.

$$x + x + 1 \quad > \quad 21$$

$2x + 1 > 21$ Combine like terms.
$2x > 20$ Subtract 1.
$x > 10$ Divide by 2.

The solution set, $x > 10$, of the open sentence has an infinite number of elements. The problem requires only one number as an answer: the smallest integer greater than 10. Thus, 11 is the required number.

Check: If $x = 11$, the consecutive integers are 11 and 12. The sum of the consecutive integers, 11 + 12 or *23, is greater than 21.* $(23 > 21)$√

Answer: The smallest possible integral value of x is 11.

3. Let x represent the larger of two integers whose sum is less than 40. The smaller integer is one-third of the larger. Find the largest possible value of x.

Solution:

Let x = the larger integer.
Then $\frac{1}{3}x$ = the smaller integer.

$$\underbrace{\text{The sum of the two integers}}_{x + \frac{1}{3}x} \; \underbrace{\text{is less than}}_{<} \; \underbrace{40.}_{40}$$

$\frac{4}{3}x < 40$	Combine like terms.
$x < 40 \div \frac{4}{3}$	Divide by $\frac{4}{3}$.
$x < 30$	$40 \div \frac{4}{3} = 40 \times \frac{3}{4} = 30$

The solution set, $x < 30$, of the open sentence has an infinite number of elements. The problem requires only one number as an answer: the largest integer less than 30 such that one-third of that integer is also an integer. Thus, 27 is the required number.

Check: If $x = 27$, the numbers are 27 and 9. The smaller number, *9*, is one-third of the larger number, $\frac{1}{3} \cdot 27$ or *9*.✓
The sum of the numbers, 27 + 9 or *36, is less than 40.* $(36 < 40)$✓

Answer: The largest possible integral value of x is 27.

4. The sum of three consecutive odd integers is at least 189. Find the smallest possible values for these numbers.

Solution:

Let x = the first of the consecutive odd integers
$x + 2$ = the second of the consecutive odd integers
$x + 4$ = the third of the consecutive odd integers.

$$\underbrace{\text{The sum of the three consecutive odd integers}}_{(x) + (x + 2) + (x + 4)} \; \underbrace{\text{is at least}}_{\geq} \; \underbrace{189.}_{189}$$

$$3x + 6 \geq 189$$
$$3x \geq 183$$
$$x \geq 61$$

The solution set, $x \geq 61$, of the open sentence has an infinite number of elements. The smallest of these values is 61. Thus, 61, 63, and 65 are the required numbers.

Check: 61, 63, and 65 are consecutive odd integers.✓
Their sum, 61 + 63 + 65, is at least 189. 61 + 63 + 65 = 189✓

Answer: The smallest possible values for three consecutive odd integers whose sum is at least 189 are 61, 63, and 65.

EXERCISES

In 1–8, only an algebraic solution will be accepted.

1. Let x represent the smaller of two integers whose sum is greater than 40. The larger integer is seven times the smaller. Find the *smallest* possible value of x.
2. Let x represent the larger of two integers whose sum is less than 30. The smaller integer is one-half of the larger. Find the *largest* possible value of x.
3. The sum of two consecutive integers is greater than 29. If x represents the first of these integers, what is the *smallest* possible value of x?
4. The sum of two consecutive even integers is greater than 18. If x represents the first of these integers, what is the *smallest* possible value of x?
5. The sum of two consecutive even integers is less than 10. If x represents the first of these integers, what is the *largest* possible value of x?
6. The sum of three consecutive odd integers is less than 15. If x represents the first of these integers, what is the *largest* possible value of x?
7. The sum of three consecutive integers is less than 39. Find the *largest* possible values for these numbers.
8. The sum of four consecutive integers is at least 114. Find the *smallest* possible values for these numbers.

4.6 SUMMARY EXERCISES

In 1–20, find the solution set of each equation and check.

1. $x - 1 = 1$
2. $5x = 0$
3. $-3y + 1 = -11$
4. $\frac{1}{2}a + 7 = 33$
5. $\dfrac{x}{2} + \dfrac{x}{6} = 8$
6. $.8z = .2z + 2.4$
7. $2\frac{3}{4}x + 24 = 3x$
8. $\frac{1}{5}x - 3 = -4$
9. $5a + 2a - 17 = 53$
10. $5a + 5 - 9a = -11$
11. $8b = 36 + 5b$
12. $2.3y + 36 = .3y$
13. $2\frac{1}{4}x = 1\frac{1}{4}x - 8$
14. $z = 9z + 72$
15. $.5x - 30 = 1.1x$
16. $7x + 8 = 6x + 1$
17. $3(x - 9) = 30$
18. $5(3x - 2) + 8 = 43$
19. $10a - (3a - 11) = 25$
20. $4(2b + 1) - 3(2b - 5) = 29$
21. Solve for h in terms of A and b: $A = bh$
22. Solve for x in terms of r, s, and t: $rx + s = t$
23. Solve for x in terms of a, b, and c: $ax = bx + c$
24. Solve for D in terms of R and T: $T = \dfrac{D}{R}$
25. Solve for C if $F = 50$: $C = \frac{5}{9}(F - 32)$

26. Solve for F if $C = 20$: $C = \frac{5}{9}(F - 32)$

27. Solve for x: $8x \leq 3x + 15$

28. The inequality $7x - 1 > x + 5$ is equivalent to
(1) $x > 0$ (2) $x < 0$ (3) $x > 1$ (4) $x < 1$

29. If the replacement set for x is $\{-2, -1, 0, 1, 2\}$, find the solution set of $2x - 1 \geq 1$.

30. The inequality $7x - (x + 1) < 5$ is equivalent to
(1) $x < \frac{2}{3}$ (2) $x < 1$ (3) $x > 1$ (4) $x > \frac{2}{3}$

31. If the replacement set for x is $\{0, 1, 2, 3, 4, 5\}$, find the solution set for $-2x > 10$.

32. A member of the solution set of $-1 \leq x < 4$ is
(1) -2 (2) -1 (3) 4 (4) 5

33. If x is an integer, the solution set of the inequality $-5 \leq x \leq -3$ is
(1) $\{-5, -4, -3\}$ (2) $\{-5, -4\}$ (3) $\{-4, -3\}$ (4) $\varnothing$

34. If x is an integer, write the solution set of the inequality $-2 \leq x < 0$.

35. If x is an integer, find the value of x that will make the following expression true: $\frac{1}{5} < \frac{x}{10} < \frac{2}{5}$

36. If x is an integer, find all the values of x that will make the following expression true: $\frac{1}{3} \leq \frac{x}{9} < \frac{2}{3}$
(1) $\{3, 4, 5, 6\}$ (2) $\{3, 4, 5\}$ (3) $\{4, 5, 6\}$ (4) $\varnothing$

37. Which solution set is represented by the accompanying graph?

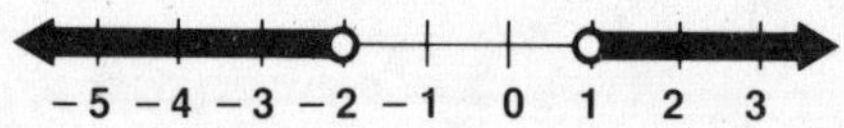

(1) $-2 < x < 1$
(2) $(x > -2) \vee (x > 1)$
(3) $(x < -2) \vee (x > 1)$
(4) $(x < -2) \vee (x < 1)$

38. Which solution set is represented by the accompanying graph?

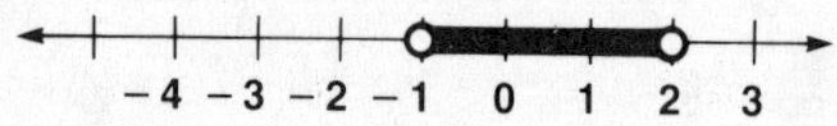

(1) $(x > -1) \wedge (x < 2)$
(2) $-1 \leq x \leq 2$
(3) $-1 \leq x < 2$
(4) $-1 < x \leq 2$

39. Which solution set is represented by the accompanying graph?

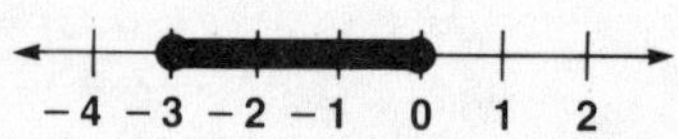

(1) $\{x \mid -3 < x < 0\}$
(2) $\{x \mid (x < -3) \wedge (x < 0)\}$
(3) $\{x \mid -3 \leq x \leq 0\}$
(4) $\{x \mid (x \leq -3) \wedge (x \leq 0)\}$

40. Which solution set does *not* represent the accompanying graph?

(1) $\{x \mid (x > -1) \wedge (x < 3)\}$
(2) $\{x \mid -1 < x < 3\}$
(3) {real numbers between -1 and 3}
(4) $\{x \mid (x > -1) \vee (x < 3)\}$

41. Which graph shows the solution to $(x < -2) \vee (x \geq 4)$?

(1) 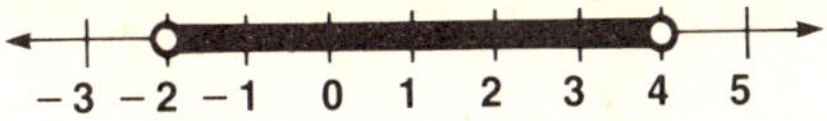

(2) 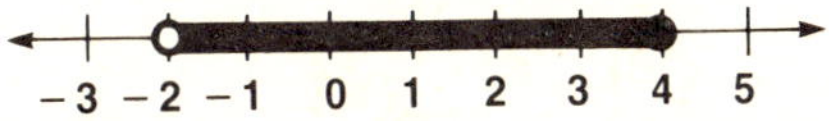

(3) 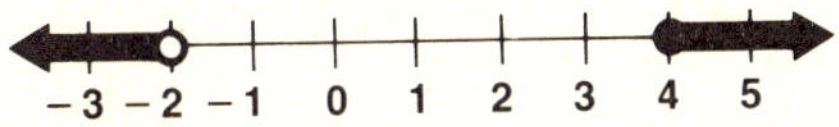

(4) 

42. Which set is shown on the accompanying graph?

(1) $\{x \mid x < -4\}$
(2) $\{x \mid x > -4\}$
(3) $\{x \mid -4 < x \leq 3\}$
(4) $\{x \mid -4 < x < 3\}$

43. Which set is represented by the accompanying graph?

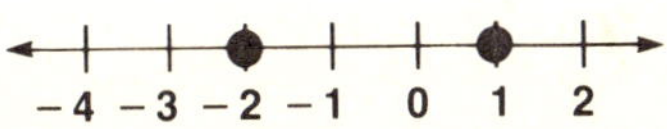

(1) $\{-2, 1\}$ (2) $\{x \mid -2 \leq x \leq 1\}$ (3) $\{x \mid -2 < x < 1\}$ (4) $\varnothing$

In 44–47, write the set that is represented by the graph.

44. 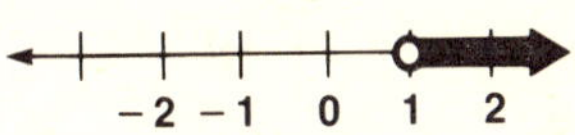

45. 

46. 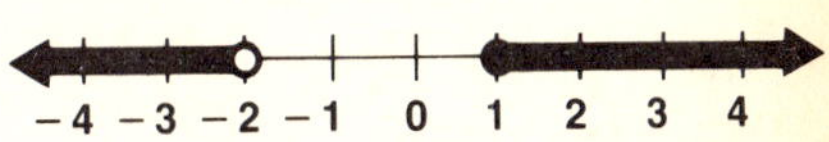

47. 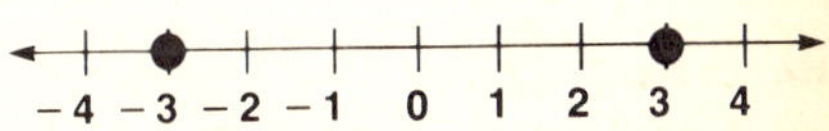

48. If eight times a number is decreased by 23, the result is 129. Find the number.

49. One number is 8 more than another. If the smaller number is multiplied by 3, the result is 18 more than the larger number. Find the numbers.

50. Three more than five times a number is less than 18. Find the largest integer for which this is true.

51. Two less than seven times a number is less than 12. Find the largest integer for which this is true.

52. If 14 is added to a certain number and the sum is doubled, the result is the same as decreasing eight times the number by 14. Find the number.

53. The sum of three numbers is 12. If the second is 1 less than the first and the third is 5 less than twice the second, find the numbers.

54. Find three consecutive integers such that the sum of the first two exceeds the third by 24.

55. Find four consecutive integers whose sum is 82.

56. Find four consecutive odd integers such that the sum of the first three is 30 more than the fourth.

57. Find four consecutive even integers such that the sum of the third and fourth is 6 less than four times the first.

58. Let x represent the larger of two integers whose sum is less than 40. The smaller integer is one-fourth of the larger. Find the largest possible value of x.

59. The sum of three consecutive odd integers is less than 30. If x represents the first of these integers, what is the largest possible value of x?

60. The sum of four consecutive even integers is at least 252. Find the smallest possible values for these numbers.

CUMULATIVE SUMMARY EXERCISES

CHAPTERS 1–4

1. Let p represent "I will go roller skating" and q represent "The sun is shining." Using p and q, write in symbolic form: "I will go roller skating if and only if the sun is shining."
2. Which is the converse of the statement $p \to \sim q$?
   (1) $p \to \sim q$ (2) $\sim p \to \sim q$ (3) $q \to \sim p$ (4) $\sim q \to p$
3. Which statement is false when p is false and q is false?
   (1) $p \wedge q$ (2) $p \to q$ (3) $\sim p \to \sim q$ (4) $p \leftrightarrow q$
4. **a.** Copy and complete the truth table for the statement:

$$\sim(p \vee q) \leftrightarrow (\sim p \wedge \sim q)$$

p	q	$p \vee q$	$\sim(p \vee q)$	$\sim p$	$\sim q$	$(\sim p \wedge \sim q)$	$\sim(p \vee q) \leftrightarrow (\sim p \wedge \sim q)$
T	T						
T	F						
F	T						
F	F						

   b. Is $\sim(p \vee q) \leftrightarrow (\sim p \wedge \sim q)$ a tautology?
   c. Justify the answer you gave in part **b.**
5. The inverse of $\sim p \to q$ is
   (1) $p \to q$ (2) $\sim p \to \sim q$ (3) $p \to \sim q$ (4) $q \to \sim p$
6. Given the true statement: "If a number is rational, then it can be expressed as the ratio of two integers." Which statement must also be true?
   (1) If a number is not rational, then it can be expressed as the ratio of two integers.
   (2) If a number cannot be expressed as the ratio of two integers, then it is rational.
   (3) If a number cannot be expressed as the ratio of two integers, then it is irrational.
   (4) If a number can be expressed as the ratio of two integers, then it is irrational.

7. Let p represent "$x - 3 = 1$" and let q represent "$\frac{1}{2}x = 2$." For what value of x is $p \wedge q$ true?

8. If $p \vee q$ is false, then
(1) p is true and q is false
(2) p is false and q is true
(3) both p and q are true
(4) both p and q are false

9. Given that $(\sim p \rightarrow q)$ and $(p \vee q)$ are logically equivalent. Which of the following statements is false?
(1) $(\sim p \rightarrow q)$ and $(p \vee q)$ have opposite truth values.
(2) $(\sim p \rightarrow q)$ and $(p \vee q)$ have the same truth values.
(3) $(\sim p \rightarrow q) \leftrightarrow (p \vee q)$ is a tautology.
(4) $(\sim p \rightarrow q) \rightarrow (p \vee q)$ is a tautology.

10. When $p \rightarrow q$ is true, which related conditional is also true?
(1) $\sim p \rightarrow q$ (2) $p \rightarrow \sim q$ (3) $q \rightarrow p$ (4) $\sim q \rightarrow \sim p$

11. Which should be the heading for column 3?
(1) $p \wedge q$
(2) $p \rightarrow q$
(3) $p \leftrightarrow q$
(4) $p \vee q$

Col. 1	Col. 2	Col. 3
p	q	
T	T	T
T	F	F
F	T	F
F	F	T

12. If p represents "x is divisible by 3" and q represents "x is divisible by 5," then which statement is true when $x = 21$?
(1) $p \wedge q$ (2) $\sim p$ (3) $\sim p \wedge q$ (4) $p \vee q$

13. **a.** Copy and complete the truth table for the statement:

$$[p \wedge (p \vee \sim q)] \rightarrow (\sim p \rightarrow q)$$

p	q	$\sim q$	$p \vee \sim q$	$p \wedge (p \vee \sim q)$	$\sim p$	$\sim p \rightarrow q$	$[p \wedge (p \vee \sim q)] \rightarrow (\sim p \rightarrow q)$

b. Is $[p \wedge (p \vee \sim q)] \rightarrow (\sim p \rightarrow q)$ a tautology?

14. Which represents the inverse of the statement $\sim p \rightarrow q$?
(1) $\sim p \rightarrow \sim q$ (2) $p \rightarrow \sim q$ (3) $q \rightarrow \sim p$ (4) $\sim q \rightarrow p$

15. Let p represent "You are correct" and let q represent "You do not win a prize." Using p and q, write in symbolic form: "If you are correct, you win a prize."

16. Which is an example of a finite set?
(1) {integers < 100,000}
(2) {integers}
(3) {positive integers < 100,000}
(4) {positive integers}

17. If A = {positive primes < 10} and B = {positive integers < 10}, which statement about $A \cap B$ is true?
(1) $A \cap B = A \cup B$
(2) $A \cap B = A$
(3) $A \cap B = B$
(4) $A \cap B = \varnothing$

18. If the universe = {whole numbers ≤ 10}, A = {even whole numbers ≤ 10}, and B = {odd whole numbers ≤ 10}, then which of the following statements is false?
(1) $A \cap B = \varnothing$ (3) $\sim A = B$
(2) $\sim A = \sim B$ (4) $\sim B = A$

19. Which is a rational number?
(1) $\sqrt{3}$ (2) $\sqrt{5}$ (3) $\sqrt{7}$ (4) $\sqrt{9}$

20. For which value(s) of x is $\frac{5}{2 + x}$ not defined?
(1) 0 (2) 2 (3) -2 (4) ± 2

21. Evaluate $|2 - 5|$.

22. The additive inverse of -7 is
(1) $-\frac{1}{7}$ (2) $\frac{1}{7}$ (3) 0 (4) 7

In 23–29, each algebraic statement is an illustration of a property listed in **a–i**. Match them.

23. $3 \cdot 5 = 5 \cdot 3$
24. $(2 + 3) + 4 = 2 + (3 + 4)$
25. $2 + (-2) = 0$
26. $3 \cdot \frac{1}{3} = 1$
27. $4 \cdot 1 = 4$
28. $(-7) + 0 = -7$
29. $5 \cdot 3 + 6 \cdot 3 = (5 + 6) \cdot 3$

a. Associative Property of Addition
b. Associative Property of Multiplication
c. Commutative Property of Addition
d. Commutative Property of Multiplication
e. Distributive Property
f. Additive Inverse Property
g. Multiplicative Inverse Property
h. Additive Identity Property
i. Multiplicative Identity Property

30. Which statement is true?
(1) $|-4| < |4|$ (3) $|+5| > |-5|$
(2) $|-2| = |+2|$ (4) $|-1| - 3 = |-3|$

31. Which number does not have a multiplicative inverse?
(1) 2 (2) $+1$ (3) -1 (4) 0

32. Write $\frac{175}{225}$ in simplest form.

33. Add $+\frac{2}{3} - \frac{1}{6} - \frac{5}{6}$ and express the answer in simplest form.

34. Multiply $(\frac{3}{7})(-\frac{14}{9})$ and express the answer in simplest form.

35. Divide $\frac{-\frac{3}{2}}{-\frac{2}{3}}$ and express the answer in simplest form.

36. Subtract $-\frac{1}{4}$ from $\frac{1}{8}$.

37. Which fraction has the greatest value?
(1) $\frac{2}{7}$ (2) $\frac{5}{7}$ (3) $\frac{1}{14}$ (4) $\frac{5}{14}$

38. Which fraction is equivalent to $-3\frac{1}{3}$?
(1) $\frac{-10}{-3}$ (2) $\frac{-10}{3}$ (3) $\frac{-8}{3}$ (4) $\frac{-8}{-3}$

39. $.64 \div .8$ is equivalent to
(1) .8 (2) .08 (3) 8 (4) 80

40. Write .16 as a fraction in lowest terms.

41. Which decimal has the greatest value?
(1) .22 (2) .201 (3) .02 (4) .002

42. Which decimal lies between .31 and .32?
(1) .301 (2) .312 (3) .322 (4) .3

43. Write the following number in symbols: eleven thousand, two hundred eight and three hundredths.

44. Write the following number in words: 1,004.1

45. Round off 736.14 to the nearest tenth.

46. Round off 849.5 to the nearest integer.

47. Express as a binomial: $7x - 2y - 3x + y$.

48. Find the sum of $3x^2 - 5x - 1$ and $2x^2 - 2x - 1$.

49. From $7x^2 - 3x - 4$ subtract $5x^2 - 3x + 5$.

50. Subtract $2a - 3b + 4c$ from $3a - 4b$.

51. The product of $5x^4$ and $4x^5$ is
(1) $9x^9$ (2) $9x^{20}$ (3) $20x^{20}$ (4) $20x^9$

52. The sum of $3x^2$ and $4x^3$ is
(1) $7x^5$ (2) $3x^2 + 4x^3$ (3) $7x^6$ (4) $12x^5$

53. When $3x^4y^3$ is multiplied by $5xy^5$, the product is
(1) $8x^5y^8$ (2) $15x^4y^8$ (3) $15x^5y^8$ (4) $15x^4y^{15}$

54. Express the product $(x - 3)(x + 1)$ as a trinomial.

55. Express the product $(2x - 5)(2x + 5)$ as a binomial.

56. When $-12x^8$ is divided by $-3x^4$, the quotient is
(1) $4x^2$ (2) $-4x^2$ (3) $4x^4$ (4) $-4x^4$

57. The quotient $\dfrac{-15x^4y^6}{3xy^2}$ is equivalent to
(1) $-5x^3y^3$ (2) $-5x^3y^4$ (3) $5x^3y^4$ (4) $5x^3y^3$

58. What is the value of $4x^4$ if $x = -1$?

59. If $x = 1$ and $y = -2$, find the value of $(x^2y)^3$.

60. If $x = \frac{1}{2}$ and $y = 2$, find the value of $(x^2y)^2$.

61. What is the value of $3(4x)^0$?
(1) 0 (2) 3 (3) 12 (4) $\frac{3}{4}$

62. The value of 4^{-2} is
(1) $\frac{1}{8}$ (2) $\frac{1}{16}$ (3) -16 (4) -8

63. The value of $3 \cdot 2^{-3}$ is
(1) $\frac{3}{8}$ (2) $\frac{3}{6}$ (3) $\frac{1}{24}$ (4) $\frac{3}{24}$

64. Arrange the polynomial $8x^2 - 3 + 5x^4$ in ascending order.

65. Arrange the polynomial $2y^2 - 3x^2 + 4xy$ in descending order of x and ascending order of y.

66. What is the degree of the polynomial $7x^3 - 2x^4 - 5$?

In 67–78, solve and check.

67. $5x + 4 = 19$

68. $\dfrac{a}{2} - 3 = 5$

69. $3x + .4 = 4$

70. $.6y + 3 = 51$

71. $0.2x + 0.5 = 9.1$

72. $7x = 24 - x$

73. $5x - 3 = 9 - x$

74. $4(x - 1) = 16$

75. $3(x - 3) - 5 = 1$

76. $\dfrac{y}{2} + \dfrac{y}{8} = 10$

77. $5(x + 2) - (x - 3) = 21$

78. $5\frac{3}{4}x - (x - \frac{1}{4}) = 5$

79. Solve for x in terms of a, b, and c: $a + bx = c$

80. Solve for d_1 in terms of A and d_2: $A = \frac{1}{2}d_1 d_2$

81. Solve for y in terms of a, c, and d: $ay = cy + d$

82. Solve for x: $2x > 5x - 9$

83. The inequality $3x + 4 > x - 2$ is equivalent to
(1) $x > -3$ (2) $x < -3$ (3) $x > -1$ (4) $x > 1$

84. If the replacement set is $\{-3, -2, -1, 0, 1, 2, 3\}$, find the solution set of $2x + 3 \leq 1$.

85. Which solution set is represented by the accompanying graph?

(1) $\{x \mid x > -2\}$ (3) $\{x \mid x \geq -2\}$
(2) $\{x \mid x < -2\}$ (4) $\{x \mid x \leq -2\}$

86. A graph of $-2x \geq 6$ is

(1) 

(3) 

(2) 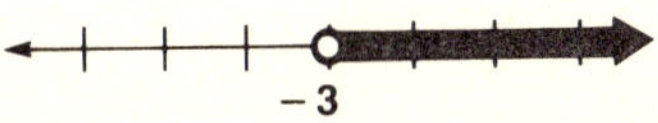

(4) 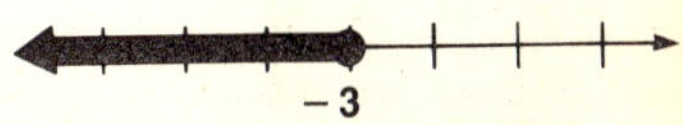

87. Which inequality is represented by the accompanying graph?
(1) $-2 \leq x \leq 3$ (3) $-2 \leq x < 3$
(2) $-2 < x < 3$ (4) $-2 < x \leq 3$

88. Which graph represents the inequality $(x < -1) \vee (x \geq 2)$?

(1) 

(2) 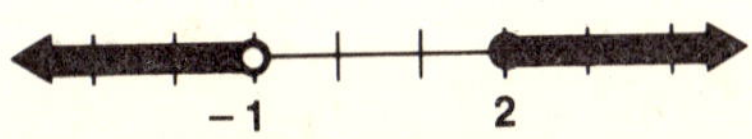

(3) 

(4) 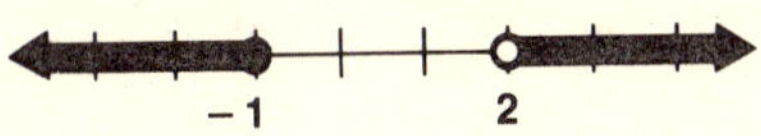

89. If x is an integer, write the solution set of $2 \leq x < 5$.

90. If the replacement set for x is $\{-3, -2, -1, 0, 1, 2, 3\}$, write the solution set for $-3x \leq 6$.

91. If x is an integer, find the value(s) of x that will make the following expression true:

$$\frac{1}{5} \leq \frac{x}{10} \leq \frac{3}{5}$$

92. If $(x \geq 2) \wedge (x \geq 5)$, which of the following may be a value of x?
(1) 2 (2) 3 (3) 5 (4) 6

In 93–100, only an algebraic solution is acceptable.

93. One number is 3 less than another number. If three times the smaller number is increased by five times the larger number, the result is 15. Find both numbers.

94. Two numbers are in the ratio 8:11 and their sum is 57. Find the numbers.

95. The sum of three numbers is 42. The second is 5 more than the first and the third is 3 less than twice the first. Find the numbers.

96. If $2n + 1$ represents an odd integer, which of the following represents the next larger odd integer?
(1) $2n - 1$ (2) $2n + 2$ (3) $2n + 3$ (4) $2n + 4$

97. Find three consecutive even integers such that twice the sum of the second and third is 10 less than five times the first.

98. Find three consecutive positive odd integers such that the largest decreased by twice the smallest is 20 less than the second.

99. Let x represent the larger of two integers whose sum is less than 30. The smaller integer is one-fifth of the larger integer. Find the largest possible value of x.

100. The sum of three consecutive odd integers is greater than 135. Find the smallest possible values for these numbers.

CHAPTER 5

Geometric Sets and Angle Measure

5.1 CLASSIFICATION OF GEOMETRIC SETS

Point, Line, Plane

To begin, some geometric terms are explained here. Also, the diagrams associated with them are shown, and the manner in which they are labeled and read is detailed.

A ***point*** is thought of as a dot. It has no ***dimensions*** (no length, no width, no height). A point is labeled by one capital letter, and is read "point P" and written P.

Point

Line

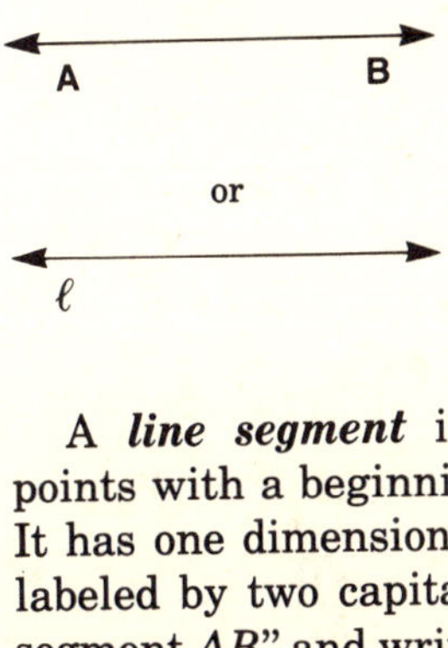

A ***line*** is an infinite collection of points with no beginning and no end, extending infinitely in both directions. It has one dimension: length. A line is labeled by two capital letters or by one small letter, and is read "line AB" and written $\overleftrightarrow{AB}$ or read "line ℓ" and written ℓ. (*Note:* A and B represent points on the line, but these points are not always shown as dots on the diagram.)

A ***line segment*** is an infinite collection of points with a beginning and an end ***(endpoints)***. It has one dimension: length. A line segment is labeled by two capital letters, and is read "line segment AB" and written $\overline{AB}$ or $\overline{BA}$.

Line Segment

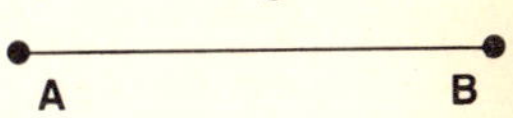

Ray

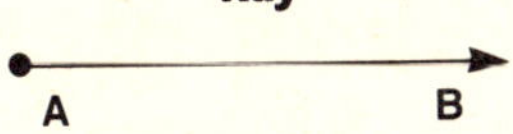

A ***ray*** is an infinite collection of points with only one endpoint. Extending infinitely in one direction, a ray has one dimension: length. A ray is labeled by two capital letters, and is read "ray AB" and written $\overrightarrow{AB}$. (*Note:* B represents a point on the ray in the direction in which the ray extends infinitely, but that point is not always shown as a dot on the diagram. A represents the endpoint from which the ray starts.)

A ***plane*** is thought of as a flat surface extending infinitely in all directions. It has two dimensions: length and width. A plane is labeled by one capital letter or by three capital letters (representing three points not on the same line), and is read "plane P" or "plane ABC." (*Note:* Only a part of a plane can be drawn since there are no boundaries.)

Plane

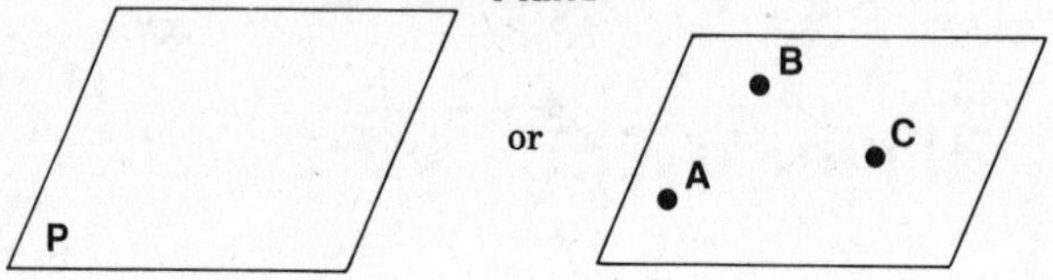

Information About Lines

Through one point in a plane, an infinite number of lines can be drawn.

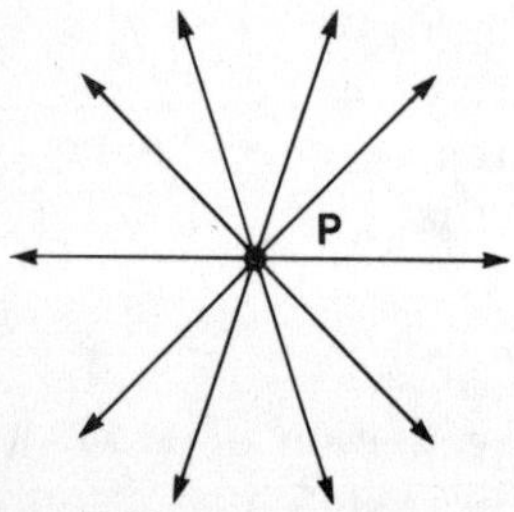

Through two points in a plane, only one line can be drawn.

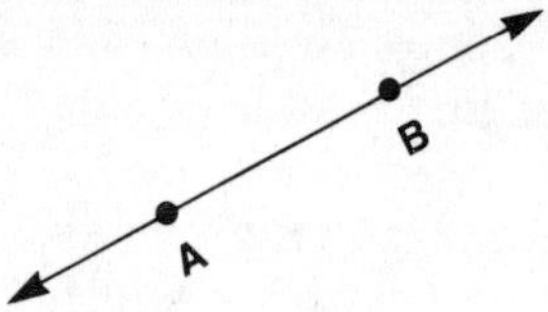

Two lines can ***intersect*** in only one point.

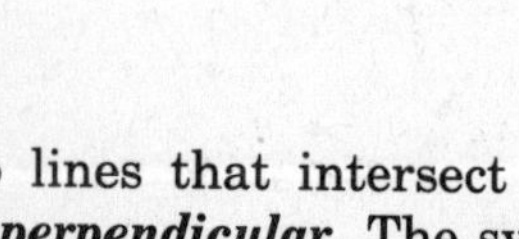

Two lines in the same plane that never intersect are called ***parallel.*** The symbol for "is parallel to" is $\parallel$. (Note the use of the same number of little "arrowheads" to show parallelism.)

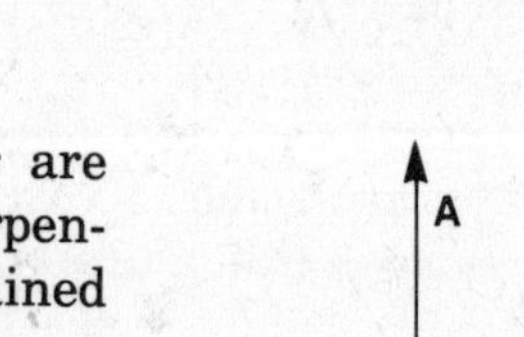

$\overleftrightarrow{AB} \parallel \overleftrightarrow{CD}$

Two lines that intersect at *right angles* are called ***perpendicular.*** The symbol for "is perpendicular to" is $\perp$. (*Note:* ***Right angles*** are explained further in an upcoming discussion.)

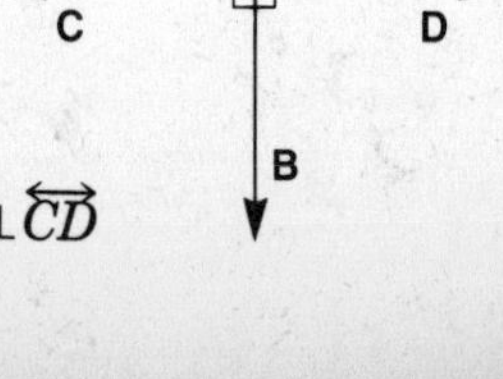

$\overleftrightarrow{AB} \perp \overleftrightarrow{CD}$

The length of a line segment can be measured with an instrument such as a ***ruler.*** When two line segments are equal in measure, they are called ***congruent.*** The symbol for "is congruent to" is $\cong$. When we speak of the measure of a line segment, we do not write a bar over the letters representing the line segment. That is, $\overline{AB}$ means line segment AB, and AB means the measure of line segment $\overline{AB}$.

The diagram shows two line segments that are equal in measure or congruent. (Note the use of the same number of little "stroke" marks to show equality.)

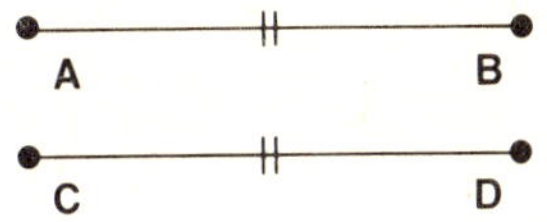

$$AB = CD \quad \text{or} \quad \overline{AB} \cong \overline{CD}$$

To ***bisect*** means to divide into two congruent parts. There is an infinite number of bisectors of $\overline{AB}$. In each case, $\overline{AM} \cong \overline{MB}$. M is the ***midpoint*** of $\overline{AB}$.

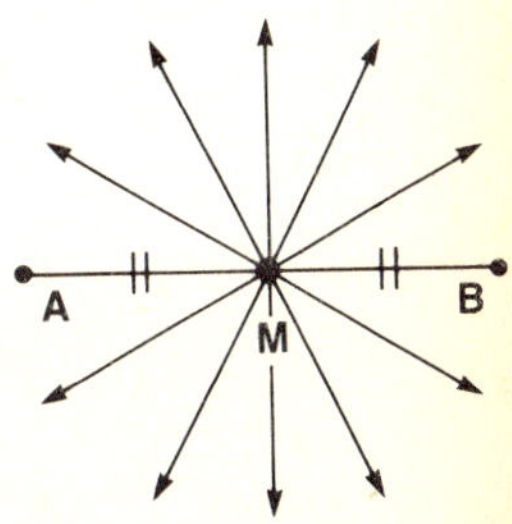

There is only one ***perpendicular bisector*** of $\overline{AB}$. $\overleftrightarrow{QR} \perp \overline{AB}$, and $\overline{AM} \cong \overline{MB}$.

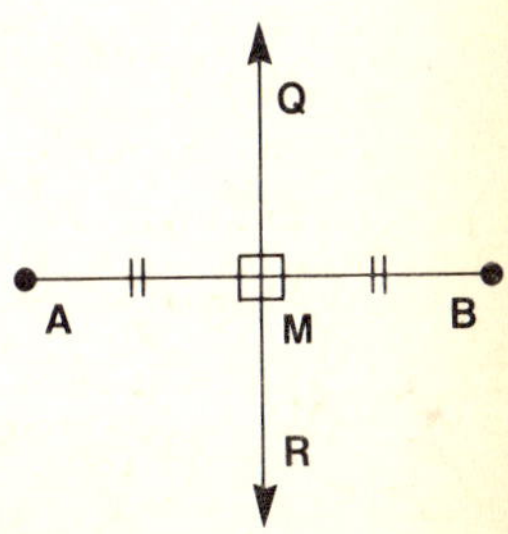

Angles

When two rays unite at their endpoints, the figure formed is called an ***angle.*** The point at which the rays unite is called the ***vertex*** of the angle, and the rays are called the ***sides*** of the angle. (*Note:* The dot for the vertex point of an angle may be omitted from the diagram as are the dots for the points in the infinite direction of each ray.)

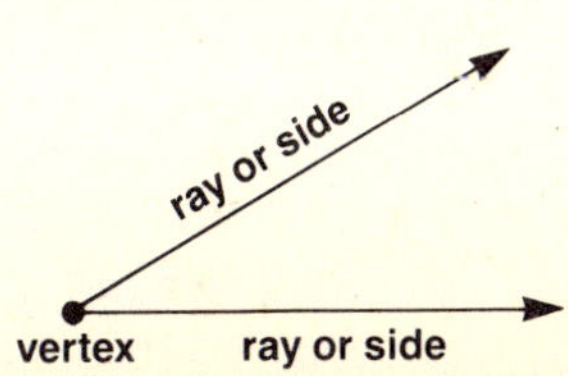

An angle can be labeled by three capital letters. The letter at the vertex *must* be read between the other two letters: angle BAC or angle CAB. The symbol for angle is $\angle$. Thus, the angle shown is $\angle BAC$ or $\angle CAB$. An angle can be labeled by one capital letter. The angle shown, $\angle BAC$ or $\angle CAB$, is also called $\angle A$.

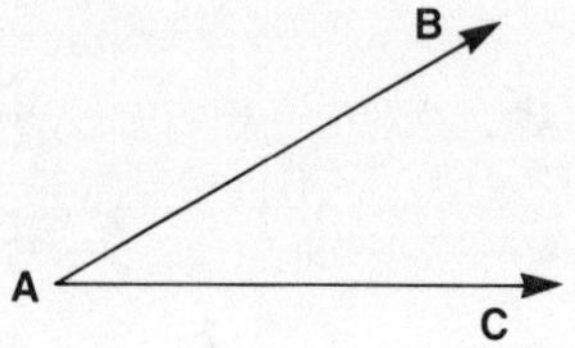

An angle can be labeled by one small letter or by a number. These labels are usually written inside the angle. The angles shown are $\angle x$ and $\angle 1$.

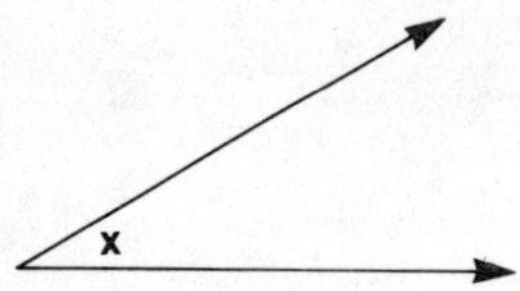

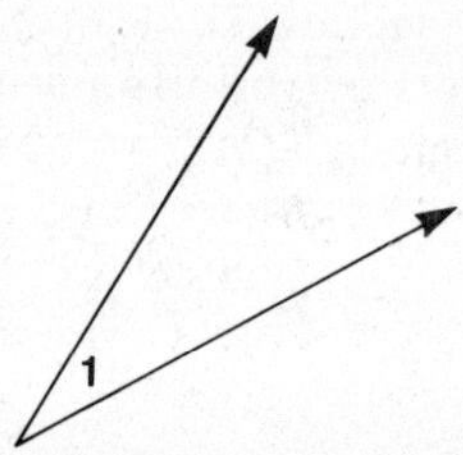

To understand what is meant by the *size of an angle,* we speak of *rotation.* As shown in the following diagrams, a *full rotation* begins when one ray is on top of the other (the rays *coincide*) and ends when the two rays again coincide.

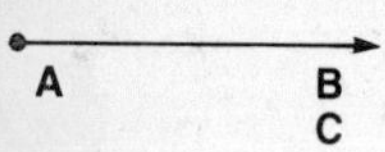

Two rays, $\overrightarrow{AB}$ and $\overrightarrow{AC}$, coincide.

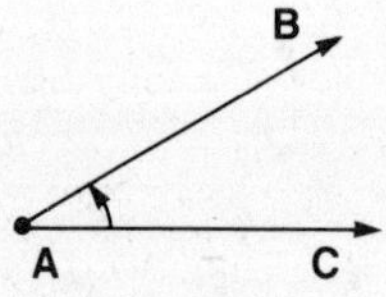

$\overrightarrow{AB}$ rotates away from $\overrightarrow{AC}$ in a *counterclockwise direction* (*opposite* from the way the hands of a clock rotate).

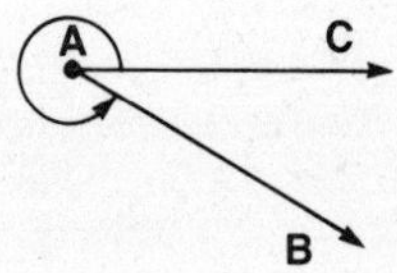

$\overrightarrow{AB}$ continues its rotation away from $\overrightarrow{AC}$.

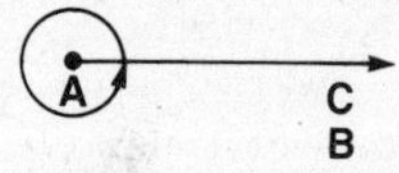

$\overrightarrow{AB}$ once again coincides with $\overrightarrow{AC}$. One full rotation has been completed.

The size of an angle is determined by the *amount of rotation* between the two rays that form the sides of the angle and not by the length of the rays (since the rays extend infinitely in length).

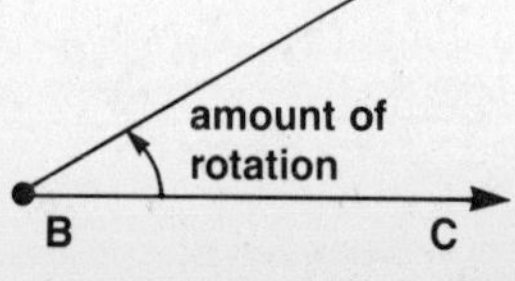

The size of an angle, or the amount of rotation contained in the angle, can be measured by an instrument called a ***protractor.*** One unit used to measure an angle is called a ***degree*** (the symbol for degree is °). One full rotation contains 360°.

When we speak of the measure of an angle we write, for example, $m\angle A$ (read "the measure of $\angle A$"). When two angles are equal in measure, the angles are congruent.

The diagram shows two angles that are equal in measure, or congruent. (Note the use of the same number of little "arc" marks to show equality.)

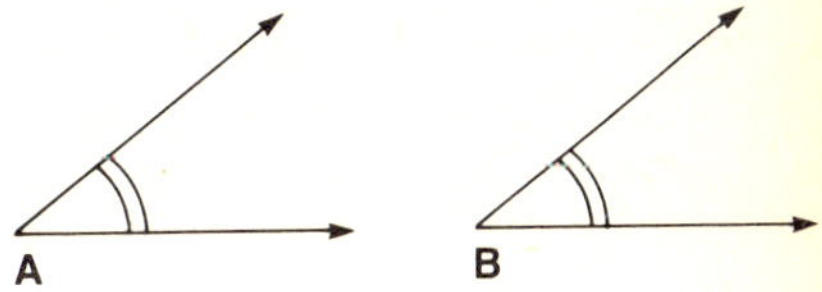

$$m\angle A = m\angle B$$
$$\angle A \cong \angle B$$

Every angle has a bisector that divides the angle into two congruent angles.

$\overrightarrow{BQ}$ is the bisector of $\angle ABC$

$$m\angle ABQ = m\angle QBC$$
$$\angle ABQ \cong \angle QBC$$

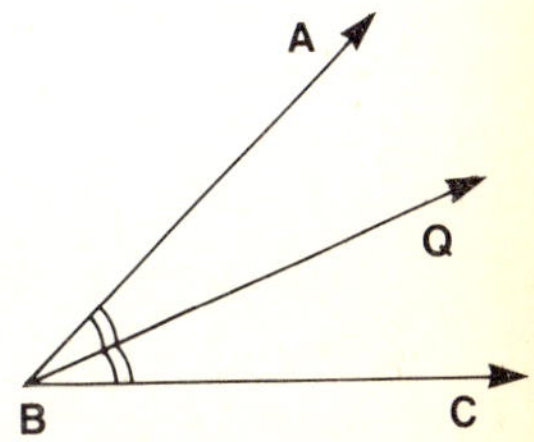

Angles can be classified by the number of degrees they contain:

Acute Angle

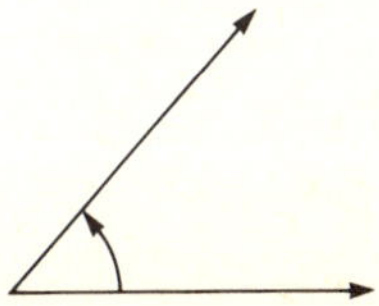

Contains between 0° and 90°.

Right Angle

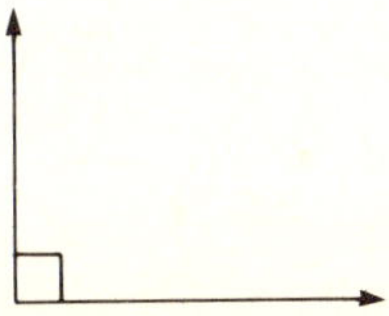

Contains exactly 90°.

Obtuse Angle

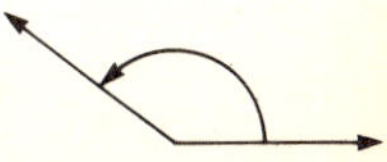

Contains more than 90° but less than 180°.

Straight Angle

Contains exactly 180°.

Reflex Angle

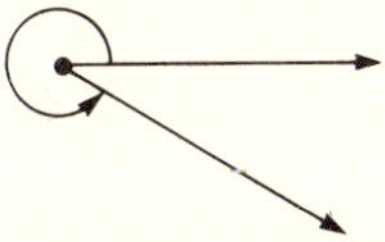

Contains more than 180° but less than 360°.

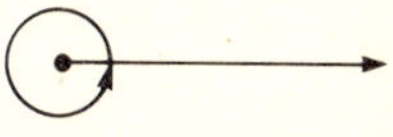

One full rotation (a ***circle***) contains exactly 360°.

Triangles

Triangles (plane figures containing three angles) can be classified by the types of angles they contain:

Acute Triangle

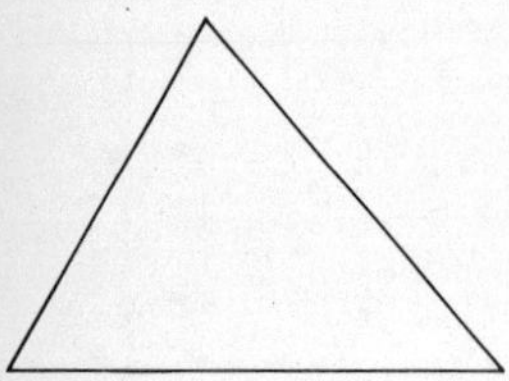

Contains three acute angles.

Right Triangle

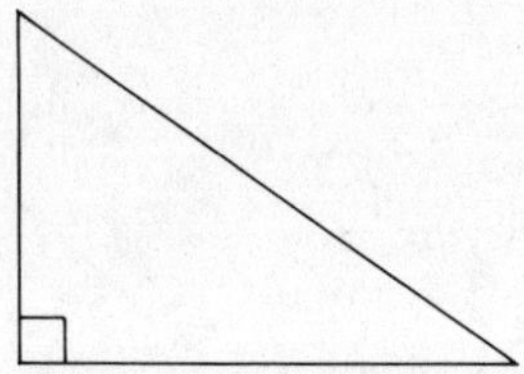

Contains one right angle and two acute angles.

Obtuse Triangle

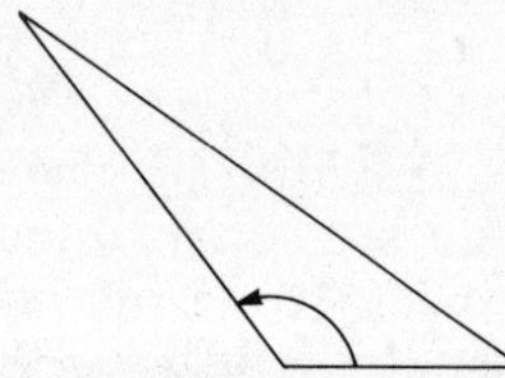

Contains one obtuse angle and two acute angles.

(*Note:* It is impossible for a triangle to contain more than one right angle or more than one obtuse angle or both a right angle and an obtuse angle.)

Triangles also contain three ***sides*** and may be classified according to the number of sides that are equal in length:

Scalene Triangle

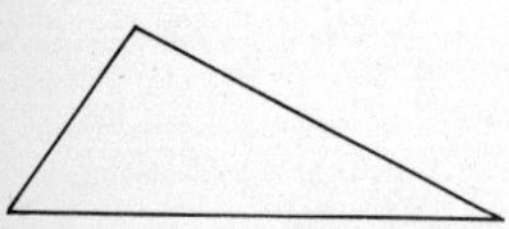

No side is equal in length to any other side. No angle is equal in measure to any other angle.

Isosceles Triangle

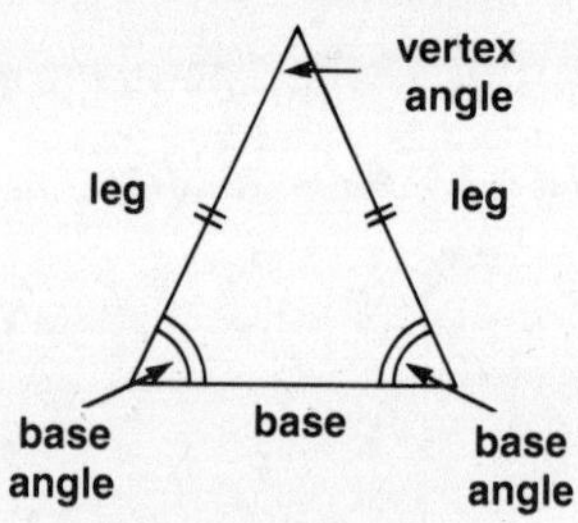

Two sides, the ***legs,*** are equal in length, or congruent. The third side is called the ***base.*** Two angles, the ***base angles,*** are equal in measure, or congruent. The angle formed by the equal legs is called the ***vertex angle.***

Equilateral Triangle

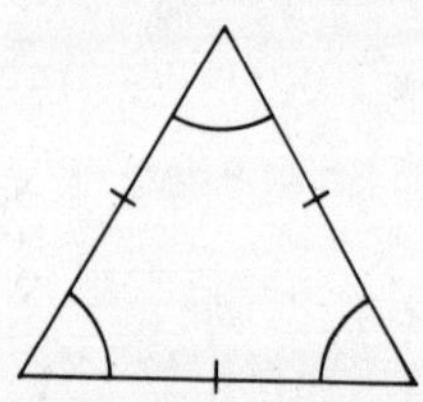

All three sides are equal in length, or congruent. All three angles are equal in measure, or congruent. An equilateral triangle is ***equiangular.***

The classification of triangles by type of angle and by equality of sides can be combined so that we can speak of:

Isosceles Right Triangle

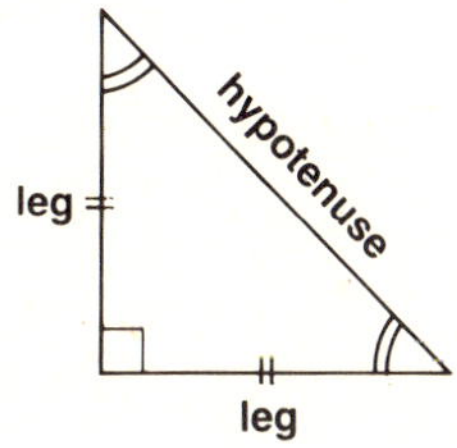

The legs are equal in length, or congruent. The side opposite the right angle, the ***hypotenuse,*** is the longest side. The two acute angles are equal in measure, or congruent.

Isosceles Obtuse Triangle

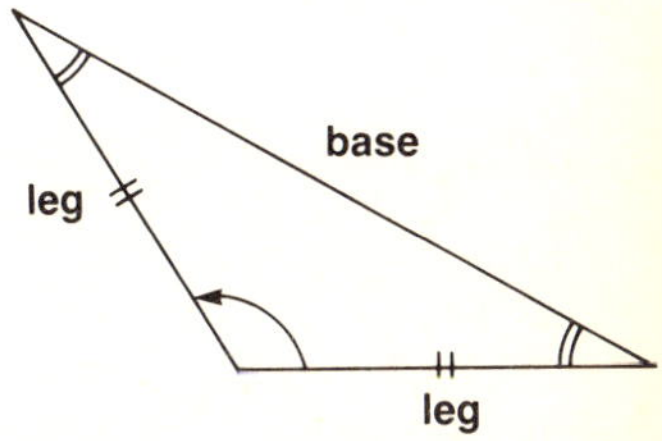

The legs are equal in length, or congruent. The side opposite the obtuse angle (the vertex angle) is the longest side. It has no special name other than the base. The two acute angles (the base angles) are equal in measure, or congruent.

Polygons

Plane figures that are *closed* and that consist of *straight sides* are called ***polygons.*** Note that the definition of a polygon is a conjunction of two conditions: (closed plane figure) $\wedge$ (sides that are straight).

A triangle is a three-sided polygon. A polygon with four sides is called a ***quadrilateral.*** The names and some properties of the special quadrilaterals are listed here:

Parallelogram

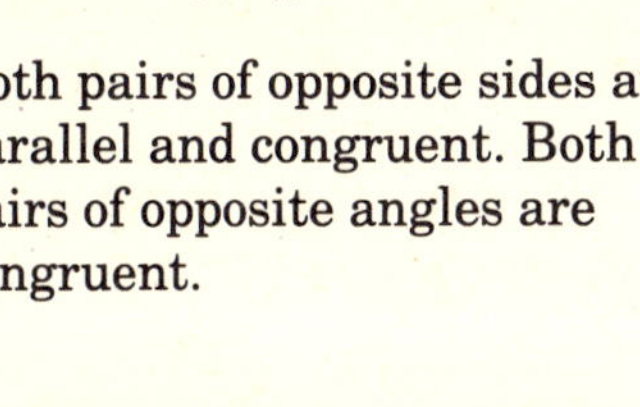

Both pairs of opposite sides are parallel and congruent. Both pairs of opposite angles are congruent.

Rectangle

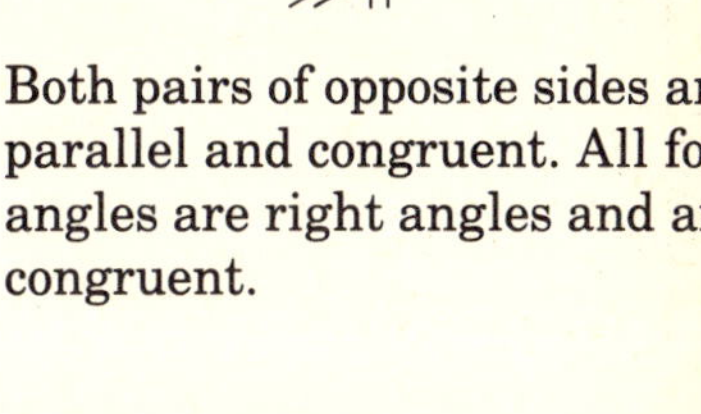

Both pairs of opposite sides are parallel and congruent. All four angles are right angles and are congruent.

Rhombus

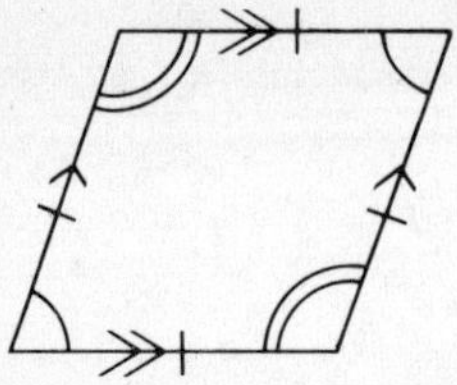

Both pairs of opposite sides are parallel. All four sides are congruent. Both pairs of opposite angles are congruent.

Square

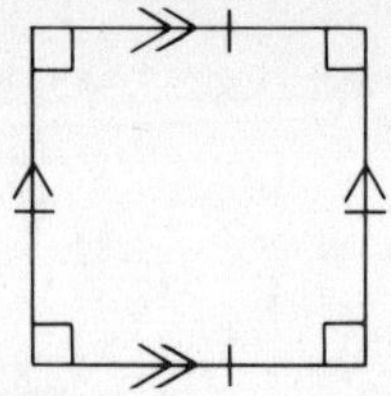

Both pairs of opposite sides are parallel. All four sides are congruent. All four angles are right angles and are congruent.

Trapezoid

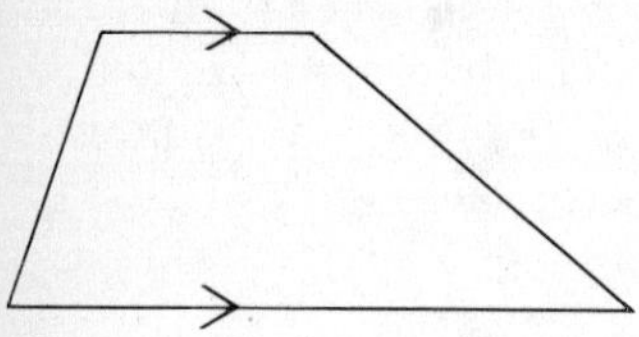

Only one pair of opposite sides is parallel (the bases). The remaining sides (the legs) are neither parallel nor congruent.

Isosceles Trapezoid

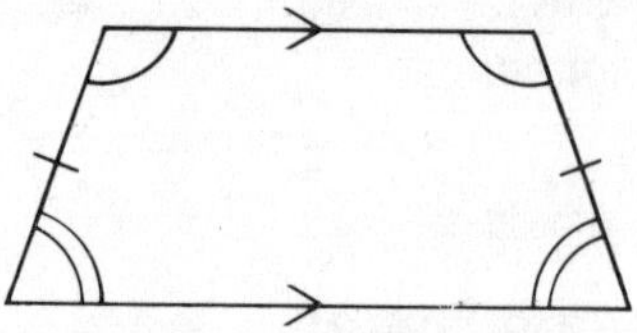

The bases are parallel. The legs are congruent. Both pairs of base angles are congruent.

The set of quadrilaterals may be arranged diagramatically to show how subsets are related:

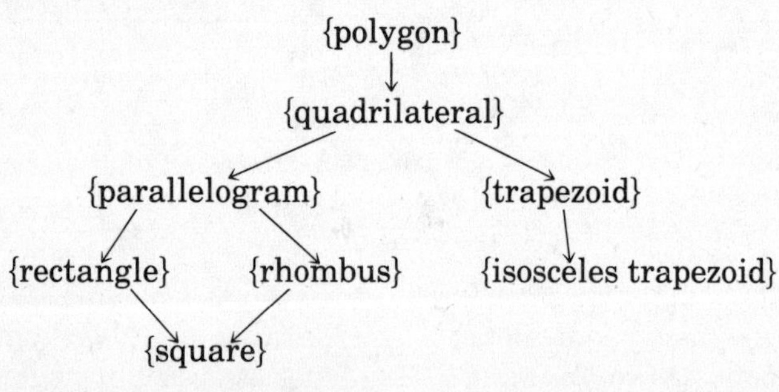

Every polygon with more than three sides contains lines called ***diagonals.*** A diagonal is a line segment that connects two nonconsecutive vertices.

Some polygons with more than four sides have special names. Some of these are:

Pentagon	**Hexagon**	**Septagon**	**Octagon**
(5 sides)	(6 sides)	(7 sides)	(8 sides)

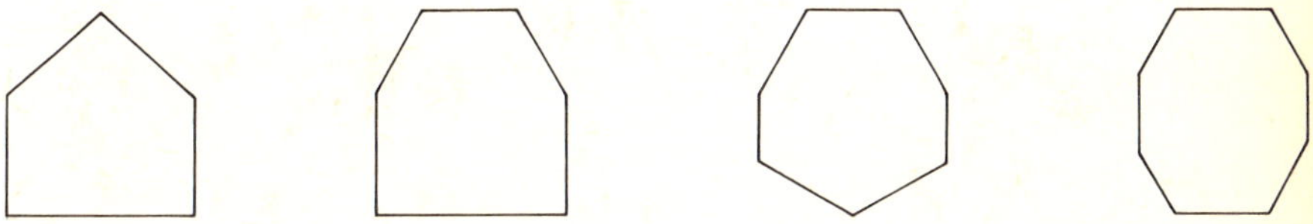

Whenever we draw a polygon of more than three sides, we usually label the vertices by letters that are in consecutive order. For example:

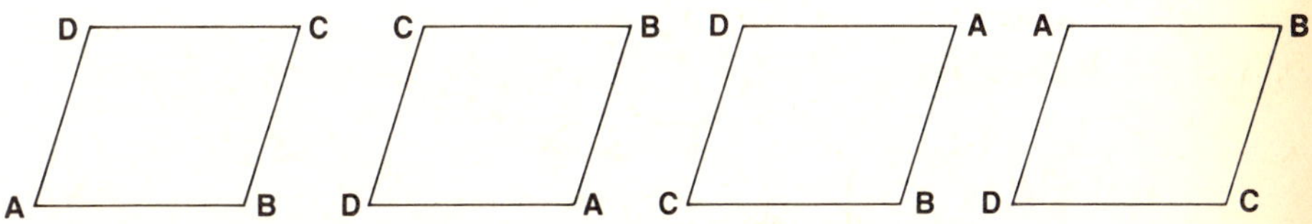

Notice that $\overline{AB}$ and $\overline{CD}$ are always opposite each other. Similarly, $\overline{BC}$ and $\overline{AD}$ are always opposite each other.

5.2 ANGLE PAIRS; ANGLES ASSOCIATED WITH LINES

Some angle pairs are given names because of their special properties:

Adjacent Angles

Adjacent angles are two angles that share the same vertex and have a common side between them.

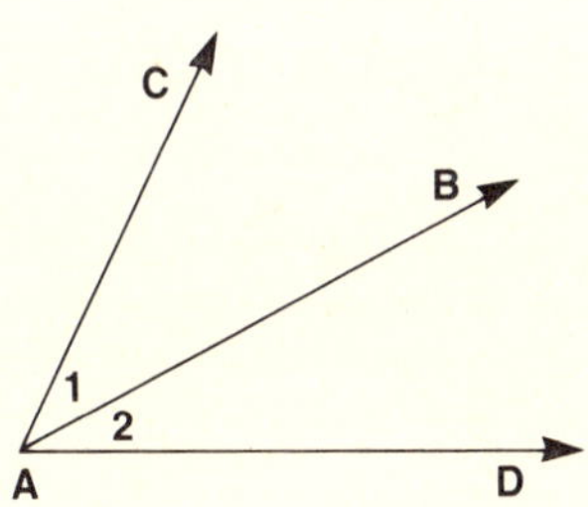

$\angle 1$ and $\angle 2$ are adjacent angles since they share the same vertex, point A, and have a common side, $\overrightarrow{AB}$, between them.

Supplementary and Complementary Angles

Supplementary angles are two angles the sum of whose measures is 180°. Numerical examples of pairs of supplementary angles are 40° and 140°, 60° and 120°, 90° and 90°, $x°$ and $(180 - x)°$.

Supplementary angles may be adjacent. (*Note:* Supplementary adjacent angles are also called a ***linear pair of angles.***)

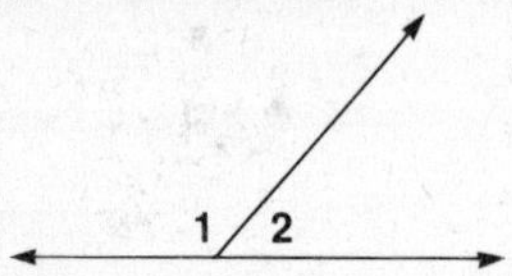

$m\angle 1 + m\angle 2 = 180°$

Supplementary angles need not be adjacent.

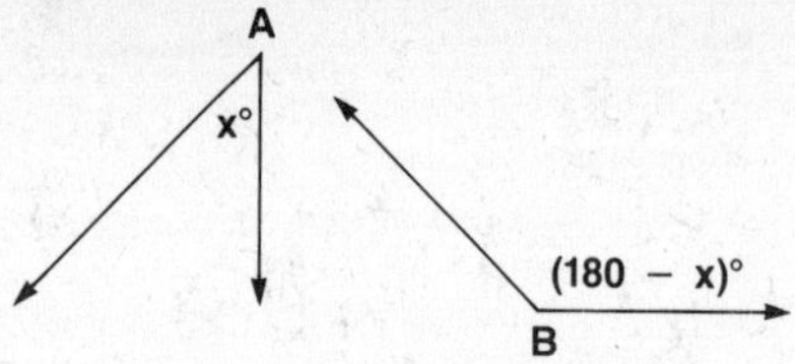

$m\angle A + m\angle B = 180°$

An important example of supplementary angles is the consecutive angles of a parallelogram.

Complementary angles are two angles the sum of whose measures is 90°. Numerical examples of pairs of complementary angles are 30° and 60°, 20° and 70°, 45° and 45°, $x°$ and $(90 - x)°$.

Complementary angles may be adjacent.

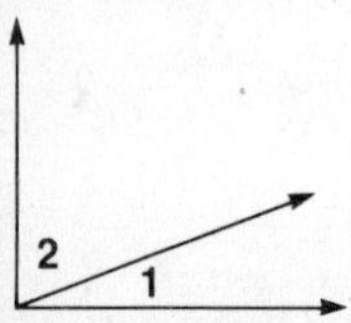

$m\angle 1 + m\angle 2 = 90°$

Complementary angles need not be adjacent.

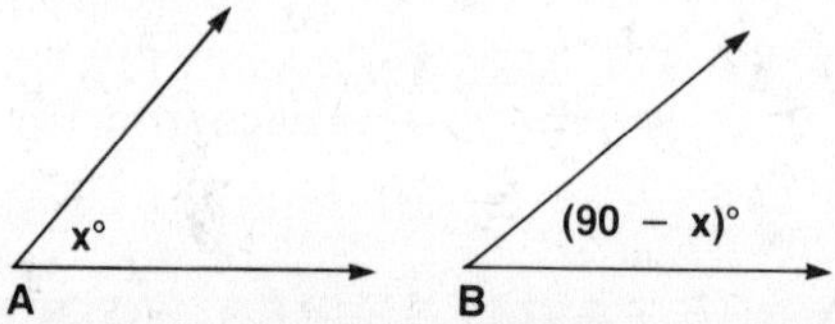

$m\angle A + m\angle B = 90°$

Geometric problems involving supplementary and complementary angles can be solved algebraically.

Examples

1. Two angles are supplementary, and the larger of the angles is three times the smaller. Find the measure of each angle.

Solution 1:

Let x = the number of degrees in the smaller angle.
Then $3x$ = the number of degrees in the larger angle.

$x + 3x = 180°$ The angles are supplementary.
$4x = 180°$
$x = 45°$
$3x = 3(45°) = 135°$

Check: The angles are supplementary. $45° + 135° \stackrel{?}{=} 180°$
$180° = 180°$ ✓

The larger angle, *135°*, is three times the smaller angle, $3 \cdot 45°$ or *135°*. ✓

Answer: The measures of the angles are 45° and 135°.

Solution 2:

Let x = the number of degrees in one angle.
Then $180 - x$ = the number of degrees in the supplementary angle.

$$\underbrace{\text{The larger of the angles}}_{x}\ \underbrace{\text{is}}_{=}\ \underbrace{\text{three times}}_{3\cdot}\ \underbrace{\text{the smaller}}_{(180-x)}.$$

$$\begin{aligned} x &= 3(180 - x) \\ x &= 3(180) - 3(x) && \text{Distributive property.} \\ x &= 540 - 3x \\ 4x &= 540 && \text{Add } 3x. \\ x &= 135^\circ && \text{Divide by 4.} \\ 180 - x &= 180^\circ - 135^\circ = 45^\circ \end{aligned}$$

Answer: 45°, 135°

Solution 3:

Since either representation can be the smaller angle, we can also use $180 - x$ as the representation for the larger angle:

$$\underbrace{\text{The larger of the angles}}_{180-x}\ \underbrace{\text{is}}_{=}\ \underbrace{\text{three times}}_{3\cdot}\ \underbrace{\text{the smaller}}_{(x)}.$$

$$\begin{aligned} 180 - x &= 3x \\ 180 &= 4x && \text{Add } x. \\ 45 &= x && \text{Divide by 4.} \\ x &= 45^\circ \\ 180 - x &= 180^\circ - 45^\circ = 135^\circ \end{aligned}$$

Answer: 45°, 135°

2. Two complementary angles are in the ratio 1:2. Find the measure of each angle.

Solution:

Let x = the common ratio factor.
Then x and $2x$ represent the number of degrees in the angles that are in the ratio 1:2.

The angles are complementary.

$$\begin{aligned} x + 2x &= 90 \\ 3x &= 90 \\ x &= 30 \\ 2x &= 60 \end{aligned}$$

Check:

The angles are complementary.

$$30^\circ + 60^\circ \stackrel{?}{=} 90^\circ$$
$$90^\circ = 90^\circ \checkmark$$

The angles are in the ratio 1:2.

$$\frac{30}{60} = \frac{1 \cdot \cancel{30}}{2 \cdot \cancel{30}} = \frac{1}{2} \checkmark$$

Answer: The measures of the angles are 30° and 60°.

Note: In an alternate solution, we would begin by representing the complementary angles as $x°$ and $(90 - x)°$. Then we would write the equation based upon the given condition that the angles are in the ratio 1:2. The solution of the type of equation that results from this set-up is not considered until Chapter 6. Thus, the alternate solution is only mentioned here.

3. In the accompanying diagram, $\overrightarrow{OA} \perp \overrightarrow{OC}$ and $\overrightarrow{OD} \perp \overrightarrow{OB}$. If $m\angle AOD = 20°$, find $m\angle COB$.

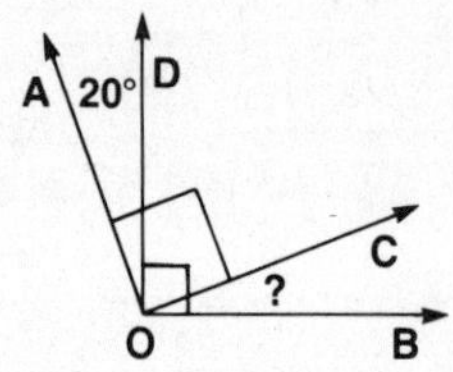

Solution:

$\angle AOC$ is a right angle $= 90°$ ($\overrightarrow{OA} \perp \overrightarrow{OC}$; perpendicular lines form right angles).

(1) First find $m\angle DOC$.

$$\begin{aligned} m\angle AOC &= m\angle AOD + m\angle DOC \\ 90 &= 20 + m\angle DOC \\ 70 &= m\angle DOC \end{aligned}$$ See diagram.

$\angle DOB$ is a right angle $= 90°$ $\quad \overrightarrow{OD} \perp \overrightarrow{OB}$

(2) Use $m\angle DOC$ to find $m\angle COB$.

$$\begin{aligned} m\angle DOB &= m\angle DOC + m\angle COB \\ 90 &= 70 + m\angle COB \\ 20 &= m\angle COB \end{aligned}$$ See diagram.

Answer: $m\angle COB = 20°$

EXERCISES

1. Two angles are supplementary and congruent. How many degrees are there in the measure of each angle?
2. Two angles are supplementary and in the ratio of 1:5. Find the measure of the *smaller* angle.
3. Two complementary angles are in the ratio of 1:4. Find the measure of the *smaller* angle.
4. If two supplementary angles are in the ratio of 1:3, find the number of degrees in the measure of the *smaller* angle.
5. Two complementary angles are in the ratio 4:5. What is the measure in degrees of the *smaller* angle?
   (1) 10 (2) 40 (3) 50 (4) 80
6. The larger of two supplementary angles is 20° less than four times the smaller. Find the measure of each angle.
7. Two angles are complementary and the measure of one angle is 30° less than twice the measure of the other. Find the measure of each angle.
8. Find the measure of the angle whose supplement is twice itself.
9. Find the measure of the angle whose complement is half of itself.

10. Find the measure of the angle whose complement equals one-third of its supplement.

11. In the figure, $\overrightarrow{OA}$ is perpendicular to $\overrightarrow{OB}$, and $\overrightarrow{OD}$ is perpendicular to $\overrightarrow{OC}$. If $m\angle 3 = 28°$, find $m\angle 1$.

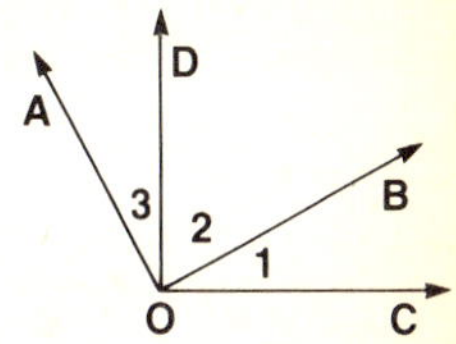

12. In the accompanying figure, $\overrightarrow{AB} \perp \overrightarrow{AE}$ and $\overrightarrow{AD} \perp \overrightarrow{AC}$. If $m\angle BAC = 160$, find $m\angle DAE$.

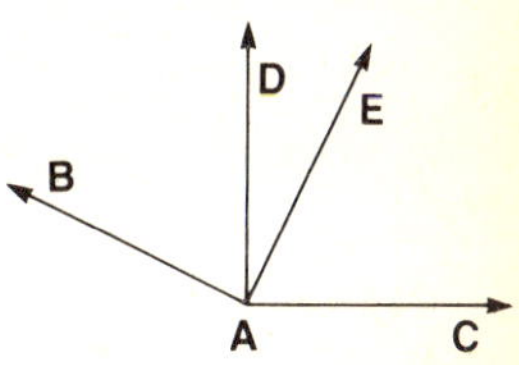

Angles Formed by Two Intersecting Lines

Another special angle pair is formed when two lines intersect. In the figure, $\angle a$ and $\angle b$ are a pair of ***vertical angles.*** Also, $\angle c$ and $\angle d$ are a pair of ***vertical angles.***

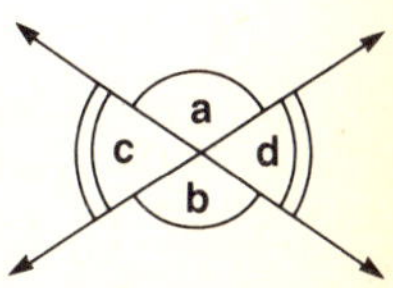

◆ ***The two angles in each pair of vertical angles are equal in measure, or congruent.***

$$m\angle a = m\angle b \text{ or } \angle a \cong \angle b$$
$$m\angle c = m\angle d \text{ or } \angle c \cong \angle d$$

◆ ***The sum of the measures of all the angles around the point of intersection of the lines is 360°*** since one complete rotation exists.

$$m\angle a + m\angle b + m\angle c + m\angle d = 360°$$

Examples

As shown in the accompanying figure, $\overleftrightarrow{AEB}$ and $\overleftrightarrow{CED}$ intersect at E. $\angle AEC$ is represented by $(3x + 20)$ and $\angle BED$ is represented by $(2x + 60)$.

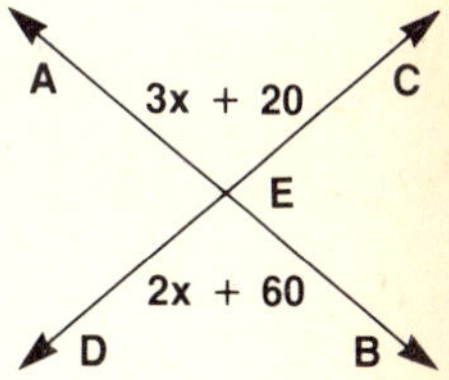

1. Find the value of x.
2. Find the measures of $\angle AEC$ and $\angle BED$.
3. Find the measure of $\angle CEB$.
4. Find the measure of $\angle AED$.

Solutions:

1. $m\angle AEC = m\angle BED$ Vertical angles are equal in measure.

$$3x + 20 = 2x + 60$$
$$x + 20 = 60$$
$$x = 40 \quad \textit{Ans.}$$

2. $m\angle AEC = 3x + 20$ $\quad$ $m\angle BED = 2x + 60$

$= 3(40) + 20$ $\quad$ $= 2(40) + 60$

$= 120 + 20$ $\quad$ $= 80 + 60$

$= 140°$ *Ans.* $\quad$ $= 140°$ *Ans.*

(*Note:* After calculating $m\ \angle AEC$, we do not need to calculate $m\ \angle BED$. Since vertical angles are equal in measure, the measures of both angles must be 140°. Doing the calculation for $m\ \angle BED$ acts as a check.)

3. $m\angle CEB + m\angle AEC = 180°$ $\quad$ $\angle CEB$ is supplementary to $\angle AEC$

$m\angle CEB + 140° = 180°$ $\quad$ since $\angle AEB$ is a straight angle.

$m\angle CEB = 40°$ *Ans.*

4. $140° + 40° + 140° + m\angle AED = 360°$

$320° + m\angle AED = 360°$

$m\angle AED = 40°$ *Ans.*

(*Note:* Since $\angle AED$ and $\angle CEB$ are vertical angles, their measures are equal.)

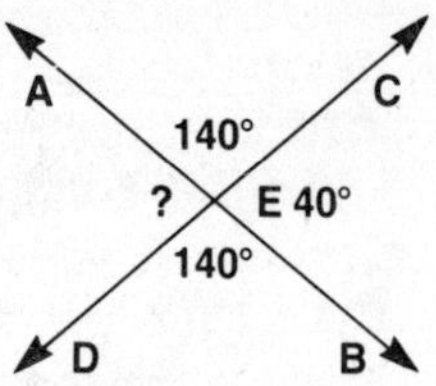

Angles Formed by Two Lines and a Transversal

When two lines are cut by a ***transversal*** (a line crossing both of the original lines), several types of angle pairs are formed. To help remember these special pairs of angles and their names, first look at the following two general diagrams of a pair of lines and a transversal. Note that there are many angles formed. We will concentrate on eight angles.

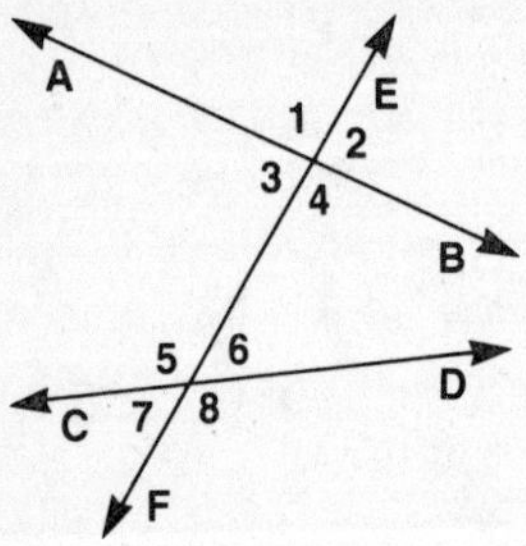

Transversal $\overleftrightarrow{EF}$ intersects $\overleftrightarrow{AB}$ and $\overleftrightarrow{CD}$. $\overleftrightarrow{AB}$ and $\overleftrightarrow{CD}$ are not parallel.

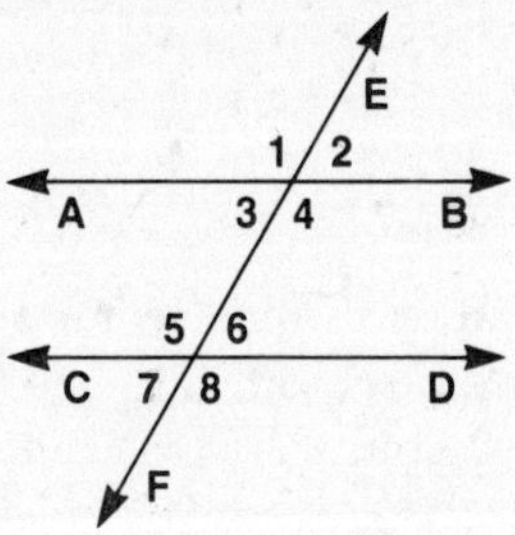

Transversal $\overleftrightarrow{EF}$ intersects $\overleftrightarrow{AB}$ and $\overleftrightarrow{CD}$. $\overleftrightarrow{AB} \parallel \overleftrightarrow{CD}$.

Four of these angles ($\angle 3$, $\angle 4$, $\angle 5$, $\angle 6$) are between or inside the two lines $\overleftrightarrow{AB}$ and $\overleftrightarrow{CD}$ and are called ***interior angles.*** Four of these angles ($\angle 1$, $\angle 2$, $\angle 7$, $\angle 8$) are outside the two lines $\overleftrightarrow{AB}$ and $\overleftrightarrow{CD}$ and are called ***exterior angles.***

The four interior angles are thought of in pairs. Two pairs of interior angles are next to each other *(adjacent)* and are *supplementary* (their sum is a straight angle, or 180°) regardless of whether the original lines are parallel or not.

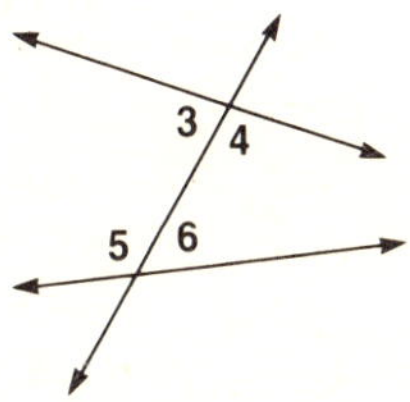

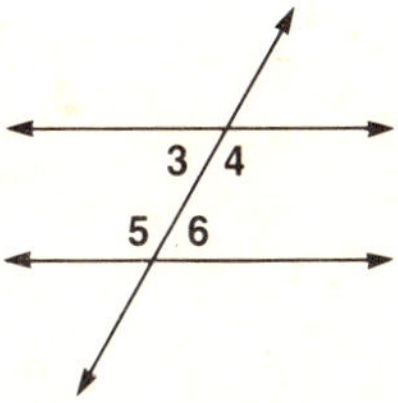

In both diagrams, $\angle 3$ and $\angle 4$ are interior angles that are supplementary. Also, $\angle 5$ and $\angle 6$ are interior angles that are supplementary.

Example

If $\angle AEF$ is represented by $(7x + 55)$ and $\angle BEF$ is represented by $(2x - 10)$, find the measures of $\angle AEF$ and $\angle BEF$.

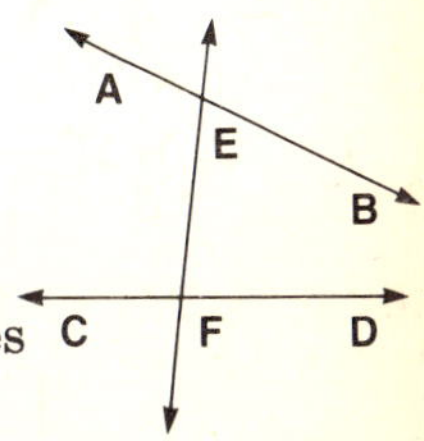

Solution:

$$\begin{aligned} m\angle AEF + m\angle BEF &= 180° && \text{Adjacent interior angles are supplementary.} \\ 7x + 55 + 2x - 10 &= 180 \\ 9x + 45 &= 180 && \text{Collect like terms.} \\ 9x &= 135 && \text{Subtract 45.} \\ x &= 15 && \text{Divide by 9.} \\ 7x + 55 &= 7(15) + 55 = 160 \\ 2x - 10 &= 2(15) - 10 = 20 \end{aligned}$$

Answer: $m\angle AEF = 160°$; $m\angle BEF = 20°$

◆ Two pairs of interior angles are called ***alternate interior angles. The measures of the alternate interior angles are equal only if the original lines are parallel.***

The diagrams below show pairs of alternate interior angles. Since the lines are not parallel, the alternate interior angles are not equal in measure.

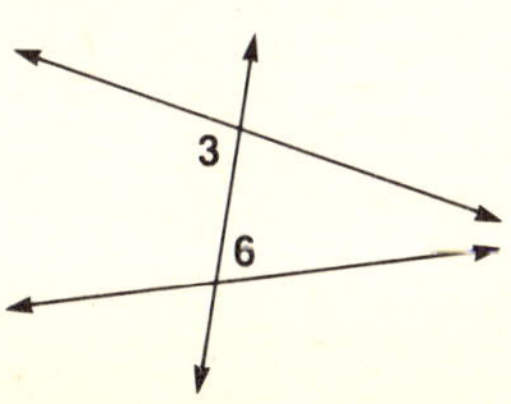

$\angle 3$ and $\angle 6$ are alternate interior angles.

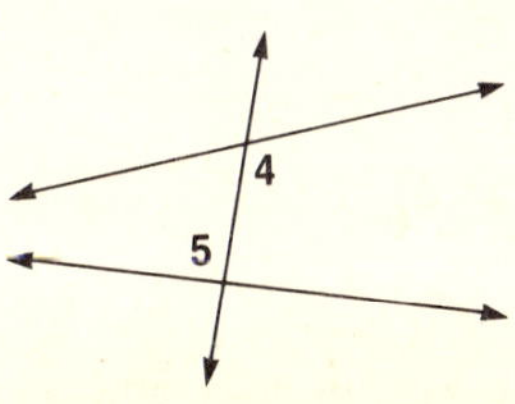

$\angle 4$ and $\angle 5$ are alternate interior angles.

The diagrams below show pairs of alternate interior angles that are equal in measure. The lines that are crossed by the transversal are parallel.

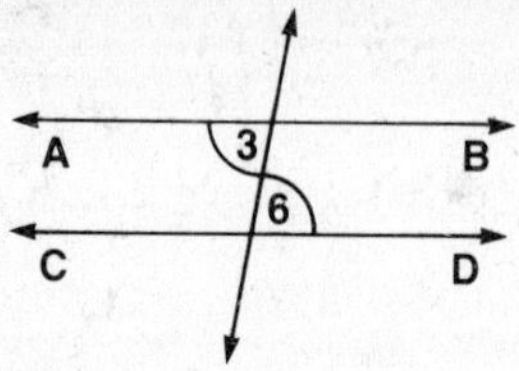

$\angle 3$ and $\angle 6$ are alternate interior angles. Since $\overleftrightarrow{AB} \parallel \overleftrightarrow{CD}$, $m\angle 3 = m\angle 6$, or $\angle 3 \cong \angle 6$.

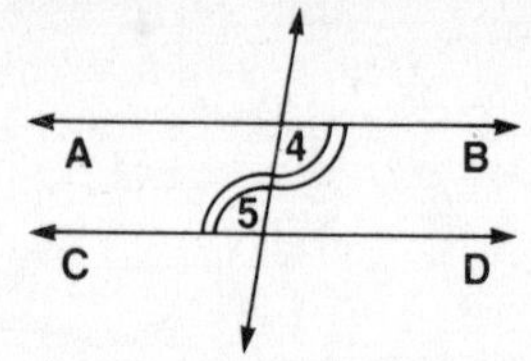

$\angle 4$ and $\angle 5$ are alternate interior angles. Since $\overleftrightarrow{AB} \parallel \overleftrightarrow{CD}$, $m\angle 4 = m\angle 5$, or $\angle 4 \cong \angle 5$.

To help remember the placement of alternate interior angles, think of the following pattern:

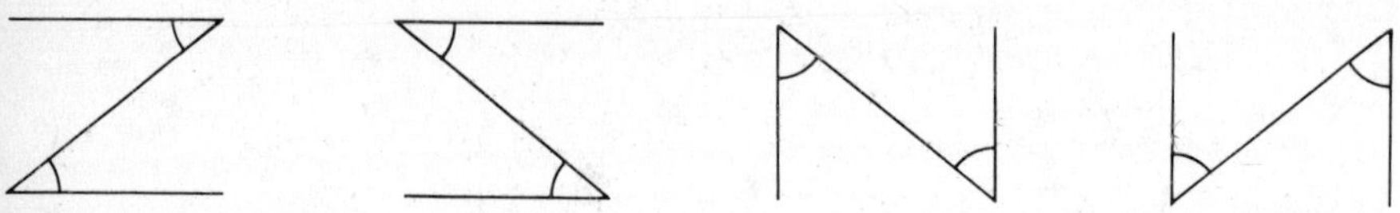

Note that for a pair of alternate interior angles, the angles are on different sides of the transversal.

Examples

As shown in the accompanying diagram, $\overleftrightarrow{AB}$ is parallel to $\overleftrightarrow{CD}$, and these lines are cut by the transversal $\overleftrightarrow{GEFH}$ at E and F, respectively. If $\angle AEF$ is represented by $(4x - 10)$ and $\angle DFE$ is represented by $(2x + 20)$, find:

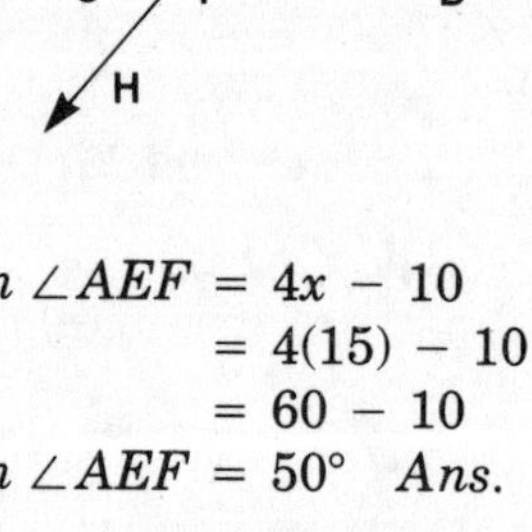

1. the value of x.
2. the measure of $\angle AEF$.
3. the measure of $\angle BEF$.

Solutions:

1. Alternate interior angles of parallel lines have equal measures.

$$\begin{aligned} m\angle AEF &= m\angle DFE \\ 4x - 10 &= 2x + 20 \\ 2x - 10 &= 20 \\ 2x &= 30 \\ x &= 15 \quad \textit{Ans.} \end{aligned}$$

2. $$\begin{aligned} m\angle AEF &= 4x - 10 \\ &= 4(15) - 10 \\ &= 60 - 10 \\ m\angle AEF &= 50^\circ \quad \textit{Ans.} \end{aligned}$$

3. $$\begin{aligned} m\angle AEF + m\angle BEF &= 180^\circ \\ 50^\circ + m\angle BEF &= 180^\circ \\ m\angle BEF &= 130^\circ \quad \textit{Ans.} \end{aligned}$$

Adjacent interior angles are supplementary.

◆ ***When two lines are parallel, the pairs of interior angles on the same side of the transversal are supplementary.***

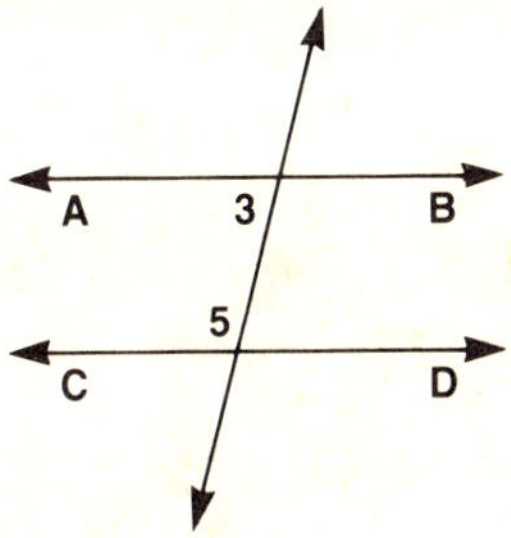

If $\overleftrightarrow{AB} \parallel \overleftrightarrow{CD}$, $\angle 3$ and $\angle 5$ are supplementary.

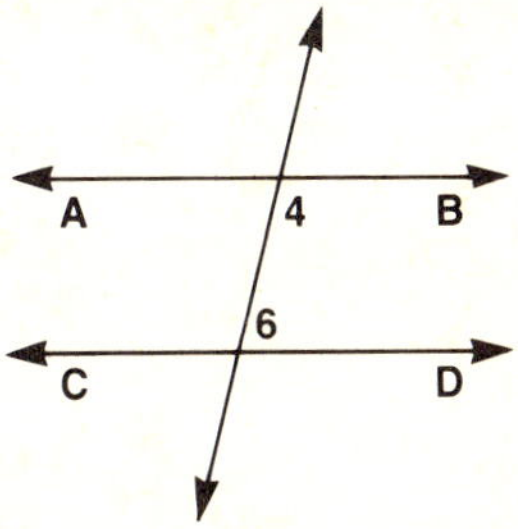

If $\overleftrightarrow{AB} \parallel \overleftrightarrow{CD}$, $\angle 4$ and $\angle 6$ are supplementary.

Examples

In the accompanying diagram, $\overleftrightarrow{AB} \parallel \overleftrightarrow{CD}$, and these lines are cut by the transversal $\overleftrightarrow{GEFH}$ at E and F, respectively. If $\angle BEF$ is represented by $(5x + 50)$ and $\angle EFD$ is represented by $(x + 10)$, find:

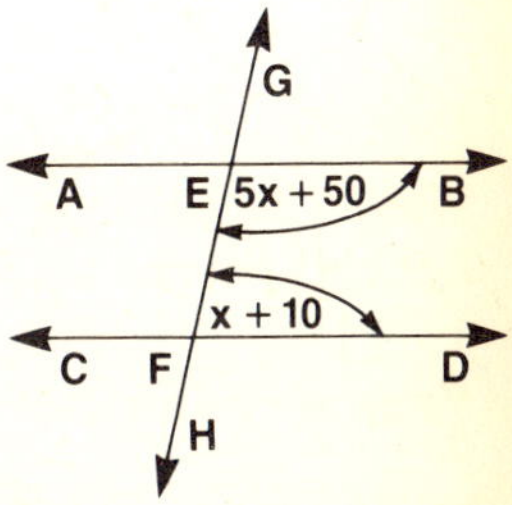

1. the value of x.
2. the measure of $\angle BEF$.
3. the measure of $\angle EFC$.
4. the measure of $\angle AEF$.

Solutions:

1. $m\angle BEF + m\angle EFD = 180°$ When two lines are parallel, the interior angles on the same side of the transversal are supplementary.

$$(5x + 50) + (x + 10) = 180$$
$$6x + 60 = 180$$
$$6x = 120$$
$$x = 20 \quad Ans.$$

2. $m\angle BEF = 5x + 50$

$$= 5(20) + 50$$
$$= 100 + 50 = 150° \quad Ans.$$

3. $m\angle EFC = m\angle BEF$ Alternate interior angles of parallel lines are equal in measure.

$m\angle EFC = 150°$ *Ans.*

4. $m\angle AEF + m\angle BEF = 180°$ Adjacent interior angles are supplementary.

$$m\angle AEF + 150 = 180$$
$$m\angle AEF = 30° \quad Ans.$$

◆ A pairing between one interior angle and one exterior angle is made. They are called ***corresponding angles*** of the lines. ***The measures of the corresponding angles are equal only if the original lines are parallel.***

Notice the placement. Corresponding angles appear when the lines and the transversal form the following pattern:

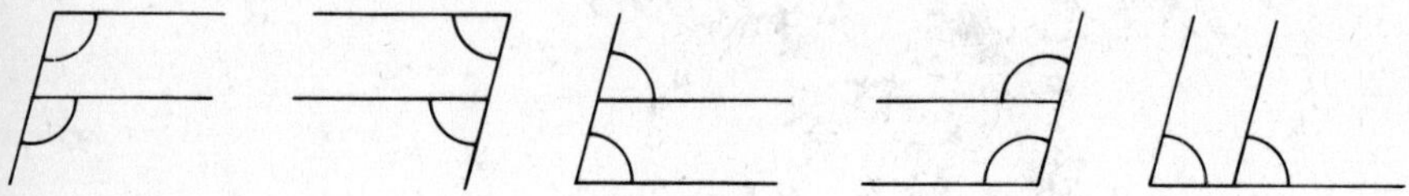

Examples

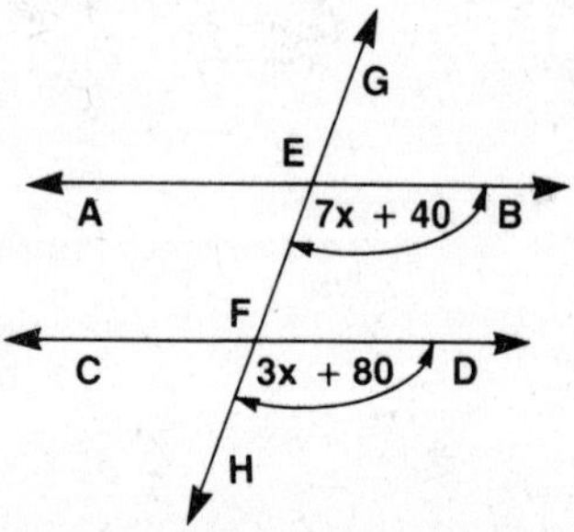

As shown in the accompanying diagram, $\overleftrightarrow{AB} \parallel \overleftrightarrow{CD}$, and these lines are cut by the transversal $\overleftrightarrow{GEFH}$ at E and F, respectively. If $\angle BEF$ is represented by $(7x + 40)$ and $\angle DFH$ is represented by $(3x + 80)$, find:

1. the value of x.
2. $m\angle BEF$.
3. $m\angle EFC$.
4. $m\angle EFD$.
5. $m\angle AEF$.

Solutions:

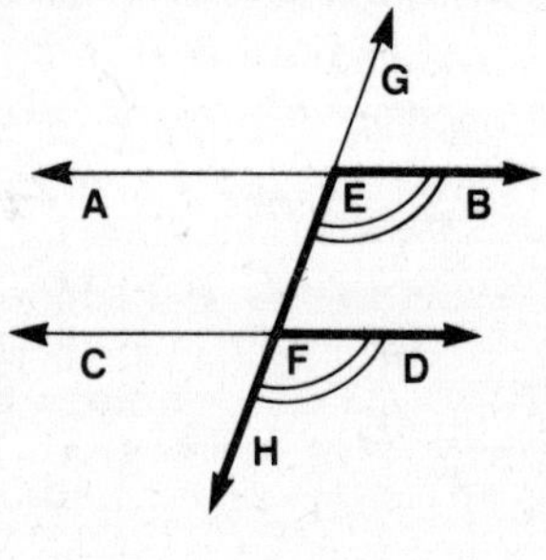

1. Corresponding angles of parallel lines have equal measures.

$$\begin{aligned} m\angle BEF &= m\angle DFH \\ 7x + 40 &= 3x + 80 \\ 4x + 40 &= 80 \\ 4x &= 40 \\ x &= 10 \quad \textit{Ans.} \end{aligned}$$

2. $$\begin{aligned} m\angle BEF &= 7x + 40 \\ &= 7(10) + 40 \\ &= 70 + 40 \\ &= 110° \quad \textit{Ans.} \end{aligned}$$

3. Alternate interior angles of parallel lines have equal measures.

$$\begin{aligned} m\angle BEF &= m\angle EFC \\ 110° &= m\angle EFC \quad \textit{Ans.} \end{aligned}$$

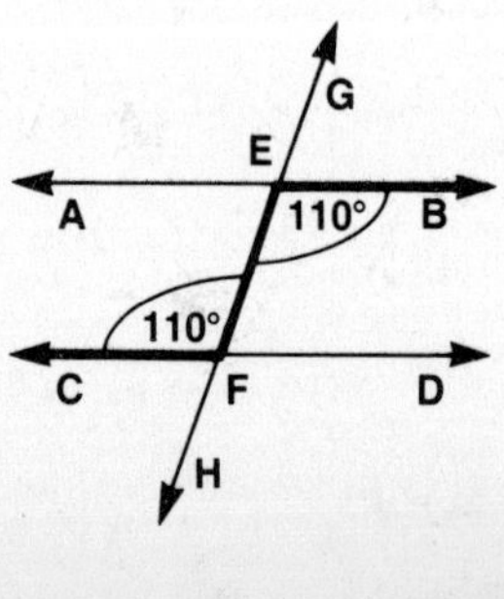

4. Adjacent interior angles are supplementary.
$m \angle EFC + m \angle EFD = 180°$
$110° + m \angle EFD = 180°$
$m \angle EFD = 70°$ *Ans.*

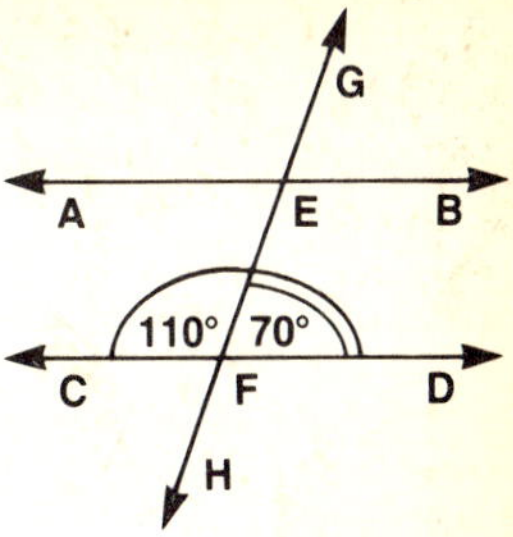

5. Alternate interior angles of parallel lines have equal measures.
$m \angle AEF = m \angle EFD$
$m \angle AEF = 70°$ *Ans.*

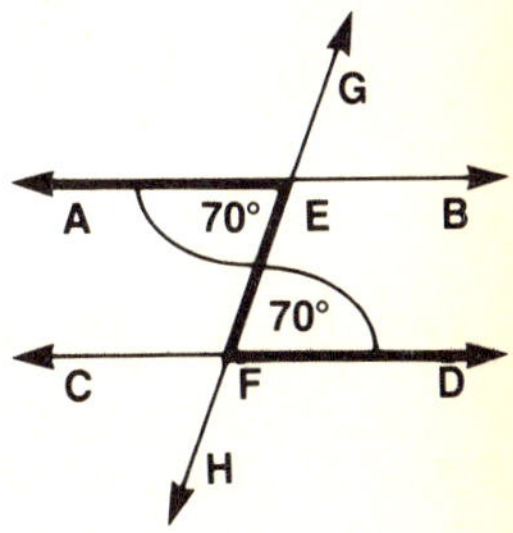

SUMMARY OF ANGLE PAIRS

Name	Description/Properties	Illustrative Diagrams
Adjacent angles	Two angles that have the same vertex and a common side between them.	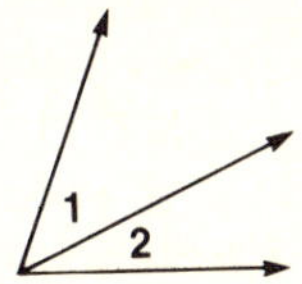 ∠1 and ∠2 are adjacent angles
Supplementary angles	Two angles whose sum is 180°.	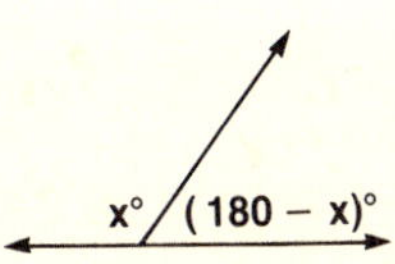 Supplementary angles that are also adjacent (a linear pair of angles) 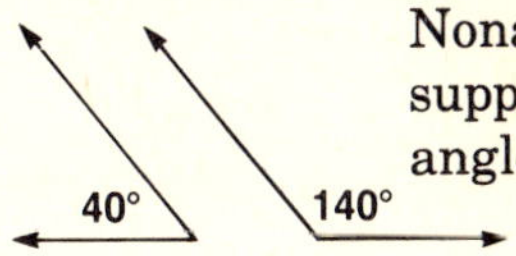 Nonadjacent supplementary angles

<table>
<tr><th>Name</th><th>Description/Properties</th><th>Illustrative Diagrams</th></tr>
<tr><td>Complementary angles</td><td>Two angles whose sum is 90°.</td><td>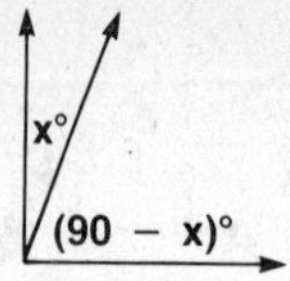

Complementary angles that are also adjacent
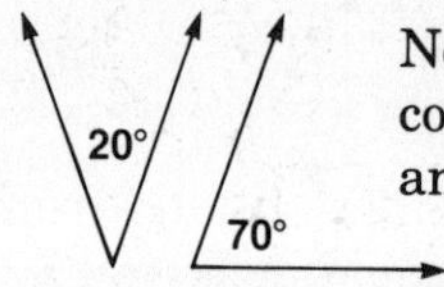

Nonadjacent complementary angles</td></tr>
<tr><td>Vertical angles</td><td>Two pairs of vertical angles are formed whenever two lines intersect. Vertical angles are always equal in measure.</td><td>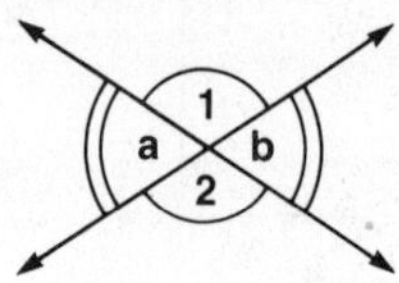

$\angle 1 \cong \angle 2$
$\angle a \cong \angle b$</td></tr>
<tr><th colspan="3">Interior Angles With Respect to Two Lines Cut by a Transversal</th></tr>
<tr><td rowspan="2">Alternate interior angles</td><td></td><td>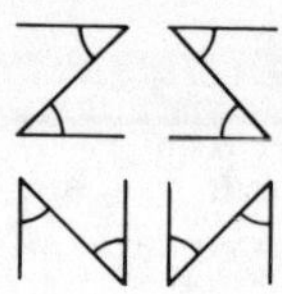
Patterns showing alternate interior angles</td></tr>
<tr><td>Alternate interior angles are equal in measure only when the lines are parallel.</td><td>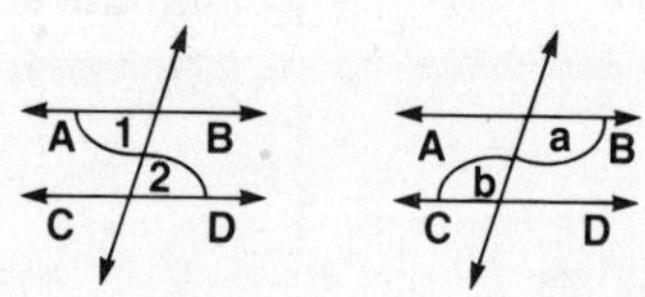

When $\overleftrightarrow{AB} \parallel \overleftrightarrow{CD}$, $\angle 1 \cong \angle 2$ and $\angle a \cong \angle b$.</td></tr>
<tr><td>Interior angles on the same side of the transversal</td><td>Interior angles on the same side of the transversal are supplementary only when the lines are parallel.</td><td>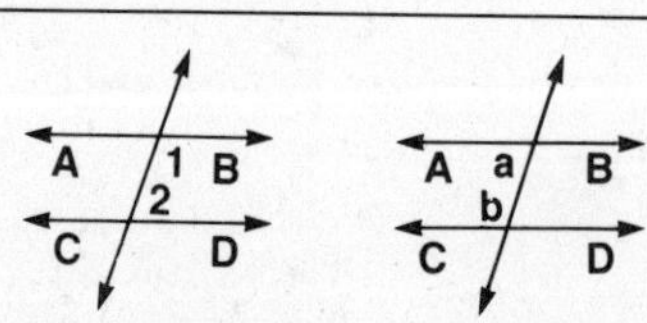

When $\overleftrightarrow{AB} \parallel \overleftrightarrow{CD}$, $\angle 1$ and $\angle 2$ are supplementary and $\angle a$ and $\angle b$ are supplementary.</td></tr>
</table>

Name	Description/Properties	Illustrative Diagrams
Adjacent interior angles	Formed by one line and the transversal, these angles are supplementary adjacent angles.	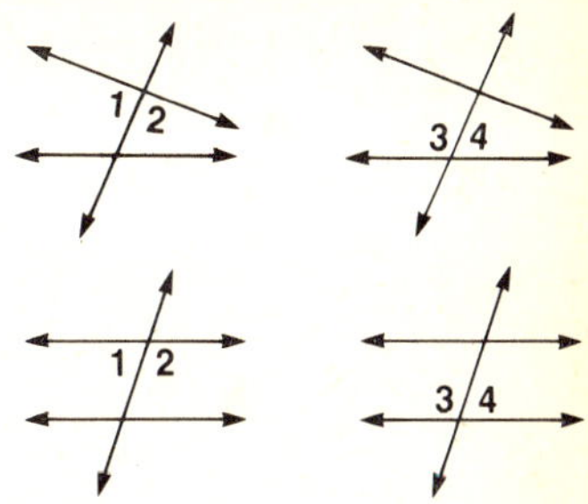 $m\,\angle 1 + m\,\angle 2 = 180°$ $m\,\angle 3 + m\,\angle 4 = 180°$ The lines may or may not be parallel.
Corresponding angles with respect to two lines cut by a transversal		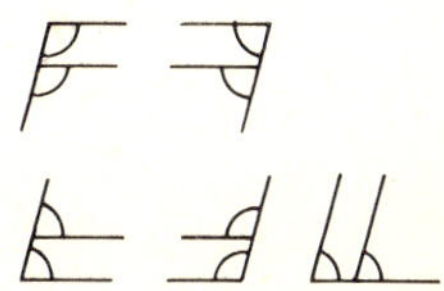 patterns
	Corresponding angles are equal in measure only when the lines are parallel.	When $\overleftrightarrow{AB} \parallel \overleftrightarrow{CD}$: 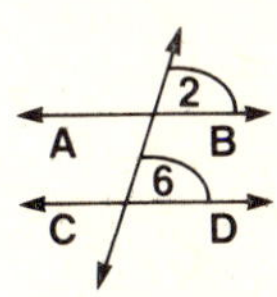 $\angle 2 \cong \angle 6$ 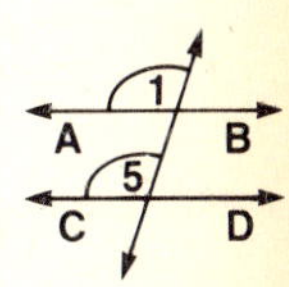 $\angle 1 \cong \angle 5$ 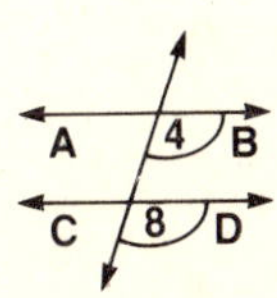 $\angle 4 \cong \angle 8$ 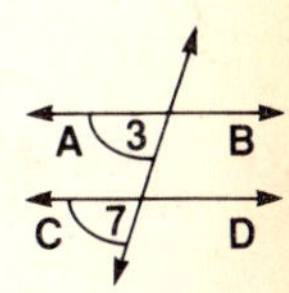 $\angle 3 \cong \angle 7$

Knowing When Two Lines Are Parallel

◆ The converse of each statement concerning the angle pairs formed by two parallel lines and a transversal is also true. That is:

1. If a pair of congruent alternate interior angles is formed, then the lines that form them are parallel.
2. If a pair of congruent corresponding angles is formed, then the lines that form them are parallel.
3. If a pair of interior angles on the same side of the transversal is supplementary, then the lines that form them are parallel.

(Note the application to logic. Since three conditional statements and their respective converses are true, there are three true biconditional statements concerning parallel lines and their special angle pairs.)

Thus, if we have information about angle pairs, we can make judgments about whether certain lines are parallel. For example, in the diagram, if $\angle 1 \cong \angle 2$, then $\overleftrightarrow{AB} \parallel \overleftrightarrow{CD}$.

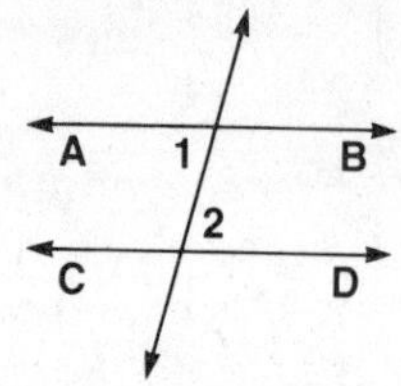

Examples

In each of the following diagrams, $\angle 1 \cong \angle 2$. Name the line segments that are therefore parallel. Give a reason for your answer.

1.

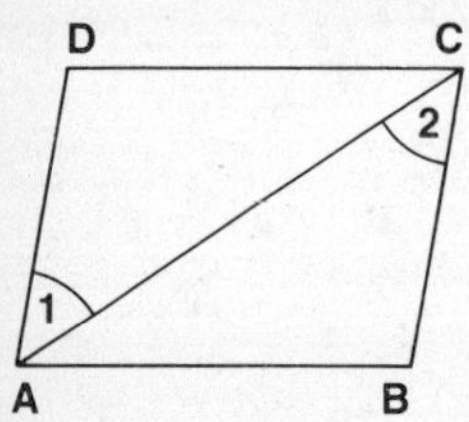

2.

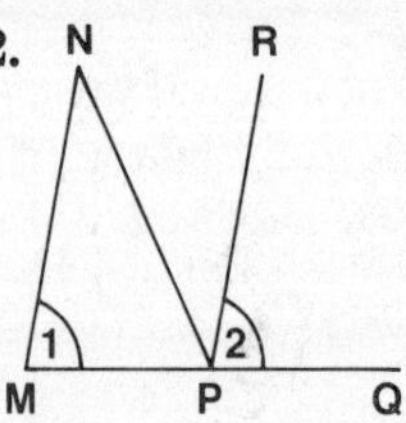

3.

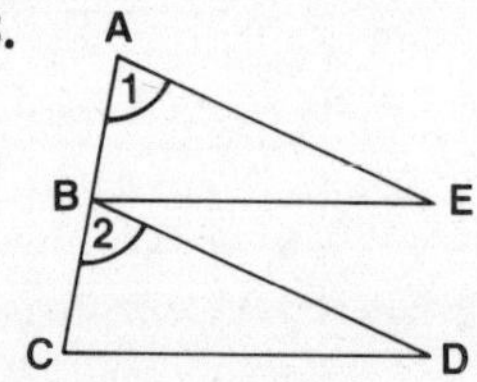

Solutions:

1. *Answer:* Since a pair of congruent alternate interior angles is formed, $\overline{AD} \parallel \overline{BC}$.

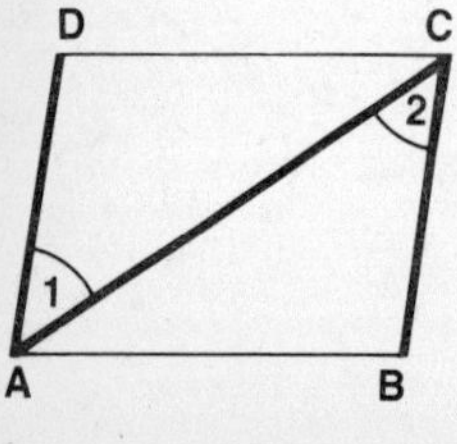

2. *Answer:* Since a pair of congruent corresponding angles is formed, $\overline{MN} \parallel \overline{PR}$.

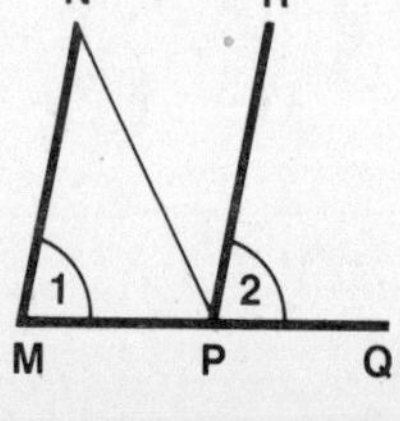

3. *Answer:* Since a pair of congruent corresponding angles is formed, $\overline{AE} \parallel \overline{BD}$.

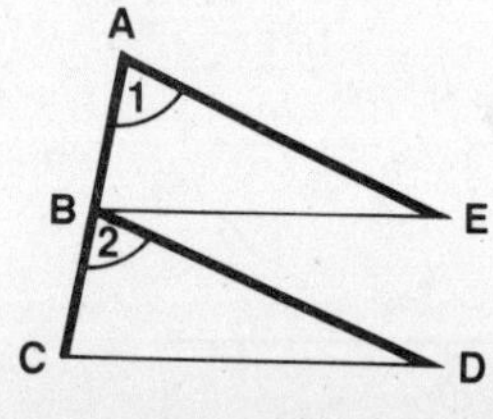

EXERCISES

1. As shown in the accompanying figure, vertical angles have degree measures of x and $(2x - 35)$. Find x.

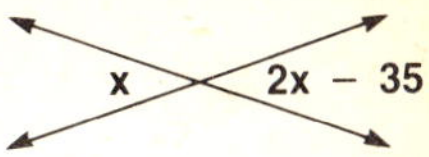

2. As shown in the accompanying diagram, vertical angles have degree measures of $(7x - 10)$ and $(6x + 15)$. Find the measure of $\angle q$.

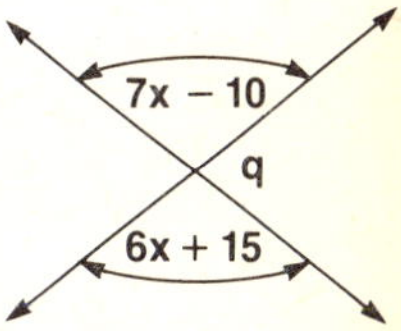

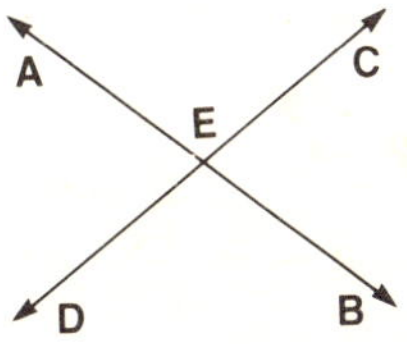

3. Use the accompanying diagram to determine which of the following statements is *false*.
   (1) $\angle AEC$ and $\angle BED$ are congruent
   (2) $\angle AED$ and $\angle CEB$ are congruent
   (3) $\angle AEC$ and $\angle AED$ are supplementary
   (4) $\angle AEC$ and $\angle AED$ are complementary

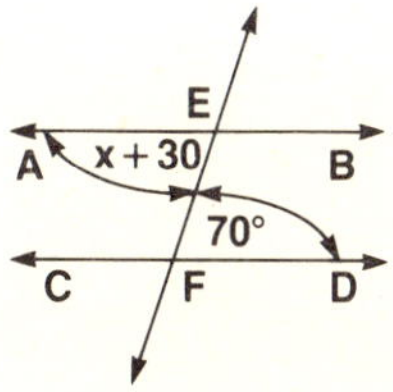

4. As shown in the accompanying figure, $\overleftrightarrow{AB}$ is parallel to $\overleftrightarrow{CD}$, and $\overleftrightarrow{AB}$ and $\overleftrightarrow{CD}$ are cut by transversal $\overleftrightarrow{EF}$ at E and F, respectively. If the measure of $\angle AEF$ equals $(x + 30)°$ and the measure of $\angle DFE$ equals 70°, find x.

5. As shown in the accompanying diagram, $\overleftrightarrow{AB}$ is parallel to $\overleftrightarrow{CD}$ and transversal $\overleftrightarrow{EF}$ intersects $\overleftrightarrow{AB}$ and $\overleftrightarrow{CD}$ at G and H, respectively. If the measure of angle CHG is $2x°$ and the measure of angle BGH is $(x + 15)°$, find the value of x.

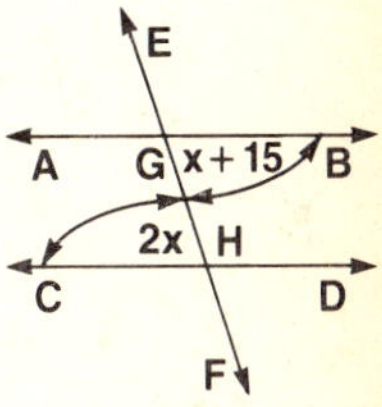

6. In the accompanying figure, $\overleftrightarrow{AB}$ is parallel to $\overleftrightarrow{CD}$, and these lines are cut by the transversal $\overleftrightarrow{EF}$ at points E and F, respectively. If the measure of $\angle AEF$ is 70°, find the number of degrees in the measure of $\angle DFE$.

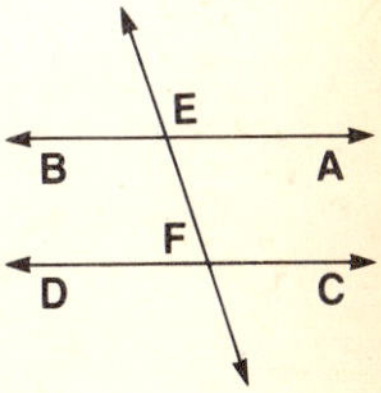

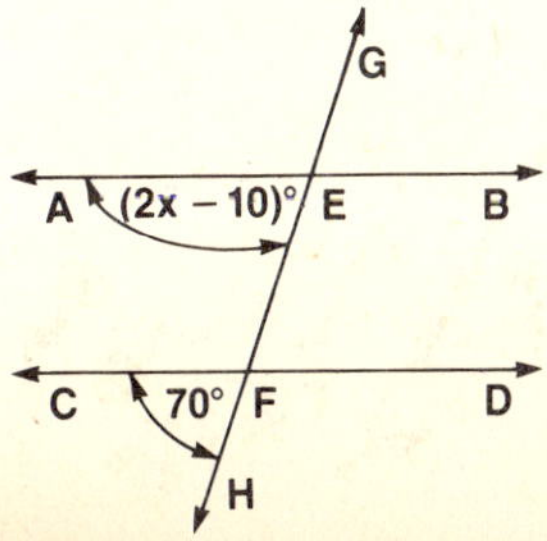

7. In the accompanying diagram, parallel $\overleftrightarrow{AB}$ and $\overleftrightarrow{CD}$ are cut by transversal $\overleftrightarrow{GH}$ at points E and F, respectively. If $m\angle AEF = (2x - 10)°$ and $m\angle CFH = 70°$, find the value of x.

8. Use the accompanying diagram to determine which of the following statements is *false*.
(1) $\angle a \cong \angle d$
(2) $\angle c \cong \angle f$
(3) $\angle d \cong \angle f$
(4) $\angle d \cong \angle h$

$\overleftrightarrow{AB} \parallel \overleftrightarrow{CD}$

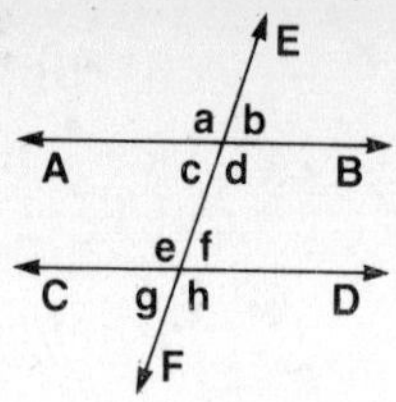

9. Use the accompanying diagram to determine which of the following statements is *false*.
(1) $\angle a$ and $\angle b$ are supplementary
(2) $\angle a$ and $\angle c$ are supplementary
(3) $\angle c$ and $\angle e$ are supplementary
(4) $\angle c$ and $\angle g$ are supplementary

$\overleftrightarrow{AB} \parallel \overleftrightarrow{CD}$

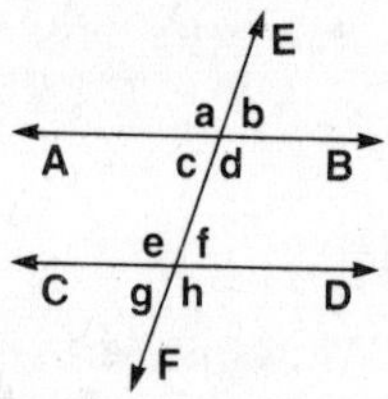

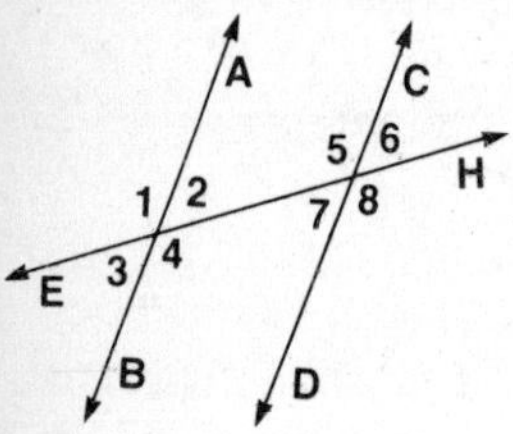

$\overleftrightarrow{AB} \parallel \overleftrightarrow{CD}$

10. Use the accompanying diagram to determine which of the following statements is *true*.
(1) $\angle 1$ is the supplement of $\angle 4$
(2) $\angle 1$ is the supplement of $\angle 3$
(3) $\angle 5$ is the supplement of $\angle 8$
(4) $\angle 3$ is the supplement of $\angle 7$

11. Use the accompanying diagram to determine which of the following statements is *true*.
(1) $\angle c \cong \angle b$
(2) $\angle c \cong \angle z$
(3) $\angle c \cong \angle x$
(4) $\angle c \cong \angle d$

$\overleftrightarrow{AB} \parallel \overleftrightarrow{CD}$

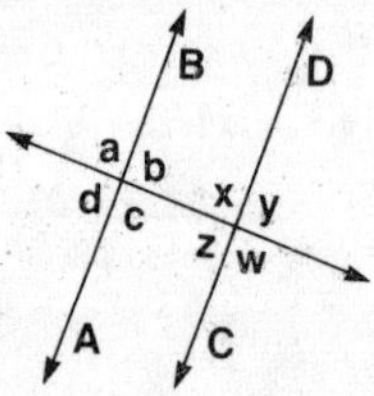

12. Use the accompanying diagram to determine which of the following statements is *false*.
(1) $\angle 1 \cong \angle 4$
(2) $\angle 1$ is the supplement of $\angle 3$
(3) $\angle 1 \cong \angle 5$
(4) $\angle 1 \cong \angle 7$

$\overleftrightarrow{AB} \parallel \overleftrightarrow{CD}$

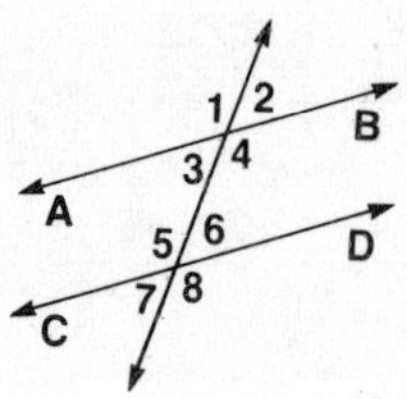

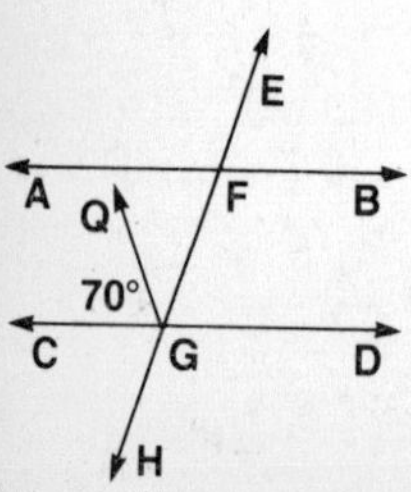

13. In the accompanying diagram, transversal $\overleftrightarrow{EFGH}$ intersects $\overleftrightarrow{AB}$ and $\overleftrightarrow{CD}$. If $\overleftrightarrow{AB} \parallel \overleftrightarrow{CD}$, $\overrightarrow{GQ}$ is the bisector of $\angle CGF$ and the measure of $\angle QGC = 70°$, find the measure of $\angle GFB$.

14. If two parallel lines are intersected by a transversal, which angles must be supplementary?
(1) a pair of alternate interior angles
(2) a pair of alternate exterior angles
(3) two corresponding angles
(4) two interior angles on the same side of the transversal

15. In the accompanying diagram, $\overleftrightarrow{AB}$ is parallel to $\overleftrightarrow{CD}$ and $\overrightarrow{FG}$ bisects $\angle EFD$. If $m\angle EFG = x$ and $m\angle FEG = 4x$, find x.

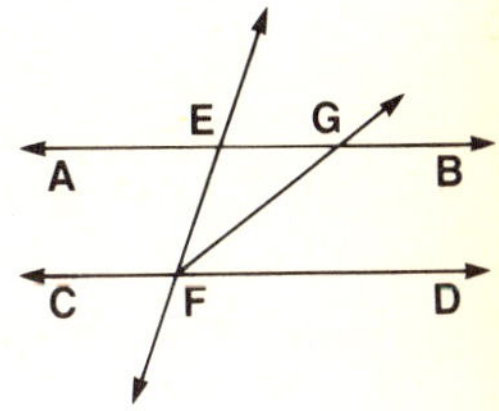

In 16–18, $\angle 1 \cong \angle 2$ in each diagram. Name the line segments that are therefore parallel. Give a reason for your answer.

16.

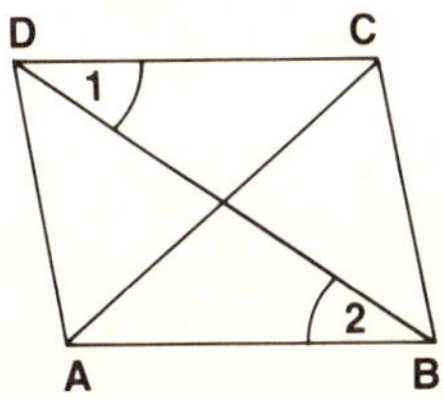

17.

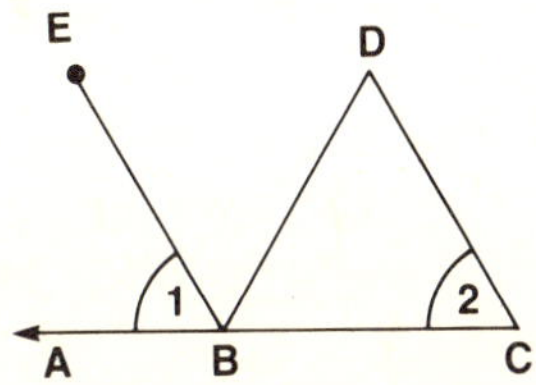

18.

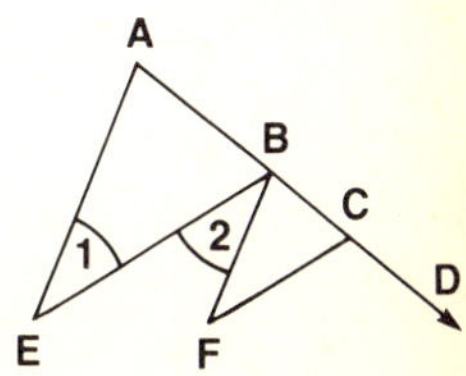

5.3 ANGLES ASSOCIATED WITH TRIANGLES

◆ A triangle (△) contains three interior angles. It can be shown that the ***sum of the measures of the three interior angles of a triangle is always 180°.***

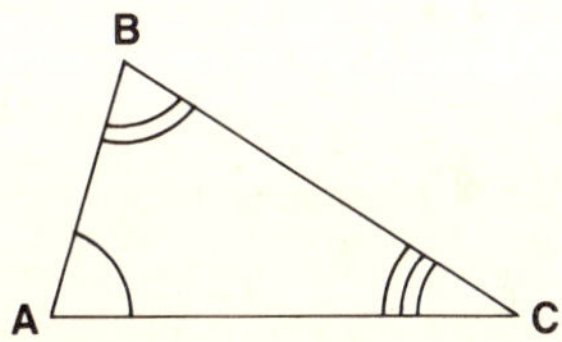

$m\angle A + m\angle B + m\angle C = 180°$ for all triangles.

Examples

1. In triangle ABC, the measures of angles A and B are in the ratio 2:5, and the measure of angle C is 72° more than the measure of angle A. Find the number of degrees in each angle.

Solution:

Let x = the common ratio factor.

Then $2x = m\angle A \quad \frac{2x}{5x} = \frac{2}{5}$ or 2:5

$5x = m\angle B$

$2x + 72 = m\angle C$

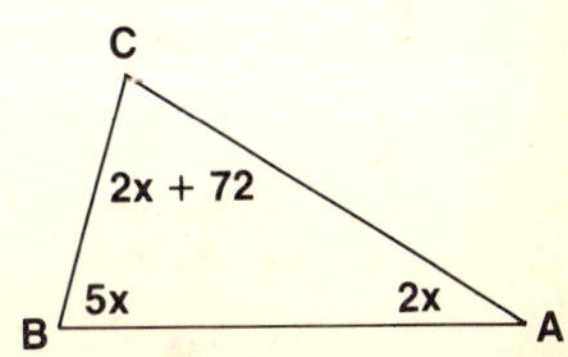

$m\angle A + m\angle B + m\angle C = 180°$ The sum of the measures of the three interior angles of a triangle is 180°.

$$\begin{aligned}(2x) + (5x) + (2x + 72) &= 180\\ 9x + 72 &= 180\\ 9x &= 108\\ x &= 12\end{aligned}$$

$m\angle A = 2x = 2(12) = 24$
$m\angle B = 5x = 5(12) = 60$
$m\angle C = 2x + 72 = 2(12) + 72 = 24 + 72 = 96$

Check:

$$\begin{aligned}m\angle A + m\angle B + m\angle C &= 180\\ 24 + 60 + 96 &\stackrel{?}{=} 180\\ 84 + 96 &\stackrel{?}{=} 180\\ 180 &= 180\checkmark\end{aligned}$$

Answer: $m\angle A = 24°$, $m\angle B = 60°$, $m\angle C = 96°$

2. The three angles of a triangle are in the ratio 1:2:3.

a. Find the number of degrees in each angle.

b. Based on your answer to part **a,** tell which is the most accurate description of the triangle.

(1) right and isosceles
(2) isosceles but not right
(3) right but not isosceles
(4) not right and not isosceles

Solution:

a. Let x = the common ratio factor.
Then x, $2x$, and $3x$ represent the measures of the angles in the ratio 1:2:3.

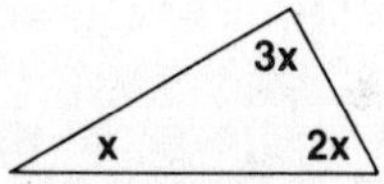

The sum of the measures of the three interior angles of a triangle is 180°.

$$\begin{aligned}x + 2x + 3x &= 180\\ 6x &= 180\\ x &= 30\\ 2x &= 2(30) = 60\\ 3x &= 3(30) = 90\end{aligned}$$

Check:

$$\begin{aligned}30 + 60 + 90 &\stackrel{?}{=} 180\\ 90 + 90 &\stackrel{?}{=} 180\\ 180 &= 180\checkmark\end{aligned}$$

Answer: The measures of the angles of the triangle are 30°, 60°, and 90°.

b. From part **a** we see that one angle of the triangle is a right angle and that no two angles of the triangle are equal in measure. Thus, the triangle is a right triangle but it is not isosceles.

Answer: (3)

3. The measure of the vertex angle of an isosceles triangle is 30° more than three times the measure of a base angle. Find the number of degrees in each angle of the triangle.

Solution:

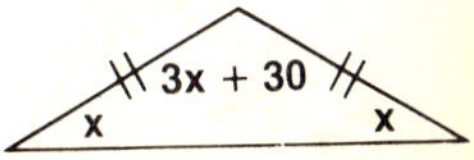

Let x = the measure of each of the two base angles.

Then $3x + 30$ = the measure of the vertex angle.

$(x) + (x) + (3x + 30) = 180$ The sum of the measures of the three
$5x + 30 = 180$ interior angles of a triangle = 180°.
$5x = 150$
$x = 30$

$3x + 30 = 3(30) + 30 = 90 + 30 = 120$

Check:

$$m \text{ base } \angle + m \text{ base } \angle + m \text{ vertex } \angle = 180$$
$$30 + 30 + 120 \stackrel{?}{=} 180$$
$$60 + 120 \stackrel{?}{=} 180$$
$$180 = 180 \checkmark$$

Answer: The measure of the vertex angle is 120° and the measure of each base angle is 30°. (*Note:* The triangle is an isosceles obtuse triangle.)

Using the fact that the sum of the three interior angles of a triangle is 180°, we can easily show each of the following:

◆ *The measure of each angle of an equilateral triangle is 60°.* (An equilateral triangle is also equiangular.)

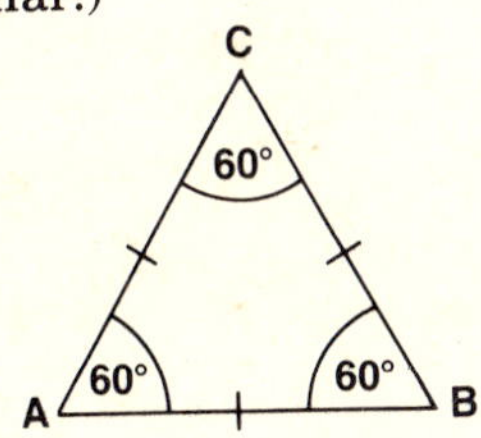

Equilateral (Equiangular) Triangle
$m\angle A = m\angle B = m\angle C = 60°$

◆ *The two acute angles of a right triangle are complementary.* (The sum of the two acute angles of a right triangle is 90°.)

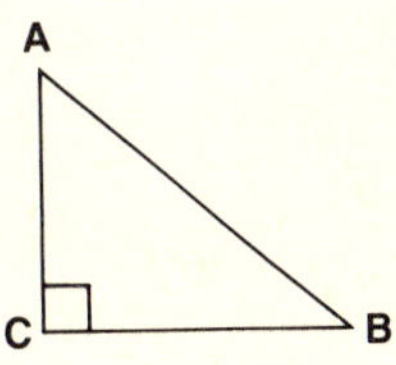

Right Triangle
$m\angle A + m\angle B = 90°$

Any side of a triangle can be extended in length. Thus, an angle is formed that is outside of the triangle—***an exterior angle.*** An exterior angle of a triangle is related to the interior angles in the following manner:

An exterior angle of a triangle is the supplement of the interior angle ***adjacent*** *to it.*

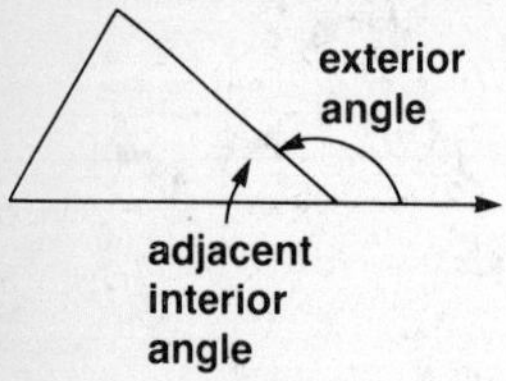

The measure of an exterior angle of a triangle is equal to the sum of the measures of the two ***remote*** *interior angles* (the two interior angles "far away from" it).

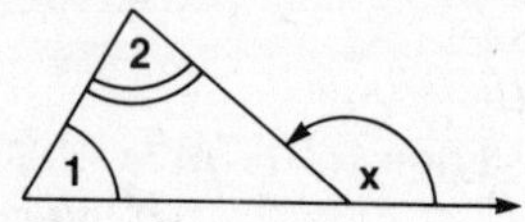

$$m\angle x = m\angle 1 + m\angle 2$$

m exterior angle + m adjacent interior angle = 180°

Examples

1. In the accompanying diagram, side $\overline{AC}$ of triangle ABC is extended to D and the measure of angle BCD is 130°. What is the measure of angle BCA?

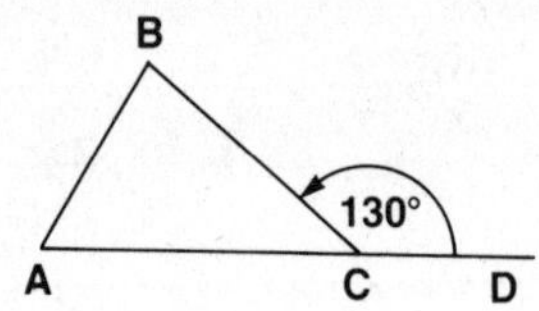

Solution:

Exterior $\angle BCD$ is the supplement of adjacent interior $\angle BCA$.

$$\begin{aligned} m\angle BCD + m\angle BCA &= 180° \\ 130 + m\angle BCA &= 180 \\ m\angle BCA &= 50° \quad \textit{Ans.} \end{aligned}$$

2. In the accompanying diagram, $\angle BCD$ is an exterior angle to $\triangle ABC$ at vertex C. If $m\angle A = 60°$ and $m\angle B = 75°$, find the number of degrees in the measure of $\angle BCD$.

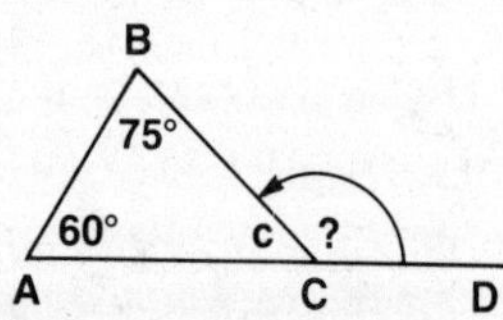

Solution:

The measure of exterior angle BCD is equal to the sum of the measures of remote, interior angles A and B.

$$\begin{aligned} m\angle BCD &= m\angle A + m\angle B \\ &= 60° + 75° \\ m\angle BCD &= 135° \quad \textit{Ans.} \end{aligned}$$

Note: Other positions for an exterior angle of a triangle are shown here. In each diagram, the exterior angle is labeled x. The interior angle that is adjacent to x is labeled a. The remote interior angles *with respect to* x are labeled 1 and 2.

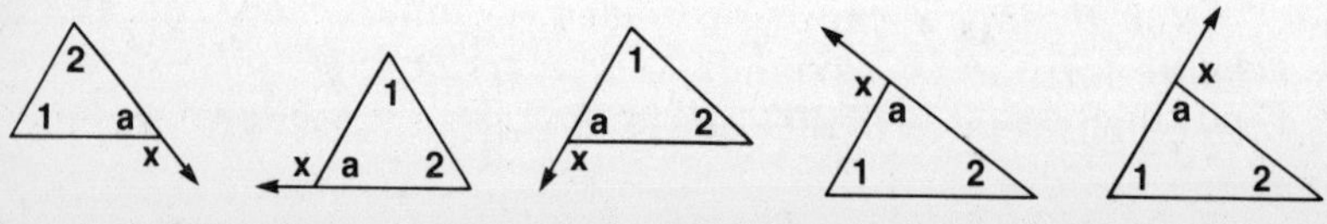

Example

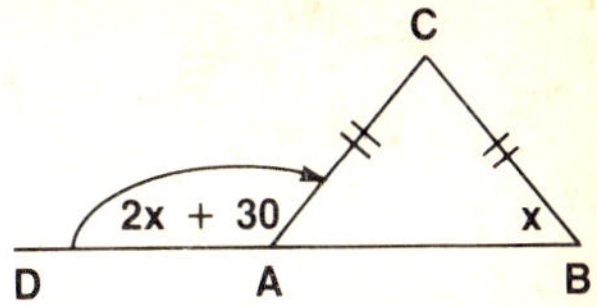

In the accompanying diagram, $\overline{AC} \cong \overline{BC}$ and $\overline{BA}$ is extended through point A to D. If $m\angle B = x$ and $m\angle CAD = (2x + 30)$, find x.

Solution:

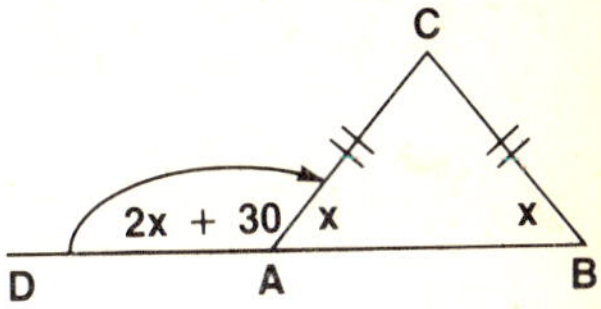

Since $\overline{AC} \cong \overline{BC}$, $\triangle ABC$ is isosceles and its base angles are congruent. Thus, $m\angle CAB = x$.

Exterior $\angle CAD$ is supplementary to adjacent interior $\angle CAB$.

$$\begin{aligned} m\angle CAD + m\angle CAB &= 180^\circ \\ m\angle CAD + x &= 180 \\ (2x + 30) + x &= 180 \\ 3x + 30 &= 180 \\ 3x &= 150 \\ x &= 50^\circ \quad \textit{Ans.} \end{aligned}$$

EXERCISES

In 1–16, only an algebraic solution is acceptable.

1. If two angles of a triangle measure 30° and 70°, what is the number of degrees in the third angle of the triangle?
2. In triangle ABC, angle B is twice as large as angle A and angle C is 20° more than angle A. How many degrees are there in each angle?
3. In triangle ABC, angle A is 30° more than angle B. Angle C equals the sum of angle A and angle B. Find the measures of each of the three angles.
4. In triangle ABC, the measures of angles A and B are in the ratio 1:3. Angle C is 5 degrees more than angle A. Find the number of degrees in each angle.
5. If the three angles of a triangle are in the ratio 1:3:5, find the number of degrees in the smallest angle.
6. If the measures of the angles of a triangle are in the ratio 2:2:5, the triangle is
   (1) a right triangle
   (2) an acute triangle
   (3) an isosceles triangle
   (4) an equilateral triangle
7. The degree measures of the three angles of a triangle are $(x + 10)$, $(2x + 20)$, and $(2x - 25)$.
   a. Find x.
   b. Find the number of degrees in each of the three angles of the triangle.

c. Based on your answer to part **b**, which is the most accurate description of the triangle?

(1) right and isosceles (3) right but not isosceles
(2) isosceles but not right (4) not right and not isosceles

8. If each base angle of an isosceles triangle measures 40°, find the number of degrees in the measure of the vertex angle.
9. If each base angle of an isosceles triangle measures 50°, find the measure in degrees of the vertex angle of the triangle.
10. Two angles of a triangle are equal in measure and the measure of the third angle is 150°. Find the number of degrees in one of the two equal angles.
11. If one base angle of an isosceles triangle measures 35°, find the measure of the vertex angle of the triangle.
12. The measure of the vertex angle of an isosceles triangle is 100°. Find the number of degrees in one of the base angles of the triangle.
13. If the vertex angle of an isosceles triangle measures 80°, find the number of degrees in the measure of one of the base angles.
14. What is the measure in degrees of one angle of an equilateral triangle?
15. If two angles of a triangle are complementary, find the number of degrees in the third angle of the triangle.
16. Find the number of degrees in each of the acute angles of an isosceles right triangle.
17. In the diagram, $\overline{BDC}$ is a segment, and $\overline{DE} \perp \overline{AEC}$. If $m\angle BDA = 80$ and $m\angle EDC = 50$, find $m\angle DAC$.

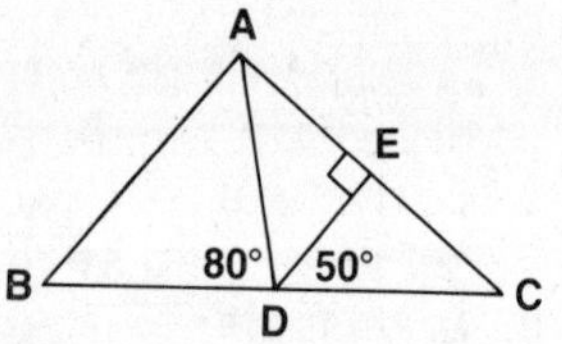

18. In the accompanying diagram, side $\overline{AC}$ of triangle ABC is extended to D and the measure of angle BCD is 110°. What is the measure in degrees of angle BCA?

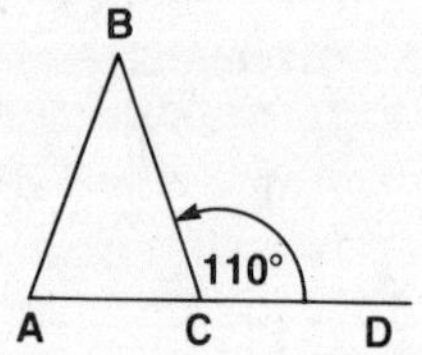

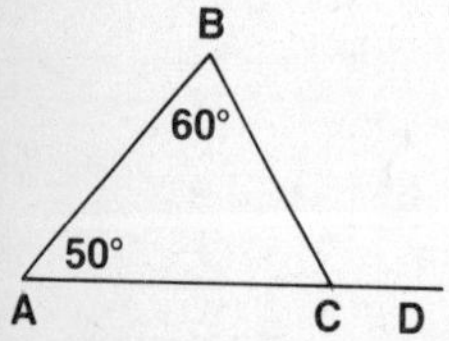

19. In the accompanying figure, $\angle BCD$ is an exterior angle to $\triangle ABC$ at vertex C. If $m\angle A = 50$ and $m\angle B = 60$, find the number of degrees in the measure of $\angle BCD$.

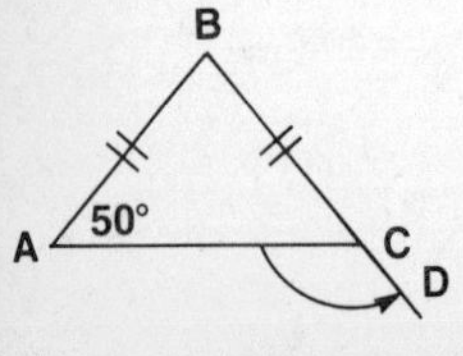

20. In the accompanying diagram, side $\overline{BC}$ of triangle ABC is extended to point D. If $\overline{AB} \cong \overline{BC}$ and $m\angle A = 50°$, find $m\angle ACD$.

21. In the accompanying diagram, $\angle QRT$ is an exterior angle to $\triangle PQR$ at vertex R. If $m\angle QRT = 115°$ and $m\angle QPR = 60°$, find $m\angle PQR$.

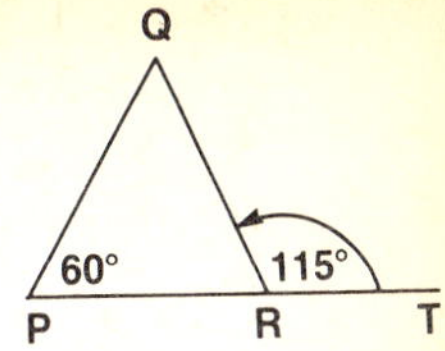

22. In the figure, $\overline{CBD}$ is a segment, and $\overline{AC} \cong \overline{BC}$ in $\triangle ABC$. If the measure of the exterior angle at C is 40°, find the number of degrees in the measure of angle x.

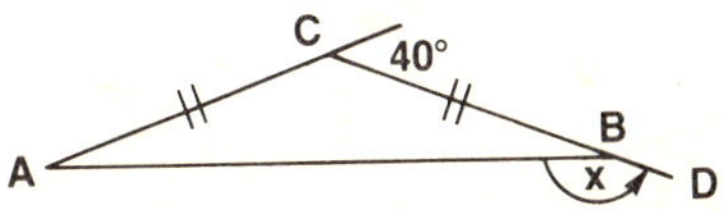

23. In the figure below, $\overleftrightarrow{AB}$ is parallel to $\overleftrightarrow{EF}$, $m\angle x = 70$, and $m\angle y = 105$. Find $m\angle z$.

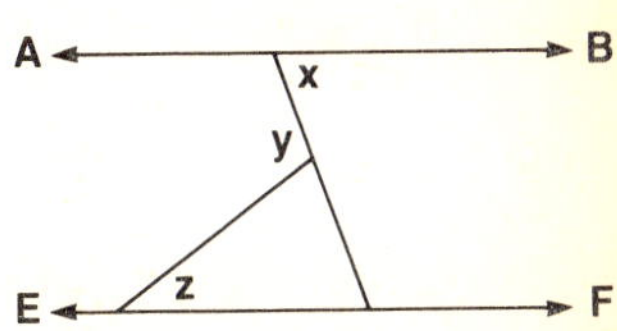

24. If an exterior angle of a triangle is congruent to the adjacent interior angle, then the remaining two interior angles must be
(1) supplementary
(2) complementary
(3) congruent
(4) obtuse

5.4 THE CIRCLE; ANGLES IN THE CIRCLE

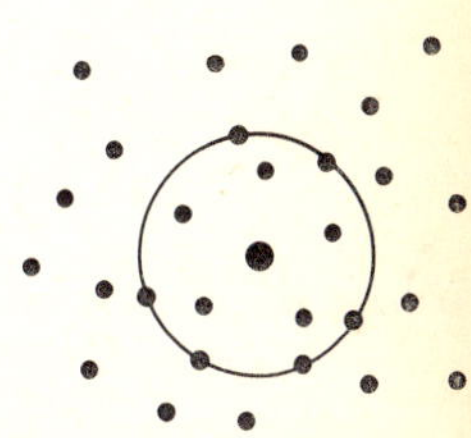

Considering a circle as a plane figure, we observe the following: There is an infinite number of points of the plane *inside* the circle, *on* the circle, and *outside* the circle.

Every point *on* the circle (on its ***circumference***) is the same distance from its ***center*** point. A line segment joining the center to any point on the circumference is called a ***radius.*** All the radii (plural of radius) of one circle are equal in length, or congruent.

A ***diameter*** of a circle is a line segment that joins two points on the circumference and passes through the center. A diameter is equal in length to two radii. All the diameters of one circle are congruent to one another.

An ***arc*** is a part of the circumference.

A ***chord*** of a circle is a line segment that joins two points on the circumference and, in general, does not pass through the center. (*Note:* A diameter is, therefore, a special chord and is also the longest chord of the circle.)

O is the center of the circle and the circle is called by its center: circle O or $\odot O$.

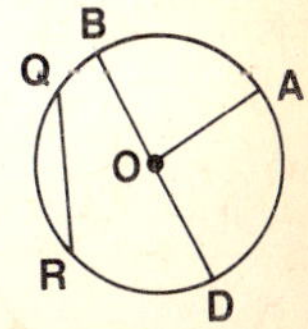

$\overline{OA}$ is a radius. $\overline{QR}$ is a chord.

$\overline{BOD}$ is a diameter. $\overset{\frown}{AB}$ is an arc.

Angle AOB is called a ***central angle*** of $\odot O$ since the vertex of the angle is at the center of the circle. The sides of central $\angle AOB$ are radii $\overline{OA}$ and $\overline{OB}$.

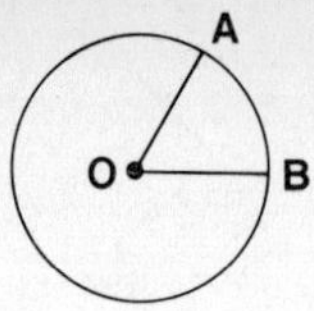

◆ ***The measure of a central angle of a circle is equal to the measure of its intercepted arc.***

Note also that the measure of a ***semicircle*** (half a circle) = 180°. An arc that contains less than 180° is called a ***minor arc,*** and an arc that contains more than 180° is called a ***major arc.***

Examples

1. In the accompanying diagram, $\overline{OA}$ is a radius and $\overline{BOD}$ is a diameter of $\odot O$. If $m\,\overset{\frown}{AB} = 80°$, find $m\,\angle AOD$.

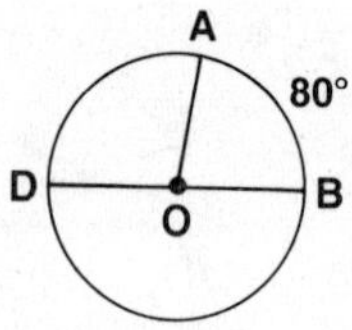

Solution 1:

$$\begin{aligned} m \text{ central } \angle AOB &= m\,\overset{\frown}{AB} \\ m\,\angle AOB &= 80° \\ \angle AOD &\text{ is the supplement of } \angle AOB \\ \text{Thus, } m\,\angle AOD + m\,\angle AOB &= 180° \\ m\,\angle AOD + 80° &= 180° \\ m\,\angle AOD &= 100° \quad \textit{Ans.} \end{aligned}$$

Solution 2:

$$\begin{aligned} m\,\overset{\frown}{AD} + m\,\overset{\frown}{AB} &= m \text{ semicircle} \\ m\,\overset{\frown}{AD} + 80° &= 180° \\ m\,\overset{\frown}{AD} &= 100° \\ m \text{ central } \angle AOD &= m\,\overset{\frown}{AD} \\ m\,\angle AOD &= 100° \quad \textit{Ans.} \end{aligned}$$

2. In $\odot O$, $\overline{OA}$ is a radius and $\overline{BOD}$ is a diameter. If $m\,\overset{\frown}{AB}$ and $m\,\overset{\frown}{AD}$ are in the ratio 1 : 2, find $m\,\angle AOB$.

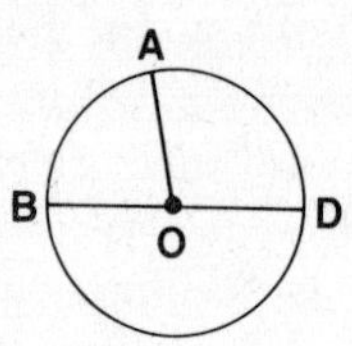

Solution:

Let x = the common ratio factor.
Then x = the number of degrees in $m\,\overset{\frown}{AB}$
and $2x$ = the number of degrees in $m\,\overset{\frown}{AD}$.

$$\begin{aligned} m\,\overset{\frown}{AB} + m\,\overset{\frown}{AD} &= m \text{ semicircle} \\ x + 2x &= 180° \\ 3x &= 180° \\ x &= 60° \\ m\,\overset{\frown}{AB} &= 60° \\ m \text{ central } \angle AOB &= m\,\overset{\frown}{AB} \\ m\,\angle AOB &= 60° \quad \textit{Ans.} \end{aligned}$$

Angle PQR is an *inscribed angle* of $\odot O$. Its vertex is on the circumference and its sides are chords.

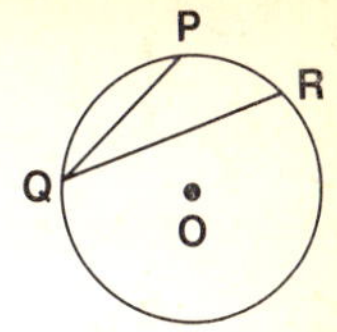

◆ ***The measure of an inscribed angle of a circle is equal to one-half the measure of its intercepted arc.***

Examples

1. In $\odot O$, $\overline{OA}$ and $\overline{OB}$ are radii, $\overline{AC}$ and $\overline{BC}$ are chords. If $m\angle AOB = 70°$, find $m\angle ACB$.

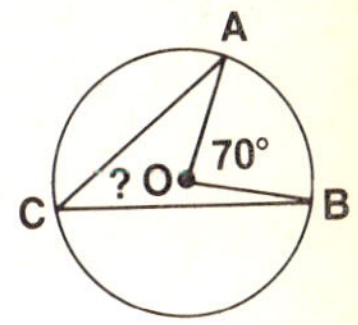

Solution:

$$\begin{aligned} m \text{ central } \angle AOB &= m\overset{\frown}{AB} \\ 70° &= m\overset{\frown}{AB} \\ m \text{ inscribed } \angle ACB &= \tfrac{1}{2} m\overset{\frown}{AB} \\ &= \tfrac{1}{2}\cdot 70° = 35° \quad \textit{Ans.} \end{aligned}$$

2. In $\odot O$, $\overline{OA}$ and $\overline{OB}$ are radii, $\overline{AC}$ is a chord, and $\overline{COB}$ is a diameter. If $m\angle ACB = 40°$, find $m\overset{\frown}{AC}$.

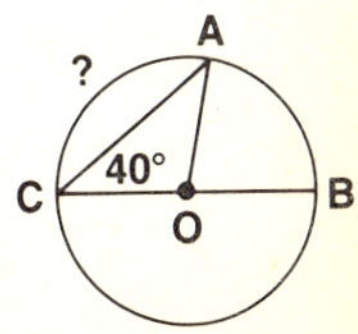

Solution:

$$\begin{aligned} m \text{ inscribed } \angle ACB &= \tfrac{1}{2} m\overset{\frown}{AB} \\ 40° &= \tfrac{1}{2} m\overset{\frown}{AB} \\ 80° &= m\overset{\frown}{AB} \\ m\overset{\frown}{AC} + m\overset{\frown}{AB} &= m \text{ semicircle} \\ m\overset{\frown}{AC} + 80° &= 180° \\ m\overset{\frown}{AC} &= 100° \quad \textit{Ans.} \end{aligned}$$

3. Isosceles $\triangle ABC$, with $\overline{AB} \cong \overline{AC}$, is inscribed in $\odot O$. If m vertex $\angle A = 70°$, find $m\overset{\frown}{AB}$.

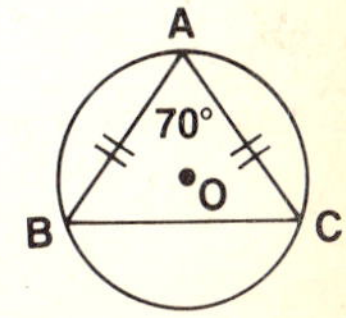

Solution:

In any triangle, the sum of the measures of the three interior angles is 180°. $\quad m\angle A + m\angle B + m\angle C = 180°$

Replace $m\angle A$ by its equal measure, 70°. $\quad 70° + m\angle B + m\angle C = 180°$

Replace $m\angle B$ by its equal measure, $m\angle C$ ($\angle B \cong \angle C$).

$$\begin{aligned} 70° + m\angle C + m\angle C &= 180° \\ 70° + 2m\angle C &= 180° \\ 2m\angle C &= 110° \\ m\angle C &= 55° \end{aligned}$$

$\angle C$, a base angle of isosceles $\triangle ABC$, is also an inscribed angle of $\odot O$, inscribed $\angle ACB$. The measure of an inscribed angle of a circle is one-half the measure of its intercepted arc.

$$\begin{aligned} m \text{ inscribed } \angle ACB &= \tfrac{1}{2} m\overset{\frown}{AB} \\ 55° &= \tfrac{1}{2} m\overset{\frown}{AB} \\ 110° &= m\overset{\frown}{AB} \\ &\textit{Ans.} \end{aligned}$$

Angle AEC is an ***angle formed by two intersecting chords*** of $\odot O$. The vertex of the angle is the point of intersection of the chords. The sides of the angle are segments of the intersecting chords. There are two intercepted arcs.

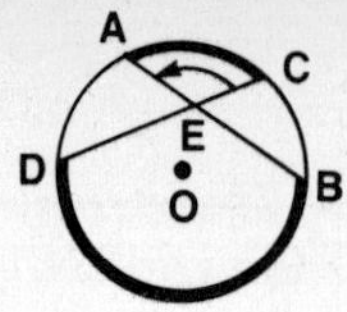

◆ ***The measure of an angle formed by two intersecting chords of a circle is one-half the sum of the measures of the intercepted arcs.***

Note: Whenever two chords intersect, four angles are formed: two pairs of vertical angles. Thus, in the preceding diagram:

$$m\angle AEC = m\angle DEB = \tfrac{1}{2}(m\overset{\frown}{AC} + m\overset{\frown}{DB}) \text{ and}$$
$$m\angle CEB = m\angle AED = \tfrac{1}{2}(m\overset{\frown}{AD} + m\overset{\frown}{CB})$$

Examples

1. Chords $\overline{LM}$ and $\overline{PQ}$ intersect at R in $\odot O$. If $m\overset{\frown}{LP} = 50°$ and $m\overset{\frown}{QM} = 120°$, find $m\angle LRP$.

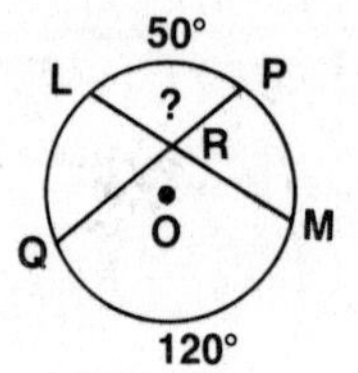

Solution:

$$\begin{aligned} m\angle LRP \text{ formed by two intersecting chords} &= \tfrac{1}{2}(m\overset{\frown}{LP} + m\overset{\frown}{QM}) \\ &= \tfrac{1}{2}(50° + 120°) \\ &= \tfrac{1}{2}(170°) \\ &= 85° \quad \textit{Ans.} \end{aligned}$$

2. Chord $\overline{CD}$ intersects diameter $\overline{AOB}$ at P. If $m\overset{\frown}{AC} = 110°$ and $m\overset{\frown}{AD} = 150°$, find $m\angle CPB$.

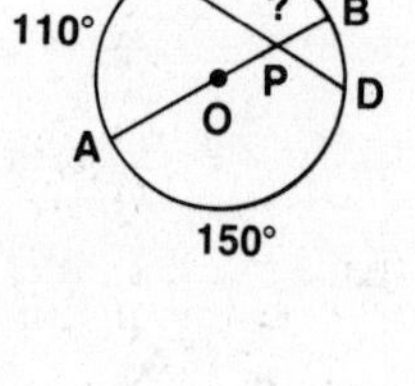

Solution:

$$\begin{aligned} m\angle CPB \text{ formed by two intersecting chords} &= \tfrac{1}{2}(m\overset{\frown}{CB} + m\overset{\frown}{AD}) \\ m\angle CPB &= \tfrac{1}{2}(m\overset{\frown}{CB} + 150°) \end{aligned}$$

$$\begin{aligned} m\overset{\frown}{CB} + m\overset{\frown}{AC} &= m \text{ semicircle} \\ m\overset{\frown}{CB} + 110° &= 180° \\ m\overset{\frown}{CB} &= 70° \end{aligned}$$

$$\begin{aligned} m\angle CPB &= \tfrac{1}{2}(70° + 150°) \\ &= \tfrac{1}{2}(220°) \\ &= 110° \quad \textit{Ans.} \end{aligned}$$

SUMMARY OF ANGLES IN A CIRCLE

Name	Description and Rule for Finding the Measure
Central angle	The vertex of the angle is at the center of the circle. The sides of the angle are radii. 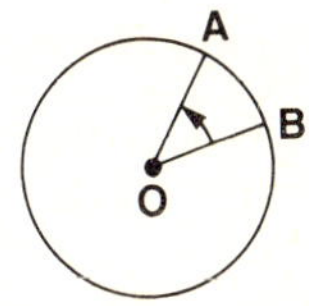 m central angle $=$ m intercepted arc $m \angle AOB = m\,\overset{\frown}{AB}$
Inscribed angle	The vertex of the angle is on the circumference of the circle. The sides of the angle are chords. 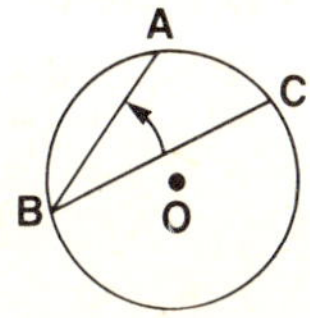 m inscribed angle $= \frac{1}{2}m$ intercepted arc $m \angle ABC = \frac{1}{2}m\,\overset{\frown}{AC}$
Angle formed by two intersecting chords	The vertex of the angle is at the point of intersection of the chords. The sides of the angle are segments of the chords. 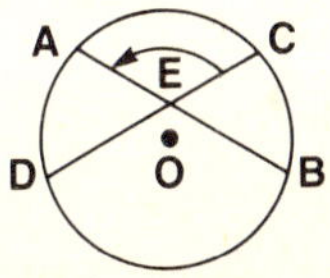 m angle formed by two intersecting chords $=$ $\frac{1}{2}$ the sum of the measures of the intercepted arcs $m \angle AEC = \frac{1}{2}(m\,\overset{\frown}{AC} + m\,\overset{\frown}{BD})$

Examples

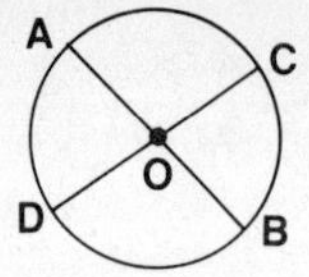

1. In the accompanying diagram, diameters $\overline{AOB}$ and $\overline{COD}$ are drawn in circle O.
If $m\,\widehat{AC}:m\,\widehat{CB}:m\,\widehat{BD}:m\,\widehat{DA} = 5:4:5:4$, find $m\angle AOC$.

Solution:

Let x = the common ratio factor.

Then $5x = m\,\widehat{AC}$ $\quad 4x = m\,\widehat{CB}$ $\quad 5x = m\,\widehat{BD}$ $\quad 4x = m\,\widehat{DA}$.

The entire circumference of the circle contains 360°.

$$\begin{aligned} 5x + 4x + 5x + 4x &= 360° \\ 18x &= 360 \\ x &= 20 \\ 5x &= 100 \\ 4x &= 80 \end{aligned}$$

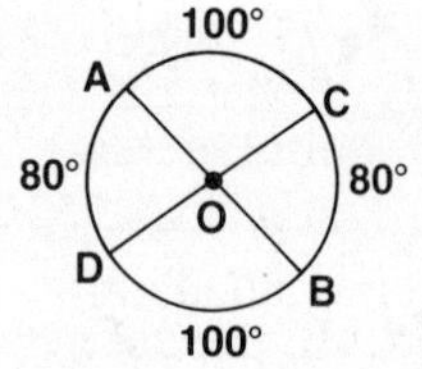

m central $\angle AOC = m\,\widehat{AC}$

$m\angle AOC = 100°$ *Ans.*

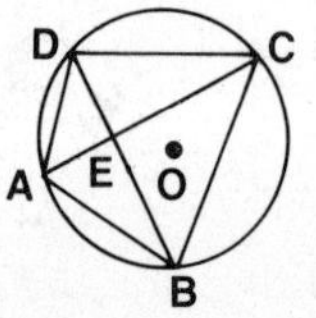

2. In the accompanying diagram, chords $\overline{AB}$, $\overline{BC}$, $\overline{CD}$, $\overline{DA}$, $\overline{AC}$, and $\overline{BD}$ are drawn as shown. If $m\,\widehat{AD} = x$, $m\,\widehat{AB} = (x + 20)$, $m\,\widehat{BC} = (2x + 10)$, and $m\,\widehat{CD} = (x + 40)$, find the measures of:
   a. $\widehat{AD}$, $\widehat{AB}$, $\widehat{BC}$, and $\widehat{CD}$.
   b. $\angle DAC$.
   c. $\angle DEC$.

Solution:

a. The entire circumference of the circle contains 360°.

$$\begin{aligned} m\,\widehat{AD} + m\,\widehat{AB} + m\,\widehat{BC} + m\,\widehat{CD} &= 360° \\ x + x + 20 + 2x + 10 + x + 40 &= 360 \\ 5x + 70 &= 360 \\ 5x &= 290 \\ x &= 58° \end{aligned}$$

$m\,\widehat{AD} = x = 58°$

$m\,\widehat{AB} = x + 20 = 58 + 20 = 78°$

$m\,\widehat{BC} = 2x + 10 = 2(58) + 10 = 126°$

$m\,\widehat{CD} = x + 40 = 58 + 40 = 98°$

Check:
58
78
126
98
360 ✓

Answer: $m\,\widehat{AD} = 58°$, $m\,\widehat{AB} = 78°$, $m\,\widehat{BC} = 126°$, $m\,\widehat{CD} = 98°$

b. m inscribed $\angle DAC = \frac{1}{2}m\,\widehat{DC}$

$= \frac{1}{2} \cdot 98°$

$= 49°$ *Ans.*

c. m angle formed by two intersecting chords $=$ $\frac{1}{2}$ the sum of the measures of the intercepted arcs

$$m\angle DEC = \tfrac{1}{2}(m\overset{\frown}{DC} + m\overset{\frown}{BA})$$
$$= \tfrac{1}{2}(98 + 78)$$
$$= \tfrac{1}{2}(176)$$
$$= 88^\circ \quad \textit{Ans.}$$

EXERCISES

1. In the accompanying diagram, radius $\overline{OA}$ and diameter $\overline{BOD}$ are drawn in $\odot O$. If $m\angle AOD = 130^\circ$, find $m\overset{\frown}{AB}$.

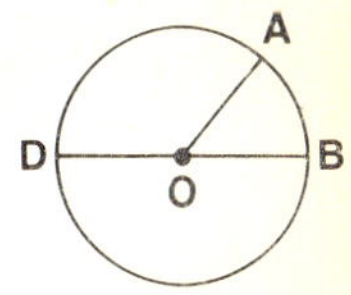

2. In the accompanying figure, the measure of inscribed angle ABC is 40°. Find the number of degrees in the measure of arc AC.

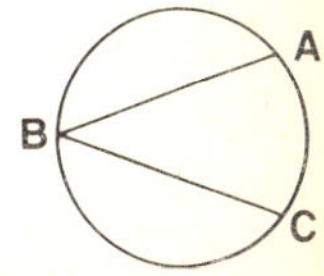

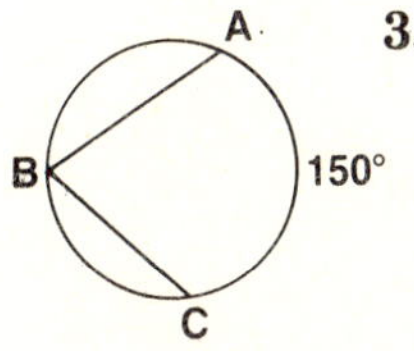

3. In the accompanying figure, arc AC measures 150°. Find the number of degrees in inscribed angle ABC.

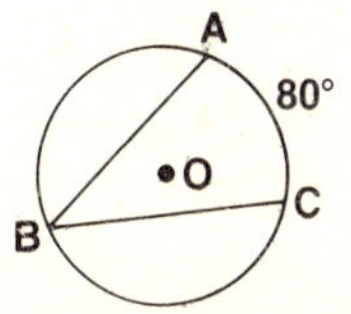

4. As shown in the accompanying diagram, angle ABC is inscribed in circle O and the measure of arc $AC = 80^\circ$. Find the number of degrees in the inscribed angle ABC.

5. In the accompanying figure, $\triangle ABC$ is inscribed in circle O as shown and $m\overset{\frown}{AB} = 80^\circ$. Find the number of degrees in $\angle ACB$.

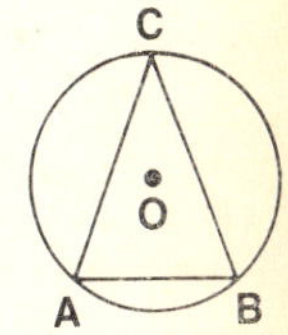

6. In the accompanying diagram, radii $\overline{OQ}$ and $\overline{OR}$ and chords $\overline{PQ}$ and $\overline{PR}$ are drawn in $\odot O$. If $m\angle QOR = 140^\circ$, find $m\angle P$.

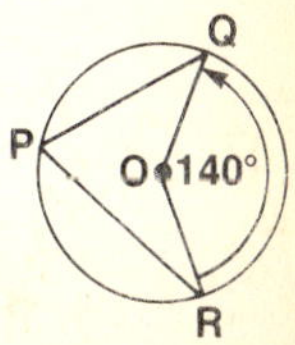

7. In the accompanying figure, $\triangle ABC$ is inscribed in circle O. Arc AB measures 120°, and arc BC measures 60°. Find the number of degrees in angle B.

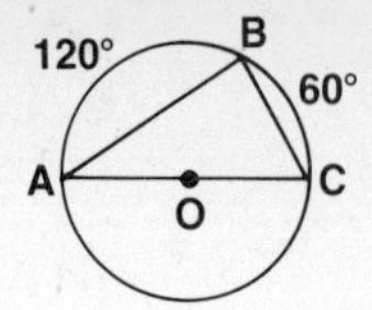

8. In the accompanying figure, equilateral triangle ABC is inscribed in circle O. Find the number of degrees in arc AB.

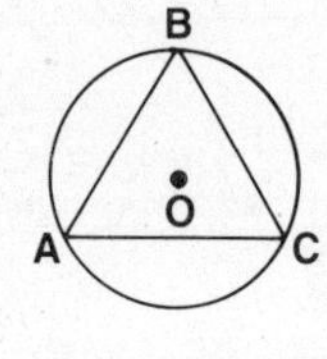

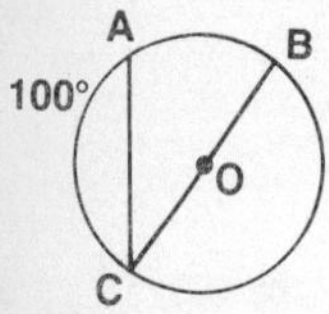

9. In the accompanying figure, chord $\overline{AC}$ and diameter $\overline{BOC}$ are drawn in $\odot O$. If $m\ \widehat{AC} = 100°$, find $m\ \angle ACB$.

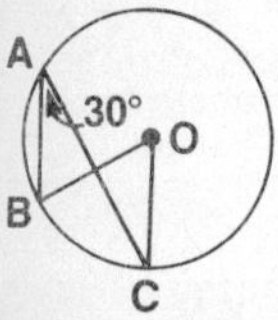

10. In the accompanying diagram, chords $\overline{AB}$ and $\overline{AC}$ and radii $\overline{OB}$ and $\overline{OC}$ are drawn in $\odot O$. If $m\ \angle BAC = 30°$, find $m\ \angle BOC$.

11. In the accompanying diagram, chords $\overline{AB}$ and $\overline{AC}$ are drawn in $\odot O$. If $m\ \widehat{AB} = 100°$ and $m\ \widehat{AC} = 140°$, find $m\ \angle BAC$.

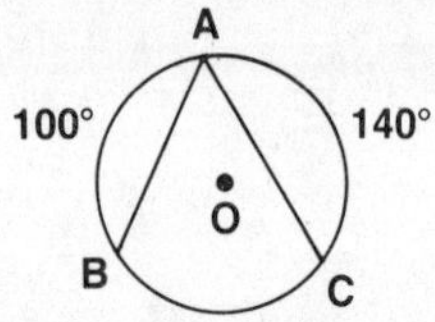

12. One circle is inside another so that they touch in only one point (point A), as shown in the accompanying diagram. Chords of the two circles are drawn as shown. If $m\ \widehat{BD} = 40°$, find $m\ \widehat{CE}$.

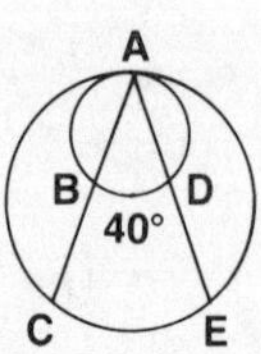

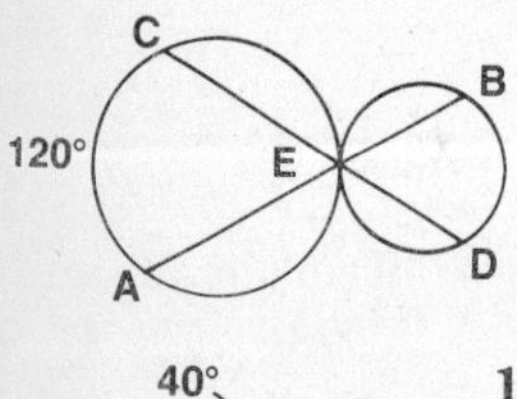

13. Two circles touch each other in only one point (point E), as shown in the accompanying diagram. $\overline{CED}$ and $\overline{BEA}$ are drawn as shown. If $m\ \widehat{AC} = 120°$, find $m\ \widehat{BD}$.

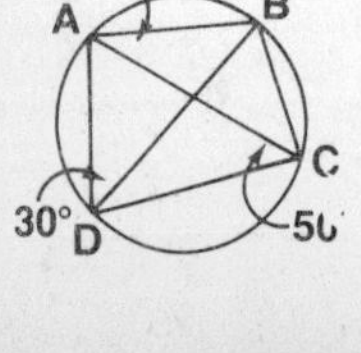

14. Six chords are drawn in a circle, as shown in the accompanying diagram. If $m\ \angle BAC = 40°$, $m\ \angle ADB = 30°$, and $m\ \angle ACD = 50°$, find $m\ \angle DBC$.

15. In the accompanying figure, $\overline{ROS}$ is a diameter of circle O, as shown. Radius $\overline{OT}$ and chords $\overline{RT}$ and $\overline{TS}$ are drawn. If $m\angle R = 50$, find $m\angle 1$.

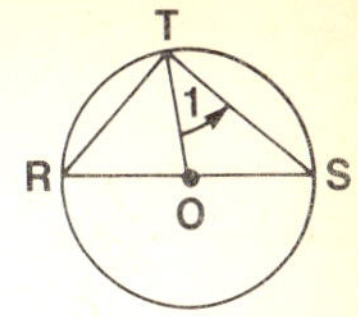

16. In the accompanying figure, O is the center of the circle, and $\overline{AEOB}$ and $\overline{ACD}$ are drawn. Radii $\overline{OC}$ and $\overline{OD}$ are drawn. If $\overline{AC} \cong \overline{CO}$ and $m\angle A = 30$, find $m\angle BOD$.

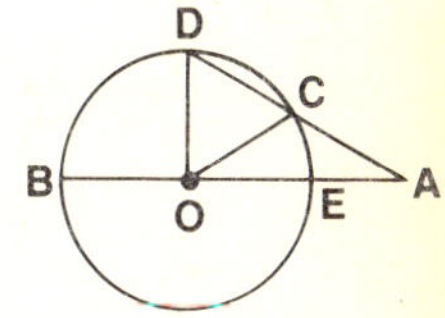

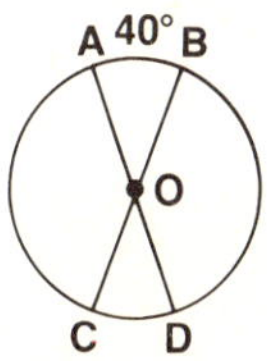

17. In the accompanying diagram, $\overline{AD}$ and $\overline{BC}$ are diameters that intersect at the center of circle O, and $m\overset{\frown}{AB} = 40°$. Find the number of degrees in angle COD.

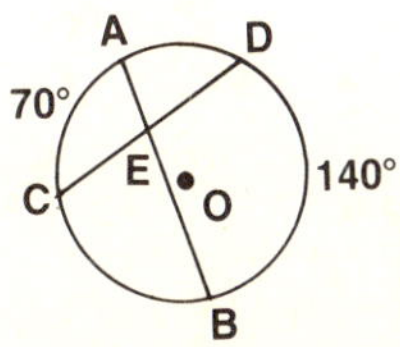

18. In the accompanying diagram, chords $\overline{AB}$ and $\overline{CD}$ intersect at E in circle O. If $m\overset{\frown}{AC} = 70°$ and $m\overset{\frown}{BD} = 140°$, find $m\angle DEB$.

19. In the accompanying diagram, chords $\overline{AB}$ and $\overline{CD}$ intersect at E in a circle. If $m\overset{\frown}{AC} = 140°$ and $m\overset{\frown}{BD} = 100°$, find $m\angle AED$.

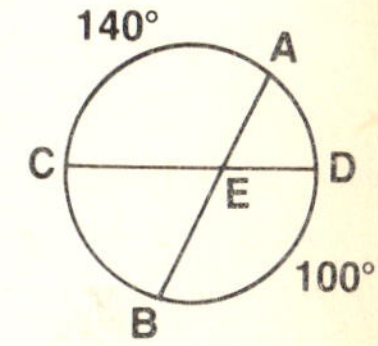

20. In the accompanying diagram, chords $\overline{AB}$ and $\overline{CD}$ intersect at E in circle O. If $m\angle AEC = 100°$ and $m\overset{\frown}{BD} = 70°$, find $m\overset{\frown}{AC}$.

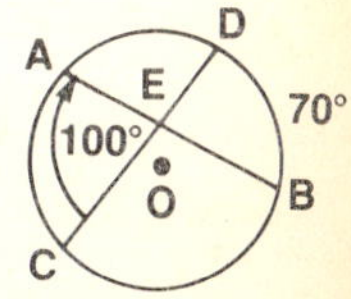

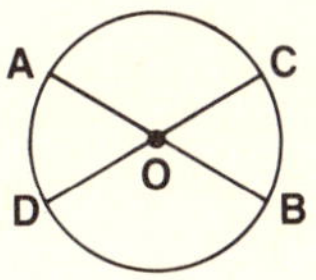

21. In the accompanying diagram, diameters $\overline{AOB}$ and $\overline{COD}$ are drawn in circle O.
If $m\overset{\frown}{AD}:m\overset{\frown}{DB}:m\overset{\frown}{BC}:m\overset{\frown}{CA} = 6:12:6:12$, find the measures of: **a.** $\overset{\frown}{AD}, \overset{\frown}{DB}, \overset{\frown}{BC}, \overset{\frown}{CA}$ **b.** $\angle AOC$

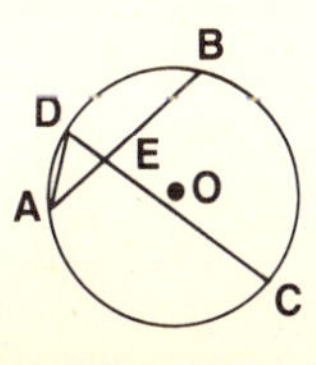

22. In the accompanying diagram, chords $\overline{AB}$, $\overline{CD}$, and $\overline{AD}$ are drawn in circle O.
If $m\overset{\frown}{AD}:m\overset{\frown}{DB}:m\overset{\frown}{BC}:m\overset{\frown}{CA} = 1:4:6:7$, find the measures of:
a. $\overset{\frown}{AD}, \overset{\frown}{DB}, \overset{\frown}{BC}, \overset{\frown}{CA}$ **b.** $\angle DAB$ **c.** $\angle DEB$

5.5 SUMMARY EXERCISES

1. Two angles are complementary and congruent. How many degrees are there in the measure of each angle?
2. $\angle C$ is the complement of $\angle 1$, and $\angle 5$ is the supplement of $\angle 1$. What is the difference in the degree measures of $\angle 5$ and $\angle C$?
3. Two angles are supplementary and in the ratio of 2:3. Find the measure of each angle.
4. Two angles are complementary and in the ratio of 1:5. Find the measure of each angle.
5. The larger of two supplementary angles is 30° less than twice the smaller. Find the measure of each angle.
6. The larger of two complementary angles exceeds three times the smaller angle by 10°. Find the measure of each angle.
7. Find the measure of the angle whose supplement is three times itself.
8. Find the measure of the angle whose complement is one-fifth of itself.
9. In the accompanying figure, $\overrightarrow{BA} \perp \overrightarrow{BE}$ and $\overrightarrow{BD} \perp \overrightarrow{BC}$. If $m\angle ABC = 150°$, find $m\angle DBE$.

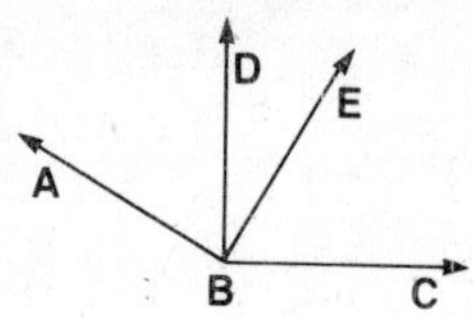

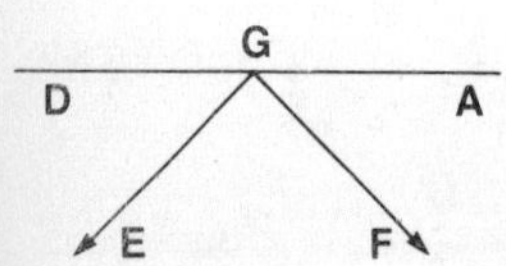

10. In the accompanying figure, $\overrightarrow{AD}$, $\overrightarrow{GE}$, and $\overrightarrow{GF}$ intersect so that $\angle DGE$ is the complement of $\angle AGF$. What is the measure of $\angle EGF$?

In 11–14, the accompanying figure shows the intersection of $\overleftrightarrow{AB}$, $\overleftrightarrow{CD}$, and $\overrightarrow{GF}$. $\overrightarrow{GF}$ bisects $\angle BGC$. If $m\angle DGB = x$ and $m\angle AGC = (3x - 280)$, find:

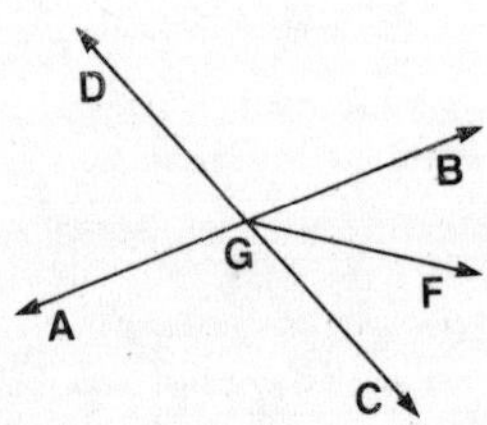

11. x.
12. $m\angle AGC$.
13. $m\angle DGA$.
14. $m\angle BGF$.

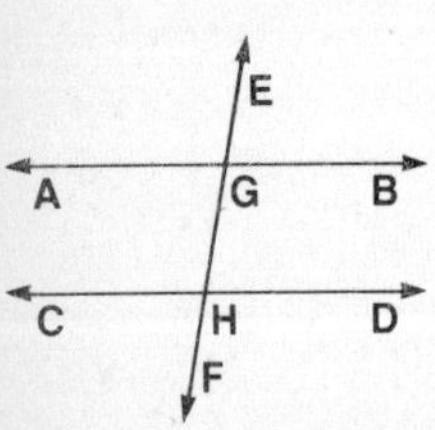

15. In the accompanying diagram, $\overleftrightarrow{AB} \parallel \overleftrightarrow{CD}$, and $\overleftrightarrow{EF}$ intersects $\overleftrightarrow{AB}$ at G and $\overleftrightarrow{CD}$ at H. If the degree measure of $\angle AGH$ is $(3x - 10)$ and the degree measure of $\angle GHD$ is 80, find the value of x.

16. In the diagram, $\overleftrightarrow{AB} \parallel \overleftrightarrow{CD}$ and $\overleftrightarrow{EF}$ intersects $\overleftrightarrow{AB}$ at G and $\overleftrightarrow{CD}$ at H. If $m\angle EGB = (2x - 10)$ and $m\angle GHD = (x + 70)$, find the value of x.

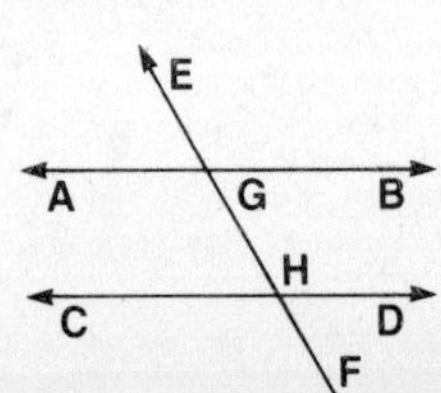

In 17–20, the diagram shows $\overleftrightarrow{AB} \parallel \overleftrightarrow{CD}$ and $\overleftrightarrow{EF}$ intersecting $\overleftrightarrow{AB}$ at G and $\overleftrightarrow{CD}$ at H. If $m\angle AGE = 100°$, find:

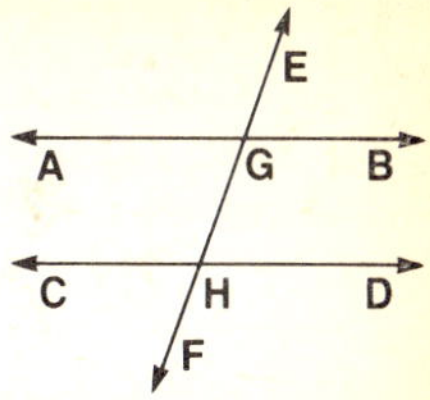

17. $m\angle AGH$ **19.** $m\angle HGB$
18. $m\angle CHG$ **20.** $m\angle FHD$

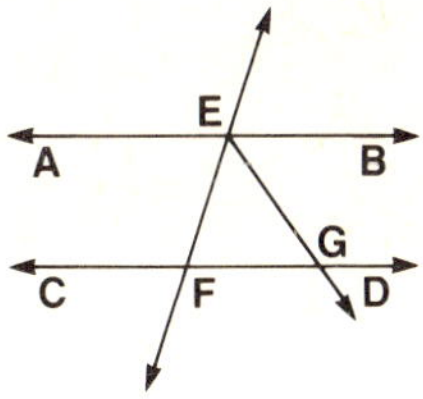

21. In the accompanying figure, $\overleftrightarrow{AB} \parallel \overleftrightarrow{CD}$, $\overleftrightarrow{EF}$ intersects $\overleftrightarrow{AB}$ and $\overleftrightarrow{CD}$, and $\overrightarrow{EG}$ bisects $\angle FEB$. If $m\angle FEG = x$ and $m\angle EFG = 3x$, find x.

22. In the accompanying figure, $\overline{AB}$, $\overline{CD}$, $\overline{BE}$, $\overline{DF}$, and $\overline{AF}$ intersect as shown. If $\angle 1 \cong \angle 2$, name the line segments that must be parallel.

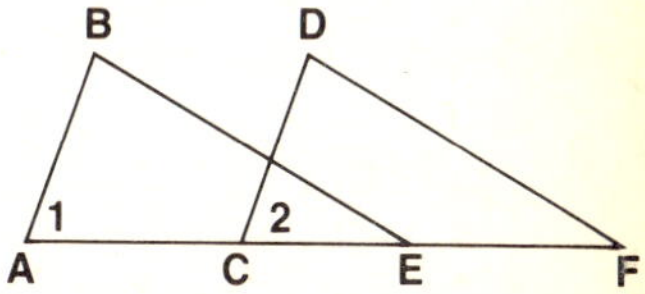

23. The three angles of a triangle are in the ratio 1:1:2.
a. Find the number of degrees in each angle.
b. Based on your answer to part **a,** tell which is the most accurate description of the triangle.
(1) isosceles but not right (3) isosceles and obtuse
(2) right and isosceles (4) not isosceles

24. If the measure of the vertex angle of an isosceles triangle is 110°, find the measure of each base angle.

25. In triangle ABC, the measure of $\angle B$ is 10° more than the measure of $\angle A$, and the measure of $\angle C$ is 30° more than twice the measure of $\angle B$. Find the measures of each of the three angles.

In 26–35, find the number of degrees in the measures of $\angle x$ and $\angle y$.

26.

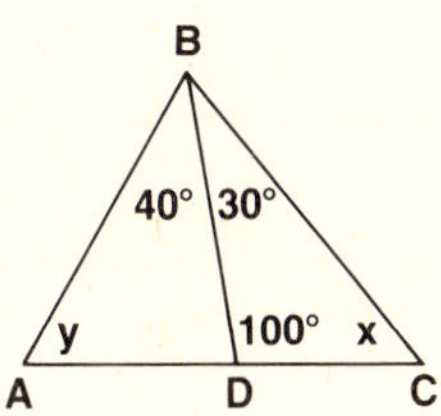

27. $\overline{AC} \cong \overline{BC}$

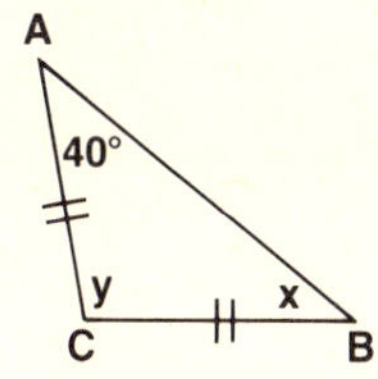

28. $\overline{CD} \perp \overline{AB}$

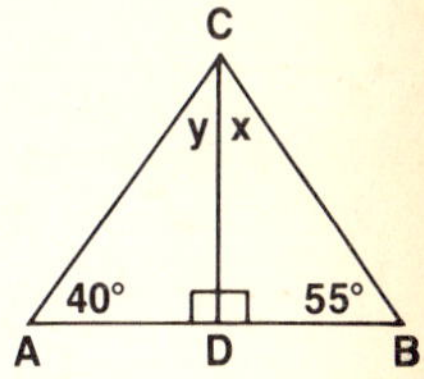

29. $\overline{AB}$ is extended to Q.
$\overline{AC}$ is extended to R.
$\overline{AC} \cong \overline{BC}$

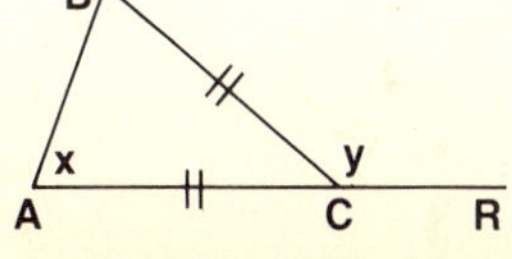

30. $\overline{AB}$ is extended to P.
$\overline{CB} \perp \overrightarrow{AP}$
$\overline{AQ}$ bisects $\angle CAB$.
$\overline{BQ}$ bisects $\angle CBA$.

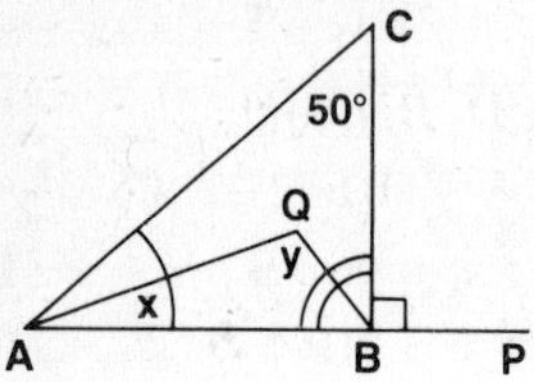

33. $\overleftrightarrow{RS} \parallel \overleftrightarrow{TV}$
$\overline{AB}$ bisects $\angle SAC$.
$\overline{CB}$ bisects $\angle ACV$.

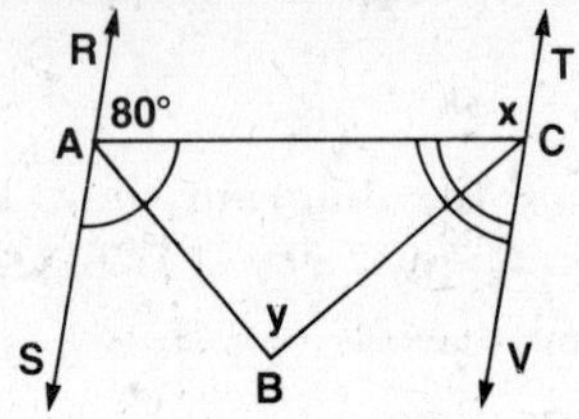

31. $\overrightarrow{BF} \parallel \overrightarrow{DA}$
$\overline{BCE}$

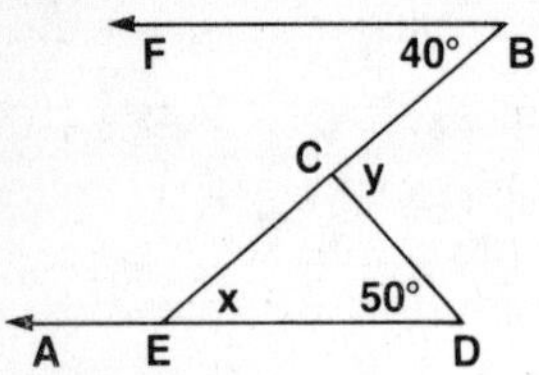

34. $\overline{AC}$, $\overline{CBD}$, $\overline{BEA}$, $\overline{DE}$

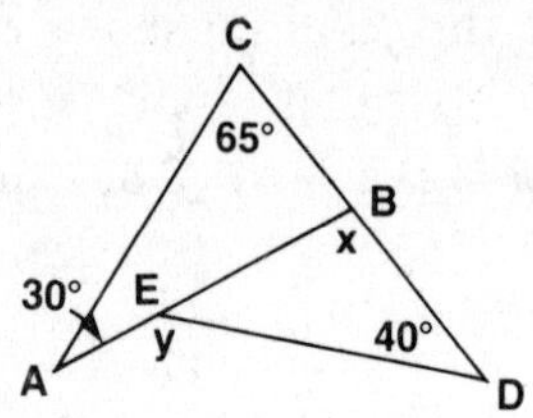

32. $\overline{AD} \cong \overline{DC} \cong \overline{CA}$
$\overline{AB} \cong \overline{BC}$
$\overline{AB} \perp \overline{BC}$

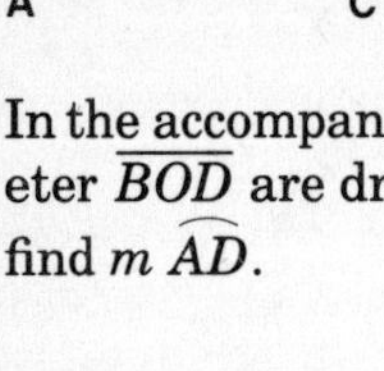

35. $\overrightarrow{BAE}$, $\overline{CDB}$
$\overline{CD} \cong \overline{DA} \cong \overline{AB}$

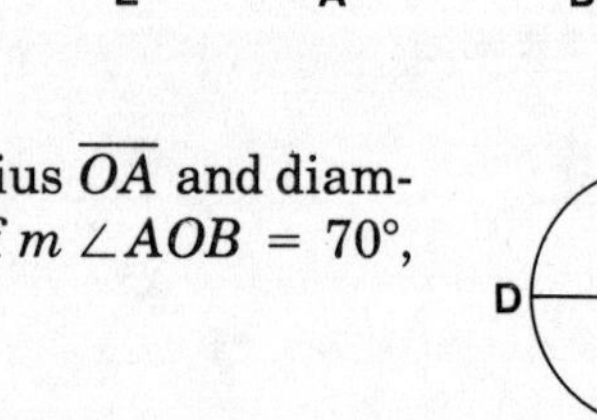

36. In the accompanying diagram, radius $\overline{OA}$ and diameter $\overline{BOD}$ are drawn in circle O. If $m\ \angle AOB = 70°$, find $m\ \widehat{AD}$.

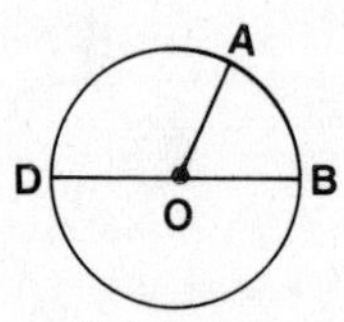

37. In the accompanying diagram, triangle ABC is inscribed in a circle as shown. If $m\ \widehat{AC} = 130°$, find $m\ \angle ABC$.

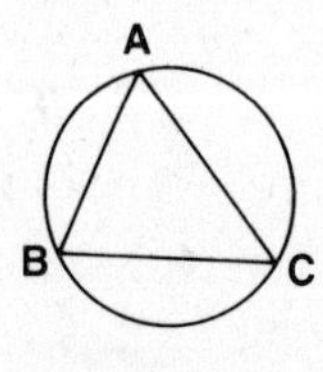

38. In the accompanying diagram, chords $\overline{AB}$ and $\overline{CD}$ intersect at E in circle O. If $m\ \widehat{AC} = 80°$ and $m\ \widehat{DB} = 160°$, find $m\ \angle AEC$.

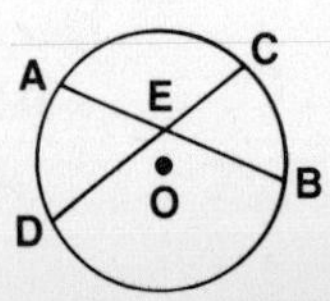

39. In the accompanying diagram, chords $\overline{AB}$ and $\overline{CD}$ intersect at E in circle O. If $m\angle CEB = 80°$ and $m\overset{\frown}{CB} = 100°$, find $m\overset{\frown}{AD}$.

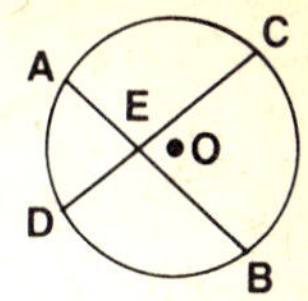

In 40–43, the diagram shows chords $\overline{AB}$, $\overline{CD}$, $\overline{BD}$, and $\overline{AD}$ in a circle. If $m\overset{\frown}{AD} : m\overset{\frown}{DB} : m\overset{\frown}{BC} : m\overset{\frown}{CA} = 8:7:10:11$, find the measures of:

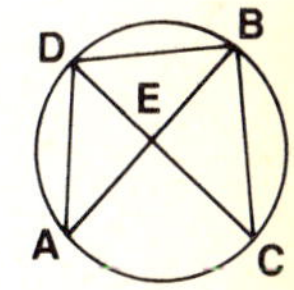

40. $\overset{\frown}{AD}$, $\overset{\frown}{DB}$, $\overset{\frown}{BC}$, and $\overset{\frown}{CA}$ **42.** $\angle BED$
41. $\angle DAB$ **43.** $\angle DEA$

In 44–47, the diagram shows chords $\overline{AC}$, $\overline{CB}$, $\overline{BD}$, $\overline{AB}$, and $\overline{CD}$ in a circle. If $m\overset{\frown}{AC} = 80$, $m\overset{\frown}{CB} = x$, $m\overset{\frown}{BD} = (2x - 8)$, and $m\overset{\frown}{DA} = (x + 32)$, find the measures of:

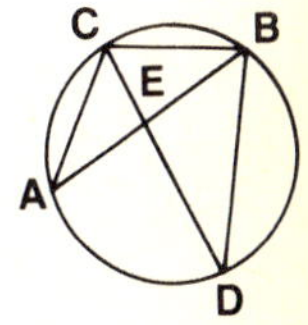

44. $\overset{\frown}{BC}$, $\overset{\frown}{BD}$, and $\overset{\frown}{DA}$ **46.** $\angle ACB$
45. $\angle ACD$ **47.** $\angle CEA$

In 48–50, use the accompanying diagram, which shows diameters $\overline{AOC}$ and $\overline{BOD}$ and chords $\overline{AB}$, $\overline{AD}$, $\overline{DC}$, and $\overline{CB}$ in circle O. If $m\overset{\frown}{AB} = 60°$, find the measures of:

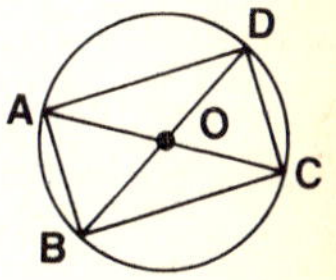

48. $\overset{\frown}{AD}$ **49.** $\angle ABD$ **50.** $\angle AOB$

CHAPTER 6

Ratio and Proportion

6.1 RATIO AS A RATE; SOLVING A PROPORTION; APPLICATIONS

We have already spoken of ***ratio*** as the comparison between two numbers by division. For example, the ratio of 2 to 3 is written 2:3 or $\frac{2}{3}$. An infinite number of numbers is in the ratio 2:3 $\left(\frac{6}{9}, \frac{8}{12}, \frac{10}{15}, \ldots, \frac{2x}{3x}\right.$ where x is the common ratio factor$\left.\right)$. We know that a ratio may be extended to involve more than two numbers in a comparison. For example, we can speak of three numbers that are in the ratio 2:3:5. We have used the concept of ratio in solving different types of problems.

The order in which a ratio is written is important for the comparison. If, for example, one woman is twice as old as another, the ratio may be thought of as $\frac{\text{younger woman}}{\text{older woman}}$ and written $\frac{1}{2}$ or as $\frac{\text{older woman}}{\text{younger woman}}$ and written $\frac{2}{1}$. You must be sure to write the ratio in the order of the comparison.

Example

A baseball team won 9 out of 16 games. The ratio of "wins" to "losses" is (1) 9:16 (2) 7:16 (3) 9:7 (4) 7:9

Solution:

We are asked to write the ratio of wins to losses. We already know the number of wins, 9. The number of losses must be 7, the total number of games minus the number of wins.

Thus, $\frac{\text{wins}}{\text{losses}} = \frac{9}{7}$ or 9:7.

Answer: (3)

Now let us investigate another aspect of ratio. Ratio is used to compare two numbers in the *same unit*. For example, if a gasoline tank holds 25 gallons of gas and there are 5 gallons of gasoline in the tank, the tank is filled to $\frac{1}{5}$ of its capacity, since $\frac{5 \text{ gallons}}{25 \text{ gallons}} = \frac{1}{5}$.

Ratio is also used to compare two numbers in *different units*. For example, if a car uses 1 gallon of gas for every 20 miles that it travels, the ***rate*** of gas consumption is 20 *miles per gallon*, since $\frac{20\text{ miles}}{1\text{ gallon}}$ is the comparison. Similarly, a plane is said to be traveling at the rate of 500 *miles per hour* $\left(\frac{500\text{ miles}}{1\text{ hour}}\right)$ if it covers a distance of 500 miles in 1 hour. Or, if a hospital ward has 60 patients and 5 nurses, we speak of 12 patients *per* nurse, since $\frac{60\text{ patients}}{5\text{ nurses}} = \frac{12\text{ patients}}{1\text{ nurse}} = 12\text{ patients}:1\text{ nurse}$.

When two ratios are equal, the equation that shows this is called a ***proportion.*** For example, $\frac{8}{10} = \frac{24}{30}$ is a proportion since the ratio $\frac{8}{10}$ is equivalent to $\frac{4}{5}$ and the ratio $\frac{24}{30}$ is equivalent to $\frac{4}{5}$. In the proportion $\frac{8}{10} = \frac{24}{30}$, there are four terms: 8, 10, 24, and 30. Two of the terms, 10 and 24, are called the ***means*** and the other two terms, 8 and 30, are called the ***extremes.***

◆ ***In a proportion, the product of the means is equal to the product of the extremes.*** For example:

$$\frac{8}{10} = \frac{24}{30}$$
$$10 \times 24 = 8 \times 30$$
$$240 = 240$$

The expression $\frac{2}{3} = \frac{6}{12}$ is not a proportion since $3 \times 6 \neq 2 \times 12$.

In general:

$$\underset{\text{mean}}{\overset{\text{extreme}}{\frac{a}{b}}} = \underset{\text{extreme}}{\overset{\text{mean}}{\frac{c}{d}}} \text{ is a proportion if and only if } \underset{\text{mean}}{b} \times \underset{\text{mean}}{c} = \underset{\text{extreme}}{a} \times \underset{\text{extreme}}{d}$$

(*Note:* The denominators cannot equal zero.)

If one term of a proportion is unknown, we can find the missing value by applying the principle of a proportion.

Examples

1. Solve for x: $\frac{3}{x} = \frac{9}{15}$

Solution:

To make this statement a proportion, we apply the principle that the product of the means is equal to the product of the extremes.

$$\frac{3}{x} = \frac{9}{15}$$
$$9 \cdot x = 3 \cdot 15 \quad \text{Cross-multiply.}$$
$$9x = 45$$
$$x = 5$$

Check:

Substitute $x = 5$ into the original proportion.

$$\frac{3}{x} = \frac{9}{15}$$

$$\frac{3}{5} \stackrel{?}{=} \frac{9}{15}$$

$$\frac{3}{5} \stackrel{?}{=} \frac{3 \cdot \cancel{3}}{5 \cdot \cancel{3}}$$

$$\frac{3}{5} = \frac{3}{5} \checkmark$$

Note that we check to see if the two ratios are equal by reducing each ratio to lowest terms. We do not cross-multiply in the check.

Answer: $x = 5$

2. Solve for x: $\dfrac{x + 2}{15} = \dfrac{42}{45}$

Solution:

$$\frac{x + 2}{15} = \frac{42}{45}$$

$15 \cdot 42 = 45(x + 2)$	Cross-multiply.
$630 = 45x + 90$	Distributive property.
$540 = 45x$	Subtract 90.
$12 = x$	Divide by 45.

Answer: $x = 12$

Check:

Substitute $x = 12$ into the original proportion.

$$\frac{x + 2}{15} = \frac{42}{45}$$

$$\frac{12 + 2}{15} \stackrel{?}{=} \frac{42}{45}$$

$$\frac{14}{15} \stackrel{?}{=} \frac{\cancel{3} \cdot 14}{\cancel{3} \cdot 15}$$

$$\frac{14}{15} = \frac{14}{15} \checkmark$$

A proportion can be used to solve verbal problems involving equal rates.

Examples

1. If Joan can read 80 pages in 2 hours, how many pages can she read in 5 hours?

Solution:

Let x = the number of pages Joan reads in 5 hours.

The assumption is that Joan reads at a constant rate. That is, Joan can always read the same number of pages in one hour.

$\dfrac{80 \text{ pages}}{2 \text{ hours}} = \dfrac{x \text{ pages}}{5 \text{ hours}}$	Write a proportion.
$2 \cdot x = 80 \cdot 5$	Cross-multiply to solve the proportion.
$2x = 400$	
$x = 200$	

Check:

Substitute $x = 200$ into the original proportion to see if we have a constant reading rate.

$$\frac{80 \text{ pages}}{2 \text{ hours}} \stackrel{?}{=} \frac{200 \text{ pages}}{5 \text{ hours}}$$

$$\frac{\not{2} \cdot 40}{\not{2} \cdot 1} \stackrel{?}{=} \frac{\not{5} \cdot 40}{\not{5} \cdot 1}$$

$$\frac{40 \text{ pages}}{1 \text{ hour}} = \frac{40 \text{ pages}}{1 \text{ hour}} \checkmark$$

Answer: Joan can read 200 pages in 5 hours.

2. If 10 dimes weigh 37 grams, find the weight in grams of 25 dimes.

Solution:

Let x = the weight in grams of 25 dimes.
The assumption is that all dimes weigh the same number of grams.

$\frac{10 \text{ dimes}}{37 \text{ grams}} = \frac{25 \text{ dimes}}{x \text{ grams}}$	Write a proportion.
$37 \times 25 = 10 \times x$	Cross-multiply to solve
$925 = 10x$	the proportion.
$92.5 = x$	Divide by 10.

Check:

Substitute $x = 92.5$ into the original proportion to see if we have the same gram weight per dime.

$$\frac{10 \text{ dimes}}{37 \text{ grams}} \stackrel{?}{=} \frac{25 \text{ dimes}}{92.5 \text{ grams}}$$

If 10 dimes weigh 37 grams, 1 dime weighs $\frac{37}{10}$ or 3.7 grams.	If 25 dimes weigh 92.5 grams, 1 dime weighs $\frac{92.5}{25}$ or:	$25\overline{)92.5}$ = 3.7

$$\text{3.7 grams per dime} = \text{3.7 grams per dime} \checkmark$$

Answer: 25 dimes weigh 92.5 grams.

3. If on a map $\frac{1}{2}$ inch represents a distance of 15 miles, how many inches will represent a distance of 50 miles?

Solution:

Let x = the number of inches that represents 50 miles.

$$\frac{\frac{1}{2}\text{ inch}}{15\text{ miles}} = \frac{x\text{ inches}}{50\text{ miles}} \qquad \text{Write a proportion.}$$

$$15 \cdot x = \tfrac{1}{2} \cdot 50 \qquad \text{Cross-multiply.}$$

$$15x = 25$$

$$x = \frac{25}{15} \qquad \text{Divide by 15.}$$

$$x = \frac{\cancel{5} \cdot 5}{\cancel{5} \cdot 3} = \frac{5}{3} \text{ or } 1\frac{2}{3} \qquad \text{Reduce.}$$

Check:

Substitute $x = \frac{5}{3}$ into the original proportion to see if we have the same scale.

$$\frac{\frac{1}{2}\text{ inch}}{15\text{ miles}} \stackrel{?}{=} \frac{\frac{5}{3}\text{ inch}}{50\text{ miles}}$$

If $\frac{1}{2}$ inch represents 15 miles, then 1 inch is represented by 2×15 or 30 miles.	If $\frac{5}{3}$ inch represents 50 miles, than 1 inch is represented by $\frac{3}{5} \times 50$ or 30 miles.
(*Note:* to change $\frac{1}{2}$ to 1, multiply by 2.)	(*Note:* to change $\frac{5}{3}$ to 1, multiply by $\frac{3}{5}$.)
1 inch represents 30 miles	= 1 inch represents 30 miles✓

Answer: $\frac{5}{3}$ inches or $1\frac{2}{3}$ inches represent a distance of 50 miles.

Note: The four terms of a proportion can be arranged in different ways to produce the same product of the means and of the extremes.

$\frac{a}{b} = \frac{c}{d}$	$\frac{a}{c} = \frac{b}{d}$	$\frac{d}{b} = \frac{c}{a}$	$\frac{d}{c} = \frac{b}{a}$
	$a \times d = c \times b$	$d \times a = b \times c$	$d \times a = c \times b$
$a \times d = b \times c$	$a \times d = b \times c$	$a \times d = b \times c$	$a \times d = b \times c$

Thus, when we form a proportion to solve a verbal problem involving equal rates, other arrangements are possible. Go back to example 1 and see that $\frac{80}{x} = \frac{2}{5}, \frac{5}{2} = \frac{x}{80}$, etc. are other ways in which the proportion could have been written. As long as 80 pages corresponds to 5 hours and x pages corresponds to 2 hours, the proportion will lead to the same rate of reading.

Another application of proportions is in the conversion of units of measure. There are different systems of measure (we work with English measures and metric measures), and there are different units of measure within each system for weight, length, area, and volume. There are also units of measure for time.

Proportions enable us to *convert from one measure to another within a system.*

Examples

1. How many inches are there in $2\frac{1}{2}$ feet?

Solution:

Let x = the number of inches in $2\frac{1}{2}$ feet. Since there are 12 inches in 1 foot, we write a proportion.

$$\frac{12 \text{ inches}}{1 \text{ foot}} = \frac{x \text{ inches}}{2\frac{1}{2} \text{ feet}}$$
$$1 \cdot x = 2\tfrac{1}{2} \cdot 12$$
$$x = \tfrac{5}{2} \cdot 12$$
$$x = 30$$

Answer: There are 30 inches in $2\frac{1}{2}$ feet.

Check:

Substitute $x = 30$ into the original proportion.

$$\frac{12}{1} = \frac{x}{2\frac{1}{2}}$$
$$\frac{12}{1} \stackrel{?}{=} \frac{30}{2\frac{1}{2}}$$
$$\frac{12}{1} \stackrel{?}{=} \frac{30}{\frac{5}{2}}$$
$$\frac{12}{1} \stackrel{?}{=} 30 \div \tfrac{5}{2}$$
$$\frac{12}{1} \stackrel{?}{=} \overset{6}{\cancel{30}} \cdot \frac{2}{\underset{1}{\cancel{5}}}$$
$$\frac{12}{1} = \frac{12}{1} \checkmark$$

2. How many pounds are there in 40 ounces?

Solution:

Let x = the number of pounds in 40 ounces. Since there are 16 ounces in 1 pound, we write a proportion.

$$\frac{16 \text{ ounces}}{1 \text{ pound}} = \frac{40 \text{ ounces}}{x \text{ pounds}}$$
$$1 \cdot 40 = 16 \cdot x$$
$$\tfrac{40}{16} = x$$
$$\tfrac{5}{2} = x$$
$$2\tfrac{1}{2} = x$$

Answer: There are $2\frac{1}{2}$ pounds in 40 ounces.

Check:

Substitute $x = 2\frac{1}{2}$ into the original proportion.

$$\frac{16}{1} = \frac{40}{x}$$
$$\frac{16}{1} \stackrel{?}{=} \frac{40}{2\frac{1}{2}}$$
$$\frac{16}{1} \stackrel{?}{=} \frac{40}{\frac{5}{2}}$$
$$\frac{16}{1} \stackrel{?}{=} 40 \div \tfrac{5}{2}$$
$$\frac{16}{1} \stackrel{?}{=} \overset{8}{\cancel{40}} \cdot \frac{2}{\underset{1}{\cancel{5}}}$$
$$\frac{16}{1} = \frac{16}{1} \checkmark$$

3. How many centimeters are there in $2\frac{1}{2}$ meters?

Solution:

Let x = the number of centimeters in $2\frac{1}{2}$ meters.
Since there are 100 centimeters in 1 meter, we write a proportion.

$$\frac{100 \text{ centimeters}}{1 \text{ meter}} = \frac{x \text{ centimeters}}{2\frac{1}{2} \text{ meters}}$$
$$1 \cdot x = 2\tfrac{1}{2} \cdot 100$$
$$x = \tfrac{5}{2} \cdot 100$$
$$x = 250$$

Check:

Substitute $x = 250$ into the original proportion.

$$\frac{100}{1} = \frac{x}{2\frac{1}{2}}$$
$$\frac{100}{1} \stackrel{?}{=} \frac{250}{2\frac{1}{2}}$$
$$\frac{100}{1} \stackrel{?}{=} \frac{250}{\frac{5}{2}}$$
$$\frac{100}{1} \stackrel{?}{=} 250 \div \tfrac{5}{2}$$
$$\frac{100}{1} \stackrel{?}{=} \overset{50}{\cancel{250}} \cdot \frac{2}{\cancel{5}_1}$$
$$\frac{100}{1} = \frac{100}{1} \checkmark$$

Answer: There are 250 centimeters in $2\frac{1}{2}$ meters.

4. How many days are there in w weeks?

Solution:

Let x = the number of days in w weeks.
Since there are 7 days in 1 week, we write a proportion.

$$\frac{7 \text{ days}}{1 \text{ week}} = \frac{x \text{ days}}{w \text{ weeks}}$$
$$1 \cdot x = 7 \cdot w$$
$$x = 7w$$

Check:

$$\frac{7}{1} = \frac{x}{w}$$
$$\frac{7}{1} \stackrel{?}{=} \frac{7\cancel{w}}{\cancel{w}_1}$$
$$\frac{7}{1} = \frac{7}{1} \checkmark$$

Answer: There are $7w$ days in w weeks.

5. How many minutes are there in 5 hours and 12 minutes?

Solution:

First find the number of minutes in 5 hours. To do this, use a proportion.

Let x = the number of minutes in 5 hours. There are 60 minutes in 1 hour.

$$\frac{60 \text{ minutes}}{1 \text{ hour}} = \frac{x \text{ minutes}}{5 \text{ hours}}$$
$$1 \cdot x = 5 \cdot 60$$
$$x = 300$$

There are 300 minutes in 5 hours.
Now add the additional number of minutes given:

$$300 \text{ minutes} + 12 \text{ minutes} = 312 \text{ minutes}$$

The check is left to the student.

Answer: There are 312 minutes in 5 hours and 12 minutes.

6. How many seconds are there in m minutes and t seconds?

Solution:

First find the number of seconds in m minutes. To do this, use a proportion.

Let x = the number of seconds in m minutes. There are 60 seconds in 1 minute.

$$\frac{60 \text{ seconds}}{1 \text{ minute}} = \frac{x \text{ seconds}}{m \text{ minutes}}$$
$$1 \cdot x = 60 \cdot m$$
$$x = 60m$$

There are $60m$ seconds in m minutes.
Now add the additional number of seconds given:

$$60m \text{ seconds} + t \text{ seconds} = (60m + t) \text{ seconds}$$

The check is left to the student.

Answer: There are $(60m + t)$ seconds in m minutes and t seconds.

Note that when we were converting from one unit to another within a system, we had to know certain facts (for example: 16 ounces = 1 pound, 7 days = 1 week, etc.). Each of these facts told us how many of a certain unit were in 1 of another unit. Following is a list of the most commonly used facts about measure:

TIME

60 seconds	=	1 minute
60 minutes	=	1 hour
24 hours	=	1 day
7 days	=	1 week
approximately 4 weeks	=	1 month
approximately 30 days	=	1 month
365 days	=	1 year
(366 days	=	1 leap year)
52 weeks	=	1 year
12 months	=	1 year

OTHER MEASUREMENTS

	English System	Metric System
Length:	12 inches = 1 foot 3 feet = 1 yard 5,280 feet = 1 mile 1,760 yards = 1 mile	basic unit = meter 100 centimeters = 1 meter 1,000 meters = 1 kilometer
Weight:	16 ounces = 1 pound 2,000 pounds = 1 ton	basic unit = gram 1,000 grams = 1 kilogram
Liquid Measures:	16 fluid ounces = 1 pint 2 pints = 1 quart 4 quarts = 1 gallon	basic unit = liter 1,000 liters = 1 kiloliter

Note: In the metric system, a series of prefixes is attached to each basic unit. Each prefix is a power of 10. The prefixes are:

milli	thousandth
centi	hundredth
deci	tenth
deka	ten
hecto	hundred
kilo	thousand

Thus, we can tell the relation between a given unit and its basic unit by the prefix. For example,

1 centimeter = $\frac{1}{100}$ of a meter
1 kilometer = 1,000 times a meter

Proportions also enable us to *convert from one system of measure to another*. To do this, we must know another set of facts relating the two systems. Some of these facts are:

1 kilometer = .6 mile
1 inch = 2.5 centimeters

1 quart = .9 liter

1 kilogram = 2.2 pounds
1 ounce = 28 grams

Example

How many centimeters are there in 3 inches?

Solution:

Let x = the number of centimeters in 3 inches.
Since there are 2.5 centimeters in 1 inch, we write a proportion.

$$\frac{2.5 \text{ centimeters}}{1 \text{ inch}} = \frac{x \text{ centimeters}}{3 \text{ inches}}$$
$$1(x) = 3(2.5)$$
$$x = 7.5$$

The check is left to the student.

Answer: There are 7.5 centimeters in 3 inches.

EXERCISES

1. If Tom "hits" the basket 7 times out of 10 throws, his ratio of "hits" to "misses" is (1) 3:7 (2) 3:10 (3) 7:3 (4) 7:10
2. A baseball team won 8 out of 11 games. The ratio of "wins" to "losses" is (1) 8:11 (2) 8:3 (3) 3:8 (4) 3:11
3. In a span of 17 days, there were 12 sunny days. The ratio of "sunny" to "non-sunny" days is
(1) 12:17 (2) 12:5 (3) 5:12 (4) 5:17
4. Sarah does a successful handstand 6 out of 13 times. The ratio of "successes" to "non-successes" is
(1) 6:13 (2) 6:7 (3) 7:6 (4) 7:13
5. A coin is tossed 40 times and it comes up "heads" 17 times. The ratio of "heads" to "tails" is
(1) 17:23 (2) 23:17 (3) 17:40 (4) 23:40

In 6–14, solve the given proportion.

6. $\frac{18}{12} = \frac{x}{2}$
7. $\frac{3}{5} = \frac{9}{x}$
8. $\frac{24}{h} = \frac{16}{4}$
9. $\frac{w}{6} = \frac{7}{2}$
10. $\frac{3}{8} = \frac{11}{x}$
11. $15:n = 12:16$
12. $\frac{x+1}{8} = \frac{9}{24}$
13. $\frac{y+1}{12} = \frac{3}{4}$
14. $\frac{2}{3} = \frac{p-2}{15}$

15. A boy takes 3 minutes to read a story of 315 words. How many minutes will it take him to read a story of 945 words at the same rate?
16. If 25 pennies weigh 42 grams, find the weight in grams of 75 pennies.
17. A farmer grew 105 bushels of grain on a 3-acre plot. How many bushels can he expect to grow under the same conditions on a 5-acre plot?
18. A copying machine can make 40 copies per minute. How many copies can the machine make in $3\frac{1}{4}$ minutes?

19. If x gumballs cost 25 cents, express in terms of x the number of gumballs that can be bought for 75 cents.
20. If a car can travel 51 kilometers on 3 liters of gasoline, how many kilometers can it travel under the same conditions on 7 liters?
21. If a car can travel 60 kilometers on 3 liters of gasoline, how many kilometers can it travel under the same conditions on 7 liters?
22. A man travels 120 miles in 3 hours. If he continues at the same rate, how many miles will he travel in the next 2 hours?
23. In 3 days a car traveled 573 miles. At the same average rate, how many miles can the car travel in 5 days?
24. On a map, 1 cm represents 50 km. If two cities are 175 km apart, how many cm apart are they on the map?
25. If on a map 1 inch represents 2 miles, how many inches will represent a distance of 3 miles?
26. On a map, 1 centimeter represents 4 kilometers. How many centimeters will represent 12 kilometers?
27. If 1 centimeter on a map represents 50 kilometers, how many centimeters will represent 225 kilometers?
28. On a map, if one inch represents 60 miles, how many miles are represented by $2\frac{1}{2}$ inches?
29. If $\frac{1}{2}$ inch represents 3 feet in a scale drawing, then how many inches will represent 24 feet?
30. If $\frac{1}{2}$ inch represents 10 feet in a scale drawing, then how many inches will represent 25 feet?

In 31–50, write the number of:

31. yards in 12 feet.
32. feet in 38 inches.
33. ounces in 4 pounds.
34. pounds in 60 ounces.
35. gallons in 12 quarts.
36. quarts in 6 gallons.
37. centimeters in 6 meters.
38. meters in 400 centimeters.
39. kilograms in 5,000 grams.
40. grams in 12 kilograms.
41. seconds in t minutes.
42. seconds in q minutes and r seconds.
43. inches in f feet.
44. inches in f feet and i inches.
45. weeks in y years.
46. weeks in y years and w weeks.
47. yards in f feet.
48. centimeters in m meters.
49. centimeters in 8 inches.
50. pounds in 40 kilograms.

6.2 PERCENT; PERCENTAGE PROBLEMS

Percent means *per hundred* (or *hundredths*) and can be thought of as the ratio of a number to the number 100. The symbol for percent is %.

A percent can be written as a fraction or as a decimal. To write a number with the symbol % as a fraction, rewrite it as the ratio of the given number to 100 by dropping the symbol % and using a fraction with a denominator of 100. The fraction with a denominator of 100 is also equivalent to a decimal number of two places. For example:

$$12\% = \frac{12}{100} = .12$$

$$4\% = \frac{4}{100} = .04$$

A fraction or a decimal can be written as a percent. Work with a denominator of 100. For example:

$$.46 = \frac{46}{100} = 46\%$$

$$\frac{3}{5} = \frac{3(20)}{5(20)} = \frac{60}{100} = 60\%$$

$$\frac{1}{8} = 8\overline{)1.000}^{\,.125} = .125 = \frac{125}{1000} = \frac{12.5}{100} = 12.5\%$$

One hundred percent, 100%, is equal to 1 since $100\% = \frac{100}{100} = 1$. Any percent greater than 100% is equal to a number greater than 1. For example:

$$200\% = \tfrac{200}{100} = 2$$

$$250\% = \tfrac{250}{100} = 2\tfrac{50}{100} = \begin{cases} 2\tfrac{1}{2} \\ 2.50 \text{ (or 2.5)} \end{cases}$$

$$375\% = \tfrac{375}{100} = 3\tfrac{75}{100} = \begin{cases} 3\tfrac{3}{4} \\ 3.75 \end{cases}$$

Percents containing fractions can be written as decimals with the fraction remaining or with the fraction changed to part of the decimal. For example:

$$17\tfrac{1}{2}\% = \frac{17\frac{1}{2}}{100} = .17\tfrac{1}{2} \qquad 4\tfrac{3}{4}\% = \frac{4\frac{3}{4}}{100} = .04\tfrac{3}{4}$$

$$\text{or } 17\tfrac{1}{2}\% = 17.5\% = \frac{17.5}{100} = .175 \qquad \text{or } 4\tfrac{3}{4}\% = 4.75\% = \frac{4.75}{100} = .0475$$

To find a certain *percent of* a number, we *multiply* the number by the percent, after changing the percent to either a decimal or a fraction. (*Remember:* Replace "of" by "multiply, ×.")

Examples

1. Find 3% of 150.

Solution 1:

$3\% \underbrace{\text{of}} 150$

$= \frac{3}{100} \times 150$

$= \frac{3}{\cancel{50} \cdot 2} \times \cancel{50} \cdot 3$

$= \frac{9}{2}$ or $4\frac{1}{2}$ *Ans.*

Solution 2:

$3\% \underbrace{\text{of}} 150$

$= .03 \times 150$

$= 4.50$ or 4.5 *Ans.*

2. Find $8\frac{1}{4}\%$ of 200.

Solution 1:

$8\frac{1}{4}\% \underbrace{\text{of}} 200$

$= 8\frac{1}{4}\% \times 200$

$= \frac{8\frac{1}{4}}{100} \times 200$

$= \frac{8\frac{1}{4}}{\cancel{100}} \times 2 \cdot \cancel{100}$

$= (8 + \frac{1}{4}) \times 2$ $\quad 8\frac{1}{4} = 8 + \frac{1}{4}$

$= 16 + \frac{1}{2}$ $\quad$ Distribute 2 and multiply.

$= 16\frac{1}{2}$ *Ans.*

Solution 2:

$8\frac{1}{4}\% \underbrace{\text{of}} 200$

$= 8\frac{1}{4}\% \times 200$

$= 8.25\% \times 200$

$= .0825 \times 200$

$= 16.5$ *Ans.*

If we know a certain percent of a number, we can find the original number by translating the sentence into symbols, replacing "of" by "multiply, ×, or •, or ( )" and "is" by "equals, =."

Examples

1. If 40% of a number is 12, find the number.

Solution:

Let x = the number.

$$40\% \underbrace{\text{of}} \underbrace{\text{a number}} \underbrace{\text{is}} 12$$
$$40\% \cdot \quad x \quad = 12$$

Write 40% as a fraction and solve for x.

$$
\begin{aligned}
40\% \cdot x &= 12 \\
\tfrac{40}{100} \cdot x &= 12 \\
\tfrac{2}{5} \cdot x &= 12 \\
x &= 12 \cdot \tfrac{5}{2} \quad \text{Multiply by } \tfrac{5}{2}. \\
x &= 30
\end{aligned}
$$

or

Write 40% as a decimal and solve for x.

$$
\begin{aligned}
40\% \cdot x &= 12 \\
.4\ (x) &= 12 \\
x &= \frac{12}{.4} \quad \text{Divide by .4.} \\
x &= 30
\end{aligned}
$$

$$.4\overline{)12.0} = 30.$$

Check:

40% of 30 is 12

$.4 \times 30 \stackrel{?}{=} 12$

$12 = 12\checkmark$

Answer: The number is 30.

2. A test was passed by 75% of a class. If 27 students passed, how many students were in the class?

Solution:

Let x = the number of students in the class.

75% of the class is 27

$75\% \cdot x = 27$

Write 75% as a fraction and solve for x.

$$
\begin{aligned}
75\% \cdot x &= 27 \\
\tfrac{75}{100} \cdot x &= 27 \\
\tfrac{3}{4} \cdot x &= 27 \\
x &= 27 \cdot \tfrac{4}{3} \quad \text{Multiply by } \tfrac{4}{3}. \\
x &= 36
\end{aligned}
$$

or

Write 75% as a decimal and solve for x.

$$
\begin{aligned}
75\% \cdot x &= 27 \\
.75\ (x) &= 27 \\
x &= \frac{27}{.75} \quad \text{Divide by .75.} \\
x &= 36
\end{aligned}
$$

Check:

75% of 36 is 27

$.75 \times 36 \stackrel{?}{=} 27$

$27 = 27\checkmark$

Answer: There were 36 students in the class.

Sometimes we want to know *how many* of a certain group did a certain thing, and the information is told to us as a percent of the original group.

Example

There are 40 students in a math class, and 85% of these students passed the last test. How many students in the class passed the test? How many students in the class failed the test?

Solution:

$$\begin{aligned} \text{number passed} &= 85\% \text{ of } 40 \\ &= .85 \times 40 \\ &= 34 \\ \text{number failed} &= \text{total number} - \text{number passed} \\ &= 40 - 34 \\ &= 6 \end{aligned}$$

Answer: Of the 40 students in the class, 34 passed the test and 6 failed the test.

Sometimes we want to know *what percent* of the whole a certain number represents.

Examples

1. In a cooking school, 18 of the 60 students are female. What percent of the student body is female?

   Solution:

Represent the part of the student body that is female as a fraction.	$\frac{18}{60}$
Reduce the fraction to lowest terms.	$\frac{3}{10}$
Change the fraction to a denominator of 100, since that is easy to do.	$\frac{3}{10} \cdot \frac{10}{10} = \frac{30}{100}$
Write as a percent by omitting the denominator of 100 and using the symbol %.	30%

   Answer: 30% of the student body is female.

2. A pamphlet contains 96 pages. If 12 of these pages have artwork, what percent of the pages in the pamphlet have artwork?

   Solution:

Represent the part of the pamphlet that has artwork as a fraction.	$\frac{12}{96}$
Reduce the fraction to lowest terms.	$\frac{1}{8}$

Since it is not immediately easy to write this fraction with a denominator of 100, change the fraction to a decimal by dividing the denominator into the numerator. $8\overline{)1.000}$ = .125

Since there are three decimal places, write the decimal as a fraction with a denominator of 1,000. $\frac{125}{1000}$

Change to a denominator of 100 by dividing the numerator and denominator by 10. $\frac{12.5}{100}$

Write as a percent: drop the denominator 100 and use %. 12.5%

Answer: 12.5% of the pages in the pamphlet have artwork.

Other types of percent problems involve taxes and discounts.

Examples

1. Ed bought supplies for \$36. If the sales tax is $7\frac{1}{2}$%, what was the total cost?

Solution:

$$\text{cost of sales tax} = 7\tfrac{1}{2}\% \text{ of } \$36 = .075 \times 36 = \$2.70$$

$$\text{total cost} = \text{cost} + \text{cost of tax} = \$36.00 + \$2.70 = \$38.70$$

Answer: The total cost was \$38.70.

2. At a sale, Dan paid \$9 for a necktie that had been reduced by 20% of the original price. What was the original price?

Solution:

Let x = the original price.

$$\begin{aligned} \text{sale price} &= \text{original price} - 20\% \text{ of original price} \\ \$9 &= x - 20\% \cdot x \\ 9 &= 1.00x - .20x \\ 9 &= .80x \\ 9 &= .8x \\ \frac{9}{.8} &= x \end{aligned}$$

$.8\overline{)9.0\,00}$ = 11.25

Answer: The original price of the tie was \$11.25.

EXERCISES

1. Write 36% as a fraction in lowest terms.
2. Write $4\frac{1}{2}$% as a decimal.
3. Write .24 as a percent.
4. Write $\frac{1}{5}$ as a percent.
5. What is 20% of 250?
6. What is 12.5% of 80?
7. What is 2.5% of 1,000?
   (1) 2,500 (2) 250 (3) 25 (4) 2.50
8. If 60% of a number is 144, what is the number?
9. If 10 percent of a number is 14, what is the number?
10. If 75% of x is 210, find the value of x.
11. If 10% of a number is 8, what is 35% of the number?
12. What percent of 200 is 14?
13. What percent of 300 is 18?
14. A test was failed by 15% of a class. If 6 students failed, how many students were in the class?
15. In a senior grade, 85% of the students applied to college. If 425 students applied, how many students were in the grade?
16. There are 2,000 students in a law school, and 48% of them are over age 25. What is the number of students over age 25?
17. The Freshman Magazine has 80 pages. Of these 80 pages, 16 are advertising. What percent of the magazine is advertising?
   (1) 5% (2) 20% (3) $\frac{1}{5}$% (4) 16%
18. If 30 students took an examination and 24 passed, what percent of the students passed the examination?
19. A booklet contains 30 pages. If 9 pages in the booklet have drawings, what percent of the pages in the booklet have drawings?
   (1) 30% (2) 9% (3) 3% (4) $\frac{3}{10}$%
20. Jorge buys a television set for $600. If the sales tax is 7%, find the sales tax in dollars.
21. If the sales tax rate is $8\frac{1}{4}$%, what is the amount of sales tax that must be paid on a $160 suit?
22. Alice bought supplies for $40.00. If the sales tax is 7%, what was the total cost?
   (1) $2.80 (2) $37.20 (3) $42.80 (4) $47.00
23. The sale price of a chair is $510 after a 15% discount has been given. Find the original price of the chair.
24. At a sale of coats, Mrs. Brown paid $360 for a coat that had been reduced by 20% of the original price. What was the original price?
25. At a sale, a scarf sold for $4. This amount was 80% of the original price. What was the original price?
26. The marked price of a skirt was $36.75, which represented 75% of the original selling price. What was the original selling price?
   (1) $27.56 (2) $42.35 (3) $45.94 (4) $49.00

6.3 SIMILARITY

Two polygons are ***similar*** if they have the *same shape*. Since the sizes of the angles in a polygon determine the shape of the polygon, the *polygons are similar if their corresponding angles are congruent (equal in measure). The corresponding sides of similar polygons are in proportion.*

$\triangle ABC$ is similar to $\triangle A'B'C'$
$\triangle ABC \quad \sim \quad \triangle A'B'C'$
$\triangle ABC$ and $\triangle A'B'C'$ have the *same shape.*

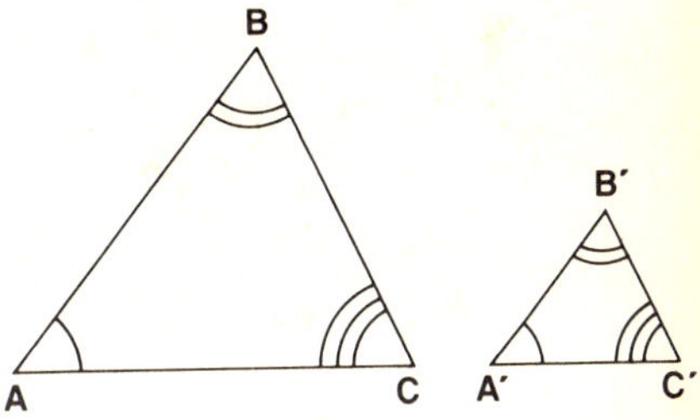

Corresponding angles are congruent:
$\angle A \cong \angle A'$
$\angle B \cong \angle B'$
$\angle C \cong \angle C'$

Corresponding sides are in proportion:
$$\frac{BC}{B'C'} = \frac{AC}{A'C'} = \frac{AB}{A'B'}$$

Note 1: We need to know only that *two* pairs of corresponding angles are congruent to say that two triangles are similar. Since the sum of the interior angles of a triangle is 180°, the third pair of corresponding angles must then be congruent.

Note 2: In the triangles, corresponding sides are opposite corresponding angles and corresponding angles are congruent (equal in measure). $\overline{BC}$ (opposite $\angle A$) corresponds to $\overline{B'C'}$ (opposite $\angle A'$) and $\angle A \cong \angle A'$; $\overline{AC}$ (opposite $\angle B$) corresponds to $\overline{A'C'}$ (opposite $\angle B'$) and $\angle B \cong \angle B'$; $\overline{AB}$ (opposite $\angle C$) corresponds to $\overline{A'B'}$ (opposite $\angle C'$) and $\angle C \cong \angle C'$.

Note 3: When we refer to a side of a triangle as, for example, $\overline{BC}$, we must include the line segment symbol. When we write a proportion, however, we are dealing with the *lengths* of the sides of the triangles. Thus, in a proportion (or in a ratio), we omit the line segment symbols.

Examples

1. In the accompanying diagram, $\triangle ABC \sim \triangle DEF$, $\angle A \cong \angle D$, $\angle B \cong \angle E$, $AB = 14$, $AC = 10$, and $DE = 7$. Find the length of $\overline{DF}$.

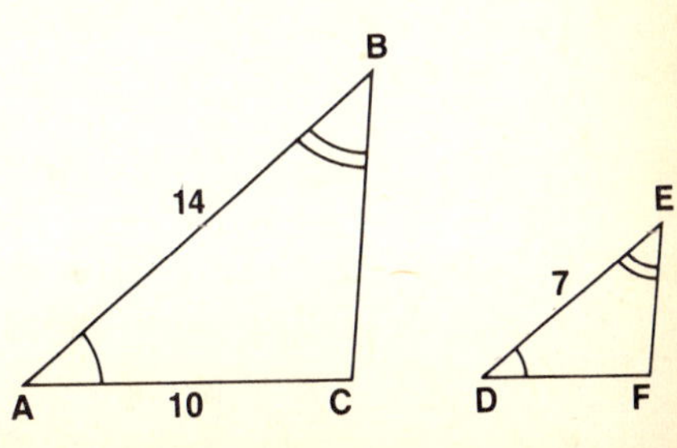

Solution:

Corresponding sides of similar triangles are in proportion.

$$\frac{AB}{AC} = \frac{DE}{DF}$$
$$\frac{14}{10} = \frac{7}{DF}$$
$$10 \cdot 7 = 14 \cdot DF \quad \text{Cross-multiply to}$$
$$70 = 14 \cdot DF \quad \text{solve the proportion.}$$
$$5 = DF$$

Answer: $DF = 5$

2. The lengths of the sides of a triangle are 5, 8, and 11. If the shortest side of a similar triangle is 10, what is the length of its longest side?

Solution:

Corresponding sides of similar triangles are in proportion.

$$\frac{\text{shortest side of } \triangle 1}{\text{shortest side of } \triangle 2} = \frac{\text{longest side of } \triangle 1}{\text{longest side of } \triangle 2}$$
$$\frac{5}{10} = \frac{11}{x}$$
$$10 \cdot 11 = 5 \cdot x \quad \text{Cross-multiply.}$$
$$110 = 5x$$
$$22 = x$$

Answer: The longest side of the second triangle is 22.

3. Right triangles ABC and DEF are similar. In $\triangle ABC$, the lengths of the legs are 5 and 12. In $\triangle DEF$, the length of the longer of the two legs is 24. What is the length of the shorter leg of $\triangle DEF$?

Solution:

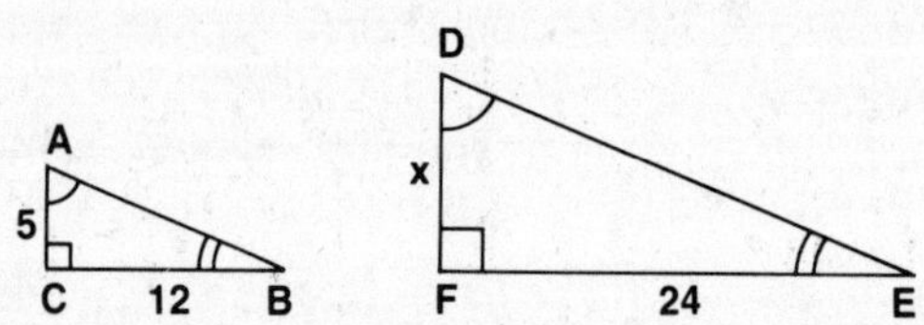

Since $\triangle ABC \sim \triangle DEF$, the corresponding sides are in proportion.

$$\frac{5}{12} = \frac{x}{24}$$
$$12x = 5 \cdot 24 \quad \text{Cross-multiply.}$$
$$12x = 120$$
$$x = 10$$

Answer: The shorter leg of right $\triangle DEF$ is 10.

Similar triangles are useful for ***indirectly measuring*** the heights of things too large to be measured directly, such as a tall building. By using the proportionality of the corresponding sides of similar triangles, the height of the building can be determined conveniently from the ground. The diagrams set up to solve ***shadow problems*** will consist of a pair of similar right triangles.

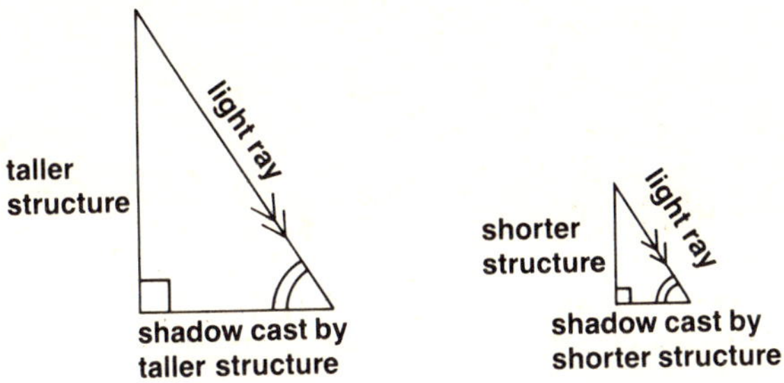

Example

A person 6 feet tall is standing near a tree. If the person's shadow is 4 feet long and the tree's shadow is 10 feet long, what is the height of the tree?

Solution:

Set up a pair of similar right triangles.

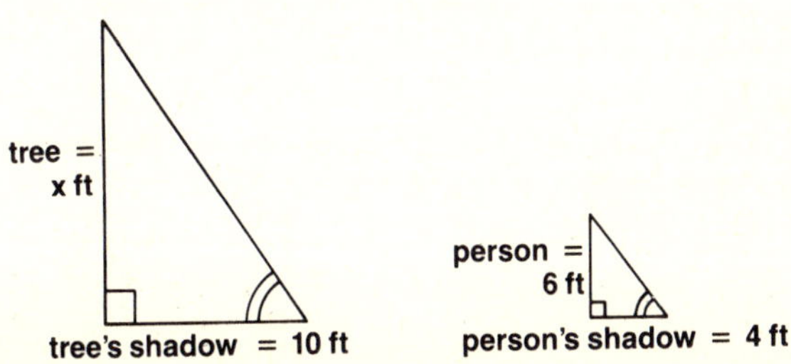

Corresponding sides of similar triangles are in proportion.

$$\frac{\text{tree}}{\text{tree's shadow}} = \frac{\text{person}}{\text{person's shadow}}$$

$$\frac{x}{10} = \frac{6}{4}$$

$$10 \cdot 6 = 4 \cdot x \quad \text{Cross-multiply.}$$

$$60 = 4x$$

$$15 = x$$

Check $x = 15$ in the proportion.

$$\frac{x}{10} = \frac{6}{4}$$

$$\frac{15}{10} \stackrel{?}{=} \frac{6}{4}$$

$$\frac{\cancel{5} \cdot 3}{\cancel{5} \cdot 2} \stackrel{?}{=} \frac{\cancel{2} \cdot 3}{\cancel{2} \cdot 2}$$

$$\frac{3}{2} = \frac{3}{2} \checkmark$$

Answer: The tree is 15 feet high.

EXERCISES

1. In the accompanying diagram of triangles ABC and DGF, $\angle A \cong \angle D$ and $\angle B \cong \angle G$. If $AC = 3$, $AB = 4$, and $DF = 9$, what is the length of $\overline{DG}$?

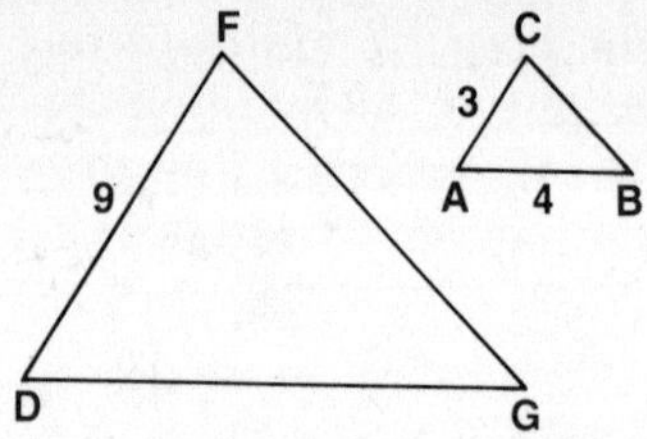

2. In the accompanying diagram, triangle ABC is similar to triangle DFE ($\angle A \cong \angle D$, $\angle B \cong \angle F$). If $AC = 8$, $AB = 12$, and $DE = 6$, find the length of $\overline{DF}$.

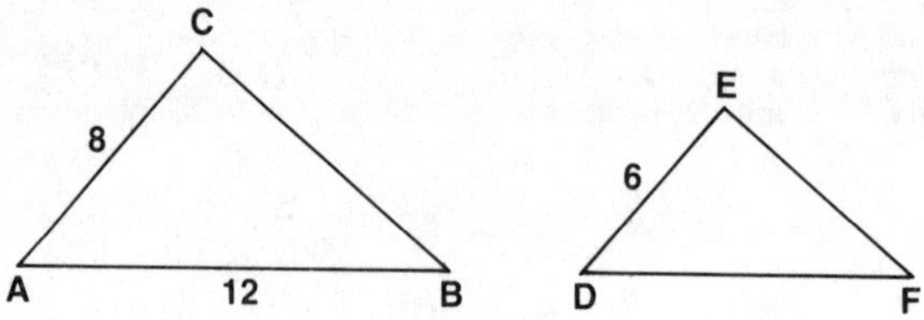

3. In the diagram of triangles ABC and DGF, $\angle A \cong \angle D$ and $\angle B \cong \angle G$. If $AC = 3$, $AB = 5$, and $DF = 9$, what is the length of $\overline{DG}$?

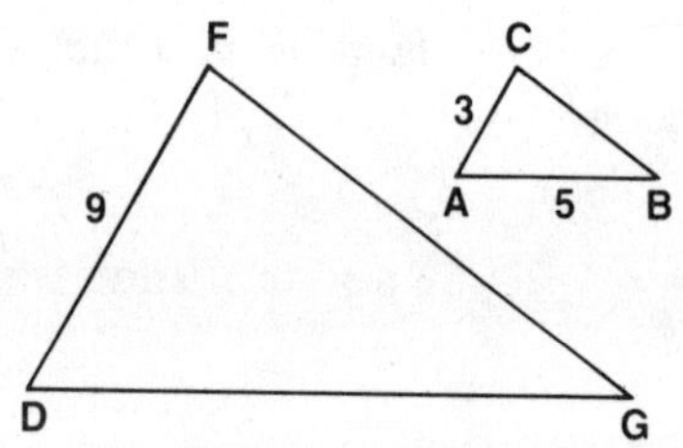

4. In the similar triangles ABC and DEF, as shown in the accompanying figure, $\angle B \cong \angle E$, $\angle C \cong \angle F$, $AC = 12$, $AB = 14$, and $DE = 42$. Find the length of $\overline{DF}$.

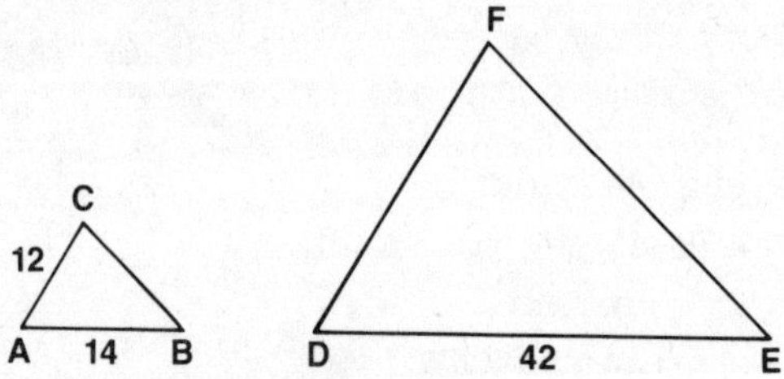

5. The lengths of the sides of $\triangle ABC$ are 5, 6, and 7. Triangle RST is similar to $\triangle ABC$. The longest side of $\triangle RST$ is 21. Find the length of the shortest side of $\triangle RST$.
6. The lengths of the sides of a triangle are 12, 8, and 6. If the shortest side of a similar triangle is 12, what is the length of its longest side?
7. The lengths of the sides of a triangle are 24, 20, and 12. If the longest side of a similar triangle is 6, what is the length of its shortest side?

8. Right triangles ABC and DEF are similar. In $\triangle ABC$, the lengths of the legs are 3 and 4. In $\triangle DEF$, the length of the longer of the two legs is 12. What is the length of the shorter leg of $\triangle DEF$?
9. A person 5 feet tall is standing near a tree 30 feet high. If the person's shadow is 4 feet long, how many feet long is the shadow of the tree?
10. A person 6 feet tall casts a shadow 15 feet long. At the same time, a nearby tower casts a shadow 100 feet long. Find the number of feet in the height of the tower.
11. What is the height, in feet, of a vertical pole that casts a shadow 8 feet long at the same time that another vertical pole 12 feet high casts a shadow 3 feet long?
12. A tree casts a shadow 20 meters long at the same time that a man 2 meters tall casts a shadow of 5 meters. How many meters are in the height of the tree?
13. A tree casts a 20-meter shadow at the same time that a 6-meter pole casts an 8-meter shadow. Find the height of the tree in meters.

6.4 CONGRUENCE

Polygons that have the *same size* AND the *same shape* are ***congruent.*** The *corresponding angles* AND the *corresponding sides of congruent polygons are congruent (equal in measure).*

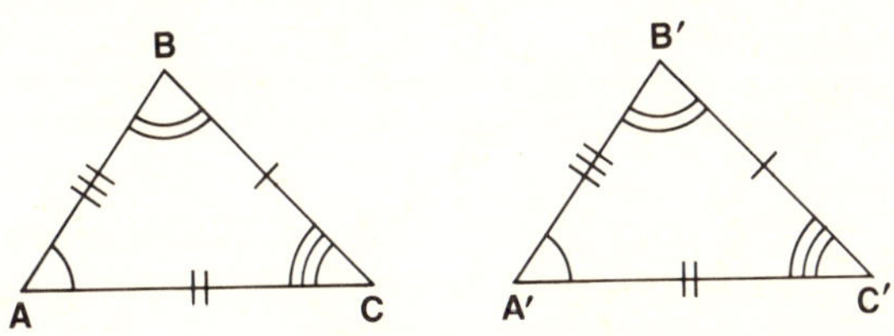

$$\triangle ABC \cong \triangle A'B'C'$$

$\triangle ABC$ and $\triangle A'B'C'$ have the *same size* and the *same shape.*

Corresponding angles are congruent:	Corresponding sides are congruent:
$\angle A \cong \angle A'$	$\overline{BC} \cong \overline{B'C'}$
$\angle B \cong \angle B'$	$\overline{AC} \cong \overline{A'C'}$
$\angle C \cong \angle C'$	$\overline{AB} \cong \overline{A'B'}$

(*Note:* In the triangles, corresponding sides are opposite the congruent corresponding angles and, conversely, corresponding angles are opposite the congruent corresponding sides.)

When we know that two triangles are congruent, we know that the three pairs of corresponding angles are congruent and that the three pairs of corresponding sides are congruent. But, if we wish to prove that two triangles are congruent, we need not first show that all six pairs of corresponding parts are congruent. It is sufficient to show that *three strategic pairs* of corresponding parts are congruent:

◆ Two triangles are congruent if:

1. Two pairs of corresponding sides and the *included* pair of corresponding angles are congruent.

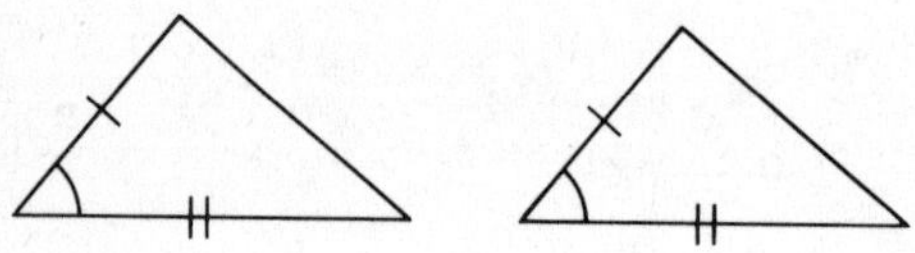

If s.a.s. ≅ s.a.s., the triangles are congruent.

2. Two pairs of corresponding angles and *any* pair of corresponding sides are congruent.

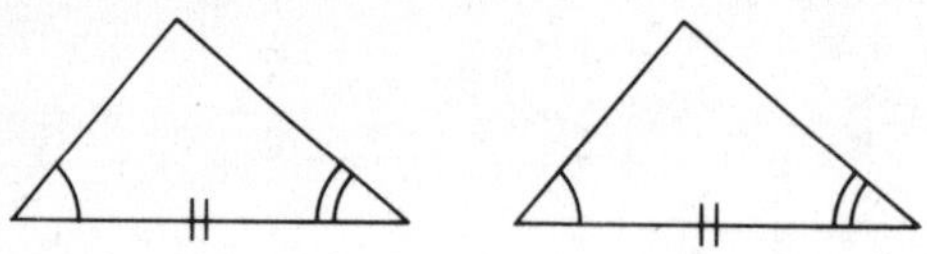

If a.s.a. ≅ a.s.a., the triangles are congruent.

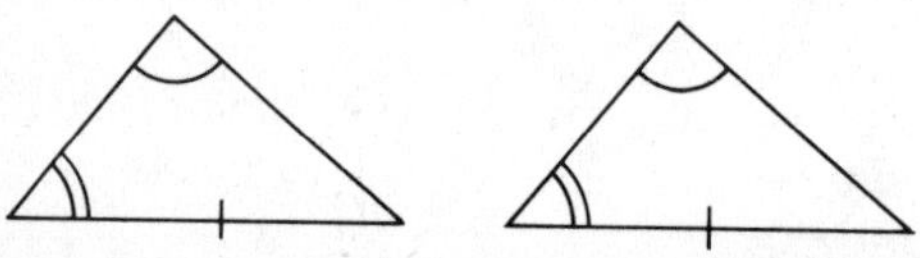

If a.a.s. ≅ a.a.s., the triangles are congruent.

3. Three pairs of corresponding sides are congruent.

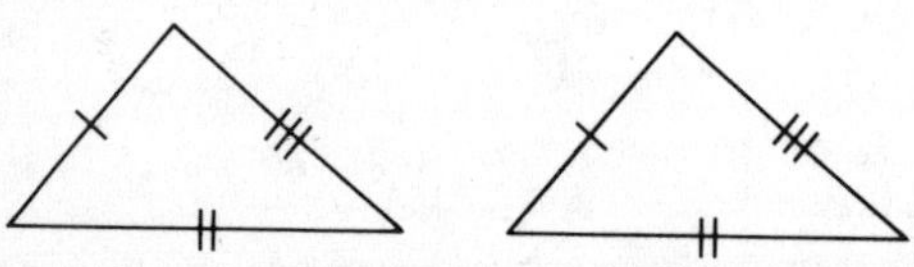

If s.s.s. ≅ s.s.s., the triangles are congruent.

(*Note* 1: It is not sufficient to show three pairs of congruent angles to establish that the triangles are congruent, as this condition will only establish that the triangles are similar. That is, a.a.a. ≅ a.a.a. is *not* a method of proving congruence.)

(*Note* 2: When we are using two pairs of sides and one pair of angles to prove congruence, the pair of angles *must be included* between the sides as shown in 1 above. That is, s.s.a. ≅ s.s.a. is *not* a method of proving congruence.)

EXERCISES

1. In the accompanying figure, triangle ABC is congruent to triangle RST, $\overline{AB} \cong \overline{RS}$, $\overline{BC} \cong \overline{ST}$, and $\overline{AC} \cong \overline{RT}$. What angle in triangle RST is congruent to $\angle BAC$?

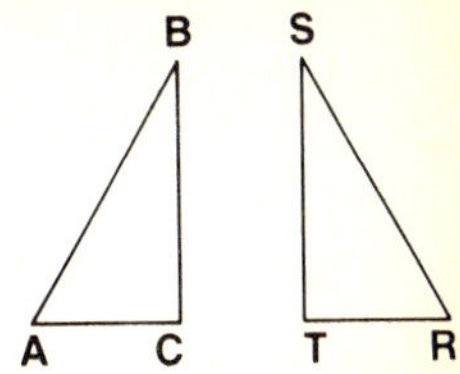

2. In the accompanying figure, triangle ABC is congruent to triangle DEF, $\overline{BA} \cong \overline{ED}$, $\angle A \cong \angle D$, and $\overline{AC} \cong \overline{DF}$. What angle in triangle DEF is congruent to $\angle CBA$?

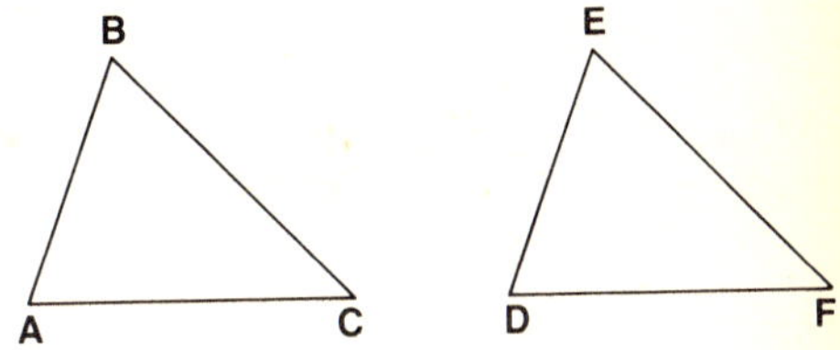

3. According to the markings shown, which pair of triangles *must* be congruent?

4. According to the markings shown, which pair of triangles *must* be congruent?

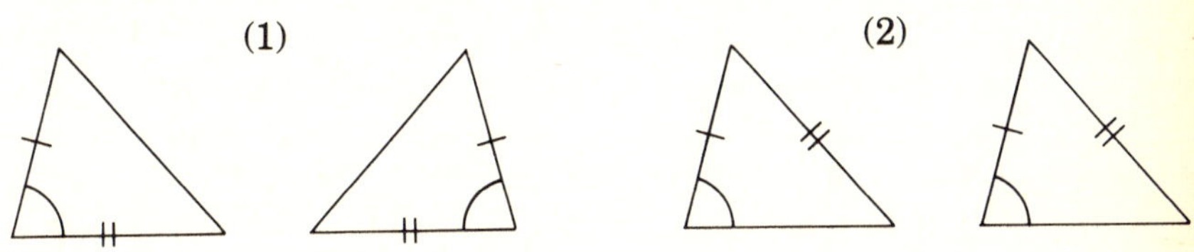

5. According to the markings shown, which pair of triangles *must* be congruent?

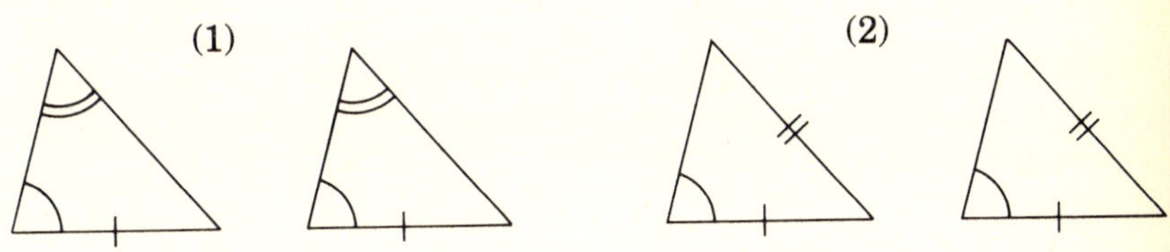

6. According to the markings shown, which pair of triangles is *not necessarily* congruent?

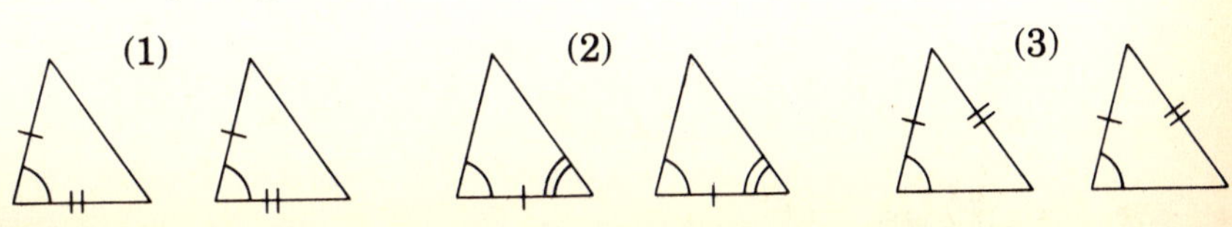

In 7–10, each diagram shows a pair of triangles that can be shown to be congruent. In each case, only two pairs of parts are given as congruent. For each case:

a. Name a third pair of parts that must be congruent because of the placement of the figures.

b. Name the method that makes the triangles congruent.

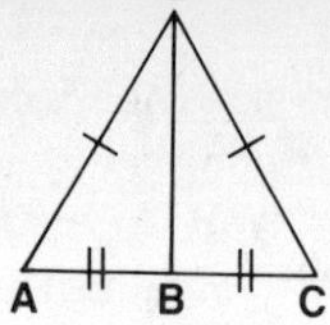

7.

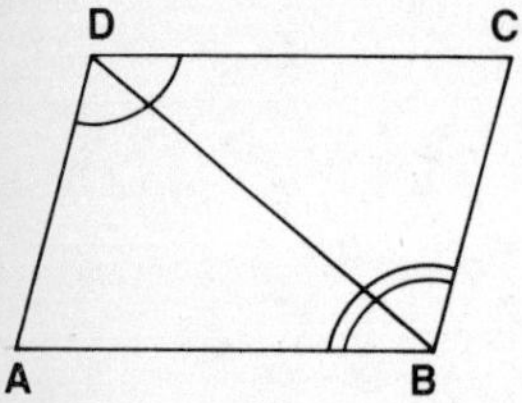

8.

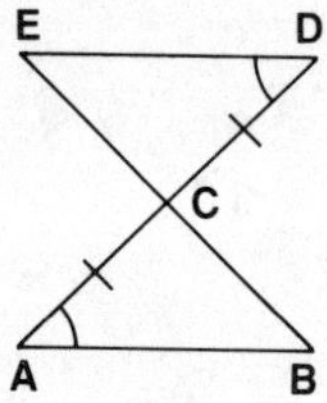

9.

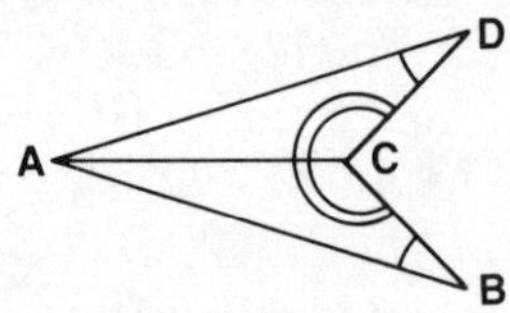

10.

6.5 CONSTRUCTIONS

When we *draw* a figure in geometry, we may use a protractor to measure angles and a ruler to measure line segments. When we *construct* figures, however, we limit our tools. In constructions, the only tools allowed are a pair of *compasses* and a *straightedge,* which is an unmarked ruler.

Some basic constructions are explained here.

To construct a line segment congruent to a given line segment. (To copy a given line segment.)

Given: line segment $\overline{AB}$

Required: To construct a line segment congruent to $\overline{AB}$ (to copy $\overline{AB}$).

Construction:

1. Draw a line ℓ on which $\overline{AB}$ will be copied. Mark a point P on ℓ.
2. Measure $\overline{AB}$ with the compasses: place the point of the compasses on A and adjust the compasses so that the point of the pencil is on B. Hold that measure on the compasses.
3. On line ℓ, place the point of the compasses on P and let the pencil make a mark. Label this mark Q.

Result: $\overline{PQ} \cong \overline{AB}$

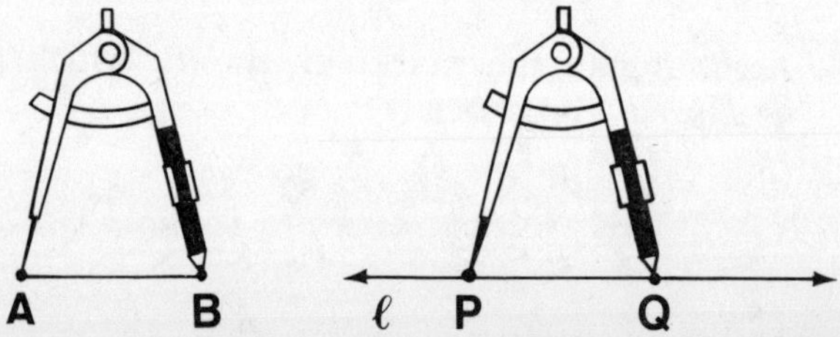

To construct a triangle given three sides.

Given: line segments *a, b,* and *c,* which are the sides of $\triangle ABC$

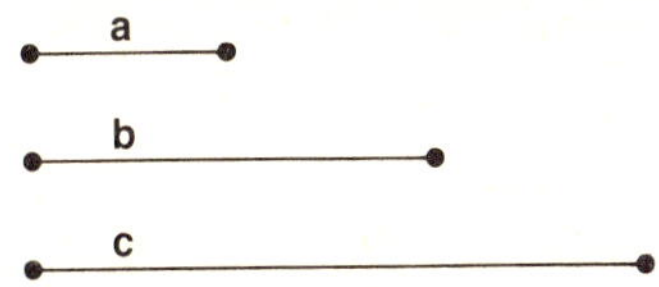

Required: To construct $\triangle ABC$.

Plan of completed triangle:

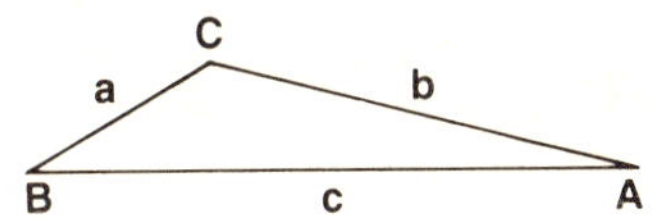

Construction:

1. Draw a line ℓ on which segment c will be copied. Label a point B on ℓ. Measure segment c with the compasses and copy it on line ℓ at B. B is the beginning of copied segment c; label the end of c by A.
2. Measure segment a with the compasses. Copy segment a at point B: put the point of the compasses on B and let the pencil make an arc whose radius is a.
3. Measure segment b with the compasses. Copy segment b at point A: put the point of the compasses on A and let the pencil make an arc whose radius is b intersecting the first arc whose radius is a.
4. Label the point of intersection C. With a straightedge, connect B to C and A to C.

Result: $\triangle ABC$ has segments a, b, and c as sides.

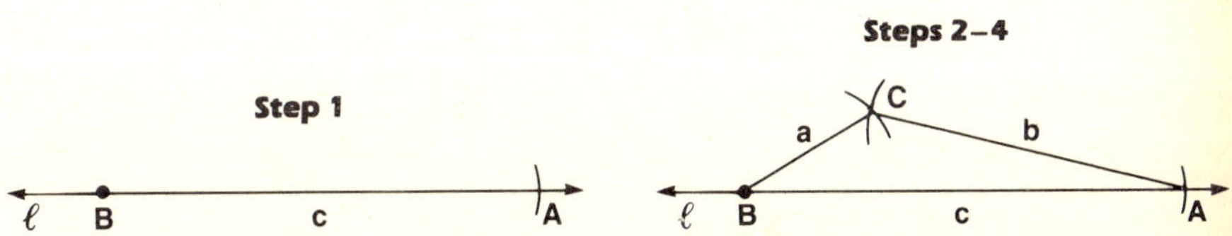

Note 1: This same construction would be used to construct a triangle congruent to a given triangle by copying the three sides of the given triangle. (That is, instead of being given three separate segments to use as sides of the constructed triangle, copy the sides of the given triangle.)

Note 2: This same construction would be used to construct an equilateral triangle given one segment length. (That is, copy the one length three times.)

Note 3: This same construction would be used to construct an isosceles triangle given the length of a leg and the length of the base. (That is, use the length of the leg twice.)

To construct an angle congruent to a given angle. (To copy a given angle.)

Given: angle A

Required: To construct an angle congruent to $\angle A$ (to copy $\angle A$).

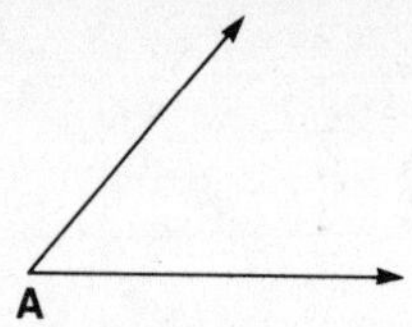

Construction:

1. Draw a line ℓ on which $\angle A$ will be copied. Mark a point P on line ℓ.
2. First construct an arc along which $\angle A$ can be measured: using any convenient radius on the compasses, put the point of the compasses on A and let the pencil make an arc that crosses both sides of $\angle A$. Label the places where the arc crosses the sides of $\angle A$ as B and C.
3. Construct this same arc at P: keeping the same radius on the compasses, put the point of the compasses on P and let the pencil make an arc one part of which crosses line ℓ. Label the point at which the arc crosses ℓ as Q.
4. Measure the arc constructed on $\angle A$: put the point of the compasses on B and adjust the compasses so that the point of the pencil is on C. Hold that measure on the compasses.
5. Copy the measure of this arc at P: go to the arc constructed at P and put the point of the compasses on Q. Let the pencil make a mark that crosses the arc. Label this point R.
6. With a straightedge, draw a ray from P through R.

Result: $\angle RPQ \cong \angle BAC$

Step 2

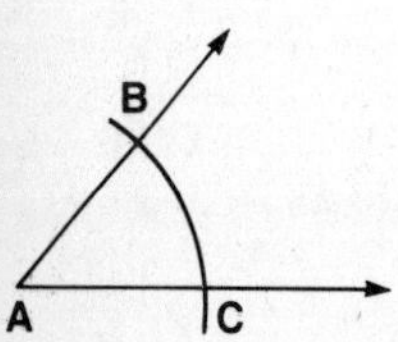

Step 3

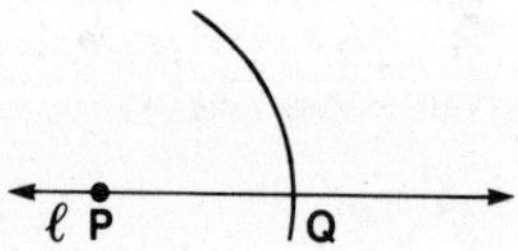

Steps 5 and 6

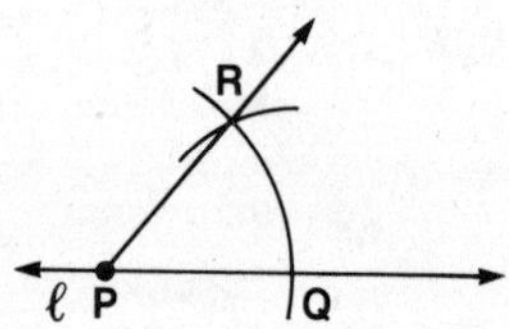

To construct a triangle similar to a given triangle.

Given: triangle ABC

Required: To construct a triangle similar to $\triangle ABC$.

Plan: Copy two angles of $\triangle ABC$.

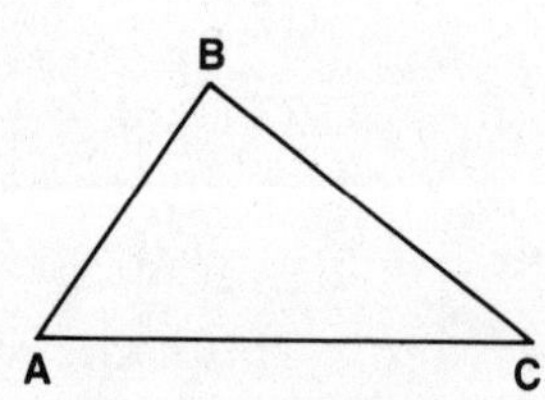

Construction:

1. Draw a line segment of any length to act as the base of the constructed triangle. Label this base $\overline{PQ}$.
2. At P, copy $\angle A$.
3. At Q, copy $\angle C$.

4. The point at which the ray from copied $\angle A$ intersects the ray from copied $\angle C$ is the third vertex of the constructed triangle. Label this point R.

Result: $\triangle PQR \sim \triangle ABC$ since two pairs of corresponding angles are congruent.

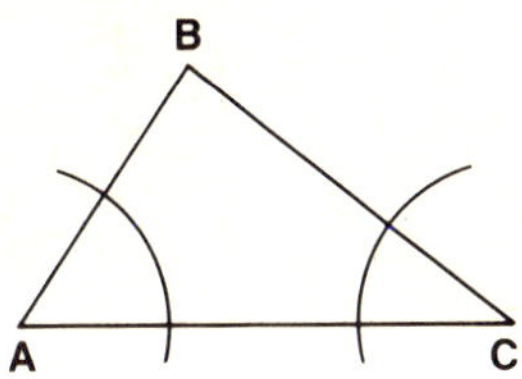

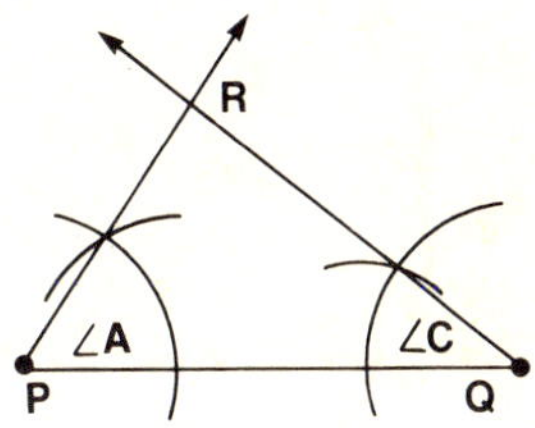

EXERCISES

1. Construct a line segment congruent to $\overline{AB}$.

2. Construct $\triangle RST$ with sides

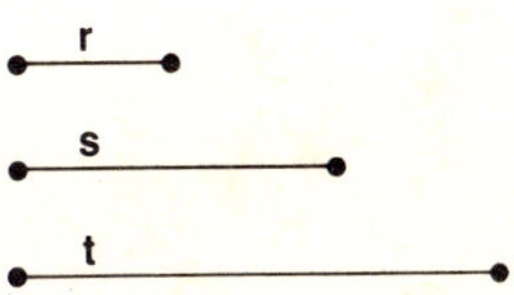

3. Construct an equilateral triangle RST with $\overline{RS}$ as the base.

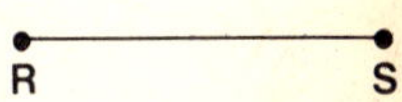

4. Using line segment $\overline{AB}$ as the base, construct an isosceles triangle whose equal sides are congruent to $\overline{CD}$.

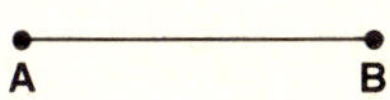

5. Using point B as the vertex, construct an angle congruent to angle A.

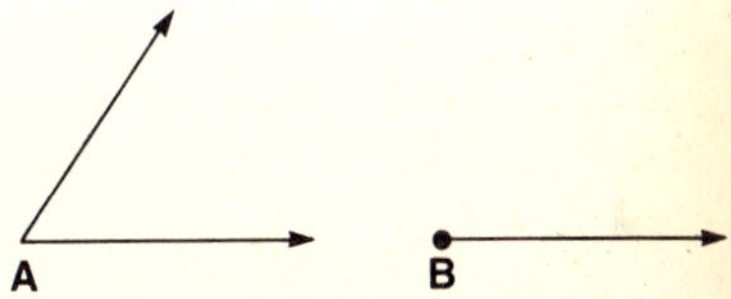

6. Construct a triangle similar to $\triangle ABC$ using $\overline{RS}$ as the base.

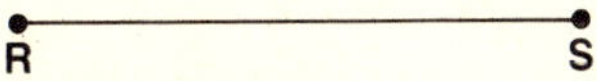

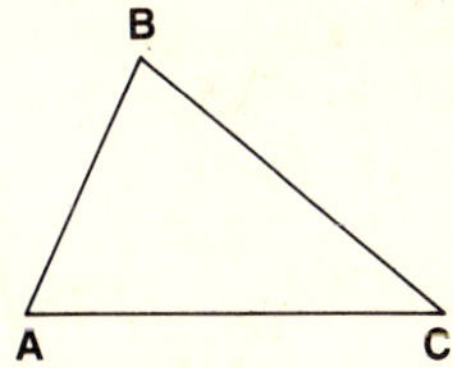

7. Construct a line segment whose measure is twice that of $\overline{AB}$.

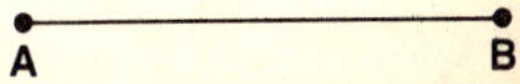

6.6 SUMMARY EXERCISES

1. A bowling team won 7 out of 9 games. The ratio of "wins" to "losses" is
(1) 7:9 (2) 2:9 (3) 2:7 (4) 7:2
2. A coin is tossed 30 times and it comes up "heads" 13 times. The ratio of "heads" to "tails" is
(1) 13:17 (2) 13:30 (3) 17:30 (4) 17:13

In 3–5, solve the given proportion.

3. $\frac{4}{9} = \frac{x}{27}$ 4. $13:26 = 39:x$ 5. $\frac{2}{5} = \frac{x-1}{20}$

6. If a car can travel 80 kilometers on 4 liters of gasoline, how many kilometers can it travel under the same conditions on 7 liters of gasoline?
7. If a disk can spin at a rate of 78 revolutions per minute, how many revolutions can it make under the same conditions in $2\frac{1}{2}$ minutes?
8. If x pencils cost 50 cents, express in terms of x the number of the same kind of pencils that can be bought for \$1.50.
9. On a map, 1 cm represents 100 km. If two cities are 225 km apart, how many cm apart are they on the map?
10. If $\frac{1}{2}$ inch represents 5 feet in a scale drawing, then how many inches will represent 15 feet?
11. How many inches are there in 4 yards?
12. How many gallons are there in 18 quarts?
13. How many centimeters are there in 8 meters?
14. How many grams are there in 5 kilograms?
15. How many minutes are there in h hours?
16. How many days are there in k weeks and 3 days?
17. How many hours are there in d days and h hours?
18. How many meters are there in c centimeters?
19. How many centimeters are there in 4 inches?
20. How many miles are there in 80 kilometers?
21. How many pounds are there in 60 kilograms?
22. Write 42% as a fraction in lowest terms.
23. Write $6\frac{1}{2}\%$ as a decimal.
24. Write .37 as a percent.
25. Write $\frac{3}{5}$ as a percent.
26. What is 40% of 375?
27. What is 10.5% of 70?
28. What is 3.5% of 1,000?
29. If 60% of a number is 180, what is the number?
30. What percent of 300 is 36?
31. If 80% of a number is 140, find the number.
32. What percent of 150 is 25?
33. If 20% of a number is 12, what is 45% of the number?

34. A test was passed by 85% of a class. If 34 students passed, how many students were in the class?

35. In a school, 75% of the freshmen take a typing course. If 225 freshmen are taking typing, how many freshmen are in the school?

36. A magazine has 120 pages. If 30 of the 120 pages are advertising, what percent of the magazine is advertising?

37. If 36 students took an examination and 24 passed, what percent of the students passed the examination?

38. A microcomputer costs $1,200. If the sales tax rate is 8%, what is the amount of money paid for tax?

39. If the sales tax rate is $7\frac{1}{2}$%, what is the amount of sales tax that must be paid on an $80 purchase?

40. Paul bought some items whose cost was $45. If the sales tax rate was 7%, what was the total cost of the purchases?

41. The sale price of a coat was $150 after a 20% discount had been given. Find the original price of the coat.

42. At a sale, a radio sold for $37.50, which represented 75% of the original selling price. What was the original selling price?

43. In the accompanying diagram of triangles ABC and XYZ, $\angle A \cong \angle X$ and $\angle B \cong \angle Y$. If $AB = 6$, $XY = 2$, and $XZ = 5$, find the length of $\overline{AC}$.

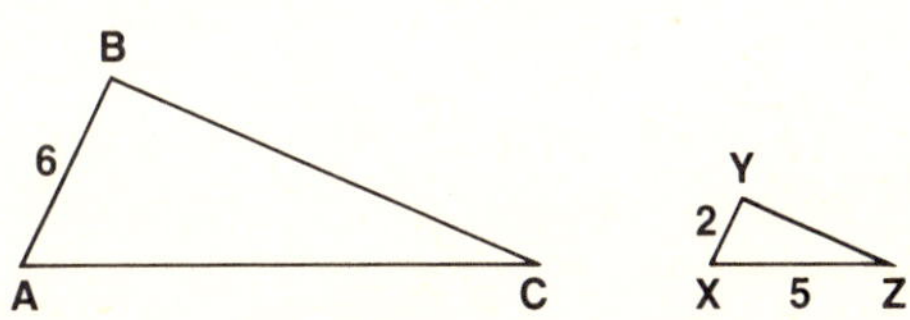

44. The lengths of the sides of a triangle are 9, 12, and 15. If the longest side of a similar triangle is 45, what is the length of its shortest side?

45. Right triangles ABC and PQR are similar. In $\triangle ABC$, the lengths of the legs are 8 and 15. In $\triangle PQR$, the length of the longer of the two legs is 30. What is the length of the shorter leg of $\triangle PQR$?

46. A person 6 feet tall casts a shadow 15 feet long. At the same time, a nearby tower casts a shadow 150 feet long. Find the number of feet in the height of the tower.

47. A building casts a shadow 30 meters long at the same time that a person 2 meters tall casts a shadow of 5 meters. How many meters are there in the height of the building?

48. A tower 40 feet tall casts a 15-foot shadow at the same time that a tree casts a 6-foot shadow. Find the number of feet in the height of the tree.

49. In the accompanying figure, triangle ABC is congruent to triangle MNO, $\overline{AB} \cong \overline{NM}$, $\overline{BC} \cong \overline{NO}$, and $\overline{CA} \cong \overline{MO}$. What angle in triangle MNO is congruent to $\angle CAB$?

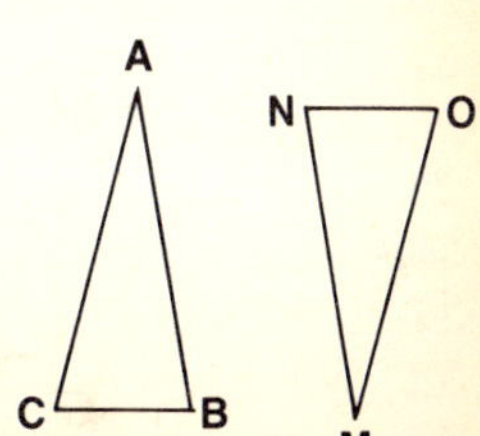

50. Which of the following is *not* a method of proving triangles congruent?

(1) s.a.s. ≅ s.a.s. (3) a.a.a. ≅ a.a.a.
(2) s.s.s. ≅ s.s.s. (4) a.a.s. ≅ a.a.s.

51. According to the markings shown, which pair of triangles must be congruent?

(1)

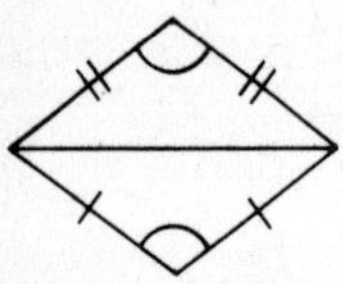

(2)

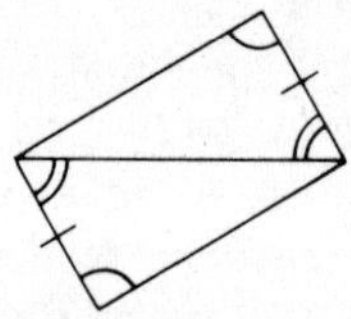

(3)

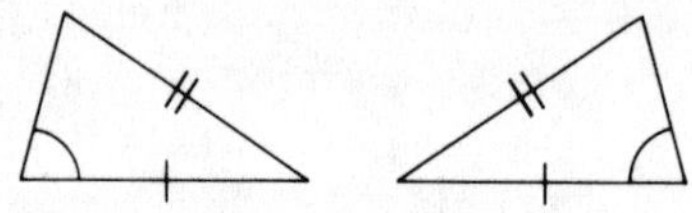

52. Construct a line segment congruent to $\overline{AB}$.

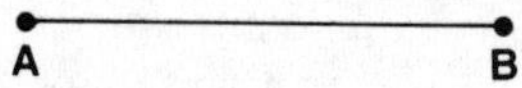

53. Construct an angle congruent to $\angle A$.

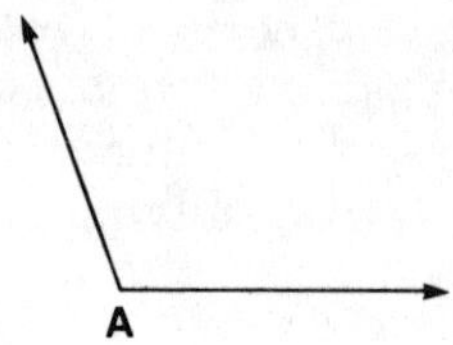

54. Construct an equilateral triangle ABC with $\overline{AB}$ as the base.

55. Construct a triangle similar to triangle ABC, using $\overline{PQ}$ as the base.

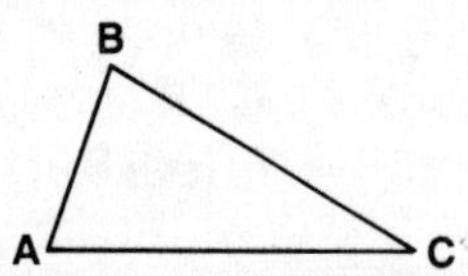

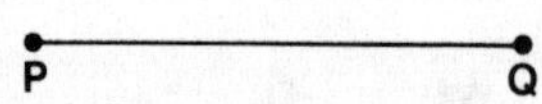

CHAPTER 7

Algebraic Procedures

7.1 FACTORING

Primes and Composites; Greatest Common Factor

When two numbers or two terms are multiplied, the answer is called the ***product.*** Each of the numbers or terms is called a ***factor.*** In our work with multiplication, we have learned to multiply two integers, two monomials, and two binomials. For example:

$$\begin{array}{ccccc} \text{Factor} & \cdot & \text{Factor} & = & \text{Product} \\ 3 & \cdot & 5 & = & 15 \\ 2x & \cdot & 3x & = & 6x^2 \\ (x+2) & \cdot & (x+3) & = & x^2+5x+6 \\ (x+4) & \cdot & (x-4) & = & x^2-16 \end{array}$$

Often we must begin with an algebraic expression and write its factors. If the expression to be factored is an integer, there may be:

1. only one way to factor, as in $15 = 5 \times 3$.
2. more than one way, as in $16 = 8 \times 2$ or $16 = 4 \times 4$.
3. no factors other than the number itself and 1, as in $7 = 7 \times 1$.

Numbers that have no factors other than themselves and 1 are called ***prime numbers.*** Integers that are not prime numbers are called ***composite numbers.*** No matter how the factoring of a composite number is begun, if the process of factoring is continued until prime factors are reached, the result will always be the same. For example:

$$\begin{aligned} 12 &= 6 \times 2 \\ &= 3 \times 2 \times 2 \end{aligned} \qquad \begin{aligned} 12 &= 4 \times 3 \\ &= 2 \times 2 \times 3 \end{aligned}$$

Thus, the combination of prime factors of 12 is $2^2 \times 3$.

$$\begin{aligned} 100 &= 10 \times 10 \\ &= 5 \times 2 \times 5 \times 2 \end{aligned} \qquad \begin{aligned} 100 &= 50 \times 2 \\ &= 25 \times 2 \times 2 \\ &= 5 \times 5 \times 2 \times 2 \end{aligned}$$

$$\begin{aligned} 100 &= 25 \times 4 \\ &= 5 \times 5 \times 2 \times 2 \end{aligned}$$

Thus, the combination of prime factors of 100 is $2^2 \times 5^2$.

Sometimes we want to know if two or more integers have a ***common factor.*** For example, 6, 12, 24, and 36 can all be divided by 2. Thus, 2 is a common factor. Also, they can all be divided by 3. Thus, 3 is a common factor. Also, they can all be divided by 6. Thus, 6 is a common factor. Note that 6 is the ***greatest common factor*** of 6, 12, 24, and 36.

Similarly, we may want to know the greatest common factor of two or more monomials.

Examples

1. What is the greatest common factor of $4x^2$, $8x^2$, $16x^3$, and $24x^4$?

Solution:

Work with the numerical coefficient first: both 2 and 4 are common factors of 4, 8, 16, and 24. The greatest common factor is 4.

Work with the letters and their exponents next: both x and x^2 are common factors of x^2, x^2, x^3, and x^4. The greatest common factor is x^2.

Thus, the greatest common factor of $4x^2$, $8x^2$, $16x^3$, and $24x^4$ is $4x^2$.

Answer: $4x^2$

2. Factor: $3x + 6x^2$

Solution:

The greatest common factor of 3 and 6 is 3. The greatest common factor of x and x^2 is x. Thus, $3x$ is the greatest common factor of $3x$ and $6x^2$.

Divide each term by the greatest common factor. That is, $\frac{3x}{3x} = 1$ and $\frac{6x^2}{3x} = 2x$.

Write the original expression as a product of the greatest common factor ($3x$) and the result of dividing each term by the greatest common factor ($1 + 2x$). That is, $3x + 6x^2 = 3x\,(1 + 2x)$. (*Note:* If the factors are multiplied, the product is the original expression.)

Answer: $3x + 6x^2 = 3x(1 + 2x)$

3. Factor: $5x - 10z$

Solution:

The greatest common factor of 5 and 10 is 5. There is no common factor of x and z. Thus, the greatest common factor of $5x$ and $10z$ is 5.

$\frac{5x}{5} = x$ and $\frac{10z}{5} = 2z$.

Answer: $5x - 10z = 5(x - 2z)$

EXERCISES

1. The prime factors of 30 are
(1) 1, 2, 15 (2) 2, 3, 5 (3) 3, 10 (4) 6, 5
2. The greatest common factor of the numbers 16, 20, and 40 is
(1) 320 (2) 2 (3) 40 (4) 4
3. The greatest common factor of the numbers 12, 30, and 60 is
(1) 360 (2) 60 (3) 3 (4) 6
4. What is the greatest common factor of $18x$ and $12x^2$?
5. What is the greatest common factor of $5x^3 + 10x$?
(1) 5 (2) x (3) $5x$ (4) $10x^3$

In 6–15, factor each expression by finding the greatest common factor.

6. $3x + 6y$
7. $6a - 9$
8. $x^2 - 7x$
9. $2x^2 + x$
10. $x^2 + 10x$
11. $4x^2 + 2x$
12. $x^3 - 5x$
13. $p + prt$
14. $\frac{1}{2}ha + \frac{1}{2}hb$
15. $\frac{1}{2}m\overset{\frown}{AB} + \frac{1}{2}m\overset{\frown}{CD}$

Factoring the Difference of Two Perfect Squares

In our work, we have seen that the product of two binomials can be a binomial or a trinomial, depending upon the form of the original binomials. (*Note:* The product of two binomials can also be more than a trinomial. This case is left for future study.)

Let us closely examine the case when the product of two binomials is a binomial. Consider $(x + 4)(x - 4) = x^2 - 16$. Recall that the product, $x^2 - 16$, is obtained either by a vertical multiplication or by the FOIL method:

$$\begin{array}{rrr} & x & +\ 4 \\ & x & -\ 4 \\ \hline x^2 & +\ 4x & \\ & -\ 4x & -\ 16 \\ \hline x^2 & & -\ 16 \end{array}$$

$$(x + \overset{+4x}{4)(x} - 4) = x^2 - 16 \qquad \text{outer: } -4x$$

Using either method of multiplication, we see that the middle term of the product is zero. The first and last terms of the product, x^2 and 16, are ***perfect squares*** (numbers whose two factors are identical, $x^2 = x \cdot x$ and $16 = 4 \cdot 4$). Also, the perfect squares are separated by a minus sign (indicating a subtraction, or *difference*). This type of product, ***the difference of two perfect squares,*** can occur only when the original binomials are identical in number but opposite in second sign; that is, $(x + 4)(x - 4)$ or $(x - 3)(x + 3)$ or $(a + b)(a - b)$.

◆ Thus, if we are asked to factor a difference of two perfect squares, $a^2 - b^2$, the factors must be $(a + b)(a - b)$, since $(a + b)(a - b) = a^2 - b^2$.

Examples

1. Factor: $x^2 - 49$

Solution:

$x^2 - 49$ is the difference of two perfect squares, since $x^2 = x \cdot x$ and $49 = 7 \cdot 7$. The factors must be identical in number and opposite in second sign. The factors are, therefore, $(x + 7)(x - 7)$.

Answer: $x^2 - 49 = (x + 7)(x - 7)$

2. Factor: $100 - x^2$

Solution:

$100 - x^2$ is the difference of two perfect squares. The factors are $(10 + x)(10 - x)$. (Note that the order of placement is important since subtraction is not commutative. That is, $(x + 10)(x - 10)$ are *not* the factors of $100 - x^2$.)

Answer: $100 - x^2 = (10 + x)(10 - x)$

3. Factor: $25z^2 - .09$

Solution:

This is, again, a difference of two perfect squares, since $25z^2 = 5z \cdot 5z$ and $.09 = .3 \times .3$.

Answer: $25z^2 - .09 = (5z + .3)(5z - .3)$

4. Factor: $36x^2 + y^2$

Solution:

This is not a *difference* of two perfect squares, since there is a + sign separating the two perfect squares. At this time, we can factor only a difference of two perfect squares, and we must consider a sum as not factorable.

Answer: $36x^2 + y^2$ is not factorable.

EXERCISES

In 1–10, factor the difference of two perfect squares where possible.

1. $x^2 - 49$
2. $x^2 - y^2$
3. $36 - x^2$
4. $4x^2 - 25$
5. $x^2 - .81$
6. $x^2 - \frac{9}{25}$
7. $9x^2 - .09$
8. $x^2 + y^2$
9. $x^2 - .1$
10. $a^2x^2 - y^4$

Factoring a Trinomial

The product of two binomials may be a trinomial, as in $(x + 2)(x + 3) = x^2 + 5x + 6$. Thus, if we are asked to factor a trinomial, we can expect two binomial factors. Examining the manner in which the product of two binomials is obtained can help us to understand how to factor a trinomial into two binomials.

Multiplication:

$$\begin{array}{r} x + 2 \\ x + 3 \\ \hline x^2 + 2x \phantom{{}+6} \\ + 3x + 6 \\ \hline x^2 + 5x + 6 \end{array} \qquad \overset{+2x}{(x + 2)(x + 3)} = x^2 + 5x + 6 \quad (\text{outer: } +3x)$$

Factoring:

$$x^2 + 5x + 6 = (x + 2)(x + 3)$$

The first term, x^2, is factored into $x \cdot x$. The last term, 6, is factored into $2 \cdot 3$. The middle term, $5x$ or $2x + 3x$, is a result of the way the first and last terms were factored. If the last term had been factored as $6 \cdot 1$, the resulting factors, $(x + 6)(x + 1)$, would have produced a middle term of $6x + x$ or $7x$, and $(x + 6)(x + 1)$ would not have been the factors of $x^2 + 5x + 6$.

◆ Thus, to factor a trinomial:

1. We expect the answer to be a product of two binomials.
2. The first term of each binomial is a result of factoring the first term of the trinomial.
3. The second term of each binomial, including signs, is a result of factoring the last term of the trinomial.
4. The only acceptable factors of the first and last terms of the trinomial (these are the terms of the binomials) are those that produce the exact middle term of the trinomial when the binomials are multiplied.

Example

Factor: $x^2 + 7x + 10$

Solution:

Factor the first term of the trinomial: $x^2 = x \cdot x$. Place these factors as the first terms of the expected binomials: $(x \quad)(x \quad)$.

Factor the last term of the trinomial, including signs. Since $+10$ is positive, its factors are either both positive or both negative. Take into account all possible ways of factoring 10, including prime factors. That is, $10 = 10 \cdot 1$ or $5 \cdot 2$. Thus:

$$+10 = (+5)(+2) = (+2)(+5) \qquad +10 = (+10)(+1) = (+1)(+10)$$
$$+10 = (-5)(-2) = (-2)(-5) \qquad +10 = (-10)(-1) = (-1)(-10)$$

[*Note:* Because of the commutative property of multiplication, $(+5)(+2)$ is equivalent to $(+2)(+5)$, $(-5)(-2)$ is equivalent to $(-2)(-5)$, $(+10)(+1)$ is equivalent to $(+1)(+10)$, and $(-10)(-1)$ is equivalent to $(-1)(-10)$.]

Try each of these eight sets of factors in the expected binomials to see which ones produce the exact middle term of the trinomial:

$$\overbrace{(x + 5)(x + 2)}^{+5x} \quad \text{outer: } +2x$$

$+5x + 2x = +7x$
This is the exact middle term of the trinomial.

$$\overbrace{(x + 2)(x + 5)}^{+2x} \quad \text{outer: } +5x$$

$+2x + 5x = +7x$
This is the exact middle term of the trinomial.

$$\overbrace{(x - 5)(x - 2)}^{-5x} \quad \text{outer: } -2x$$

$-5x - 2x = -7x$
This is not the exact middle term of the trinomial.

$$\overbrace{(x - 2)(x - 5)}^{-2x} \quad \text{outer: } -5x$$

$-2x - 5x = -7x$
This is not the exact middle term of the trinomial.

$$\overbrace{(x + 10)(x + 1)}^{+10x} \quad \text{outer: } +1x$$

$+10x + 1x = +11x$
This is not the exact middle term of the trinomial.

$$\overbrace{(x + 1)(x + 10)}^{+1x} \quad \text{outer: } +10x$$

$+1x + 10x = +11x$
This is not the exact middle term of the trinomial.

$$\overbrace{(x - 10)(x - 1)}^{-10x} \quad \text{outer: } -1x$$

$-10x - 1x = -11x$
This is not the exact middle term of the trinomial.

$$\overbrace{(x - 1)(x - 10)}^{-1x} \quad \text{outer: } -10x$$

$-1x - 10x = -11x$
This is not the exact middle term of the trinomial.

Thus, $(x + 5)(x + 2)$ or its equivalent $(x + 2)(x + 5)$ is the factored form of $x^2 + 7x + 10$.

Answer: $x^2 + 7x + 10 = (x + 5)(x + 2)$ or $(x + 2)(x + 5)$

Considering all the possible sets of factors that are equivalent because of the commutative property leads to much extra work. We can cut back on our work by not actually writing or testing all those sets. In the example just worked out, it would be sufficient to test $(+5)(+2)$, $(-5)(-2)$, $(+10)(+1)$, and $(-10)(-1)$. Thus, we have cut back the work from testing eight sets to testing only four sets. We must remember, though, that an answer of factors always has an equivalent form because of the commutative property, and this equivalent form is considered to be the same answer. In the example, $(x + 5)(x + 2)$ and $(x + 2)(x + 5)$ are considered the same answer.

Let us look at some examples to see how the procedure for factoring trinomials works. Study each example with its explanation to see how different signs, primes, and composites each may affect what you must consider.

Examples

1. Factor: $x^2 - 4x + 3$

Solution:

Factor the first term of the trinomial, and place the resulting factors as the first terms of the expected binomials: $x^2 = x \cdot x$ or $(x \quad)(x \quad)$.

Factor the last term of the trinomial, including signs. Take into account all possible numerical factors including primes: since $+3$ is positive, its factors are either both positive or both negative. Since 3 is a prime, its only factors are $3 \cdot 1$. Thus, the possibilities to be tested are:

$$+3 = (+3)(+1)$$
$$+3 = (-3)(-1)$$

Test the possibilities as the second terms of the expected binomials to see which set of factors results in the exact middle term of the trinomial:

$$(x + 3)(x + 1) \qquad\qquad (x - 3)(x - 1)$$

inner: $+3x$, outer: $+1x$ $\qquad$ inner: $-3x$, outer: $-1x$

$$+3x + 1x = +4x \qquad\qquad -3x - 1x = -4x$$

no $\qquad\qquad$ yes

Answer: $x^2 - 4x + 3 = (x - 3)(x - 1)$

2. Factor: $x^2 - 11x + 18$

Solution:

Factor the first term of the trinomial, and place the resulting factors as the first terms of the expected binomials: $x^2 = x \cdot x$ or $(x \quad)(x \quad)$.

Factor the last term of the trinomial including signs. Take into account all possible numerical factors, including primes: since +18 is positive, its factors are either both positive or both negative. The factors of 18 are $1 \cdot 18$, $2 \cdot 9$, and $3 \cdot 6$. Thus, the possibilities to be tested are:

$+18 = (+1)(+18)$ $+18 = (+2)(+9)$ $+18 = (+3)(+6)$
$+18 = (-1)(-18)$ $+18 = (-2)(-9)$ $+18 = (-3)(-6)$

Test the possibilities as the second terms of the binomials to see which set of factors results in the exact middle term of the trinomial:

$+1x$
$(x + 1)(x + 18)$
$+18x$
$+1x + 18x = +19x$
no

$-1x$
$(x - 1)(x - 18)$
$-18x$
$-1x - 18x = -19x$
no

$+2x$
$(x + 2)(x + 9)$
$+9x$
$+2x + 9x = +11x$
no

$-2x$
$(x - 2)(x - 9)$
$-9x$
$-2x - 9x = -11x$
yes

$+3x$
$(x + 3)(x + 6)$
$+6x$
$+3x + 6x = +9x$
no

$-3x$
$(x - 3)(x - 6)$
$-6x$
$-3x - 6x = -9x$
no

Answer: $x^2 - 11x + 18 = (x - 2)(x - 9)$

Note that you must think of *all* the possible factors to be tested. In more complicated examples in later studies, there may be many factors, and they can be difficult. The need for you actually to write all the possibilities depends upon how much you can "see." When testing the possibilities, you may stop as soon as a pair "works," since there is only one possible answer.

For example, in regard to Example 2, look over all the listed factors, $1 \cdot 18$, $2 \cdot 9$, and $3 \cdot 6$, to see which pair is the most likely. Note that when the factors are added their sum must be the numerical coefficient of the middle term of $x^2 - 11x + 8$, which is -11. Since $1 + 18 = 19$ and $3 + 6 = 9$, we can eliminate these factors at once and not bother testing them. This leaves only $+2 \cdot +9$ and $-2 \cdot -9$. Since the sum of any two positive numbers is a positive number, we can eliminate $+2 \cdot +9$. This leaves only $-2 \cdot -9$, whose sum is -11.

In the following examples, all the possibilities are written and all are tested.

3. Factor: $x^2 + 10x - 11$

Solution:

Factor the first term of the trinomial, and place the resulting factors as the first terms of the expected binomials: $x^2 = x \cdot x$ or $(x \quad)(x \quad)$.

Factor the last term of the trinomial, including signs. Take into account all possible numerical factors, including primes: since -11 is negative, it has one positive factor and one negative factor. Since 11 is a prime, its only factors are $11 \cdot 1$. Thus, the possibilities to be tested are:

$$-11 = (+11)(-1)$$
$$-11 = (-11)(+1)$$

Test the possibilities as the second terms of the expected binomials to see which set of factors results in the exact middle term of the trinomial:

$$\begin{array}{cc} +11x & -11x \\ (x + 11)(x - 1) & (x - 11)(x + 1) \\ -1x & +1x \\ +11x - 1x = +10x & -11x + 1x = -10x \\ \text{yes} & \text{no} \end{array}$$

Answer: $x^2 + 10x - 11 = (x + 11)(x - 1)$

4. Factor: $x^2 - 2x - 15$

Solution:

Factor the first term of the trinomial, and place the resulting factors as the first terms of the expected binomials: $x^2 = x \cdot x$ or $(x \quad)(x \quad)$.

Factor the last term of the trinomial, including signs. Take into account all possible numerical factors, including primes: since -15 is negative, it has one positive factor and one negative factor. The factors of 15 are $1 \cdot 15$ and $3 \cdot 5$. Thus, the possibilities to be tested are:

$$-15 = (+1)(-15) \qquad -15 = (+3)(-5)$$
$$-15 = (-1)(+15) \qquad -15 = (-3)(+5)$$

Test the possibilities as the second terms of the expected binomials to see which set of factors results in the exact middle term of the trinomial:

$$(x + 1)(x - 15): \quad +1x - 15x = -14x \quad \text{no}$$

$$(x - 1)(x + 15): \quad -1x + 15x = +14x \quad \text{no}$$

$$(x + 3)(x - 5): \quad +3x - 5x = -2x \quad \text{yes}$$

$$(x - 3)(x + 5): \quad -3x + 5x = +2x \quad \text{no}$$

Answer: $x^2 - 2x - 15 = (x + 3)(x - 5)$

5. Factor: $2x^2 - 5x - 3$

Solution:

Factor the first term of the trinomial, and place the resulting factors as the first terms of the expected binomials: $2x^2 = 2x \cdot x$ or $(2x \quad)(x \quad)$.

Factor the last term of the trinomial, including signs. Take into account all possible numerical factors, including primes: since -3 is negative, it has one positive factor and one negative factor. Since 3 is a prime, its only factors are $3 \cdot 1$. Thus, the possibilities to be tested are:

$$-3 = (+3)(-1)$$
$$-3 = (-3)(+1)$$

Test the possibilities as the second terms of the expected binomials to see which set of factors results in the exact middle term of the trinomial:

$$(2x + 3)(x - 1): \quad +3x - 2x = +x \quad \text{no}$$

$$(2x - 3)(x + 1): \quad -3x + 2x = -x \quad \text{no}$$

Notice that we have not come up with the answer because we have not yet taken into account all the possibilities for the binomial factors now that the first term of the trinomial has a coefficient other than 1.

The other sets of possible binomial factors are:

$$\underbrace{(2x + \overset{+1x}{\overline{1)(x}} - 3)}_{-6x} \qquad \underbrace{(2x - \overset{-1x}{\overline{1)(x}} + 3)}_{+6x}$$

$+1x - 6x = -5x$ yes $\qquad -1x + 6x = +5x$ no

Answer: $2x^2 - 5x - 3 = (2x + 1)(x - 3)$

(*Note:* Detailed work involving trinomials where the coefficient of x^2 is other than 1 is left for future study. You will be able to factor simple cases if you follow the general procedure for factoring trinomials and remember to include all possible binomial factors for testing.)

EXERCISES

In 1–15, factor each trinomial.

1. $x^2 + 8x + 15$
2. $x^2 + 13x + 12$
3. $x^2 + 2x - 3$
4. $x^2 - 11x + 24$
5. $a^2 - 5a - 24$
6. $a^2 - 4a - 21$
7. $a^2 - 4a + 4$
8. $x^2 + 27x + 50$
9. $x^2 + 51x + 50$
10. $x^2 - 37x + 36$
11. $2x^2 + 7x + 3$
12. $3x^2 + 4x + 1$
13. $2x^2 + 11x + 5$
14. $5x^2 + 14x - 3$
15. $4x^2 - 9x + 2$

Combining Types of Factoring

Combinations of types of factoring can be applied in some problems. In these cases, first look for a greatest common factor. After the greatest common factor has been removed, the remaining factor may still be factorable, either as a difference of two perfect squares or as a trinomial.

Examples

1. Factor completely: $2x^2 - 50$

Solution:

$2x^2 - 50 = 2(x^2 - 25)$	The greatest common factor is 2.
$= 2(x + 5)(x - 5)$	$x^2 - 25$ is a difference of two perfect squares.

Answer: $2x^2 - 50 = 2(x + 5)(x - 5)$

2. Factor completely: $ax^2 + 7ax + 12a$

Solution:

$ax^2 + 7ax + 12a = a(x^2 + 7x + 12)$ — The greatest common factor is a.

$= a(x + 4)(x + 3)$ — The trinomial $x^2 + 7x + 12$ can be factored.

Answer: $ax^2 + 7ax + 12a = a(x + 4)(x + 3)$

3. Factor completely: $3x^4 - 243$

Solution:

$3x^4 - 243 = 3(x^4 - 81)$ — The greatest common factor is 3.

$= 3(x^2 + 9)(x^2 - 9)$ — $x^4 - 81$ is a difference of two perfect squares.

$= 3(x^2 + 9)(x + 3)(x - 3)$ — $x^2 - 9$ is a difference of two perfect squares.

Answer: $3x^4 - 243 = 3(x^2 + 9)(x + 3)(x - 3)$

EXERCISES

In 1–20, factor completely.

1. $3x^2 - 48$
2. $18x^2 - 8$
3. $ax^2 - 49a$
4. $\pi x^2 - \pi y^2$
5. $12x^2 - 27y^2$
6. $x^3 - x$
7. $4x^3 - 49x$
8. $x^4 - 9x^2$
9. $5x^4 - 80$
10. $x^6 - x^2$
11. $ax^2 - 2ax - 3a$
12. $4x^2 - 4x - 48$
13. $3x^2 + 6x + 3$
14. $4x^2 + 16x - 48$
15. $2ax^2 - 2ax - 12a$
16. $x^3 + 8x^2 + 12x$
17. $8x^2 - 2x - 6$
18. $4x^2 - 6x - 4$
19. $x^4 - 10x^2 + 9$
20. $x^4 + x^2 - 2$

7.2 OPERATIONS WITH ALGEBRAIC FRACTIONS

We are already familiar with rational numbers, which include the fractions of arithmetic. In Chapter 2, the operations with fractions were reviewed. Now we will work with algebraic fractions, which are fractions that contain bases and exponents as well as numerical coefficients. We will see how the operations with fractions can be extended to include algebraic fractions.

Reducing Fractions

Fractions can be reduced to *simplest form,* or *lowest terms,* by factoring and cancelling common factors. Try to eliminate the greatest common factor at the beginning. If the greatest common factor is not eliminated at the beginning, the process of factoring will have to be carried out more than once.

Examples

Reduce each fraction to simplest form:

1. $\frac{18}{72}$

Solution 1:

$$\frac{18}{72} = \frac{\cancel{18} \cdot 1}{\cancel{18} \cdot 4} = \frac{1}{4} \quad \textit{Ans.}$$

Solution 2:

$$\frac{18}{72} = \frac{\cancel{9} \cdot 2}{\cancel{9} \cdot 8} = \frac{\cancel{2} \cdot 1}{\cancel{2} \cdot 4} = \frac{1}{4} \quad \textit{Ans.}$$

2. $\frac{12x^7}{4x^2}$

Solution 1:

$$\frac{12x^7}{4x^2} = \frac{\cancel{4} \cdot 3 \cdot \cancel{x^2} \cdot x^5}{\cancel{4} \cdot 1 \cdot \cancel{x^2} \cdot 1}$$
$$= \frac{3x^5}{1} = 3x^5 \quad \textit{Ans.}$$

Solution 2:

$$\frac{12x^7}{4x^2} = (12x^7) \div (4x^2)$$
$$= (12 \div 4) \cdot (x^7 \div x^2)$$
$$= 3 \cdot x^{7-2}$$
$$= 3x^5 \quad \textit{Ans.}$$

(*Note:* In Solution 2, the fraction $\frac{12x^7}{4x^2}$ is thought of as the quotient of two monomials.)

3. $\frac{5x^2 - 10x}{x - 2}$

Solution:

$$\frac{5x^2 - 10x}{x - 2} = \frac{5x\cancel{(x - 2)}}{\cancel{(x - 2)}} = 5x \quad \textit{Ans.}$$

4. $\frac{x^2 + 2x - 3}{x^2 - 1}$

Solution:

$$\frac{x^2 + 2x - 3}{x^2 - 1} = \frac{(x + 3)(x - 1)}{(x + 1)(x - 1)}$$

Factor the trinomial.

Factor the difference of two perfect squares.

$$= \frac{(x + 3)\cancel{(x - 1)}}{(x + 1)\cancel{(x - 1)}}$$

Cancel common factors.

$$= \frac{x + 3}{x + 1} \quad \textit{Ans.}$$

(*Note:* The x's here cannot be cancelled because they are not factors: they are not attached to the other numbers by multiplication.)

EXERCISES

1. The expression $\frac{15x^6}{3x^3}$ is equivalent to
   (1) $5x^2$ (2) $5x^3$ (3) $12x^2$ (4) $12x^3$
2. The expression $\frac{8x^8}{4x^4}$ is equivalent to
   (1) $2x^2$ (2) $4x^2$ (3) $2x^4$ (4) $4x^4$
3. The expression $\frac{-12x^6y^2}{3x^2y}$ is equivalent to
   (1) $9x^3y^2$ (2) $4x^4y$ (3) $-4x^3y$ (4) $-4x^4y$
4. Express in lowest terms: $\frac{8x^4y^3}{2xy^3}$
5. Express $\frac{x^2 + 3x}{2(x + 3)}$ as a fraction in lowest terms.

In 6–8, express in lowest terms.

6. $\frac{2x - 10}{x^2 - 25}$ 7. $\frac{x^2 - 4}{x + 2}$ 8. $\frac{x + 3}{x^2 - 2x - 15}$

9. The expression $\frac{x^2 + 2x - 3}{x^2 + 3x}$ is equivalent to
   (1) $\frac{x - 1}{x}$ (2) $\frac{2x - 3}{3x}$ (3) $-\frac{1}{3}$ (4) $\frac{3}{x}$
10. Express the fraction $\frac{n^2 - n - 6}{4n - 12}$ in its lowest terms.

Combining Fractions by Addition or Subtraction

CASE 1:

If the denominators are alike, keep the denominator and add or subtract the numerators. After combining the given fractions, reduce the single fraction as needed.

Examples

In each example, write the sum or difference of the fractions as a single fraction in simplest form:

1. $\frac{3}{7} + \frac{2}{7} = \frac{3 + 2}{7} = \frac{5}{7}$ *Ans.* 2. $\frac{3x}{7} + \frac{2x}{7} = \frac{3x + 2x}{7} = \frac{5x}{7}$ *Ans.*

3. $\frac{3}{7x} + \frac{2}{7x} = \frac{3 + 2}{7x} = \frac{5}{7x}$ *Ans.*

4. $\frac{7}{2x^2} - \frac{5}{2x^2} = \frac{7-5}{2x^2} = \frac{2}{2x^2} = \frac{\not{2} \cdot 1}{\not{2} \cdot 1 \cdot x^2} = \frac{1}{x^2}$ *Ans.*

5. $\frac{7x}{y} - \frac{5x}{y} = \frac{7x - 5x}{y} = \frac{2x}{y}$ *Ans.*

6. $\frac{8}{x+y} - \frac{2}{x+y} = \frac{8-2}{x+y} = \frac{6}{x+y}$ *Ans.*

7. $\frac{9}{x^2-4} + \frac{3}{x^2-4} = \frac{9+3}{x^2-4} = \frac{12}{x^2-4}$ *Ans.*
(*Note:* No reduction can be done in Example 7 because the 4 in the denominator is not a factor.)

8. $\frac{3x}{x+y} + \frac{2x}{x+y} = \frac{3x+2x}{x+y} = \frac{5x}{x+y}$ *Ans.*
(*Note:* No reduction can be done in Example 8 because the x in the denominator is not a factor.)

CASE 2:

If the denominators are different, rewrite each fraction so that the denominators are alike. Use the lowest common denominator, or L.C.D. To determine the L.C.D., use the principle of prime factors. First, we will practice finding L.C.D.'s.

Examples

1. Find the L.C.D. of 12 and 16.

Solution:

Factor 12 and 16 into primes:

$$12 = 3 \cdot 2 \cdot 2 \text{ or } 3 \cdot 2^2$$
$$16 = 2 \cdot 2 \cdot 2 \cdot 2 \text{ or } 2^4$$

The L.C.D. is the product of the highest powers of the different prime factors. There are two different primes, 3 and 2. Write the product of their highest powers: $3 \cdot 2^4$.

Thus, the L.C.D. of 12 and 16 is $3 \cdot 2^4$ or 48. *Ans.*

2. Find the L.C.D. of x^2 and x^5.

Solution:

There is only one kind of prime involved here, x. Take the highest power, x^5. The L.C.D. of x^2 and x^5 is x^5. *Ans.*

To rewrite a fraction, we must multiply both its numerator and denominator by the same factor. To determine the factor by which to multiply the numerator and denominator, divide the original denominator into the new denominator you wish to use.

Examples

1. Write $\frac{3}{7}$ with a denominator of 14.

Solution:

Divide 7 into 14 to determine the factor by which you will multiply the numerator and denominator. Since $14 \div 7 = 2$, multiply the numerator and denominator by 2.

$$\frac{3}{7} = \frac{3 \cdot 2}{7 \cdot 2} = \frac{6}{14} \quad Ans.$$

2. Write $\frac{3}{5x}$ with a denominator of $10x^2$.

Solution:

Since $10x^2 \div 5x = 2x$, $\frac{3}{5x} \cdot \frac{2x}{2x} = \frac{6x}{10x^2}$ *Ans.*

Now that we have reviewed finding an L.C.D. and rewriting a fraction, let us look at examples of combining fractions whose denominators are different.

Examples

1. Combine into a single fraction: $\frac{3}{14} + \frac{8}{21}$

Solution:

Determine the L.C.D. of 14 and 21.

$$14 = 7 \cdot 2$$
$$21 = 7 \cdot 3$$
$$\text{L.C.D.} = 2 \cdot 3 \cdot 7 \text{ or } 42$$

Rewrite each fraction with a denominator of 42.

$$42 \div 14 = 3 \qquad 42 \div 21 = 2$$
$$\frac{3}{14} \cdot \frac{3}{3} = \frac{9}{42} \qquad \frac{8}{21} \cdot \frac{2}{2} = \frac{16}{42}$$

Thus, the solution may be written:

$$\frac{3}{14} + \frac{8}{21}$$
$$= \frac{3}{14} \cdot \frac{3}{3} + \frac{8}{21} \cdot \frac{2}{2} = \frac{9}{42} + \frac{16}{42} = \frac{9 + 16}{42} = \frac{25}{42} \quad Ans.$$

2. Combine into a single fraction: $\frac{10}{x} - \frac{3}{x^2}$

Solution:

The L.C.D. of x and x^2 is x^2. $x^2 \div x = x$ and $x^2 \div x^2 = 1$.

Thus:

$$\frac{10}{x} - \frac{3}{x^2}$$
$$= \frac{10}{x} \cdot \frac{x}{x} - \frac{3}{x^2} \cdot \frac{1}{1}$$
$$= \frac{10x}{x^2} - \frac{3}{x^2} = \frac{10x - 3}{x^2} \quad \textit{Ans.}$$

(*Note:* Since $10x$ and -3 are unlike terms, $10x - 3$ cannot be combined. Since x is not a common factor in the numerator, no x can be cancelled.)

3. Express as a single fraction in lowest terms: $\frac{x + 5}{3} + \frac{x - 2}{2}$

Solution:

The L.C.D. of 3 and 2 is 6.

$$\frac{x + 5}{3} + \frac{x - 2}{2}$$
$$= \frac{x + 5}{3} \cdot \frac{2}{2} + \frac{x - 2}{2} \cdot \frac{3}{3}$$
$$= \frac{2(x + 5)}{6} + \frac{3(x - 2)}{6}$$
$$= \frac{2(x + 5) + 3(x - 2)}{6}$$
$$= \frac{2x + 10 + 3x - 6}{6} = \frac{5x + 4}{6} \quad \textit{Ans.}$$

4. Express as a single fraction in lowest terms: $\frac{x + 7}{3} - \frac{2x - 3}{5}$

Solution:

The L.C.D. of 3 and 5 is 15.

$$\frac{x + 7}{3} - \frac{2x - 3}{5}$$
$$= \frac{x + 7}{3} \cdot \frac{5}{5} - \frac{2x - 3}{5} \cdot \frac{3}{3}$$
$$= \frac{5(x + 7)}{15} - \frac{3(2x - 3)}{15}$$
$$= \frac{5(x + 7) - 3(2x - 3)}{15}$$
$$= \frac{5x + 35 - 6x + 9}{15} = \frac{-x + 44}{15} \quad \textit{Ans.}$$

EXERCISES

1. Express the sum of $\frac{x}{3}$ and $\frac{x}{2}$ as a single fraction in simplest form.

2. The expression $\frac{x}{3} + \frac{x}{5}$ is equivalent to

   (1) $\frac{8}{15}$ (2) $\frac{8x}{15}$ (3) $\frac{2x}{15}$ (4) $\frac{2x}{8}$

3. Express the sum of $\frac{2x}{3}$ and $\frac{3x}{4}$ as a single fraction in lowest terms.

4. Express as a single fraction: $\frac{3x}{2} + \frac{4x}{3}$

5. Express the sum of $\frac{5x}{3}$ and $\frac{x}{6}$ as a single fraction in lowest terms.

6. Express the sum of $\frac{3x}{8}$ and $\frac{5x}{12}$ as a single fraction in lowest terms.

7. The expression $\frac{x}{8} - \frac{x}{2}$ is equivalent to

   (1) $\frac{x}{6}$ (2) 0 (3) $\frac{-3x}{8}$ (4) $\frac{-3x}{-8}$

8. Which is equivalent to $\frac{8}{x} - \frac{3}{x}$?

   (1) 5 (2) $5x$ (3) $\frac{5}{x}$ (4) $-\frac{5}{x}$

In 9–20, combine and express as a single fraction in simplest form.

9. $\frac{3}{x} + \frac{5}{2x}$

10. $\frac{8}{x} + \frac{2}{x^2}$

11. $\frac{3}{x^2} + \frac{6}{x^4}$

12. $\frac{4}{5x} - \frac{1}{10x}$

13. $\frac{8}{x} - \frac{5}{x^3}$

14. $\frac{1}{x} + \frac{2}{x^2} - \frac{3}{x}$

15. $\frac{x+1}{2} + \frac{x}{3}$

16. $\frac{x-3}{3} + \frac{x+1}{6}$

17. $\frac{3x-5}{4} + \frac{2x+3}{6}$

18. $\frac{3x-2}{5} - \frac{2x+3}{15}$

19. $\frac{3x-4}{5} - \frac{x-2}{4}$

20. $\frac{x-1}{2} - \frac{x-5}{8}$

Multiplying Fractions

Fractions can be multiplied by multiplying the numerators and multiplying the denominators. The multiplication is made easier by first factoring and cancelling common factors. Cancellation of common factors also avoids having to reduce later.

Examples

Multiply in each example and write the answer in simplest form:

1. $\frac{9}{8} \cdot \frac{16}{27} = \frac{\cancel{9} \cdot 1}{\cancel{8} \cdot 1} \cdot \frac{\cancel{8} \cdot 2}{\cancel{9} \cdot 3} = \frac{2}{3}$ *Ans.*

2. $\frac{3x + 3}{x - 1} \cdot \frac{x^2 - 1}{x + 1} = \frac{3\cancel{(x + 1)}}{\cancel{(x - 1)}} \cdot \frac{(x + 1)\cancel{(x - 1)}}{\cancel{(x + 1)}} = 3(x + 1)$
or $3x + 3$ *Ans.*

3. $\frac{x^2 - x - 6}{x - 3} \cdot \frac{2}{x + 2} = \frac{\cancel{(x + 2)}\cancel{(x - 3)}}{\cancel{(x - 3)}} \cdot \frac{2}{\cancel{(x + 2)}} = \frac{2}{1}$ or 2 *Ans.*

EXERCISES

1. Express the product $\frac{(x + y)^2}{4} \cdot \frac{2}{x + y}$ in simplest form.
2. Express as a single fraction in lowest terms: $\left(\frac{y^2 - 16}{2y + 6}\right)\left(\frac{y + 3}{y - 4}\right)$
3. The result of multiplying $\frac{x^2 - 1}{x}$ by $\frac{4x^2}{x + 1}$ is
   (1) $\frac{x - 1}{4x^3}$ (2) $\frac{(x^2 - 1)(x + 1)}{4x^3}$ (3) $4x(x + 1)$ (4) $4x(x - 1)$

In 4–10, multiply and express the answer in simplest form.

4. $\frac{3x + 6}{x^2 - 7x + 10} \cdot \frac{x - 5}{2 + x}$
5. $\frac{x - 3}{x^2 - 4x} \cdot \frac{x^2 - 16}{x^2 + x - 12}$
6. $\frac{x^2 - 25}{3x + 9} \cdot \frac{(x + 3)}{x^2 + 10x + 25}$
7. $\frac{a^2 - 25}{5a^2} \cdot \frac{a^2 - 5a}{a^2 - 10a + 25}$
8. $\frac{(x - 3)^2}{x^2 - x - 6} \cdot \frac{x + 2}{x - 3}$
9. $\frac{x^2 + 7x + 12}{x^2 - 4} \cdot \frac{2x + 4}{x + 3}$
10. $\frac{x^2 + 8x + 12}{x^2 - 8x + 16} \cdot \frac{x^2 + x - 20}{x^2 + 11x + 30}$

Dividing Fractions

A division problem can be turned into a multiplication problem by inverting the second fraction. That is, division by a fraction is equivalent to multiplication by the multiplicative inverse, or reciprocal, of that fraction.

Examples

Divide in each example and write the answer in simplest form.

1. $\frac{5}{8} \div \frac{15}{24} = \frac{5}{8} \cdot \frac{24}{15} = \frac{\cancel{5}^1}{\cancel{8}_1} \cdot \frac{\cancel{8}^1 \cdot \cancel{3}^1}{\cancel{5}_1 \cdot \cancel{3}_1} = 1$ *Ans.*

2. $\frac{x^2 - 9}{x^2 + 5x - 6} \div \frac{x^2 + x - 12}{3x - 3}$

$$= \frac{x^2 - 9}{x^2 + 5x - 6} \cdot \frac{3x - 3}{x^2 + x - 12}$$

$$= \frac{(x + 3)\cancel{(x - 3)}}{(x + 6)\cancel{(x - 1)}} \cdot \frac{3\cancel{(x - 1)}}{(x + 4)\cancel{(x - 3)}}$$

$$= \frac{3(x + 3)}{(x + 6)(x + 4)} \text{ Ans.} \qquad = \frac{3x + 9}{x^2 + 10x + 24} \text{ Equivalent Answer}$$

EXERCISES

In 1–5, perform the indicated division and express the result in lowest terms.

1. $\frac{x^2 - 25}{18} \div \frac{x - 5}{36}$
2. $\frac{a}{b^2 - 49} \div \frac{4a^3}{2b + 14}$
3. $\frac{6x + 12}{2x + 6} \div \frac{x^2 - 4}{7x + 21}$
4. $\frac{r^2 - 25}{5r^2} \div \frac{(r - 5)^2}{r^2 - 5r}$
5. $\frac{(x - 2)^2}{x^2 - 16} \div \frac{x^2 + 2x - 8}{x^2 + 6x + 8}$

7.3 SOLVING FRACTIONAL EQUATIONS

Equations that contain fractions are called ***fractional equations.*** The variable may appear in the numerators, in the denominators, or in both the numerators and the denominators. Regardless of where the variable appears, the idea for solving fractional equations is always the same: *clear of fractions by multiplying each term of the equation by the L.C.D. of all the denominators.*

◆ In general, the procedure for solving a fractional equation is:

1. Determine the L.C.D. of *all* the denominators in the equation.
2. Multiply *each* term of the equation by the L.C.D.
3. Cancel common factors and multiply.
4. The resulting equation has no fractions and may be solved as an ordinary equation.

5. Check in the *original* fractional equation by doing the arithmetic on each side of the equals sign. Do not repeat the procedure for eliminating the fractions.

The following examples illustrate this procedure.

Examples

In each of the following equations, solve for x and check:

1. $\frac{x}{5} - \frac{x}{7} = 2$

Solution:

$$\frac{x}{5} - \frac{x}{7} = 2$$

$$35 \cdot \frac{x}{5} - 35 \cdot \frac{x}{7} = \frac{2}{1} \cdot 35$$ Multiply each term of the equation by the L.C.D. of all the denominators, 35. (*Note:* $2 = \frac{2}{1}$.)

$$\overset{7}{\cancel{35}} \cdot \frac{x}{\underset{1}{\cancel{5}}} - \overset{5}{\cancel{35}} \cdot \frac{x}{\underset{1}{\cancel{7}}} = 2 \cdot 35$$ Cancel common factors.

$$7x - 5x = 70$$ Multiply.

$$2x = 70$$ Combine like terms.

$$x = 35$$ Divide by 2.

Check:

Substitute $x = 35$ into the original equation, and perform the arithmetic on each side of $\stackrel{?}{=}$ separately. Do not repeat the procedure with the L.C.D.

$$\frac{x}{5} - \frac{x}{7} = 2$$

$$\frac{35}{5} - \frac{35}{7} \stackrel{?}{=} 2$$

$$7 - 5 \stackrel{?}{=} 2$$

$$2 = 2 \checkmark$$

Answer: $x = 35$ or $\{35\}$ is the solution set.

2. $\frac{25}{x} = \frac{10}{3}$

Solution 1:

$$\frac{25}{x} = \frac{10}{3}$$

$$3x \cdot \frac{25}{x} = \frac{10}{3} \cdot 3x$$ Multiply each term of the equation by the L.C.D. $= 3x$.

$$3\cancel{x} \cdot \frac{25}{\underset{1}{\cancel{x}}} = \frac{10}{\underset{1}{\cancel{3}}} \cdot \cancel{3}x$$ Cancel common factors.

$$75 = 10x$$ Multiply.

$$7.5 = x$$ Divide by 10.

Solution 2:

Regard $\frac{25}{x} = \frac{10}{3}$ as a proportion. In a proportion, the product of the means equals the product of the extremes. Thus:

$$\begin{aligned} 10(x) &= 25(3) \\ 10x &= 75 \\ x &= 7.5 \end{aligned}$$

(*Note:* The check for both methods of solution is the same. Substitute $x = 7.5$ into the original equation, and work on each side of $\stackrel{?}{=}$. Do not repeat either method of solution as a check.)

Check:

$$\begin{aligned} \frac{25}{x} &= \frac{10}{3} \\ \frac{25}{7.5} &\stackrel{?}{=} \frac{10}{3} \\ \frac{250}{75} &\stackrel{?}{=} \frac{10}{3} \\ \frac{\cancel{25} \times 10}{\cancel{25} \times 3} &\stackrel{?}{=} \frac{10}{3} \\ \frac{10}{3} &= \frac{10}{3} \checkmark \end{aligned}$$

Answer: $x = 7.5$ or $\{7.5\}$ is the solution set.

3. $\frac{2x}{3} - 5 = \frac{3x}{4}$

Solution:

$\frac{2x}{3} - \frac{5}{1} = \frac{3x}{4}$ — $5 = \frac{5}{1}$

$12 \cdot \frac{2x}{3} - 12 \cdot \frac{5}{1} = \frac{3x}{4} \cdot 12$ — Multiply each term of the equation by the L.C.D. of all the denominators, 12.

$\overset{4}{\cancel{12}} \cdot \frac{2x}{\underset{1}{\cancel{3}}} - 12 \cdot \frac{5}{1} = \frac{3x}{\underset{1}{\cancel{4}}} \cdot \overset{3}{\cancel{12}}$ — Cancel common factors.

$8x - 60 = 9x$ — Multiply.

$-60 = x$ — Subtract $8x$.

Check:

Substitute $x = -60$ into the original equation.

$$\begin{aligned} \frac{2x}{3} - 5 &= \frac{3x}{4} \\ \frac{2(\overset{(-20)}{\cancel{-60}})}{\underset{1}{\cancel{3}}} - 5 &\stackrel{?}{=} \frac{3(\overset{(-15)}{\cancel{-60}})}{\underset{1}{\cancel{4}}} \\ -40 - 5 &\stackrel{?}{=} -45 \\ -45 &= -45 \checkmark \end{aligned}$$

Answer: $x = -60$ or $\{-60\}$ is the solution set.

4. $\frac{4}{x} + \frac{2}{7} = \frac{8}{x}$

Solution:

$$\frac{4}{x} + \frac{2}{7} = \frac{8}{x}$$

$$7x \cdot \frac{4}{x} + 7x \cdot \frac{2}{7} = \frac{8}{x} \cdot 7x$$ Multiply each term by the L.C.D. = $7x$.

$$7\cancel{x} \cdot \frac{4}{\cancel{x}_1} + \cancel{7}x \cdot \frac{2}{\cancel{7}_1} = \frac{8}{\cancel{x}_1} \cdot 7\cancel{x}$$ Cancel common factors.

$$28 + 2x = 56$$ Multiply.

$$2x = 28$$ Subtract 28.

$$x = 14$$ Divide by 2.

Check:

Substitute $x = 14$ into the original equation.

$$\frac{4}{x} + \frac{2}{7} = \frac{8}{x}$$

$$\frac{4}{14} + \frac{2}{7} \stackrel{?}{=} \frac{8}{14}$$

$$\frac{4}{14} + \frac{2}{7} \cdot \frac{2}{2} \stackrel{?}{=} \frac{8}{14}$$

$$\frac{4}{14} + \frac{4}{14} \stackrel{?}{=} \frac{8}{14}$$

$$\frac{4+4}{14} \stackrel{?}{=} \frac{8}{14}$$

$$\frac{8}{14} = \frac{8}{14} \checkmark$$

Answer: $x = 14$ or $\{14\}$ is the solution set.

5. $\frac{x+1}{3} + \frac{x+3}{4} = \frac{1}{2}$

Solution:

$$\frac{x+1}{3} + \frac{x+3}{4} = \frac{1}{2}$$

$$12\left(\frac{x+1}{3}\right) + 12\left(\frac{x+3}{4}\right) = \frac{1}{2} \cdot 12$$ Multiply by the L.C.D. = 12.

$$\overset{4}{\cancel{12}}\left(\frac{x+1}{\cancel{3}_1}\right) + \overset{3}{\cancel{12}}\left(\frac{x+3}{\cancel{4}_1}\right) = \frac{1}{\cancel{2}_1} \cdot \overset{6}{\cancel{12}}$$ Cancel common factors.

$$4(x+1) + 3(x+3) = 6$$ Multiply.

$$4x + 4 + 3x + 9 = 6$$ Distributive property.

$$7x + 13 = 6$$ Combine like terms.

$$7x = -7$$ Subtract 13.

$$x = -1$$ Divide by 7.

Check:

Substitute $x = -1$ into the original equation.

$$\frac{x+1}{3} + \frac{x+3}{4} = \frac{1}{2}$$

$$\frac{-1+1}{3} + \frac{-1+3}{4} \stackrel{?}{=} \frac{1}{2}$$

$$\frac{0}{3} + \frac{2}{4} \stackrel{?}{=} \frac{1}{2}$$

$$0 + \frac{1}{2} \stackrel{?}{=} \frac{1}{2}$$

$$\frac{1}{2} = \frac{1}{2} \checkmark$$

Answer: $x = -1$ or $\{-1\}$ is the solution set.

EXERCISES

In 1–15, solve the equation and check.

1. $\frac{a}{2} + \frac{a}{6} = 2$
2. $\frac{x}{3} - \frac{x}{4} = 1$
3. $\frac{x}{4} - \frac{x}{5} = 2$
4. $\frac{y}{3} + 2 = 5$
5. $\frac{4m}{3} = \frac{m}{6} + \frac{7}{2}$
6. $\frac{3}{5} = \frac{9}{x}$
7. $\frac{24}{h} = \frac{16}{4}$
8. $\frac{x}{3} - \frac{5}{3} = \frac{2x}{3}$
9. $\frac{5y}{3} - \frac{3y}{2} = 2$
10. $\frac{2x}{5} - 4 = \frac{2x}{3}$
11. $n - 4 = \frac{n-1}{4}$
12. $\frac{y+1}{12} = \frac{3}{4}$
13. $\frac{c+2}{2} + \frac{c-1}{3} = 4$
14. $\frac{y+2}{3} - \frac{y+3}{4} = \frac{1}{2}$
15. $12 - \frac{2x+9}{5} = \frac{x+3}{7}$

7.4 USING FRACTIONAL EQUATIONS TO SOLVE PROBLEMS

Many verbal problems lead to fractional equations. To solve these problems, follow the general procedure for solving verbal problems and apply the techniques used to solve fractional equations.

Examples

1. One-third of a number increased by 2 is the same as one-half that number decreased by 3.

Solution: Let x = the number.

One third of a number	increased by two	is the same as
$\frac{1}{3}x$	$+2$	$=$

one-half that number	decreased by three.
$\frac{1}{2}x$	-3

$\frac{1}{3}x + 2 = \frac{1}{2}x - 3$	
$6 \cdot \frac{1}{3}x + 6 \cdot 2 = 6 \cdot \frac{1}{2}x - 6 \cdot 3$	Multiply by the L.C.D. = 6.
$\overset{2}{\cancel{6}} \cdot \frac{1}{\cancel{3}_1}x + 6 \cdot 2 = \overset{3}{\cancel{6}} \cdot \frac{1}{\cancel{2}_1}x - 6 \cdot 3$	Cancel common factors.
$2x + 12 = 3x - 18$	Multiply.
$12 = x - 18$	Subtract $2x$.
$30 = x$	Add 18.

Check:
One-third of a number, $\frac{1}{3} \cdot 30$ or 10, increased by 2, 10 + 2 or *12*, is the same as one-half that number, $\frac{1}{2} \cdot 30$ or 15, decreased by 3, 15 − 3 or *12*.√

Answer: The number is 30.

2. The numerator of a fraction is 8 less than the denominator. The value of the fraction is $\frac{4}{5}$. Find the fraction.

Solution:
Let x = the denominator of the fraction.
Then $x - 8$ = the numerator of the fraction.
And $\frac{x-8}{x}$ = the fraction.

$\frac{x-8}{x} = \frac{4}{5}$	The value of the fraction is $\frac{4}{5}$.
$5x\left(\frac{x-8}{x}\right) = \frac{4}{5} \cdot 5x$	Multiply by the L.C.D. = $5x$.
$5\cancel{x}\left(\frac{x-8}{\cancel{x}_1}\right) = \frac{4}{\cancel{5}_1} \cdot \cancel{5}x$	Cancel common factors.
$5(x - 8) = 4x$	Multiply.
$5x - 40 = 4x$	Distributive property.
$x - 40 = 0$	Subtract $4x$.
$x = 40$	Add 40.
$x - 8 = 32$	
$\frac{x-8}{x} = \frac{32}{40}$	

Check:
The numerator of the fraction 32, is 8 less than the denominator, 40.√ The value of the fraction is $\frac{4}{5}$: $\frac{32}{40} = \frac{\cancel{8} \cdot 4}{\cancel{8} \cdot 5} = \frac{4}{5}$√

Answer: The fraction is $\frac{32}{40}$.

3. The denominator of a fraction is 3 less than the numerator. If 4 is added to both the numerator and denominator, the resulting fraction is equal to $\frac{3}{2}$. Find the original fraction.

Solution:

Let x = the original numerator.
Then $x - 3$ = the original denominator.
And $\frac{x}{x-3}$ = the original fraction.

$(x) + 4$ = the new numerator $= x + 4$
$(x - 3) + 4$ = the new denominator $= x + 1$
the new fraction $= \frac{x+4}{x+1}$

$\frac{x+4}{x+1} = \frac{3}{2}$ — The new fraction $= \frac{3}{2}$.
$3(x + 1) = 2(x + 4)$ — Cross-multiply.
$3x + 3 = 2x + 8$ — Distributive property.
$x + 3 = 8$ — Subtract $2x$.
$x = 5$ — Subtract 3.

The original fraction is $\frac{x}{x-3} = \frac{5}{5-3} = \frac{5}{2}$.

The new fraction is $\frac{x+4}{x+1} = \frac{5+4}{5+1} = \frac{9}{6}$.

Check:

The denominator of a fraction, 2, is 3 less than the numerator, 5.✓ If 4 is added to both the numerator and denominator, the resulting fraction, $\frac{5+4}{2+4}$ or $\frac{9}{6}$, is equal to $\frac{3}{2}$.✓

Answer: The original fraction is $\frac{5}{2}$.

If there is a remainder when a division is performed, that situation can be written in the form of an equation. For example, $15 \div 2$ has a quotient of 7 and a remainder of 1.

We can write this as an equation: $\frac{15}{2} = 7 + \frac{1}{2}$. Note that the remainder is written as a fraction, $\frac{\text{remainder}}{\text{divisor}}$.

Example

The larger of two numbers is 5 less than six times the smaller. When the larger number is divided by the smaller number, the quotient is 5 and the remainder is 2. Find the numbers.

Solution:

Let x = the smaller number.
Then $6x - 5$ = the larger number.

$\dfrac{6x-5}{x} = 5 + \dfrac{2}{x}$	$\dfrac{\text{Dividend}}{\text{Divisor}} = \text{Quotient} + \dfrac{\text{Remainder}}{\text{Divisor}}$
$x\left(\dfrac{6x-5}{x}\right) = x\left(\dfrac{5}{1}\right) + x\left(\dfrac{2}{x}\right)$	Multiply by the L.C.D. = x.
$\overset{1}{\cancel{x}}\left(\dfrac{6x-5}{\underset{1}{\cancel{x}}}\right) = x\left(\dfrac{5}{1}\right) + \overset{1}{\cancel{x}}\left(\dfrac{2}{\underset{1}{\cancel{x}}}\right)$	Cancel common factors.
$6x - 5 = 5x + 2$	Multiply.
$x - 5 = 2$	Subtract $5x$.
$x = 7$	Add 5.

The smaller number is $x = 7$.
The larger number is $6x - 5 = 6(7) - 5 = 37$.

Check:

Divide the larger number, 37, by the smaller number, 7, and see if the quotient is 5 and the remainder is 2.

$$\begin{array}{r l} 5 & \text{quotient} \\ 7\overline{)37} & \\ \underline{35} & \\ 2 & \text{remainder} \end{array}$$

Answer: The smaller number is 7 and the larger number is 37.

EXERCISES

In 1–10, only an algebraic solution will be accepted.

1. If five times a number is increased by one-third of itself, the result is 32. Find the number.
2. Find two consecutive integers such that if 7 more than the first is divided by 9 less than the second, the quotient is 6.
3. The denominator of a fraction is 8 more than the numerator. If 5 is added to both the numerator and the denominator, the result is equal to $\frac{1}{2}$. Find the original fraction.
4. The denominator of a fraction is 1 more than twice the numerator. If 2 is added to both the numerator and the denominator, the value of the fraction is $\frac{3}{5}$. Find the original fraction. Check.
5. The numerator of a fraction is 5 less than the denominator. If the numerator is doubled and the denominator is increased by 7, the value of the resulting fraction is $\frac{2}{3}$. Find the original fraction. Check.

6. The numerator of a fraction is 7 less than the denominator. If the numerator is decreased by 2 and the denominator is increased by 3, the value of the new fraction is $\frac{1}{3}$. Find the original fraction.
7. The numerator of a fraction is 7 less than the denominator. If 3 is added to the numerator and 9 is subtracted from the denominator, the resulting fraction has a value of $\frac{3}{2}$. Find the value of the original fraction. Check.
8. The larger of two numbers is 3 less than seven times the smaller. If the larger number is divided by the smaller number, the quotient is 6 and the remainder is 2. Find the numbers.
9. The larger of two numbers is one less than eight times the smaller. When the larger number is divided by the smaller number, both the quotient and the remainder are 7. Find the numbers.
10. Find two consecutive odd integers such that if seven times the second is divided by the first, the quotient is 8 and the remainder is 5.

7.5 SUMMARY EXERCISES

1. Factor 40 into primes.
   (1) $4 \cdot 10$ (2) $8 \cdot 5$ (3) $2 \cdot 3 \cdot 5$ (4) $2^3 \cdot 5$
2. Which of the following statements is true?
   (1) 12 and 25 have no prime factors in common.
   (2) 12 and 25 have the same prime factors.
   (3) All prime numbers are odd.
   (4) There is more than one prime number that is even.

In 3–6, find the greatest common factor of each set of numbers.

3. 8, 12, and 36
4. 25, 50, and 100
5. x^2 and x^3
6. x^3, x^5, and x^7

In 7–25, factor as completely as possible.

7. $3x - 9$
8. $x^2 - y^2$
9. $x^2 + x - 2$
10. $ax^2 + 7ax + 12a$
11. $x^2 - 4x$
12. $2x^2 + 4x + 14$
13. $81 - x^2$
14. $x^2 + 8x + 15$
15. $a^2 - 2a - 8$
16. $a^2 + b^2$
17. $x^2 - .64$
18. $t^2 - 12t - 13$
19. $2a^2 + 11a + 5$
20. $3y^2 + 10y - 8$
21. $15 - 8x + x^2$
22. $x^4 - 9$
23. $625 - x^4$
24. $\frac{1}{4}ax + \frac{1}{4}ay - \frac{1}{4}az$
25. $a(x + y) + b(x + y)$

In 26–29, express each fraction in lowest terms.

26. $\dfrac{18x^4}{9x}$ 27. $\dfrac{10x^5y}{2xy}$ 28. $\dfrac{x^2 - 9}{x + 3}$ 29. $\dfrac{x^2 + 2x - 3}{3x + 9}$

In 30–39, express the sum or difference of the given fractions as a single fraction in lowest terms.

30. $\frac{x}{3} - \frac{x}{6}$

31. $\frac{x}{2} + \frac{x+1}{5}$

32. $\frac{x+7}{3} + \frac{x-2}{4}$

33. $\frac{2x+1}{8} - \frac{x+2}{6}$

34. $\frac{10}{x} - \frac{3}{x}$

35. $\frac{3}{2x} + \frac{5}{x}$

36. $\frac{4}{x} + \frac{3}{x^2}$

37. $\frac{2}{x+y} + \frac{3}{x+y}$

38. $\frac{x+1}{3} - \frac{2x-1}{4}$

39. $\frac{x-1}{x} - \frac{x-1}{x}$

In 40–47, multiply or divide as indicated and express the answer in lowest terms.

40. $\frac{2}{x} \cdot \frac{x^3}{8}$

41. $\frac{3}{x^2} \div \frac{27}{x}$

42. $\frac{x^2 - 100}{5} \cdot \frac{25x}{5x - 50}$

43. $\frac{x^2 - 16}{x - 3} \cdot \frac{x^2 - 9}{x + 4}$

44. $\frac{x^2 - 25}{5x + 25} \cdot \frac{10}{(x - 5)^2}$

45. $\frac{x^2 - 7x + 10}{x^2 - 25} \cdot \frac{1}{x^2 - 4}$

46. $\frac{x}{x^2 - 49} \div \frac{4x^2}{2x - 14}$

47. $\frac{(x - 6)^2}{x^2 - 81} \div \frac{x^2 - 10x + 24}{x^2 - 10x + 9}$

In 48–56, solve the equation and check.

48. $\frac{x}{3} + \frac{x}{9} = 36$

49. $\frac{x}{2} - \frac{x}{5} = 1$

50. $\frac{x}{2} + 3 = 11$

51. $\frac{4}{7} = \frac{12}{x}$

52. $\frac{4x}{3} = 18 - \frac{x}{6}$

53. $\frac{x+1}{8} = \frac{3}{4}$

54. $\frac{2x}{3} - 7 = \frac{x}{5}$

55. $\frac{x-3}{2} + \frac{x+3}{5} = \frac{1}{2}$

56. $\frac{5x-3}{4} - \frac{3x+5}{8} = 3$

57. If six times a number is decreased by one-third of itself, the result is 153. Find the number.

58. The denominator of a fraction is 8 more than the numerator. If the numerator is decreased by 10 and the denominator is unchanged, the value of the resulting fraction is $\frac{3}{5}$. Find the original fraction.

59. The denominator of a fraction is 3 less than the numerator. If 3 is added to the numerator and 3 is subtracted from the denominator, the value of the resulting fraction is $\frac{5}{2}$. Find the original fraction.

60. The larger of two numbers is 2 less than seven times the smaller. If the larger number is divided by the smaller number, the quotient is 6 and the remainder is 5. Find the numbers.

CHAPTER 8

Further Aspects of Algebra

8.1 MORE ABOUT REAL NUMBERS

The Inverse Operation of Raising to a Power

We have said that each operation performed on numbers has an inverse operation, an operation that "undoes" the operation performed. Addition and subtraction are inverse operations of each other; multiplication and division are inverse operations of each other.

We have also used a special form of multiplication: raising to a power. This form of multiplication may itself be thought of as an operation, and as such it must have an inverse operation. Consider the following:

Raising to a power: 2^3 means use the base 2 as a factor three times to obtain the result 8 ($2^3 = 2 \cdot 2 \cdot 2 = 8$).

The inverse of this operation means that we start with the result 8 and look for the factor that has been used three times to produce 8. This procedure is written as $\sqrt[3]{8} = 2$ and is read as "the cube root of 8 is 2."

Similarly, $2^4 = 2 \cdot 2 \cdot 2 \cdot 2 = 16$ and $\sqrt[4]{16} = 2$ (the fourth root of 16 is 2).
Also, $5^2 = 5 \cdot 5 = 25$ and $\sqrt[2]{25} = 5$ (the square root of 25 is 5).

◆ In general, *the operation of raising to a power has an inverse operation called* ***extracting a root.*** These operations may be summarized as follows:

Symbols	Meaning
$b^n = N$	Use the base b as a factor n times to produce the number N. Raise b to the nth power.
$\sqrt[n]{N} = b$	Find the factor b that has been used n times to produce the number N. Find the nth root of N.

The names associated with this symbolism are:

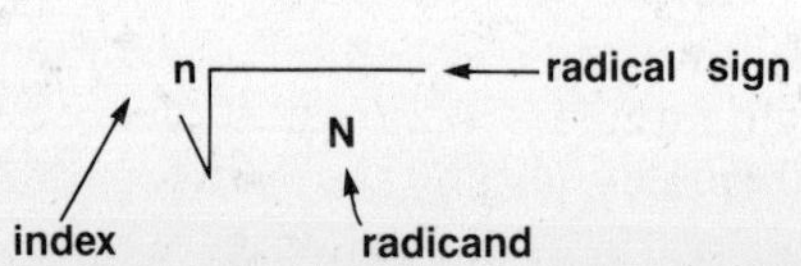

For example, in $\sqrt[3]{8}$, 3 is the ***index*** and, together with the ***radical sign*** $\sqrt{\ }$, stands for "the cube root of." The number under the radical sign, 8, is the ***radicand.***

In algebra, we agree on certain omissions in symbolism for items that are frequently used. We have already agreed, for example, that 3 means $+3$. That is, the sign may be omitted if the number is positive. However, the sign must be written if the number is negative. Also, x means $1x^1$. That is, a coefficient and exponent of 1 may be omitted. All other coefficients and exponents must be written.

Similarly, we agree that an index of 2 may be omitted. When we see, for example, $\sqrt{16}$, we know to read this as "the square root of 16." All other indexes must be written.

Another agreement is made. Consider $x^2 = 16$. These symbols can be interpreted as asking the question: "What number squared equals 16?" or "What number used as a factor two times equals 16?" The answer to that question is that there are two such numbers, since $(+4)^2 = (+4)(+4) = 16$ and $(-4)^2 = (-4)(-4) = 16$. We will agree that $\sqrt{16}$ refers only to the positive value, or ***principal square root.*** That is, $\sqrt{16} = 4$ only. If we wish to refer to both the positive and negative square roots of 16, we must write $\pm\sqrt{16}$. Whenever the index is an even number, this situation arises. For example:

$$\sqrt{16} = 4 \qquad \pm\sqrt{16} = \pm 4$$
$$\sqrt[4]{16} = 2 \qquad \pm\sqrt[4]{16} = \pm 2$$
$$\sqrt[6]{64} = 2 \qquad \pm\sqrt[6]{64} = \pm 2$$

The domain for an index (the numbers that may be used) is the set of positive integers, since the index refers to how many times something has been done (that is, how many times a factor has been used). It would not make sense to say that a factor has been used $4\frac{1}{2}$ times or -5 times, just as it does not make sense to say that a light switch has been turned on $4\frac{1}{2}$ times or -5 times.

Let us now discuss the domain for a radicand. First, let us consider what happens when the *radicand is any positive integer:*

1. If the radicand is a positive integer that is a ***perfect square*** (a number whose two factors are equal), the square root is an integer. (Examples: $\sqrt{4} = 2$, $\sqrt{9} = 3$, $\sqrt{16} = 4$, $\sqrt{25} = 5$) These numbers, the square roots of perfect squares, belong to the set of Rational Numbers.

   If the radicand is not a perfect square, the square root is not an integer. (Examples: $\sqrt{3}$, $\sqrt{5}$, $\sqrt{6}$, $\sqrt{7}$, $\sqrt{8}$, $\sqrt{10}$) These numbers, the square roots of nonperfect squares, belong to the set of Irrational Numbers.
2. Similarly, if the radicand is a positive integer that is a ***perfect cube*** (a number whose three factors are equal), the cube root is an integer. (Examples: $\sqrt[3]{8} = 2$, $\sqrt[3]{27} = 3$, $\sqrt[3]{64} = 4$, $\sqrt[3]{125} = 5$) These numbers, the cube roots of perfect cubes, belong to the set of Rational Numbers.

If the radicand is not a perfect cube, the cube root is not an integer. (Examples: $\sqrt[3]{5}$, $\sqrt[3]{6}$, $\sqrt[3]{10}$, $\sqrt[3]{28}$) These numbers, the cube roots of nonperfect cubes, belong to the set of Irrational Numbers.

3. In general, the nth roots of perfect positive powers n belong to the set of Rational Numbers and the nth roots of nonperfect positive powers n belong to the set of Irrational Numbers.
4. Note that 0 and 1 are special in that $\sqrt{0} = \sqrt[3]{0} = \sqrt[4]{0} = \sqrt[5]{0} = \ldots \sqrt[n]{0} = 0$ and $\sqrt{1} = \sqrt[3]{1} = \sqrt[4]{1} = \sqrt[5]{1} = \ldots \sqrt[n]{1} = 1$. Both 0 and 1 are integers and, therefore, belong to the set of Rational Numbers.

Second, let us consider what happens when the *radicand is any negative integer*. When the radicand is a negative integer, the index can be either odd or even. (*Note:* We have already said that the index can only be a positive integer. Now, we are considering two types of positive integers: odd and even.)

1. If the radicand is a negative integer and the index is odd, the values:
   a. will always be negative (Examples: $\sqrt[3]{-8}$, $\sqrt[3]{-9}$, $\sqrt[5]{-32}$, $\sqrt[5]{-33}$)
   b. will be either an integer and rational (Examples: $\sqrt[3]{-8}$, $\sqrt[5]{-32}$) or not an integer and irrational (Examples: $\sqrt[3]{-9}$, $\sqrt[5]{-33}$)

   The following diagram summarizes these statements:

   Radicand = negative integer
   Index = positive odd integer

Rational Number	**Irrational Number**
$\sqrt[3]{-\text{ perfect cube}} = -$ integer	$\sqrt[3]{-\text{ nonperfect cube}} = -$ noninteger
$\sqrt[5]{-\text{ perfect 5}^{th}\text{ power}} = -$ integer	$\sqrt[5]{-\text{ nonperfect 5}^{th}\text{ power}} = -$ noninteger
$\sqrt[7]{-\text{ perfect 7}^{th}\text{ power}} = -$ integer	$\sqrt[7]{-\text{ nonperfect 7}^{th}\text{ power}} = -$ noninteger
. . .	. . .
odd ↳ $\sqrt[n]{-\text{ perfect } n^{th}\text{ power}} = -$ integer	odd ↳ $\sqrt[n]{-\text{ nonperfect } n^{th}\text{ power}} = -$ noninteger

   All of these numbers belong to the set of Real Numbers.

2. If the radicand is negative and the index is even, we cannot produce a real number because:
   a. There is no real number whose square is negative.

   $$(+)^2 = (+)(+) = + \quad \text{and} \quad (-)^2 = (-)(-) = +$$

   b. There is no real number whose 4th power is negative.

   $$(+)^4 = (+)(+)(+)(+) = + \quad \text{and} \quad (-)^4 = (-)(-)(-)(-) = +$$

   c. In general:

   $$\left.\begin{aligned}(-)^{\text{even}} &= +\\ \text{and } (+)^{\text{even}} &= +\end{aligned}\right\}\text{always positive}$$

Thus, numbers in the form $\sqrt[\text{even}]{\text{any negative}}$ are not real numbers. When we are working with an even index, we must be especially careful to look at where the negative sign is, because:

a. If the negative sign is inside the radical sign, we do not have a real number. For example, $\sqrt{-16}$ is not a real number.

b. If the negative sign is outside the radical sign, we do have a real number. For example, $-\sqrt{16} = -4$, a real number.

When we are working with an odd index, we always have a real number whether the negative sign is inside or outside the radical sign. For example, $\sqrt[3]{-64}$ and $-\sqrt[3]{64}$ are both real numbers. (In fact, $\sqrt[3]{-64} = -\sqrt[3]{64} = -4$.)

Summary of the Types of Numbers Contained in the Real Number System

{Real Numbers}

The set of Real Numbers contains all the numbers in the sets below. The only numbers not contained are those of the form $\sqrt[\text{even}]{\text{negative}}$.

{Rational Numbers}

This set contains all numbers that can be expressed as the ratio of two integers. That is:

1. {Integers}
2. decimals that end
3. decimals that repeat
4. numbers of the form $\sqrt[n]{\text{perfect } n^{\text{th}} \text{ power}}$ excluding numbers of the form $\sqrt[\text{even}]{\text{negative}}$

{Irrational Numbers}

This set contains only those numbers that cannot be expressed as the ratio of two integers. That is:

1. decimals that neither end nor repeat
2. numbers of the form $\sqrt[n]{\text{nonperfect } n^{\text{th}} \text{ power}}$ excluding numbers of the form $\sqrt[\text{even}]{\text{negative}}$
3. π and others

EXERCISES

In 1–10, answer true or false.

1. Every rational number is a real number.
2. Every irrational number is not a real number.
3. Every integer is a rational number.
4. There are some numbers that are both rational numbers and irrational numbers.

5. An even root of a negative number is not a real number.
6. An irrational number can be expressed as the ratio of two integers.
7. A rational number can be expressed as a decimal that ends or as a decimal that repeats.
8. An irrational number can be expressed as a decimal that neither ends nor repeats.
9. 0 is an irrational number.
10. The principal square root of a real number is always positive.

In 11–13, choose the number in the given set that is a rational number.

11. (1) $\sqrt{2}$ (2) $\sqrt{3}$ (3) $\sqrt{9}$ (4) $\sqrt{5}$
12. (1) $\sqrt{8}$ (2) $\sqrt[3]{9}$ (3) $6\sqrt{2}$ (4) $5\sqrt{1}$
13. (1) $\sqrt{32}$ (2) $\sqrt[3]{32}$ (3) $\sqrt[4]{32}$ (4) $\sqrt[5]{32}$

In 14 and 15, choose the number in the given set that is not a rational number.

14. (1) $\sqrt{2}$ (2) $\sqrt{4}$ (3) $\frac{1}{2}$ (4) 0
15. (1) $\sqrt{\frac{9}{25}}$ (2) $\sqrt{.81}$ (3) $\sqrt{9}$ (4) $\sqrt{3}$

16. Which value of x will make the expression $\sqrt{29 - x}$ a positive integer? (1) 1 (2) 29 (3) 0 (4) 4
17. Which value of x will make the expression $\sqrt{100 - x}$ an irrational number? (1) 0 (2) 100 (3) 25 (4) 75
18. For which value of x is the expression $\frac{2}{\sqrt{x - 1}}$ undefined?

    (1) 2 (2) 1 (3) -1 (4) 0

In 19 and 20, choose the number in the given set that is not a real number.

19. (1) $-\sqrt{25}$ (2) $\sqrt{-25}$ (3) -5 (4) $\sqrt[3]{-5}$
20. (1) $\sqrt{-2}$ (2) $-\sqrt{2}$ (3) $\sqrt[3]{-2}$ (4) $-\sqrt[3]{2}$

8.2 APPROXIMATING SQUARE ROOTS

We have seen that numbers of the form $\sqrt[n]{\text{nonperfect } n^{\text{th}} \text{ powers}}$ are irrational. These numbers are equivalent to decimals that do not end, and we can find only ***approximations*** of their values. That is, we stop the never-ending decimal at some place of our choice. The cutoff place depends upon how accurate the approximation must be. We will consider here only approximations for square roots of irrational numbers.

Suppose we wish to know the approximate value of $\sqrt{65}$ correct to the nearest integer. Since 65 is very close to the perfect square 64, $\sqrt{65} \approx 8$. (*Note:* The symbol $\approx$ means "is approximately equal to.")

◆ To approximate the integral value of the square root of a nonperfect square, we find the closest perfect square and use its square root.

Example

Approximate $\sqrt{94}$ correct to the nearest integer.

Solution:

94 is between the perfect squares 81 and 100.
94 is closer to 100 than it is to 81 $(81 \underbrace{\ \ }_{13} 94 \underbrace{\ \ }_{6} 100)$
Thus, $\sqrt{94}$ is close to $\sqrt{100}$.

Answer: $\sqrt{94} \approx 10$

To obtain more accurate approximations of the square root of a nonperfect square, we refine the process of fitting the nonperfect square between the two closest perfect squares. Three different methods are shown here.

Method 1: Try Selected Values

Examples

1. Approximate $\sqrt{17}$ correct to the nearest tenth.

Solution:

Place 17 between the two closest consecutive perfect squares: 17 is between 16 and 25, or $16 < 17 < 25$.

Thus, $\sqrt{17}$ is placed between the square roots of those perfect squares:

$$\sqrt{16} < \sqrt{17} < \sqrt{25}$$
$$\text{or} \quad 4 < \sqrt{17} < 5$$
$$\text{or } \sqrt{17} \text{ is between 4 and 5}$$

Decide whether $\sqrt{17}$ is closer to 4 or to 5 by comparing 17 with 16 and 25. Since 17 is closer to 16 than it is to 25, $\sqrt{17}$ is closer to 4 than it is to 5. That is, $\sqrt{17} < 4.5$.

Now the choices for trial are 4.1, 4.2, 4.3, 4.4. Since 17 is only one unit greater than 16, try the lowest of the choices. That is, calculate $(4.1)^2$ and compare that value to 17: $(4.1)^2 = 16.81$. This value, 16.81, is .19 less than 17.

Since we have a value, 16.81, that is just less than 17, we now need a value just greater than 17. Calculate $(4.2)^2$ and compare that value to 17: $(4.2)^2 = 17.64$. This value, 17.64, is .64 greater than 17.

Thus, 17 is closer to 16.81 (.19 < .64), and the approximate value of $\sqrt{17}$ correct to the nearest tenth is 4.1.

Answer: $\sqrt{17} \approx 4.1$

2. Approximate $\sqrt{32}$ correct to the nearest tenth.

Solution:

$25 < 32 < 36$ — 25 and 36 are the consecutive perfect squares closest to 32.

$\sqrt{25} < \sqrt{32} < \sqrt{36}$ — $\sqrt{32}$ is between 5 and 6.

$5 < \sqrt{32} < 6$ — Since 32 is closer to 36 than it is to 25,

$\sqrt{32} \geq 5.5$ — $\sqrt{32}$ is closer to 6 than it is to 5.

Try 5.7 to begin. — The difference between 32 and 36, 4, is about half the difference between 32 and 25, 7. 5.7 is a middle value of the choices 5.5, 5.6, 5.7, 5.8, 5.9.

Calculate $(5.7)^2 = 32.49$. — This value is .49 greater than 32.

Calculate $(5.6)^2 = 31.36$. — Since we already had a value just greater than 32, we have to examine the value just less than 32. This value, 31.36, is .64 less than 32. Thus, 32 is closer to 32.49 than it is to 31.36 (.49 < .64).

Answer: $\sqrt{32} \approx 5.7$

Method 2: Divide and Average

This method depends upon beginning with an estimate of the square root to be found and refining it. The closer the initial estimate is to the actual answer, the more quickly the refinement takes place.

Example

Approximate $\sqrt{32}$.

Solution:

Estimate $\sqrt{32}$ and divide 32 by that estimate: $\sqrt{32} \approx 5.6$, since 32 is a little more than halfway between 25 ($\sqrt{25} = 5$) and 36 ($\sqrt{36} = 6$). *Divide* 32 by 5.6:

```
           5.71
   5.6.)32.0.00
        28 0
         4 0 0
         3 9 2
             80
             56
             24
```

Carry the division to two decimal places, one more than that of the divisor.

The actual answer is between the original guess of 5.6 and the quotient of 5.71. *Average* the divisor 5.6 and the quotient 5.71 to obtain what will be a new divisor: $\frac{5.6 + 5.71}{2} = \frac{11.31}{2} = 5.65$.

Divide 32 by the new divisor 5.65:

```
         5.663     Carry the division to
5.65.)32.00.000    three decimal places, one
      28 25        more than that of the
       3 75 0      divisor.
       3 39 0
         36 00
         33 90
          2 100
          1 695
            405
```

The actual answer is between the refined divisor of 5.65 and the quotient of 5.663. *Average* the divisor 5.65 and the quotient 5.663 to obtain what will be a new divisor: $\frac{5.65 + 5.663}{2} = \frac{11.313}{2} = 5.6565$.

This process can be continued infinitely. So far, the approximation of $\sqrt{32}$ is 5.6565. We can round off this answer to any of three decimal places:

$\sqrt{32} \approx 5.7$ to the nearest tenth
$\sqrt{32} \approx 5.66$ to the nearest hundredth
$\sqrt{32} \approx 5.657$ to the nearest thousandth

If we had been concerned with an approximation to the nearest tenth to begin with, we would have stopped this process after the first average.

Method 3: Use an Algorithm

An ***algorithm*** is the name for a special process that is used to solve a problem. Usually, an algorithm involves a repetitive calculation. That is, once the procedure is established, it continues to repeat itself. Thus, Method 2 was a type of algorithm. Another algorithm is shown now. Since the reasoning behind this algorithm is beyond the scope of this course, only the method is shown. This algorithm may be used to compute a square root to any number of places.

Examples

1. Approximate $\sqrt{32}$ correct to the nearest tenth.

Solution:

The answer will be here.

$$\sqrt{\overset{\frown}{32}.\overset{\frown}{00}\;\overset{\frown}{00}}$$

Starting at the decimal point and moving left, group the digits of the number in pairs.

Do the same groupings to the right of the decimal point. (*Note:* Four zeros have been added so that we can have two pairs of digits to the right of the decimal point. Each pair of zeros will produce one decimal place in the answer. We need two decimal places to round off to the nearest tenth.)

Place the decimal point in the answer directly above the decimal point in the radicand.

In all further operations, we can ignore the decimal point and proceed as though we are working with whole numbers.

```
    5.
  √32.00 00
   25
```

Below the first group, write the largest perfect square in that group. Write the square root of this perfect square in the answer, directly above the first group.

```
    5.
  √32.00 00
   25 ↓
    7 00
```

Subtract the perfect square from the first group. Bring down the next group of digits.

```
       5.
     √32.00 00
      25 ↓
10?  | 7 00
↑
└── new divisor
```

Form a new divisor by doubling what is already in the answer ($5 \times 2 = 10$). Write that number, 10, in the spot for the new divisor and add another place to it (indicated by "?"). The reason for the "?" is that the new divisor will be divided into the remainder, 10?)700, and must have a digit for the "?" that will allow for this division:

```
       6
  106)700
      636
```

```
            5. 6
         √32.00 00
           25
106       | 7 00
```

Write 6, the digit decided upon to replace the "?," in two places: (1) in the "?" place and (2) in the answer above the second group.

```
            5. 6
         √32.00 00
           25
106       | 7 00
            6 36
              64
```

Multiply the new divisor, 106, by the digit, 6, that was just entered in the answer. Write the product, 636, under the existing remainder, 700. Subtract. (*Note:* The new remainder must be less than the divisor used to find it. That is, $64 < 106$.)

That is the entire process. Now the process may be repeated as often as desired. We will do it one more time. Note that since we already have a remainder, we begin with the step of creating a new divisor after we have brought down the next group of digits.

```
            5. 6
         √32.00 00
           25
106       | 7 00
            6 36
112?      |   64 00
```

Form a new divisor by doubling what is already in the answer ($56 \times 2 = 112$). Write that number, 112, in the spot for the new divisor and add another place to it (indicated by "?").

Decide what number is to replace the "?".

```
112?)6400

         5
1125)6400
     5625
```

```
            5. 6  5
         √32.00 00
           25
106       | 7 00
            6 36
1125      |   64 00
              56 25
               7 75
```

Write 5, the digit decided upon for the "?," in two places: (1) in the "?" place and (2) in the answer above the third group.

Multiply the new divisor, 1125, by the digit, 5, that was just entered in the answer. Write the product, 5625, under the existing remainder, 6400. Subtract. (*Note:* The new remainder must be less than the divisor used to find it. That is, $775 < 1125$.)

$\sqrt{32} \approx 5.65$ or
$\sqrt{32} \approx 5.7$

Round off to the nearest tenth.

Answer: $\sqrt{32} \approx 5.7$

2. Approximate $\sqrt{94}$ correct to the nearest tenth.

Solution:

```
      .
  √94.00 00
```

Place the decimal point. Mark off groups of two digits (four zeros were added).

```
      9.
  √94.00 00
    81
    13
```

Below the first group, write the largest perfect square in the group. Write the square root of this perfect square in the answer, directly above the first group. Subtract.

```
      9.
  √94.00 00
    81 ↓
18? |13 00
```

Bring down the next group of digits. Form a new divisor by doubling what is in the answer and adding a "?". (*Note:* The "?" may be added mentally.)

```
      9. 6
  √94.00 00
    81 ↓
186 |13 00
     11 16
      1 84
```

Decide what number is to be used for the "?". Use 6, and enter it in two places. Multiply 186 by 6. Write this product, 1116, under 1300. Subtract. (*Note:* $184 < 186$).

```
       9. 6
  √94.00 00
    81 ↓   |
186 |13 00 |
     11 16 ↓
192? | 1 84 00
```

Bring down the next group. Form a new divisor ($96 \times 2 = 192$). "Think" the "?".

```
       9. 6  9
  √94.00 00
    81 ↓   |
186 |13 00 |
     11 16 ↓
1929 | 1 84 00
       1 73 61
         10 39
```

Decide the replacement for the "?". Use 9, and enter it in two places. Multiply 1929 by 9. Write the product, 17361, under 18400. Subtract. (*Note:* $1039 < 1929$.)

$\sqrt{94} \approx 9.69$ or
$\sqrt{94} \approx 9.7$

Round off to the nearest tenth.

Answer: $\sqrt{94} \approx 9.7$

EXERCISES

In 1–5, approximate the square root correct to the nearest integer.

1. $\sqrt{10}$ 2. $\sqrt{53}$ 3. $\sqrt{20}$ 4. $\sqrt{75}$ 5. $\sqrt{130}$

In 6–15, approximate the square root correct to the nearest tenth.

6. $\sqrt{86}$ 7. $\sqrt{62}$ 8. $\sqrt{23}$ 9. $\sqrt{42}$ 10. $\sqrt{72}$
11. $\sqrt{39}$ 12. $\sqrt{19}$ 13. $\sqrt{51}$ 14. $\sqrt{60}$ 15. $\sqrt{112}$

In 16–18, approximate the square root correct to the nearest hundredth.

16. $\sqrt{46}$ 17. $\sqrt{53}$ 18. $\sqrt{29}$

8.3 OPERATIONS WITH RADICALS

Simplifying Radicals

We have just seen how to obtain approximate values for the square roots of nonperfect squares. Sometimes, we will need to work with these irrational numbers in ***radical form*** (that is, for example, $\sqrt{8}$ and not its approximation 2.8). When an irrational number is in radical form, it is in ***simplest radical form*** if it cannot be reduced. (*Note:* We used a similar idea in working with fractions. For example, $\frac{6}{10}$ can be reduced to $\frac{3}{5}$ by dividing the numerator and denominator by the common factor 2. We say that $\frac{3}{5}$ is in simplest form because it cannot be reduced.)

Reducing to simplest radical form also involves the removal of a factor. The factor to be removed is a perfect n^{th} power, and it is removed from the radical sign. Look at the following examples to see how this is done.

Examples

Write each of the following irrational numbers in *simplest radical form.*

1. $\sqrt{8}$

Solution:

$\sqrt{8} = \sqrt{4 \cdot 2}$	Factor the radicand so that one factor is the largest perfect square contained in it.
$= \sqrt{4} \cdot \sqrt{2}$	Write the square root of the product of the factors as the product of the square roots of the factors.
$= 2\sqrt{2}$	Write the square root of the perfect square factor as an integer. Note that the radicand no longer contains a perfect square factor.

Answer: $2\sqrt{2}$

2. $\sqrt{32}$

Solution:

$$\begin{aligned}\sqrt{32} &= \sqrt{16 \cdot 2}\\ &= \sqrt{16} \cdot \sqrt{2}\\ &= 4\sqrt{2} \quad \textit{Ans.}\end{aligned}$$

If the *largest* perfect square in 32 is not one of the original factors, the procedure has to be performed more than once.

$$\begin{aligned}\sqrt{32} &= \sqrt{4 \cdot 8}\\ &= \sqrt{4} \cdot \sqrt{8}\\ &= 2\sqrt{8}\end{aligned}$$

This is not in simplest form because 8 still contains a perfect square factor.

$$\begin{aligned}&= 2\sqrt{4 \cdot 2}\\ &= 2\sqrt{4} \cdot \sqrt{2}\\ &= 2 \cdot 2 \cdot \sqrt{2}\\ &= 4\sqrt{2} \quad \textit{Ans.}\end{aligned}$$

3. $5\sqrt{12}$

Solution:

$$\begin{aligned}5\sqrt{12} &= 5\sqrt{4 \cdot 3}\\ &= 5\sqrt{4} \cdot \sqrt{3}\\ &= 5 \cdot 2 \cdot \sqrt{3}\\ &= 10\sqrt{3} \quad \textit{Ans.}\end{aligned}$$

EXERCISES

1. The expression $\sqrt{300}$ is equivalent to
(1) $50\sqrt{6}$ (2) $12\sqrt{5}$ (3) $3\sqrt{10}$ (4) $10\sqrt{3}$

2. Which expression is equivalent to $\sqrt{48}$?
(1) $16\sqrt{3}$ (2) $4\sqrt{12}$ (3) $4\sqrt{3}$ (4) $2\sqrt{3}$

3. The expression $\sqrt{90}$ is equivalent to
(1) $9\sqrt{10}$ (2) $6\sqrt{15}$ (3) $3\sqrt{10}$ (4) $10\sqrt{3}$

4. The expression $3\sqrt{18}$ is equivalent to
(1) $9\sqrt{2}$ (2) $3\sqrt{2}$ (3) $3\sqrt{6}$ (4) $9\sqrt{3}$

5. Which is equivalent to $5\sqrt{3}$?
(1) $\sqrt{15}$ (2) $\sqrt{53}$ (3) $\sqrt{75}$ (4) 225

6. The expression $\frac{1}{2}\sqrt{28}$ is equivalent to
(1) $\sqrt{14}$ (2) $2\sqrt{7}$ (3) $\sqrt{7}$ (4) 7

7. Which is *not* equal to $\sqrt{200}$?
(1) $10\sqrt{2}$ (2) $8\sqrt{5}$ (3) $5\sqrt{8}$ (4) $2\sqrt{50}$

In 8–10, write each expression in simplest radical form.

8. $2\sqrt{45}$ **9.** $3\sqrt{300}$ **10.** $\frac{1}{4}\sqrt{48}$

Combining Radicals by Addition or Subtraction

When irrational numbers are in radical form, they can be combined by addition or subtraction only if they are alike. (*Note:* We used a similar idea in combining monomials. Only like monomials can be combined by addition or subtraction. For example, $2x^2 + 3x^2 = 5x^2$, but $2x^2 + 3x$ cannot be combined.)

◆ *Numbers in radical form are alike if the radical parts are the same in index and in radicand.*

For example:

$\sqrt{3}$ and $2\sqrt{3}$ are alike.
$\sqrt{3}$ and $\sqrt{5}$ are not alike (different radicands).
$\sqrt{3}$ and $\sqrt[3]{3}$ are not alike (different indexes).

When numbers in radical form are alike, they are combined by adding or subtracting the coefficients of the radical parts (the number before the radical sign).

Examples

Combine like terms.

1. $2\sqrt{3} + 3\sqrt{3} = (2 + 3)\sqrt{3} = 5\sqrt{3}$ *Ans.*
2. $5\sqrt{10} - \sqrt{10} = (5 - 1)\sqrt{10} = 4\sqrt{10}$ *Ans.*
3. $5\sqrt[3]{2} + 2\sqrt[3]{2} = (5 + 2)\sqrt[3]{2} = 7\sqrt[3]{2}$ *Ans.*

Sometimes numbers in radical form do not look alike, but after they have been simplified they are alike.

Examples

Combine like terms.

1. $\sqrt{3} + \sqrt{75}$

Solution:

$\sqrt{3} + \sqrt{75}$	These terms are not alike and cannot be combined as they are.
$= \sqrt{3} + \sqrt{25 \cdot 3}$	Simplify $\sqrt{75}$.
$= \sqrt{3} + \sqrt{25} \cdot \sqrt{3}$	
$= \sqrt{3} + 5\sqrt{3}$	These terms are alike.
$= (1 + 5)\sqrt{3}$	Add coefficients of the like terms.
$= 6\sqrt{3}$ *Ans.*	

2. $14\sqrt{2} + 5\sqrt{3} - 3\sqrt{8}$

Solution:

$14\sqrt{2} + 5\sqrt{3} - 3\sqrt{8}$	No two of these terms are alike.
$= 14\sqrt{2} + 5\sqrt{3} - 3\sqrt{4 \cdot 2}$	Simplify $\sqrt{8}$.
$= 14\sqrt{2} + 5\sqrt{3} - 3\sqrt{4} \cdot \sqrt{2}$	
$= 14\sqrt{2} + 5\sqrt{3} - 3 \cdot 2 \cdot \sqrt{2}$	
$= 14\sqrt{2} + 5\sqrt{3} - 6\sqrt{2}$	The first and third terms are alike.
$= (14 - 6)\sqrt{2} + 5\sqrt{3}$	Combine the like terms.
$= 8\sqrt{2} + 5\sqrt{3}$ *Ans.*	

EXERCISES

1. Combine like terms: $2\sqrt{5} + 3\sqrt{2} - 7\sqrt{5}$

2. When $2\sqrt{5}$ is added to $\sqrt{45}$, the sum is
(1) $5\sqrt{5}$ (2) $5\sqrt{10}$ (3) $3\sqrt{50}$ (4) $6\sqrt{5}$

3. The sum of $6\sqrt{6}$ and $\sqrt{54}$ is
(1) $3\sqrt{6}$ (2) $6\sqrt{60}$ (3) $9\sqrt{6}$ (4) $15\sqrt{6}$

4. The expression $\sqrt{48} - \sqrt{12}$ is equivalent to
(1) 6 (2) 2 (3) $3\sqrt{2}$ (4) $2\sqrt{3}$

5. The expression $5\sqrt{2} - \sqrt{32}$ is equivalent to
(1) $\sqrt{18}$ (2) $\sqrt{2}$ (3) $9\sqrt{2}$ (4) $2\sqrt{3}$

6. The sum of $\sqrt{18}$ and $15\sqrt{2}$ is
(1) $18\sqrt{2}$ (2) $24\sqrt{2}$ (3) $15\sqrt{20}$ (4) $16\sqrt{20}$

7. Combine into a single term: $3\sqrt{12} + 2\sqrt{3}$

8. The expression $4\sqrt{2} - \sqrt{18}$ is equivalent to
(1) 1 (2) $\sqrt{2}$ (3) $7\sqrt{2}$ (4) $-5\sqrt{2}$

9. The expression $\sqrt{125} - \sqrt{20}$ is equivalent to
(1) $5\sqrt{3}$ (2) $3\sqrt{5}$ (3) $21\sqrt{5}$ (4) $\sqrt{105}$

10. The sum of $\sqrt{18}$ and $15\sqrt{2}$ is
(1) $18\sqrt{2}$ (2) $24\sqrt{2}$ (3) $15\sqrt{20}$ (4) $16\sqrt{20}$

11. The expression $\sqrt{27} + \sqrt{12}$ is equivalent to
(1) $\sqrt{39}$ (2) $13\sqrt{3}$ (3) $5\sqrt{6}$ (4) $5\sqrt{3}$

12. The expression $\sqrt{50} + 3\sqrt{2}$ can be written in the form $x\sqrt{2}$. Find x.

13. The expression $\sqrt{128} + \sqrt{2}$ is equivalent to
(1) $9\sqrt{2}$ (2) $\sqrt{130}$ (3) $4\sqrt{34}$ (4) $8\sqrt{2}$

14. The sum of $2\sqrt{18}$ and $\sqrt{50}$ is
(1) $10\sqrt{2}$ (2) $11\sqrt{2}$ (3) $2\sqrt{68}$ (4) $3\sqrt{68}$

15. The difference between $\sqrt{125}$ and $\sqrt{80}$ is
(1) 1 (2) $9\sqrt{5}$ (3) $\sqrt{45}$ (4) $\sqrt{5}$

16. Simplify: $\sqrt{50} + 3\sqrt{2}$

In 17 and 18, simplify and combine.

17. $5\sqrt{27} - \sqrt{108} + 3\sqrt{75}$ **18.** $7\sqrt{18} - \frac{3}{4}\sqrt{32} + \frac{1}{2}\sqrt{8}$

Multiplying and Dividing Radicals

Any two numbers in radical form can be multiplied or divided as long as the *indexes are the same*. This is done by multiplying or dividing the coefficients and then by multiplying or dividing the radicals. Simplify answers when needed.

Examples

Multiply and write the answers in simplest form.

1. $4\sqrt{2} \cdot 3\sqrt{5} = 4 \cdot 3\sqrt{2 \cdot 5} = 12\sqrt{10}$ *Ans.*

2. $6\sqrt{5} \cdot 2\sqrt{15} = 6 \cdot 2\sqrt{5 \cdot 15} = 12\sqrt{75}$
$= 12\sqrt{25 \cdot 3}$
$= 12\sqrt{25} \cdot \sqrt{3}$
$= 12 \cdot 5 \cdot \sqrt{3}$
$= 60\sqrt{3}$ *Ans.*

3. $3\sqrt{8} \cdot 5\sqrt{2} = 3 \cdot 5\sqrt{8 \cdot 2} = 15\sqrt{16}$
$= 15 \cdot 4 = 60$ *Ans.*

Divide and write the answers in simplest form.

4. $8\sqrt{20} \div 4\sqrt{2} = \frac{8}{4}\sqrt{\frac{20}{2}} = 2\sqrt{10}$ *Ans.*

5. $16\sqrt{125} \div 8\sqrt{5} = \frac{16}{8}\sqrt{\frac{125}{5}} = 2\sqrt{25}$
$= 2 \cdot 5 = 10$ *Ans.*

EXERCISES

1. Express the product of $2\sqrt{3}$ and $4\sqrt{2}$ as a single term in radical form.
2. Express the product of $3\sqrt{3}$ and $4\sqrt{3}$ as an integer.
3. The expression $(\sqrt{5})(\sqrt{12})$ is equivalent to
(1) $\sqrt{7}$ (2) $2\sqrt{15}$ (3) $\sqrt{17}$ (4) $4\sqrt{15}$
4. The expression $(3\sqrt{18})(4\sqrt{2})$ is equivalent to
(1) 12 (2) 72 (3) $36\sqrt{2}$ (4) $\sqrt{72}$
5. The expression $(6\sqrt{5})(2\sqrt{15})$ is equivalent to
(1) 120 (2) 180 (3) $60\sqrt{3}$ (4) $\sqrt{180}$

In 6–8, express each quotient as an integer.

6. $10\sqrt{75} \div 2\sqrt{3}$ 7. $9\sqrt{200} \div 3\sqrt{2}$ 8. $16\sqrt{125} \div 4\sqrt{5}$

In 9 and 10, divide and simplify.

9. $\frac{21\sqrt{180}}{7\sqrt{2}}$ 10. $\frac{3\sqrt{54}}{6\sqrt{3}}$

8.4 SOLVING QUADRATIC EQUATIONS

So far in our work, we have solved only equations of the first degree (the highest power of the variable is 1). In general, for equations of the first degree there is one answer, or ***root***, or one element in the solution set.

◆ Now we will turn to equations of the second degree, some of which are called ***quadratic equations***. Here are some facts about quadratic equations:

1. A quadratic equation is of degree 2, and, in general, has two answers, or roots, or two elements in the solution set.
2. The ***standard form*** of a quadratic equation is $ax^2 + bx + c = 0$. ($a \neq 0$ since there must be an x^2 term; b or c or both may $= 0$.)
3. The procedure for solving those types of quadratic equations considered here is based on factoring.
4. In the procedure, we apply the principle that if the product of two factors is 0, at least one of the factors *must* be 0.

Examples

Solve each of the following quadratic equations:

1. $x^2 + 2x - 8 = 0$

Solution:

$x^2 + 2x - 8 = 0$	The equation is in standard form.
$(x + 4)(x - 2) = 0$	Factor the trinomial member of the equation.
$x + 4 = 0 \mid x - 2 = 0$ $x = -4 \mid x = 2$	Set each factor = 0, and solve each first-degree equation.

Check each of the values in the *original* equation.

$x = -4$:
$$\begin{aligned} x^2 + 2x - 8 &= 0 \\ (-4)^2 + 2(-4) - 8 &\stackrel{?}{=} 0 \\ 16 - 8 - 8 &\stackrel{?}{=} 0 \\ 8 - 8 &\stackrel{?}{=} 0 \\ 0 &= 0 \checkmark \end{aligned}$$

$x = 2$:
$$\begin{aligned} x^2 + 2x - 8 &= 0 \\ (+2)^2 + 2(+2) - 8 &\stackrel{?}{=} 0 \\ 4 + 4 - 8 &\stackrel{?}{=} 0 \\ 8 - 8 &\stackrel{?}{=} 0 \\ 0 &= 0 \checkmark \end{aligned}$$

Answer: $x = -4$ or $x = +2$; the solution set is $\{-4, +2\}$.

2. $x^2 - 8x = -15$

Solution:

$x^2 - 8x = -15$	This equation is not in standard form.
$x^2 - 8x + 15 = 0$	Write the equation in standard form by adding 15 to both sides of the equation.
$(x - 5)(x - 3) = 0$	Factor the trinomial.
$x - 5 = 0 \mid x - 3 = 0$	Set each factor = 0.
$x = 5 \mid x = 3$	Solve for x.

Check each value in the original equation.

$$x = 5: \quad x^2 - 8x = -15 \qquad x = 3: \quad x^2 - 8x = -15$$
$$5^2 - 8(5) \stackrel{?}{=} -15 \qquad 3^2 - 8(3) \stackrel{?}{=} -15$$
$$25 - 40 \stackrel{?}{=} -15 \qquad 9 - 24 \stackrel{?}{=} -15$$
$$-15 = -15\checkmark \qquad -15 = -15\checkmark$$

Answer: $x = 5$ or $x = 3$; the solution set is $\{3, 5\}$.

3. $x^2 + 4x = 0$

Solution:

$x^2 + 4x = 0$	This equation is in standard form. The constant term is missing ($c = 0$).
$x(x + 4) = 0$	Since the constant term is missing, there is a common factor.
$x = 0 \mid x + 4 = 0$ $\quad\;\; \mid \; x = -4$	Set each factor $= 0$, and solve for x. Note that one value is $x = 0$.

Check each value in the original equation.

$$x = 0: \quad x^2 + 4x = 0 \qquad x = -4: \quad x^2 + 4x = 0$$
$$0^2 + 4(0) \stackrel{?}{=} 0 \qquad (-4)^2 + 4(-4) \stackrel{?}{=} 0$$
$$0 + 0 \stackrel{?}{=} 0 \qquad +16 - 16 \stackrel{?}{=} 0$$
$$0 = 0\checkmark \qquad 0 = 0\checkmark$$

Answer: $x = 0$ or $x = -4$; the solution set is $\{0, -4\}$.

4. $3x^2 = 3x$

Solution:

$3x^2 = 3x$	This equation is not in standard form.
$3x^2 - 3x = 0$	Write the equation in standard form by subtracting $3x$ from both sides of the equation.
$3x(x - 1) = 0$	Since the constant term is missing ($c = 0$), there is a common factor.
$3x = 0 \mid x - 1 = 0$ $x = 0 \mid x = 1$	Set each factor $= 0$, and solve for x. In $3x = 0$, divide by 3. Note that one value is $x = 0$.

Check each value in the original equation.

$$x = 0: \quad 3x^2 = 3x \qquad x = 1: \quad 3x^2 = 3x$$
$$3(0)^2 \stackrel{?}{=} 3(0) \qquad 3(1)^2 \stackrel{?}{=} 3(1)$$
$$3(0) \stackrel{?}{=} 0 \qquad 3(1) \stackrel{?}{=} 3$$
$$0 = 0\checkmark \qquad 3 = 3\checkmark$$

Answer: $x = 0$ or $x = 1$; the solution set is $\{0, 1\}$.

5. $x^2 - 16 = 0$

Solution:

$x^2 - 16 = 0$	This equation is in standard form. The x term is missing ($b = 0$).
$(x + 4)(x - 4) = 0$	Since the x term is missing, factor the difference of two perfect squares.
$x + 4 = 0 \mid x - 4 = 0$ $x = -4 \mid x = 4$	Set each factor $= 0$, and solve for x.

Check each value in the original equation.

Answer: $x = 4$ or $x = -4$; $x = \pm 4$; the solution set is $\{\pm 4\}$.

Alternate Solution:

$x^2 - 16 = 0$	
$x^2 = 16$	Add 16.
$x = \pm\sqrt{16}$	Take the square root of both sides of the equation.
$x = \pm 4$ *Ans.*	

The alternate solution allows us to obtain answers for this form of the quadratic equation ($b = 0$) even if the constant term is not a perfect square. See Example 1 below. The alternate form is helpful at other times as well. See examples 2 and 3 below.

Examples

1. $x^2 - 17 = 0$

Solution:

$x^2 - 17 = 0$
$x^2 = 17$
$x = \pm\sqrt{17}$ *Ans.*

2. $3x^2 = 75$

Solution:

$3x^2 = 75$
$x^2 = 25$
$x = \pm\sqrt{25}$
$x = \pm 5$ *Ans.*

3. $x^2 + 2 = 38$

Solution:

$x^2 + 2 = 38$
$x^2 = 36$
$x = \pm\sqrt{36}$
$x = \pm 6$ *Ans.*

EXERCISES

1. The solution set for $x^2 - 7x + 12 = 0$ is
(1) $\{3, -4\}$ (2) $\{-3, 4\}$ (3) $\{3, 4\}$ (4) $\{-3, -4\}$

2. Which is the solution set of $x^2 - 3x - 10 = 0$?
(1) $\{-5, 2\}$ (2) $\{5, -2\}$ (3) $\{2, 5\}$ (4) $\{-2, -5\}$

3. The solution set of $x^2 - x - 6 = 0$ is
(1) $\{1, -6\}$ (2) $\{-3, 2\}$ (3) $\{3, -2\}$ (4) $\{5, 1\}$

4. What is the positive root of the equation $(x - 4)(x + 3) = 0$?
5. Write the solution set of $x^2 - 2x - 35 = 0$.
6. The solution set of $x^2 - 2x = 0$ is
(1) $\{0\}$ (2) $\{2\}$ (3) $\{0, 2\}$ (4) $\{0, -2\}$
7. The solution set of $x^2 - 5x = 0$ is
(1) $\{0, -5\}$ (2) $\{0, 5\}$ (3) $\{0\}$ (4) $\{5\}$
8. The solution set of the equation $x^2 - 3x = 0$ is
(1) $\{3\}$ (2) $\{-3\}$ (3) $\{3, -3\}$ (4) $\{0, 3\}$
9. The solution set of $x^2 = 6x$ is
(1) $\{6\}$ (2) $\{0, 6\}$ (3) $\{0, -6\}$ (4) $\{-6\}$
10. The solution set of $2x^2 - 8x = 0$ is
(1) $\{0, 4\}$ (2) $\{0, -4\}$ (3) $\{2, 4\}$ (4) $\{2, -4\}$
11. What is the positive root of the equation $x^2 - 49 = 0$?
12. Solve for the positive value of x: $x^2 - 81 = 0$
13. The solution set of $x^2 - 25 = 0$ is
(1) $\{5\}$ (2) $\{-5\}$ (3) $\{\pm 5\}$ (4) $\{0, 5\}$
14. Find the positive root of $2x^2 = 72$.
15. The solution set of $2x^2 + 2 = 52$ contains two members. What is the larger member?

8.5 USING QUADRATIC EQUATIONS TO SOLVE PROBLEMS

Many types of verbal problems lead to quadratic equations. We will consider some of these problems here and others in Chapter 9. In using quadratic equations to solve these problems, we will apply the general procedure for solving verbal problems.

Examples

1. One positive number is 3 more than another. The sum of the squares of the two numbers is 17. Find the numbers.

Solution:

Let x = the smaller positive number.
Then $x + 3$ = the larger positive number.

$x^2 + (x + 3)^2 = 17$	The sum of the squares is 17.
$x^2 + (x + 3)(x + 3) = 17$	$(x + 3)^2 = (x + 3)(x + 3)$
$x^2 + x^2 + 6x + 9 = 17$	Multiply to remove parentheses.
$2x^2 + 6x + 9 = 17$	Combine x^2-terms to obtain a quadratic equation.
$2x^2 + 6x - 8 = 0$	Subtract 17 to put the quadratic equation in standard form.
$x^2 + 3x - 4 = 0$	Divide by 2 to simplify factoring.

$(x + 4)(x - 1) = 0$		Factor the trinomial.
$x + 4 = 0$	$x - 1 = 0$	Set each factor = 0.
$x = -4$	$x = 1$	Solve for x.

Reject $x = -4$ as a possible answer, since the problem calls for positive values.

$x = 1$	Smaller positive number.
$x + 3 = 1 + 3 = 4$	Larger positive number.

Check in the words of the problem. One positive number, 4, is 3 more than the other, 1.√ The sum of their squares, $4^2 + 1^2$ or $16 + 1$, is *17*.√

Answer: The numbers are 1 and 4.

2. One number is 11 more than another number. If the product of the two numbers is -24, find both numbers.

Solution:

Let x = the smaller number.
Then $x + 11$ = the larger number.

$(x) \cdot (x + 11) = -24$		The product of the two numbers is -24.
$x^2 + 11x = -24$		Distributive property.
$x^2 + 11x + 24 = 0$		Add 24 to put in form.
$(x + 8)(x + 3) = 0$		Factor the trinomial.
$x + 8 = 0$	$x + 3 = 0$	Set each factor = 0,
$x = -8$	$x = -3$	and solve for x.
$x + 11 = -8 + 11 = 3$	$x + 11 = -3 + 11 = 8$	Evaluate $x + 11$.

There are two possible sets of answers: -8 and $+3$ or -3 and $+8$. Check each set in the words of the problem.

$-8, +3$: One number, $+3$, is 11 more than the other, -8.

$$+3 \stackrel{?}{=} -8 + 11$$
$$+3 = +3 \checkmark$$

The product of the numbers, $(-8)(+3)$, is -24.√

$-3, +8$: One number, $+8$, is 11 more than the other, -3.

$$+8 \stackrel{?}{=} -3 + 11$$
$$+8 = +8 \checkmark$$

The product of the numbers, $(-3)(+8)$, is -24.√

Answer: The numbers are $-8, +3$ or $-3, +8$.

3. Find three consecutive positive odd integers such that the square of the smallest exceeds twice the largest by 7.

Solution:

Let x = the first, the smallest, of the consecutive positive odd integers
$x + 2$ = the second of the consecutive positive odd integers
$x + 4$ = the third, the largest, of the consecutive positive odd integers.

The square of the smallest (exceeds) twice the largest (by 7.)

$$(x)^2 = 2(x + 4) + 7$$

$x^2 = 2x + 8 + 7$	Distributive property.
$x^2 = 2x + 15$	Combine like terms.
$x^2 - 2x = 15$	Subtract $2x$. } For standard form.
$x^2 - 2x - 15 = 0$	Subtract 15. } For standard form.
$(x - 5)(x + 3) = 0$	Factor the trinomial.
$x - 5 = 0 \mid x + 3 = 0$	Set each factor = 0, and solve for x.
$x = 5 \mid x = -3$	

Reject $x = -3$, since the problem calls for positive numbers.

$x = 5$	Smallest 1st	consecutive positive odd integers
$x + 2 = 5 + 2 = 7$	2nd	
$x + 4 = 5 + 4 = 9$	Largest 3rd	

Check 5, 7, and 9 in the words of the problem. The square of the smallest, 5^2 or *25*, exceeds twice the largest by 7 (is 7 more than $2 \cdot 9$, is 7 more than 18, $18 + 7 = 25$).✓

Answer: The three consecutive positive odd integers are 5, 7, and 9.

4. When the first of three consecutive integers is multiplied by the third, the result is 1 less than six times the second. Find the integers.

Solution:

Let x = the first of the consecutive integers
$x + 1$ = the second of the consecutive integers
$x + 2$ = the third of the consecutive integers.

When the first is multiplied by the third, the result is

$$(x) \cdot (x + 2) =$$

(one less than) 6 times the second.

$$6(x + 1) \quad -1$$

	$x(x + 2) = 6(x + 1) - 1$	
	$x^2 + 2x = 6x + 6 - 1$	Distributive property.
	$x^2 + 2x = 6x + 5$	Combine like terms.
	$x^2 - 4x = 5$	Subtract $6x$. } For standard
	$x^2 - 4x - 5 = 0$	Subtract 5. } form.
	$(x - 5)(x + 1) = 0$	Factor the trinomial.
$x - 5 = 0$	$x + 1 = 0$	Set each factor = 0, and
$x = 5$	$x = -1$	solve for x.
$x + 1 = 5 + 1 = 6$	$x + 1 = -1 + 1 = 0$	Evaluate $x + 1$ and
$x + 2 = 5 + 2 = 7$	$x + 2 = -1 + 2 = 1$	$x + 2$.

There are two possible sets of answers: 5, 6, 7 or −1, 0, +1. Check each set in the words of the problem.

5, 6, 7: When the first, 5, is multiplied by the third, 7, the result, $5 \cdot 7$ or *35*, is one less than six times the second (one less than $6 \cdot 6$, $36 - 1 = 35$).✓

−1, 0, +1: When the first, −1, is multiplied by the third, +1, the result $(-1)(+1)$ or *−1*, is one less than six times the second (one less than $6 \cdot 0$ or 0, $0 - 1 = -1$).✓

Answer: The three consecutive integers are 5, 6, 7 or −1, 0, +1.

5. In an auditorium, 750 seats are arranged in rows with the same number of seats in each row. If the number of seats in each row is 5 more than the number of rows, find the number of rows.

Solution:

Let x = the number of rows.
Then $x + 5$ = the number of seats in each row.

the number of rows	•	the number of seats in each row	=	the total number of seats
x	•	$(x + 5)$	=	750

$$x^2 + 5x = 750$$
$$x^2 + 5x - 750 = 0$$
$$(x + 30)(x - 25) = 0$$

$x + 30 = 0$	$x - 25 = 0$
$x = -30$	$x = 25$
Reject.	$x + 5 = 25 + 5 = 30$

Check:

number of rows	•	number of seats in each row	=	total number of seats
25	•	30	$\stackrel{?}{=}$	750
		750	=	750✓

Answer: There are 25 rows in the auditorium.

EXERCISES

For each of the following problems, only an algebraic solution is acceptable.

1. One positive number is 4 more than another. The sum of their squares is 40. Find the numbers.
2. One number is 10 more than another number. If the product of the two numbers is -21, find both numbers.
3. If the square of a positive number is added to five times this number, the result is 36. Find the number.
4. The sum of a positive number and the square of its additive inverse is 30. What is the number?
5. One of two positive integers is 5 less than the other. If the product of the two integers is 36, find the integers.
6. The square of a positive number is 42 more than the number itself. What is the number?
7. The sum of the squares of two positive consecutive odd integers is 74. What are the integers?
8. Two numbers are positive consecutive integers. The sum of their squares is 25. Find the two integers.
9. Find two consecutive positive integers such that the square of the first decreased by 25 equals three times the second.
10. Find two consecutive integers such that the square of the larger diminished by five times the smaller is equal to 71.
11. Find three consecutive positive integers such that the product of the first integer and the third integer is 143. Check.
12. Find three consecutive positive integers such that the product of the second integer and the third integer is 30.
13. Find the smallest of three consecutive positive integers such that the product of the two smaller integers is 2 more than twice the largest integer.
14. Find three consecutive positive odd integers such that the square of the smallest exceeds twice the largest by 7.
15. Find three consecutive positive integers such that the square of the smallest integer exceeds the largest integer by 10.
16. Find three consecutive positive odd integers such that the product of the two smaller exceeds the largest by 8.
17. In an orchard, 160 trees are planted in rows, the same number of trees in each row. If the number of rows is 6 less than the number of trees in a row, find the number of rows.
18. There are 240 seats in the balcony of a theater. The number of seats in each row is 14 more than the number of rows. Find the number of rows. Check.

8.6 SUMMARY EXERCISES

In 1–3, answer true or false.

1. {Integers} is a subset of {Rational Numbers}.
2. Every irrational number is a real number.
3. Negative numbers are not real numbers.

4. Which of the following numbers is a rational number?
   (1) $\sqrt{1}$ (2) $\sqrt{2}$ (3) $\sqrt{3}$ (4) $\sqrt{5}$
5. Which of the following numbers is an irrational number?
   (1) $\sqrt{\frac{16}{81}}$ (2) $\sqrt{.09}$ (3) $\sqrt{30}$ (4) .333 . . .
6. Which value of x will make the expression $\sqrt{38 - x}$ a positive integer?
   (1) 1 (2) 38 (3) 0 (4) 2
7. For which value of x is the expression $\frac{7}{\sqrt{x + 1}}$ undefined?
   (1) 0 (2) 1 (3) -1 (4) 7
8. Which of the following numbers is not a real number?
   (1) $-\sqrt{9}$ (2) $\sqrt{-9}$ (3) -9 (4) $\sqrt[3]{-9}$

In 9–11, approximate the square root to the nearest tenth.

9. $\sqrt{94}$ **10.** $\sqrt{47}$ **11.** $\sqrt{115}$

12. The expression $\sqrt{400}$ is equivalent to
    (1) $10\sqrt{20}$ (2) $40\sqrt{10}$ (3) $4\sqrt{10}$ (4) 20
13. The expression $\frac{1}{5}\sqrt{500}$ is equivalent to
    (1) $2\sqrt{5}$ (2) 10 (3) 20 (4) 100
14. Which of the following is equivalent to $4\sqrt{3}$?
    (1) $\sqrt{12}$ (2) $\sqrt{24}$ (3) $\sqrt{48}$ (4) $\sqrt{144}$

In 15–17, write each expression in simplest radical form.

15. $\sqrt{54}$ **16.** $3\sqrt{128}$ **17.** $\frac{1}{4}\sqrt{96}$

18. Combine: $3\sqrt{2} - \sqrt{3} - \sqrt{2}$
19. The sum of $\sqrt{3}$ and $\sqrt{27}$ is
    (1) $\sqrt{30}$ (2) $10\sqrt{3}$ (3) $3\sqrt{3}$ (4) $4\sqrt{3}$
20. The sum of $3\sqrt{15}$ and $2\sqrt{60}$ is
    (1) $5\sqrt{75}$ (2) $7\sqrt{15}$ (3) $9\sqrt{15}$ (4) $11\sqrt{15}$
21. The difference between $\sqrt{80}$ and $\sqrt{45}$ is
    (1) $\sqrt{35}$ (2) $7\sqrt{5}$ (3) $5\sqrt{5}$ (4) $\sqrt{5}$
22. The difference between $2\sqrt{72}$ and $\sqrt{8}$ is
    (1) $10\sqrt{2}$ (2) $8\sqrt{2}$ (3) $6\sqrt{2}$ (4) $\sqrt{2}$
23. Express in simplest form: $3\sqrt{28} - 2\sqrt{63}$
24. Express the product of $5\sqrt{2}$ and $3\sqrt{2}$ as an integer.

25. Express the product of $\sqrt{72}$ and $\sqrt{2}$ as an integer.

26. The expression $(3\sqrt{5})(4\sqrt{15})$ is equivalent to
(1) $7\sqrt{20}$ (2) $12\sqrt{20}$ (3) $60\sqrt{3}$ (4) $300\sqrt{3}$

27. Express the quotient of $8\sqrt{300}$ and $2\sqrt{3}$ as an integer.

28. Divide and simplify: $\dfrac{10\sqrt{180}}{2\sqrt{3}}$

In 29–32, simplify and combine wherever possible.

29. $3\sqrt{27} + 4\sqrt{48} - \sqrt{75}$

30. $(8\sqrt{6})(5\sqrt{54})$

31. $\dfrac{75\sqrt{108}}{3\sqrt{3}}$

32. $5\sqrt{2}(6\sqrt{8} + \sqrt{20})$

In 33–42, find the solution set for each quadratic equation.

33. $x^2 - 2x - 15 = 0$

34. $x^2 + 4x = 12$

35. $x^2 - 6x = 0$

36. $x^2 = 8x$

37. $2x^2 - 16x = 0$

38. $x^2 + 26x + 25 = 0$

39. $x^2 - 64 = 0$

40. $x^2 - 10 = 0$

41. $3x^2 - 75 = 0$

42. $2x^2 - 5x - 3 = 0$

In 43–50, write an algebraic solution for each problem. Check.

43. The sum of a number and the square of its additive inverse is 72. Find the number.

44. One positive number is 4 less than another. The sum of the squares of the two numbers is 58. Find the numbers.

45. One number is 9 more than another. If the product of the two numbers is -20, find the numbers.

46. Find three consecutive positive even integers such that the square of the smallest exceeds three times the largest by 6.

47. When the first of three consecutive integers is multiplied by the third, the result is 1 less than nine times the second. Find the integers.

48. Find three consecutive positive integers such that twice the product of the first and third is 2 less than twenty times the second.

49. In a lecture hall, there are 180 seats. The number of seats in each row is 8 more than the number of rows. Find the number of rows.

50. A farmer set out 600 tomato plants. The number of plants in each row was 10 less than twice the number of rows. How many plants were there in each row?

CUMULATIVE SUMMARY EXERCISES

CHAPTERS 1–8

1. Let p represent "The polygon has exactly 3 sides," and let q represent "All angles of the polygon are right angles." Which statement is true if the polygon is a rectangle?
(1) $p \wedge q$ (2) $p \vee q$ (3) p (4) $\sim q$
2. What is the inverse of the statement "If x is an even integer, then it is divisible by 2"?
(1) If x is an odd integer, then it is divisible by 2.
(2) If x is an odd integer, then it is not divisible by 2.
(3) If x is an even integer, then it is not divisible by 2.
(4) If x is divisible by 2, then it is an even integer.
3. **a.** Copy and complete the truth table for the statement:

$$[q \wedge \sim(p \wedge q)] \leftrightarrow (q \wedge \sim p)$$

p	q	$p \wedge q$	$\sim(p \wedge q)$	$q \wedge \sim(p \wedge q)$	$\sim p$	$q \wedge \sim p$	$[q \wedge \sim(p \wedge q)] \leftrightarrow (q \wedge \sim p)$

b. Is $[q \wedge \sim(p \wedge q)] \leftrightarrow (q \wedge \sim p)$ a tautology?
4. If $p \rightarrow q$ is true, which related conditional must also be true?
(1) $q \rightarrow p$ (2) $\sim p \rightarrow \sim q$ (3) $\sim q \rightarrow \sim p$ (4) $\sim q \rightarrow p$
5. If p represents "$a \cdot b = 0$," and q represents "either $a = 0$ or $b = 0$," write in symbolic form, using p and q: "$a \cdot b = 0$ if and only if either $a = 0$ or $b = 0$."
6. Given the true statement "A square has four right angles." Which statement must also be true?
(1) If a polygon is not a square, then it does not have four right angles.
(2) If a polygon does not have four right angles, then it is not a square.
(3) If a polygon has four right angles, then it is a square.
(4) If a triangle has three right angles, then it is equiangular.
7. Which is the inverse of the statement "If $2x \geq 6$, then $x \geq 3$"?
(1) If $2x < 6$, then $x \geq 3$.
(2) If $2x < 6$, then $x < 3$.
(3) If $x \geq 3$, then $2x \geq 6$.
(4) If $2x \geq 6$, then $x < 3$.

8. **a.** Copy and complete the truth table for the statement:

$$\sim(p \wedge q) \leftrightarrow [(\sim p) \vee (\sim q)]$$

p	q	$p \wedge q$	$\sim(p \wedge q)$	$\sim p$	$\sim q$	$(\sim p) \vee (\sim q)$	$\sim(p \wedge q) \leftrightarrow [(\sim p) \vee (\sim q)]$

b. Determine if the statement is a tautology.

c. Let p represent "I am tired." Let q represent "I do sleep." Which sentence is logically equivalent to $\sim(p \wedge q)$?
(1) I am tired or I do sleep.
(2) I am not tired or I do not sleep.
(3) I am not tired and I do sleep.
(4) I am not tired and I do not sleep.

9. Which integer, when substituted for x, makes the following statement true? $(x > -3) \wedge (x < -1)$

10. If p represents "She lives in the USA," and q represents "She lives in the western hemisphere," write in symbolic form, using p and q: "If she does not live in the western hemisphere, then she does not live in the USA."

11. Let p represent "x is a prime number," and let q represent "x is an even number." For what value of x is $p \wedge q$ true?
(1) 1 (2) 2 (3) 3 (4) 4

12. What is the inverse of the statement "If x is a prime, then it has no factors other than x and 1"?
(1) If x is a composite, then it has factors other than x and 1.
(2) If x is a composite, then it has no factors other than x and 1.
(3) If x has no factors other than x and 1, then x is a prime.
(4) If x has factors other than x and 1, then x is a prime.

13. **a.** Construct a truth table for the statement $(p \rightarrow q) \leftrightarrow (p \vee q)$.
b. Is the statement in part **a** a tautology?

In 14–24, fill in the word "positive," "negative," or "zero."

14. The additive inverse of a positive number is _____.
15. _____ has no multiplicative inverse.
16. The sum of two positive numbers is _____.
17. The product of two numbers with like signs is _____.
18. The sum of two negative numbers is _____.
19. The quotient of two negative numbers is _____.
20. The absolute value of a nonzero number is _____.
21. The principal square root of a nonzero number is _____.
22. An even root of a _____ number is not a real number.
23. The absolute value of a negative number is _____.
24. The additive inverse of zero is _____.

25. Which is an irrational number?
(1) .125 (2) .131313 . . . (3) $\sqrt{25}$ (4) $\sqrt{3}$

26. Which is not a real number?
(1) -25 (2) $-\sqrt{25}$ (3) $\sqrt{-25}$ (4) $\sqrt[3]{-25}$

27. Which of the following is an infinite set?
(1) {positive integers less than 1,000}
(2) {integers less than 1,000}
(3) {positive odd integers less than 1,000}
(4) {positive even integers less than 1,000}

28. Evaluate: $\frac{|16 - 2|}{|-7|}$

29. For what value of x is $\frac{10}{x - 3}$ undefined?

For each statement in 30–33, write the number of the property of the real number system, chosen from the list below, that justifies the statement.

Properties

(1) Additive inverse property
(2) Multiplicative inverse property
(3) Additive identity property
(4) Multiplicative identity property

30. $2 + (-2) = 0$ **31.** $2 \cdot \frac{1}{2} = 1$ **32.** $2 \cdot 1 = 2$ **33.** $2 + 0 = 2$

For each statement in 34–36, write the number of the property of the real number system, chosen from the list below, that justifies the statement.

Properties

(1) Associative property of addition
(2) Commutative property of addition
(3) Associative property of multiplication
(4) Commutative property of multiplication
(5) Distributive property of multiplication over addition

34. $(5 + 6) + 7 = 5 + (6 + 7)$ **35.** $5(6 + 7) = 5(6) + 5(7)$
36. $5 \cdot 6 = 6 \cdot 5$

37. Which fraction has the greatest value?
(1) $\frac{1}{9}$ (2) $\frac{2}{9}$ (3) $\frac{1}{6}$ (4) $\frac{1}{3}$

38. Which of the following terms is a monomial?
(1) $3xy^2$ (2) $\frac{3x}{y^2}$ (3) $3x + y^2$ (4) $3x - y^2$

39. Express the sum of $7x^2 - 3x + 4$ and $-2x^2 + 8x - 1$ as a trinomial.

40. From the sum of $x + 2y$ and $x - y$, subtract $-x + y$.

41. When $6x^2y^2$ is multiplied by $3xy^4$, the product is
(1) $18x^2y^8$ (2) $18x^2y^6$ (3) $18x^3y^6$ (4) $9x^3y^6$

In 42–44, multiply each pair of binomials.

42. $(x - 3)(x + 4)$ **43.** $(x - 2)(x + 2)$ **44.** $(2x - 5)(x - 1)$

45. Divide and express the result in simplest form: $\dfrac{-20x^3yz^2}{4xyz}$

In 46–48, if $x = -2$ and $y = -3$, find the numerical value.

46. x^2y **47.** $-xy^2$ **48.** $-x + (xy)^2$

49. What is the value of $7x^0$?
50. The value of 5^{-2} is (1) -10 (2) -25 (3) $\frac{1}{25}$ (4) $-\frac{1}{25}$
51. What is the degree of the polynomial $8 - 7x^4$?
52. Write the polynomial $3x - 4 + 7x^2 - x^4$ in descending order.

In 53–55, solve and check.

53. $\dfrac{x}{4} + 7 = -9$ **54.** $.3x + .1 = 6.7$ **55.** $3(x - 4) - (x + 1) = 11$

56. Solve for x in terms of a, b, and c: $ax - bx = c$
57. Solve for x: $3x < 6x + 9$
58. If the replacement set is $\{-2, -1, 0, 1, 2\}$, find the solution set of $3x - 2 \leq 1$.

59. Which solution set is represented by the accompanying graph?
(1) $\{x \mid -2 < x \leq 1\}$
(2) $\{x \mid -2 \leq x < 1\}$
(3) $\{x \mid -2 < x < 1\}$
(4) $\{x \mid -2 \leq x \leq 1\}$

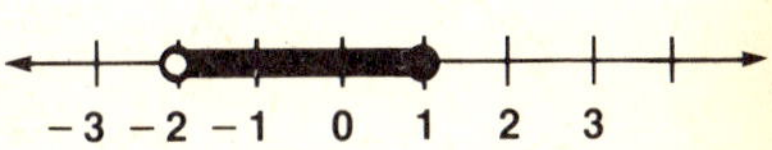

60. Which graph represents the inequality $(x < -2) \vee (x \geq 1)$?

(1) 

(3) 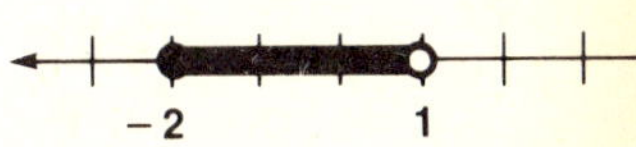

(2) 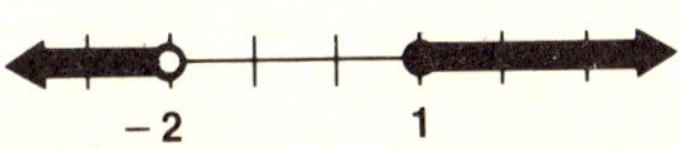

(4) 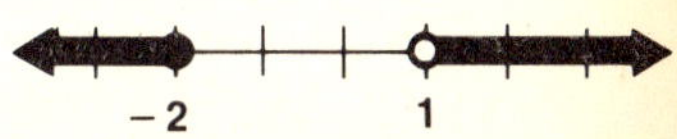

61. If $(x \geq 1) \wedge (x > 3)$, which of the following may be a value of x?
(1) 1 (2) 2 (3) 3 (4) 4
62. If two supplementary angles are in the ratio 1:8, find the number of degrees in the smaller angle.
63. Which of the following is a true relation between the supplement and complement of any acute angle?
(1) The measure of the supplement is equal to the measure of the complement.
(2) The sum of the measures of the supplement and complement is 180°.
(3) The sum of the measures of the supplement and complement is 90°.
(4) The difference between the measures of the supplement and complement is 90°.

64. The larger of two complementary angles is 10° more than three times the smaller. Find the measure of each angle.

65. Which statement is always true?
(1) A rhombus is a square.
(2) A rectangle is a parallelogram.
(3) A parallelogram is a square.
(4) A parallelogram is a rectangle.

66. In the accompanying diagram, $\overrightarrow{BCD}$, $\overrightarrow{BEF}$; $m\angle 1 = 90$, and $m\angle 2 = 30$. Find the number of degrees in the measure of angle 4.

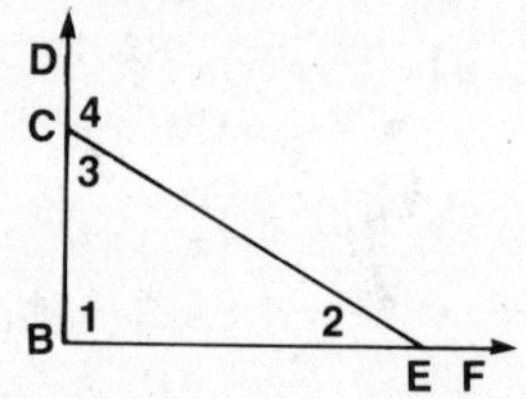

67. The degree measures of the three interior angles of a triangle are $3x$, $4x - 10$, and $18x - 60$.
a. Find the value of x.
b. Find the number of degrees in each of the interior angles of the triangle.
c. Based on your answer to part **b**, which is the most accurate description of the triangle?
(1) obtuse and isosceles
(2) isosceles but not obtuse
(3) obtuse but not isosceles
(4) not obtuse and not isosceles

68. In the diagram of triangle ABC, $\overline{BD}$ bisects $\angle ABC$ and $\overline{CDE}$ bisects $\angle ACB$. If $m\angle BDE = 70$ and $m\angle BCE = 40$, find $m\angle A$.

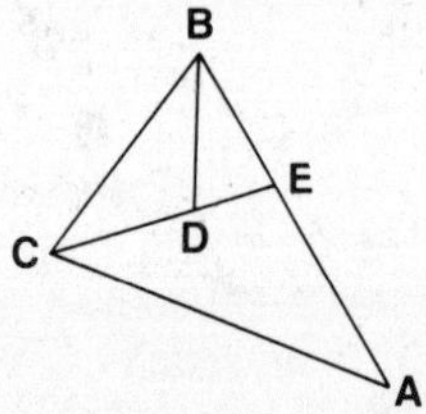

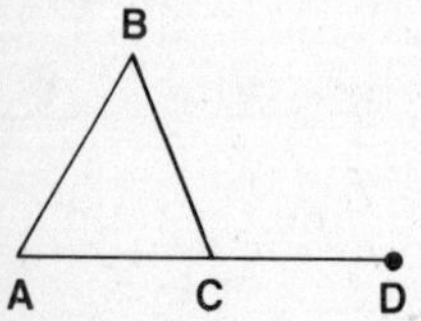

69. In isosceles triangle ABC, $\overline{AC}$ is extended to D, $\overline{AB} \cong \overline{BC}$, and $m\angle B = 40$. What is the number of degrees in the measure of exterior angle BCD?

70. In triangle ABC, $\angle B$ is 10° more than $\angle A$ and $\angle C$ is 20° less than $\angle B$. Find the measures of each of the three angles.

71. As shown in the accompanying diagram $\overleftrightarrow{AB}$ and $\overleftrightarrow{CD}$ intersect at E. If $m\angle AED = 2x + 25$ and $m\angle CEB = 4x - 25$, find the measure of $\angle AEC$.

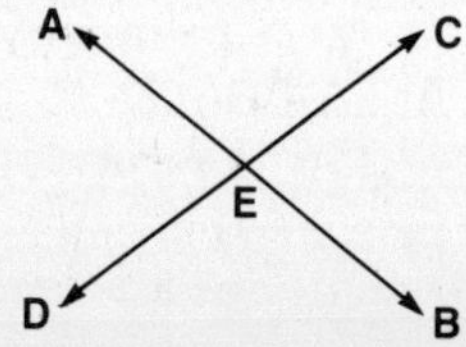

In 72–75, use the accompanying diagram that shows $\overleftrightarrow{AB} \parallel \overleftrightarrow{CD}$ and transveral $\overleftrightarrow{EH}$ intersects $\overleftrightarrow{AB}$ at F and $\overleftrightarrow{CD}$ at G. If $m\angle BFG = 3x - 10$ and $m\angle CGF = x + 50$, find the measure of

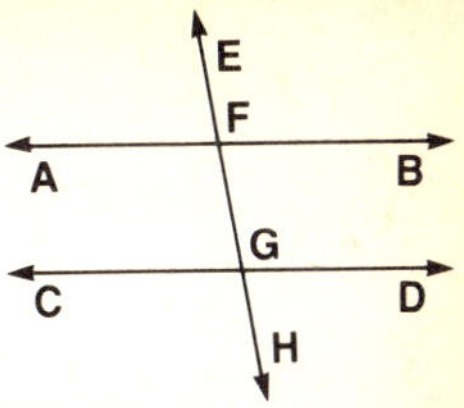

72. $\angle BFG$ **74.** $\angle DGH$

73. $\angle AFG$ **75.** $\angle AFE$

76. In the diagram, $\overleftrightarrow{AB} \parallel \overleftrightarrow{CD}$, $\overline{AE} \perp \overline{CB}$, $m\angle BCD = 2x$, and $m\angle BAE = 3x$. Find the value of x.

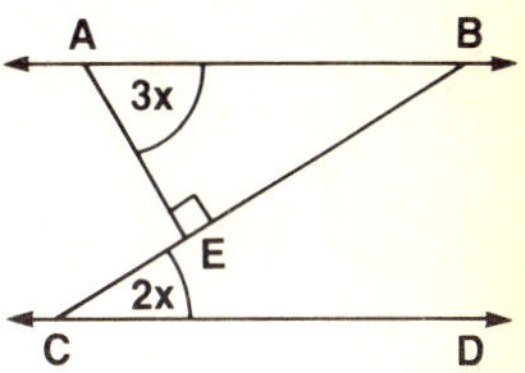

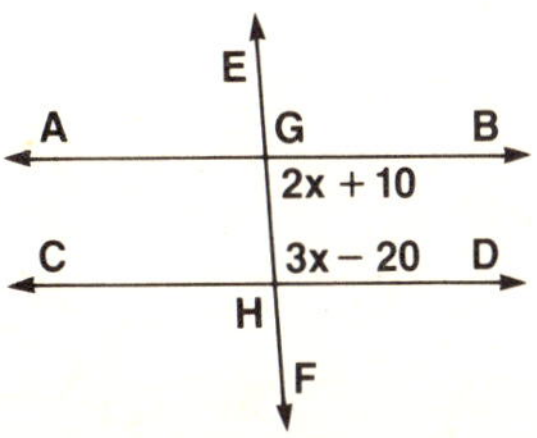

77. In the figure, two parallel lines $\overleftrightarrow{AB}$ and $\overleftrightarrow{CD}$ are cut by transversal $\overleftrightarrow{EF}$ intersecting $\overleftrightarrow{AB}$ at G and $\overleftrightarrow{CD}$ at H. If $m\angle BGH = (2x + 10)$ and $m\angle GHD = (3x - 20)$, find the value of x.

In 78–82, use the accompanying diagram that shows $\overleftrightarrow{AB} \parallel \overleftrightarrow{CD}$ and transversal $\overleftrightarrow{EH}$ intersects $\overleftrightarrow{AB}$ at F and $\overleftrightarrow{CD}$ at G. $\overrightarrow{FQ}$ and $\overrightarrow{GR}$ are the respective angle bisectors of $\angle BFG$ and $\angle DGH$. If $m\angle HGR = 80°$, find the measure of

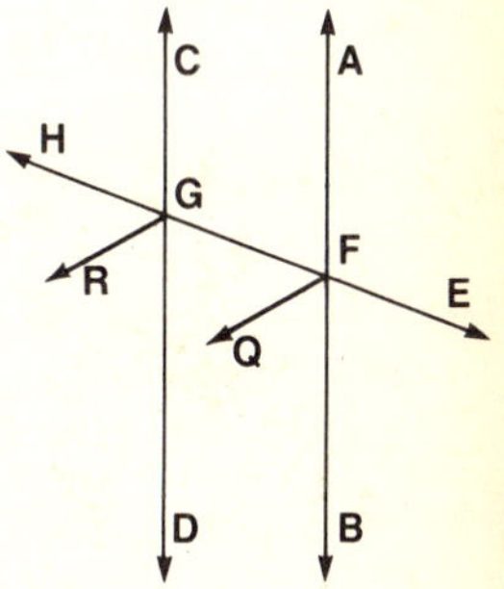

78. $\angle HGD$ **81.** $\angle CGH$

79. $\angle QFB$ **82.** $\angle AFG$

80. $\angle AFE$

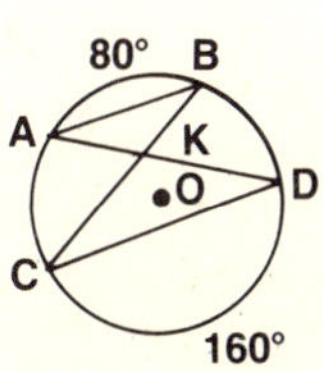

83. As shown in the accompanying diagram, chords $\overline{AB}$, $\overline{CD}$, $\overline{AD}$, and $\overline{BC}$ are drawn in circle O. If $m\overset{\frown}{AB} = 80°$ and $m\overset{\frown}{CD} = 160°$, find the measure of $\angle AKB$.

In 84–87, use the accompanying diagram that shows diameters $\overline{AB}$ and $\overline{CD}$ and chord $\overline{AC}$ drawn in circle O. If $m\overset{\frown}{AC} = 60°$, find the measures of

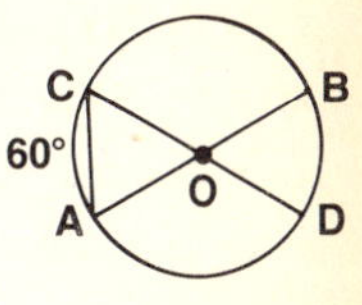

84. $\angle COA$ **85.** $\overset{\frown}{BD}$ **86.** $\overset{\frown}{AD}$ **87.** $\angle ACD$

In 88–90, use the accompanying diagram that shows chords $\overline{FC}$, $\overline{FD}$, and $\overline{CE}$ drawn in circle O. If $m\overset{\frown}{DE} : m\overset{\frown}{EF} : m\overset{\frown}{FC} : m\overset{\frown}{CD} = 6:9:12:9$, find the measures of

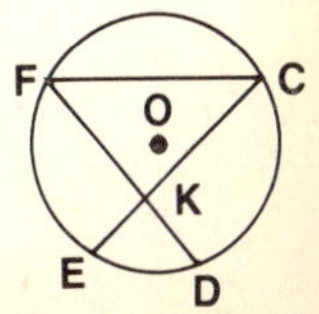

88. $\overset{\frown}{DE}, \overset{\frown}{EF}, \overset{\frown}{FC}, \overset{\frown}{CD}$ **89.** $\angle CFD$ **90.** $\angle FKC$

91. A football team won 11 out of 15 games. The ratio of "wins" to "losses" is (1) 11:15 (2) 4:11 (3) 11:4 (4) 4:15

92. Solve for x: $\frac{1}{5} = \frac{x-2}{15}$

93. On a map, 1 cm represents 50 km. If two cities are 275 km apart, how many cm apart are they on the map?

94. How many minutes are there in h hours and m minutes?

95. Write $8\frac{1}{4}\%$ as a decimal.

96. If 40% of a number is 70, what is the number?

97. A test was passed by 75% of a class. If 27 students passed the test, how many students were in the class?

98. If 42 students took a test and 35 passed, what percent of the students passed the test?

99. If the sales tax rate is 8%, what is the amount of tax that must be paid on an item that costs $6.50?

100. The sale price of an article was $102 after a 15% discount had been given. What was the original cost of the article?

101. In the accompanying diagram, $\triangle ABC \sim \triangle DEF$. If $\angle A \cong \angle D$, $\angle B \cong \angle E$, $AB = 10$, $AC = 3$, and $DE = 5$, find the length of DF.

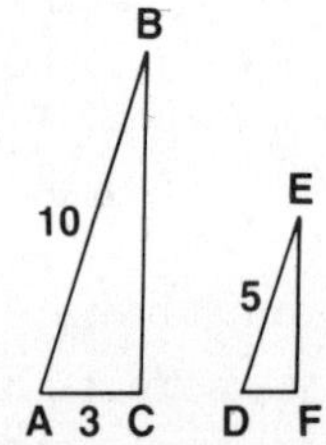

102. Right triangles ABC and PQR are similar. In $\triangle ABC$, the lengths of the legs are 5 and 12. In $\triangle PQR$, the length of the shorter of the two legs is 15. Find the length of the longer leg of $\triangle PQR$.

103. A tree casts a shadow 18 feet long at the same time that a person 6 feet tall casts a shadow 12 feet long. Find the height of the tree.

104. In the accompanying diagram, triangle ABC is congruent to triangle QPR, $\overline{AB} \cong \overline{QP}$, $\angle B \cong \angle P$, and $\overline{CB} \cong \overline{RP}$. What angle of $\triangle QPR$ is congruent to $\angle ACB$?

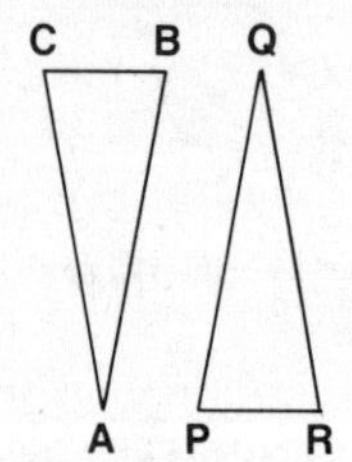

105. Which of the following is not a method of proving triangles congruent?

(1) a.a.s. ≅ a.a.s. (3) s.s.s. ≅ s.s.s.
(2) a.s.a. ≅ a.s.a. (4) s.s.a. ≅ s.s.a.

106. Construct a triangle similar to $\triangle ABC$. Use $\overline{PQ}$ as a base.

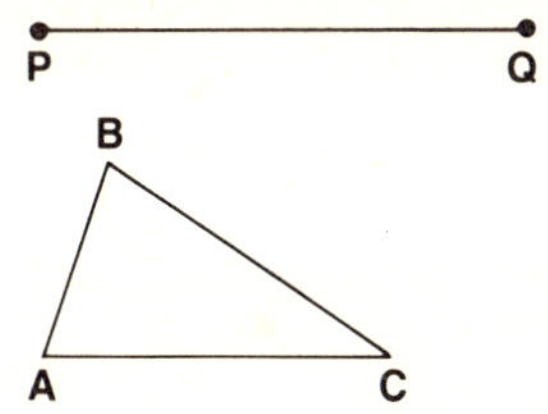

107. Construct a triangle congruent to $\triangle ABC$.

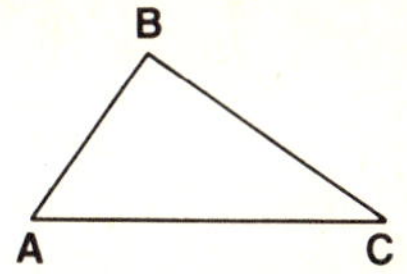

108. Factor 100 into primes.
(1) $50 \cdot 2$ (2) $25 \cdot 4$ (3) $2 \cdot 3 \cdot 5$ (4) $2^2 \cdot 5^2$

109. Find the greatest common factor of 12, 24, and 38.

In 110–115, factor as completely as possible.

110. $7x - 14xy$
111. $x^3 - 5x^2$
112. $x^2 - 81$
113. $ax^2 + 2ax - 8a$
114. $100 - x^2$
115. $3x^2 + 9x + 6$

In 116–118, express each fraction in lowest terms.

116. $\frac{27x^6}{3x^2}$ **117.** $\frac{x^2 - 25}{x + 5}$ **118.** $\frac{x^2 - 2x - 3}{x - 3}$

In 119–122, express the sum or difference of the fractions as a single fraction in lowest terms.

119. $\frac{11}{x} - \frac{3}{x}$
120. $\frac{x}{5} + \frac{x}{20}$
121. $\frac{x + 1}{4} + \frac{x - 3}{12}$
122. $\frac{x + 3}{3} - \frac{x - 1}{5}$

In 123–126, multiply or divide as indicated and express the answer in lowest terms.

123. $\frac{3}{x^2} \cdot \frac{x}{6}$
124. $\frac{5}{x^2} \div \frac{25}{x^4}$
125. $\frac{(x - 3)^2}{x^2 - 9} \cdot \frac{5x + 25}{x + 5}$
126. $\frac{x}{x^2 - 49} \div \frac{5x^3}{x^2 + 9x + 14}$

In 127–129, solve the equation and check.

127. $\frac{x}{2} - \frac{x}{8} = 3$ **128.** $\frac{5x}{3} = 5 - \frac{5x}{6}$ **129.** $\frac{x + 1}{7} = \frac{1}{14}$

130. If eight times a number is decreased by one-fifth of itself, the result is 117. Find the number.

131. The larger of two numbers is 1 less than five times the smaller. If the larger number is divided by the smaller number, the quotient is 4 and the remainder is 5. Find the numbers.

132. Which of the following numbers is not a real number?
(1) $\sqrt[3]{-27}$ (2) $-\sqrt[3]{27}$ (3) $-\sqrt{25}$ (4) $\sqrt{-25}$

133. Approximate $\sqrt{85}$ to the nearest tenth.

134. Which of the following is equivalent to $3\sqrt{5}$?
(1) $\sqrt{15}$ (2) $\sqrt{30}$ (3) $\sqrt{45}$ (4) $\sqrt{60}$

135. The sum of $\sqrt{5}$ and $\sqrt{125}$ is
(1) $\sqrt{130}$ (2) $26\sqrt{5}$ (3) $5\sqrt{5}$ (4) $6\sqrt{5}$

136. Write $(5\sqrt{3})^2$ in simplest form.

137. Express the product of $3\sqrt{50}$ and $2\sqrt{2}$ in simplest form.

138. Divide and simplify: $\dfrac{24\sqrt{150}}{8\sqrt{3}}$

139. Simplify and combine: $5\sqrt{27} + 3\sqrt{48} - \sqrt{108}$

In 140–144, find the solution set for each quadratic equation.

140. $x^2 - 11x + 28 = 0$

141. $x^2 - 3x = 0$

142. $x^2 + 6x = 160$

143. $x^2 - 121 = 0$

144. $2x^2 - 4x - 6 = 0$

145. When the first of three consecutive integers is multiplied by the third, the result is 8 more than eight times the second. Find the integers.

CHAPTER 9

Measurement of Geometric Figures

9.1 PERIMETER OF A POLYGON

The ***perimeter*** of a figure is the sum of the lengths of its sides. If the figure is a polygon, all the sides are straight line segments.

In general, the sides of a polygon are of different lengths. If two sides of a polygon, such as a triangle and a trapezoid, are equal in length, the polygon is called *isosceles*. If all sides of a polygon are equal in length, the polygon is called *equilateral*. A ***regular polygon*** is both equilateral and equiangular.

Examples

1. The perimeter of a triangle is 28. If the lengths of the sides are represented by x, $x - 2$, and $x + 3$, find the length of the longest side.

Solution:

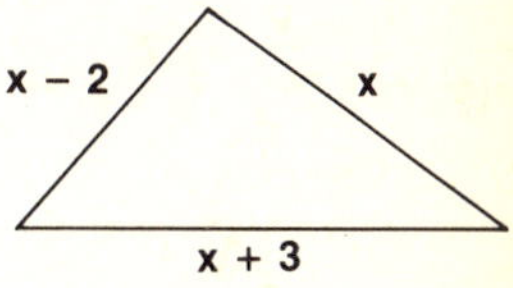

The perimeter of a triangle is equal to the sum of the lengths of the three sides.

$$(x) + (x - 2) + (x + 3) = 28$$
$$3x + 1 = 28$$
$$3x = 27$$
$$x = 9$$
$$x - 2 = 7$$
$$x + 3 = 12$$

Check:

$$9 + 7 + 12 \stackrel{?}{=} 28$$
$$16 + 12 \stackrel{?}{=} 28$$
$$28 = 28 \checkmark$$

Answer: The length of the longest side of the triangle is 12.

2. If each side of a square is multiplied by 4, the perimeter of the square
   (1) remains the same
   (2) is increased by 4
   (3) is multiplied by 4
   (4) is multiplied by 16

Solution:

If a side of the square is s, the perimeter is $s + s + s + s$ or $P = 4s$.
Replace s by $4s$: $P = 4s = 4(4s) = 16s$
Thus, the new perimeter, $16s$, is four times the old perimeter, $4s$.

Answer: (3)

3. An architect wishes to design a rectangular room so that its length is 6 feet more than its width, and its perimeter is greater than 52 feet. If each of the dimensions of the room must be a whole number of feet, what are the smallest possible measures in feet of the length and width?

Solution:

Let x = the number of feet in the width of the rectangular room.
Then $x + 6$ = the number of feet in the length of the rectangular room.

$$\underbrace{\text{the perimeter}} \quad \underbrace{\text{is greater than}} \quad \underbrace{52}$$

$$(x) + (x + 6) + (x) + (x + 6) > 52$$
$$4x + 12 > 52$$
$$4x > 40$$
$$x > 10$$

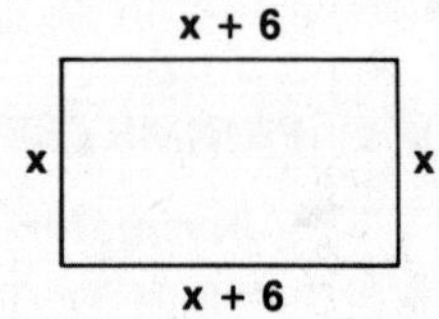

The smallest whole number > 10 is 11. Thus, the width is 11 feet and the length is 11 + 6 or 17 feet.

Check:

$$11 + 17 + 11 + 17 \overset{?}{>} 52$$
$$28 + 28 \overset{?}{>} 52$$
$$56 > 52 \checkmark$$

Answer: The smallest possible integral (whole-number) measures are 11 feet for the width and 17 feet for the length.

4. The length of a rectangle is twice the length of a side of an equilateral triangle. The width of the rectangle is 1 more than a side of the equilateral triangle. If the perimeter of the rectangle is 11 more than the perimeter of the equilateral triangle, find the length of a side of the equilateral triangle.

Solution:

Let x = the length of a side of the equilateral triangle.
Then $2x$ = the length of the rectangle.
And, $x + 1$ = the width of the rectangle.

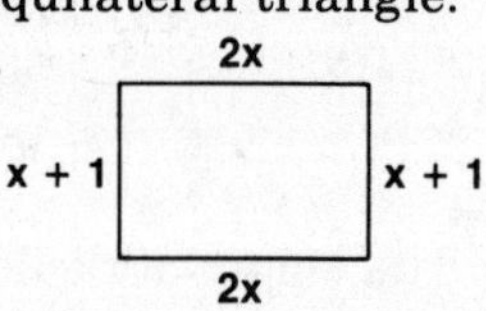

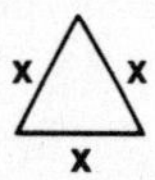

$$P_{\text{rectangle}} = P_{\text{equilateral triangle}} + 11$$
$$(x + 1) + (2x) + (x + 1) + 2x = (x) + (x) + (x) + 11$$
$$6x + 2 = 3x + 11$$
$$3x + 2 = 11$$
$$3x = 9$$
$$x = 3$$
$$2x = 6$$
$$x + 1 = 4$$

Check:

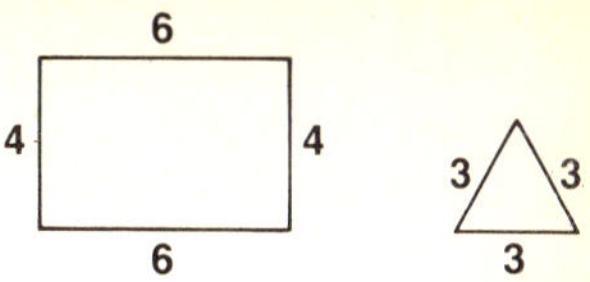

$$(4) + (6) + (4) + (6) \stackrel{?}{=} (3) + (3) + (3) + 11$$
$$10 + 10 \stackrel{?}{=} 6 + 14$$
$$20 = 20\checkmark$$

Answer: The length of a side of the equilaterial triangle is 3.

Some figures may be more complicated in structure. In these cases, the perimeter of the figure (the distance around) is calculated by studying the figure, breaking it down into components, and adding all the actual sides.

Example

A garden is in the shape of four equal squares joined as shown. If the length of the side of each square is 10 feet, how many feet of fencing are required for the outer perimeter of the garden?

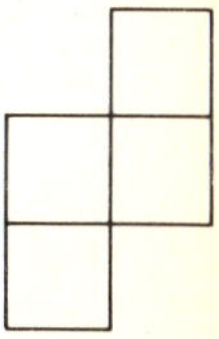

Solution:

The solid lines represent the fence around the garden. The perimeter is 10 lengths of 10, or 100 feet.

Answer: 100 feet of fencing.

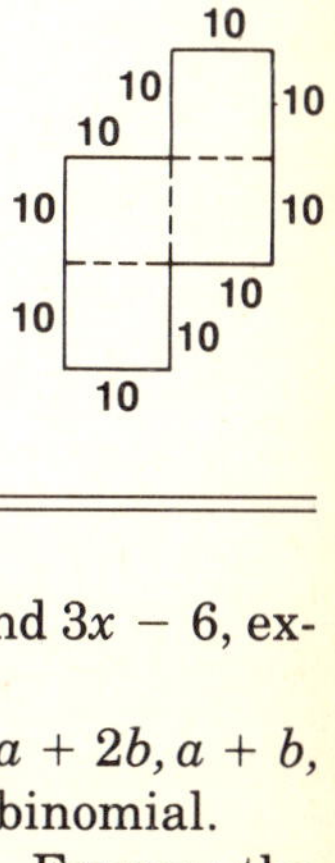

EXERCISES

1. If the sides of a triangle are represented by $2x$, $x + 5$, and $3x - 6$, express the perimeter of the triangle in terms of x.
2. The lengths of the sides of a triangle are represented by $a + 2b$, $a + b$, and $2a - b$. Express the perimeter of the triangle as a binomial.
3. The perimeter of a square is represented by $12x - 4$. Express the length of one side of the square in terms of x.
4. One side of a square is represented by $x + 3$. Express the perimeter of the square in terms of x.
5. A side of an equilateral triangle is represented by $x + 3$. The perimeter of the triangle is represented by
   (1) $3x + 9$ (2) $3x + 3$ (3) $3x + 6$ (4) $9x$
6. If each side of regular pentagon $ABCDE$ is represented by $(3x + 2)$, which expression represents the perimeter of the pentagon?
   (1) $15x + 10$ (2) $15x + 2$ (3) $8x + 7$ (4) $25x$

7. Each side of regular hexagon $ABCDEF$ is represented by $(2x + 3)$. Which expression represents the perimeter of the hexagon?
(1) $12x + 18$ (2) $12x + 3$ (3) $8x + 9$ (4) $30x$
8. The perimeter of a square is 36. What is the length of one side of the square?
9. If each side of a square is tripled, the perimeter of the square
(1) remains the same (3) is multiplied by 3
(2) is increased by 3 (4) is multiplied by 9
10. If each side of a square is doubled, then its perimeter
(1) is doubled (3) is halved
(2) is multiplied by 4 (4) stays the same
11. If the length of a rectangle is doubled and the width is also doubled, then the perimeter must be
(1) doubled (3) increased by 4
(2) increased by 2 (4) multiplied by 4

In 12–19, only an algebraic solution is acceptable.

12. The length of a rectangle is 1 centimeter less than twice the width. If the perimeter of the rectangle is 76 centimeters, find the number of centimeters in each dimension of the rectangle.
13. The length of a rectangle is 2 less than twice the width. If the perimeter of the rectangle is 92, find the numerical value of both the length and the width of the rectangle.
14. The base of an isosceles triangle is 2 less than the sum of the lengths of the two legs. If the perimeter of the triangle is 22, find the length of the base of the triangle.
15. Two sides of a triangle are equal. The length of the third side exceeds the length of each of the equal sides by 3 inches. The perimeter of the triangle is 21 inches. Find the number of inches in the length of each of the equal sides.
16. A side of a square is 7 feet longer than a side of an equilateral triangle. The perimeter of the square is twice the perimeter of the triangle. Find the length of a side of the triangle.
17. The length of a rectangle is twice the length of a side of a square and the width of the rectangle is 1 less than the side of the square. If the perimeter of the rectangle is 8 more than the perimeter of the square, find the length of a side of the square.
18. An architect wants to design a rectangular room so that its length is 8 meters more than its width, and its perimeter is greater than 56 meters. If each of the dimensions of the room must be a whole number of meters, what are the smallest possible measures in meters of the length and width?
19. Each leg of an isosceles triangle is 5 more than the base of the triangle. If the perimeter of the triangle is greater than 70, what are the smallest possible integral measures of the sides of the triangle?

20. A design is composed of two equilateral triangles and three squares joined as shown. A side of each equilateral triangle and of each square is 3 inches. If the outer perimeter of the design is to be trimmed in red tape, how many inches of tape are required?

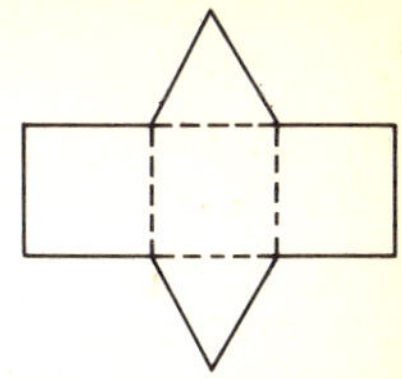

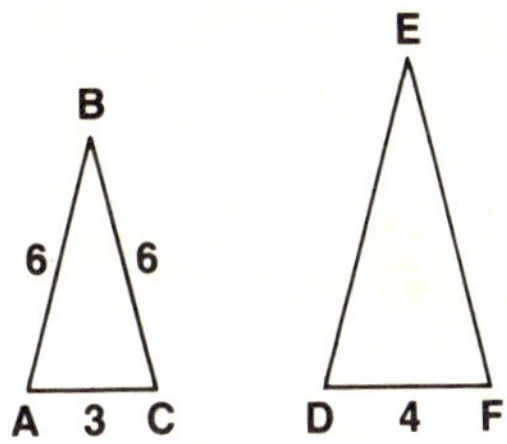

21. In the accompanying diagram, $\triangle ABC$ is similar to $\triangle DEF$, $AC = 3$, and $AB = BC = 6$. If $DF = 4$, what is the perimeter of $\triangle DEF$?
(1) 8 (3) 20
(2) 18 (4) 24

9.2 AREA OF SPECIAL QUADRILATERALS AND OF A TRIANGLE

The ***area*** of a plane figure is the portion of the plane enclosed by that figure.

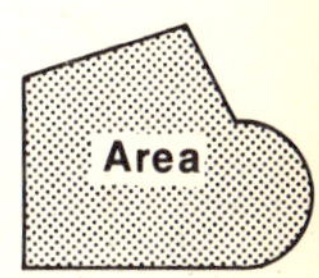

To give a numerical value to the portion of the plane enclosed by various figures, we begin with the simplest figure, a square whose side is 1 unit long. The area enclosed by this square is called ***one square unit.***

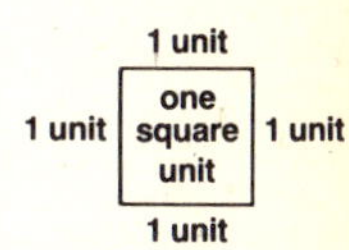

All areas are now based on this definition of *one square unit*. When we ask the question "What is the area of any plane figure?" we are really asking "How many *one square units* does that plane figure contain?"

For example, a square whose side is 3 units in length encloses 9 *one square units.*

Thus, we say that the area of a square whose side is 3 units is 9 square units.

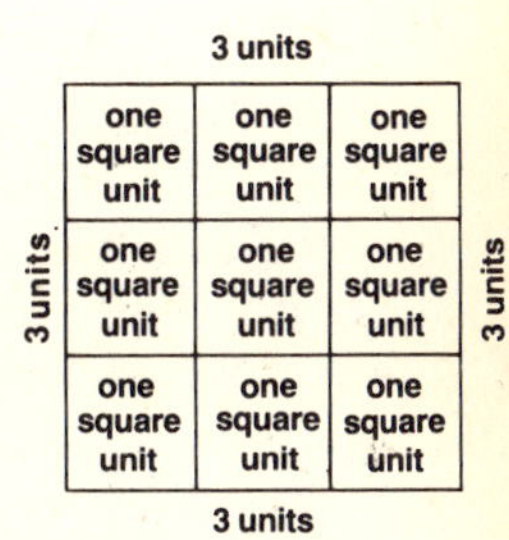

We need not draw a detailed diagram each time to count the number of *one square units* contained in a square whose side is other than 1 unit in length. Instead, we can generalize from the example. That is, since $3 \cdot 3$ or $3^2 = 9$, the **area of a square whose side is *s* units long is *s* square units.**

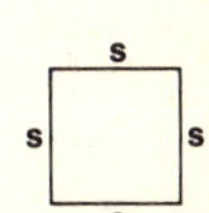

$$A_{\text{square}} = s^2$$

The unit may be any measure of length: inches, feet, miles, centimeters, meters, etc. The area will then be in square inches, square centimeters, etc.

Similarly, the area of a rectangle can be calculated by counting the number of one square units enclosed. As shown in the figure, a rectangle of base 6 units and height 3 units encloses 18 *one square units.*

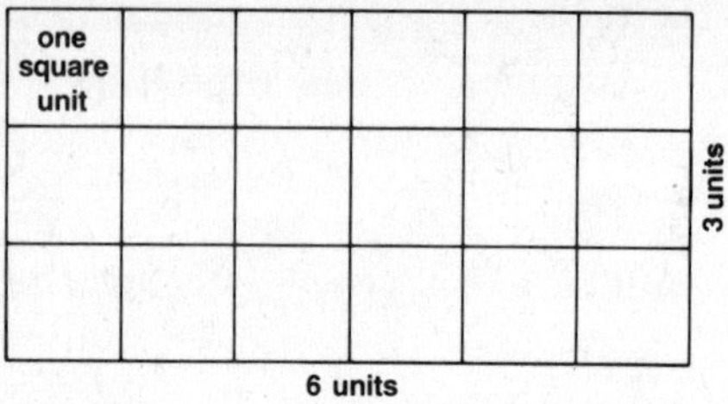

We can generalize from the example. That is, since $6 \cdot 3 = 18$, the **area of a rectangle of base b units and altitude h units is $b \cdot h$ square units.**

The base (b) of a rectangle is also called its length (ℓ) and the height (h) is also called its width (w). Thus, an alternate formula for the area of a rectangle is shown at the right.

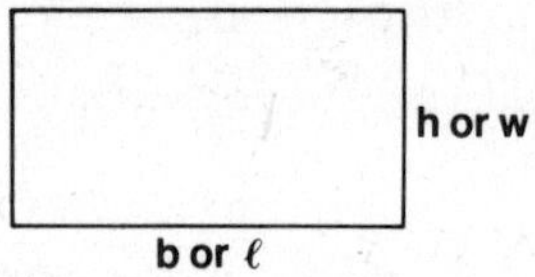

$$A_{\text{rectangle}} = \ell \cdot w \text{ or } b \cdot h$$

It was easy to see how the area of a rectangle could be built from squares. We can continue to build, now using the rectangle.

A parallelogram of base b and height h can be transformed into a rectangle of the same dimensions by cutting off the triangle at the left and placing it at the right. Thus, the **area of a parallelogram is equal to the area of a rectangle of the same dimensions.**

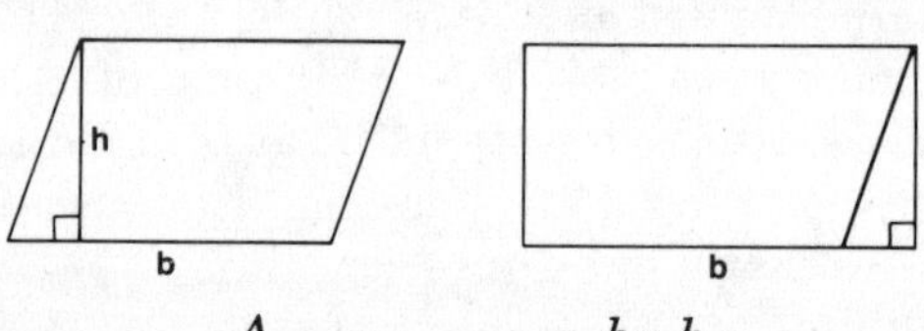

$$A_{\text{parallelogram}} = b \cdot h$$

Continuing still, we can cut a parallelogram into two equal or congruent triangles, each of base b and altitude h. Thus, the **area of a triangle is one-half the area of a parallelogram of the same dimensions.**

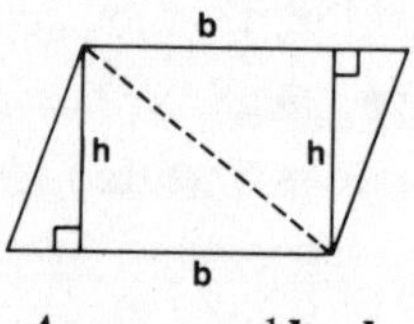

$$A_{\text{triangle}} = \tfrac{1}{2}b \cdot h$$

Finally, a trapezoid can be cut into two triangles. This time, however, the triangles are not equal. Thus, the **area of a trapezoid is the sum of the areas of the two different triangles.** The lower triangle has a base of b_2 and a height of h. Its area is $\frac{1}{2}b_2 \cdot h$. The upper triangle has a base of b_1. If its height is drawn, it is seen to be equal to h.

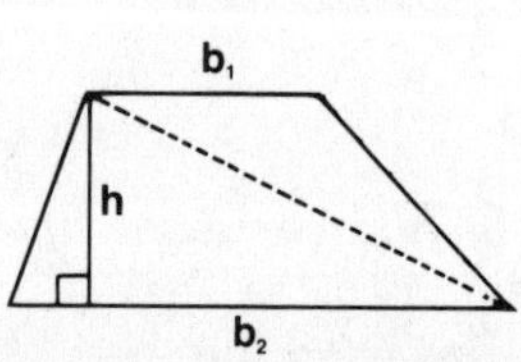

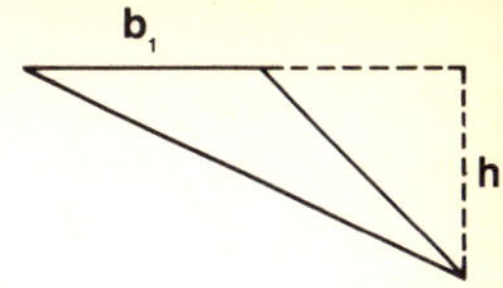

(*Note:* To draw the height, h, of this triangle, we must extend the base, b_1, since the height falls outside of the triangle. The area of this triangle is $\frac{1}{2}b_1 \cdot h$.)

Thus, the area of the trapezoid is

$$\tfrac{1}{2}b_1 \cdot h + \tfrac{1}{2}b_2 \cdot h \text{ or } \tfrac{1}{2}h\,(b_1 + b_2)$$

In words, we say that the **area of a trapezoid is the product of one-half its altitude and the sum of its bases.**

$$\boldsymbol{A_{\text{trapezoid}} = \tfrac{1}{2}h\,(b_1 + b_2)}$$

We have built a series of formulas for calculating areas based on one square unit. These formulas are summarized here.

Summary of Area Formulas

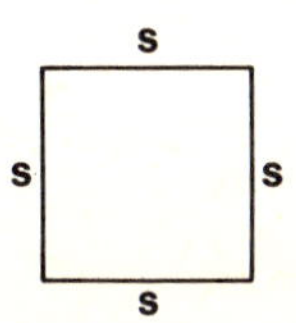

$A_{\text{square}} = s^2$

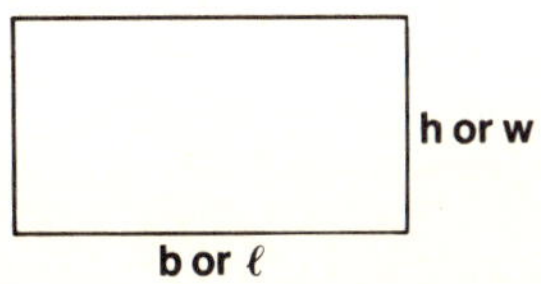

$A_{\text{rectangle}} = b \cdot h \text{ or } \ell \cdot w$

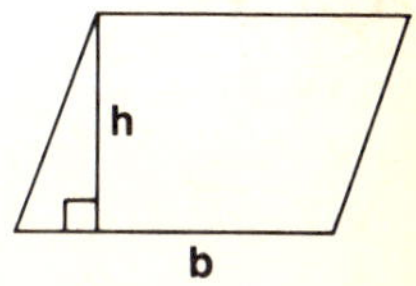

$A_{\text{parallelogram}} = b \cdot h$

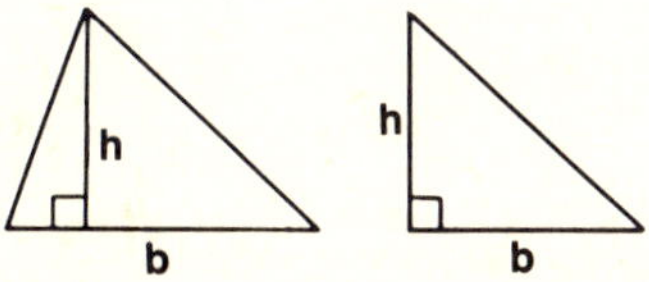

$A_{\text{triangle}} = \tfrac{1}{2}b \cdot h$

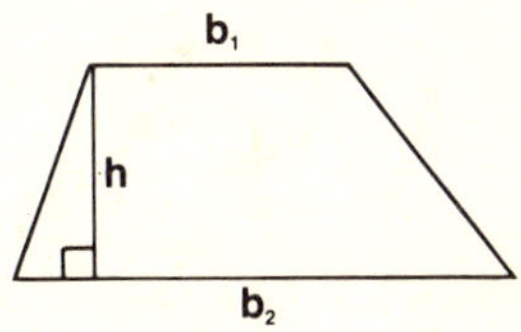

$A_{\text{trapezoid}} = \tfrac{1}{2}h\,(b_1 + b_2)$

◆ Many different types of problems involve area. To solve any area problem:

1. Read the problem carefully to determine what kind of figure is mentioned. Determine which values are known and which values are to be found. (*Note:* There can be only one unknown. Sometimes, one part has to be expressed in terms of another so that there is only one unknown. Use the words of the problem to do this.)
2. Select the appropriate area formula.
3. Substitute all the values in the proper places in the area formula.
4. Solve for the unknown value. Use this value to find all the missing parts.
5. Check.

Examples

1. Find the area of a square whose perimeter is 48 inches.

Solution:

$$A_{\text{square}} = s^2$$

We must find the value of s. Use the information about the perimeter of the square to do this.

$$P_{\text{square}} = 4s$$
$$48 = 4s$$
$$12 = s$$

Return to the area formula with the information that $s = 12$.

$$A_{\text{square}} = s^2 = 12^2 = 144$$

Answer: $A_{\text{square}} = 144$ square inches.

2. The length of a rectangle exceeds its width by 6 meters. If the area of the rectangle is 16 square meters, find the dimensions of the rectangle.

Solution:

Let x = the width of the rectangle.
Then $x + 6$ = the length of the rectangle.

$$A_{\text{rectangle}} = \ell \cdot w$$
$$16 = (x + 6)x$$
$$16 = x^2 + 6x$$
$$x^2 + 6x - 16 = 0$$
$$(x + 8)(x - 2) = 0$$

$x + 8 = 0$	$x - 2 = 0$
$x = -8$	$x = 2$ width
Reject.	$x + 6 = 8$ length

Check:

$$A = \ell \cdot w$$
$$16 \stackrel{?}{=} 8 \cdot 2$$
$$16 = 16 \checkmark$$

Answer: The width is 2 meters and the length is 8 meters.

3. If the base and height of a rectangle are tripled, then the area

(1) remains the same
(2) is multiplied by 3
(3) is increased by 9
(4) is multiplied by 9

Solution:

$$A_{\text{rectangle}} = b \cdot h$$

Replace b by $3b$ and replace h by $3h$:

$$A_{\text{rectangle}} = (3b)(3h) = 9bh$$

Thus, the new area, $9bh$, is nine times the original area, bh.

Answer: (4)

4. The area of a rectangle is 12 square units more than the area of a square. If the base of the rectangle is 4 units more than a side of the square and the height of the rectangle is 2 units less than a side of the square, find the length of a side of the square.

Solution:

Let x = the length of a side of the square.

Then $x + 4$ = the length of the base of the rectangle.

And $x - 2$ = the length of the height of the rectangle.

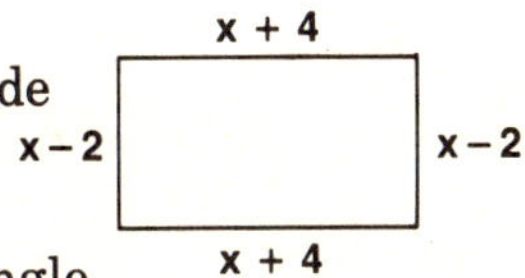

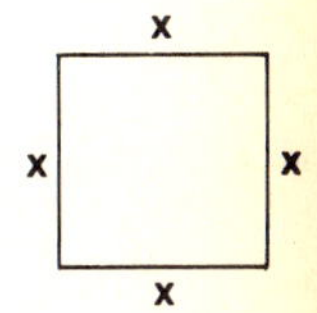

$$\begin{aligned} A_{\text{rectangle}} &= A_{\text{square}} + 12 \\ b \cdot h &= s^2 + 12 \\ (x + 4)(x - 2) &= x^2 + 12 \\ x^2 + 2x - 8 &= x^2 + 12 \\ 2x - 8 &= 12 \quad \text{Subtract } x^2. \\ 2x &= 20 \\ x &= 10 \quad \text{side of square} \\ x + 4 &= 14 \quad \text{base of rectangle} \\ x - 2 &= 8 \quad \text{height of rectangle} \end{aligned}$$

Check:

$$\begin{aligned} A_{\text{rectangle}} &= A_{\text{square}} + 12 \\ b \cdot h &= s^2 + 12 \\ 14 \cdot 8 &\stackrel{?}{=} 10^2 + 12 \\ 112 &\stackrel{?}{=} 100 + 12 \\ 112 &= 112 \checkmark \end{aligned}$$

Answer: The length of a side of the square is 10 units.

5. Find the area of the parallelogram to the nearest square unit.

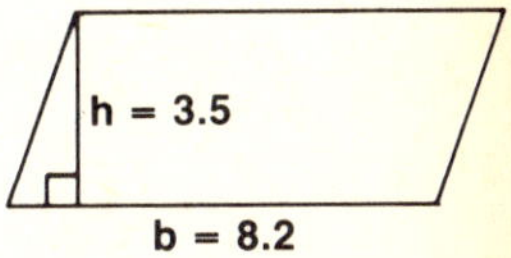

Solution:
$$\begin{aligned} A_{\text{parallelogram}} &= b \cdot h \\ &= (8.2)(3.5) \\ &= 28.70 \approx 29 \text{ square units} \quad \textit{Ans.} \end{aligned}$$

6. The area of a triangle is 40 square centimeters. If the length of the base is 10 centimeters, find the length of the altitude.

Solution:
$$\begin{aligned} A_{\text{triangle}} &= \tfrac{1}{2}b \cdot h \\ 40 &= \tfrac{1}{2}(10) \cdot h \\ 40 &= 5h \\ 8 &= h \quad \textit{Ans.} \end{aligned}$$

7. If the base of a triangle is tripled and the altitude remains the same, then the area (1) remains the same (2) is increased by 3 (3) is multiplied by 3 (4) is multiplied by 6

Solution: $A_{\text{triangle}} = \frac{1}{2}b \cdot h$

Replace b by $3b$; h remains the same:
$$\begin{aligned} A_{\text{triangle}} &= \tfrac{1}{2}(3b)h \\ &= 3(\tfrac{1}{2}bh) \end{aligned}$$

Thus, the new area, $3(\frac{1}{2}bh)$, is three times the original area, $\frac{1}{2}bh$.

Answer: (3)

8. In the accompanying diagram, $\overline{CA} \perp \overline{AB}$ and $\overline{DE} \perp \overline{AB}$.
   a. Tell why $\triangle ABC \sim \triangle EBD$.
   b. If $DE = 10$, $CA = 20$, and $BE = 4$, find BA.
   c. Find the area of $\triangle ABC$.
   d. Find the area of $\triangle DBE$.

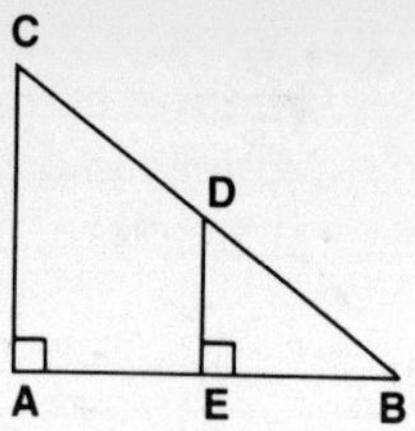

Solution:

a. Since two pairs of corresponding angles are congruent, the triangles are similar. *Ans.*

(*Note:* $\angle B$ is in each triangle.)

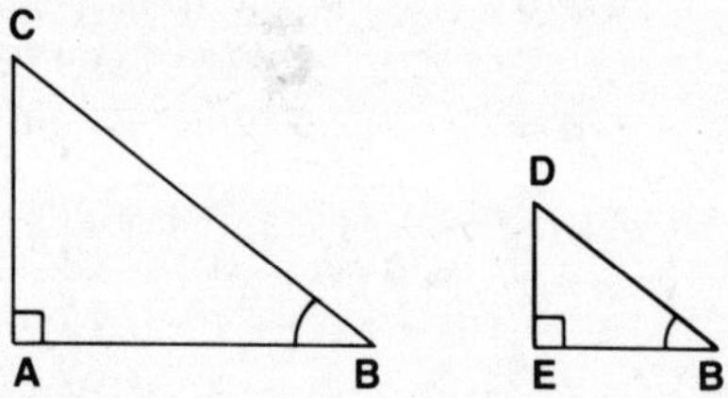

b. Corresponding sides of similar triangles are in proportion.

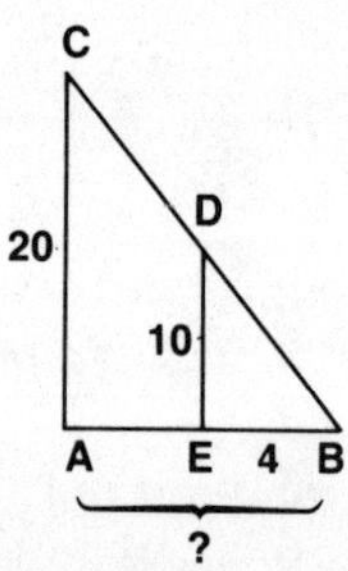

$$\frac{DE}{CA} = \frac{BE}{BA}$$
$$\frac{10}{20} = \frac{4}{BA}$$
$$10(BA) = 4(20)$$
$$10(BA) = 80$$
$$BA = 8 \quad \textit{Ans.}$$

c. $A_{\triangle ABC} = \frac{1}{2}b \cdot h$
$= \frac{1}{2}(8)(20)$
$A_{\triangle ABC} = 80$ *Ans.*

d. $A_{\triangle DBE} = \frac{1}{2}b \cdot h$
$= \frac{1}{2}(4)(10)$
$A_{\triangle DBE} = 20$ *Ans.*

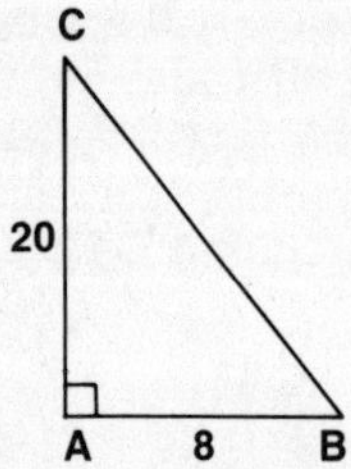

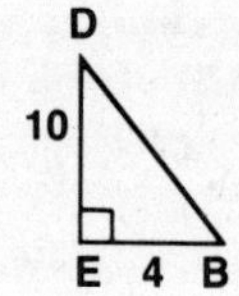

9. Find the area of a trapezoid whose altitude is 10 inches and whose bases are 14 and 21 inches.

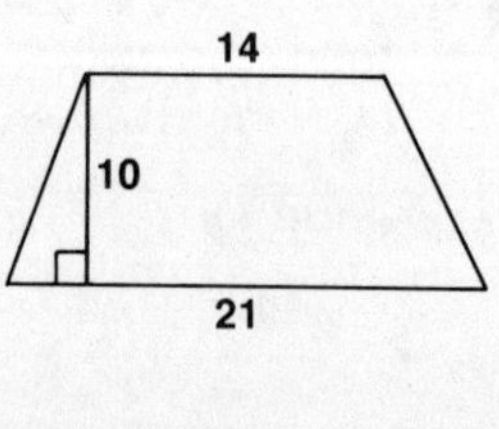

Solution:

$$A_{\text{trapezoid}} = \tfrac{1}{2}h(b_1 + b_2)$$
$$= \tfrac{1}{2} \cdot 10(14 + 21)$$
$$= 5(35)$$
$$= 175 \text{ square inches} \quad \textit{Ans.}$$

10. The area of a trapezoid is 51 square inches. If the length of the altitude is 6 inches and the length of the shorter base is 8 inches, find the length of the longer base.

Solution:

$$A_{\text{trapezoid}} = \tfrac{1}{2}h(b_1 + b_2)$$
$$51 = \tfrac{1}{2}(6)(8 + b_2)$$
$$51 = 3(8 + b_2)$$
$$51 = 24 + 3b_2$$
$$27 = 3b_2$$
$$9 = b_2$$

Check:

$$A = \tfrac{1}{2}h(b_1 + b_2)$$
$$51 \stackrel{?}{=} \tfrac{1}{2}(6)(8 + 9)$$
$$51 \stackrel{?}{=} 3(17)$$
$$51 = 51 \checkmark$$

Answer: The length of the longer base is 9 inches.

EXERCISES

1. If the length of each side of a square is represented by $2x$, which expression represents its area?
(1) $8x$ (2) $4x$ (3) $4x^2$ (4) $2x^2$

2. If the perimeter of a square is 24, find the area of the square.

3. If the perimeter of a square is 20, find the area of the square.

4. If the area of a square is 36, find the perimeter of the square.

5. If the area of a square is 49, what is its perimeter?

6. A rectangle has an area of 3.6 square meters. If the width is 1.2 meters, what is the length, in meters, of the rectangle?

7. If the length of a rectangle is represented by $2x - 3$ and its width is represented by $3x$, what is the area of the rectangle in terms of x?
(1) $6x^2 - 9x$ (2) $6x^2 - 3$ (3) $10x - 6$ (4) $5x - 3$

8. Which represents the area of a rectangle whose length is $x + 1$ and whose width is $x - 1$?
(1) $x^2 + 1$ (2) $2x$ (3) $x^2 - 1$ (4) $4x$

9. The length of a certain rectangle is 5 inches greater than a side of a square. The width of the rectangle is 2 inches less than a side of the square. The area of the rectangle is 8 square inches greater than the area of the square.

a. Find the number of inches in the length of a side of the square.

b. Find the number of square inches in the area of the rectangle.

10. Given: rectangle *ABCD* with square *ARST* as shown. The area of the rectangle is 143. If $RB = 8$ and $TD = 6$, find the length of $\overline{AT}$, a side of square *ARST*.

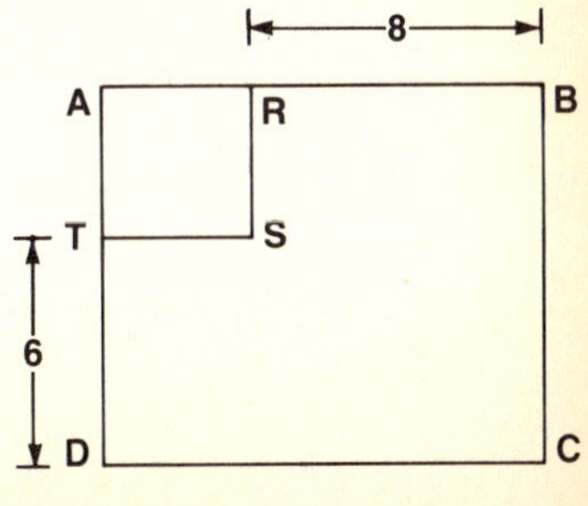

11. Three congruent squares are arranged in a row as shown in the accompanying figure. If the perimeter of $ABCD$ is 80, then the area of $ABCD$ is
(1) 64 (2) 100 (3) 193 (4) 300

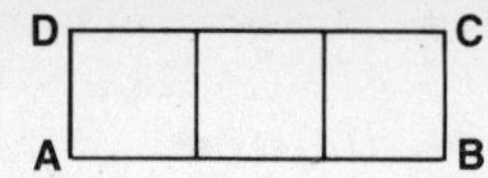

12. The area of square $ABCD$, as shown in the accompanying diagram, is 121. $\overline{EKF}$ is drawn parallel to $\overline{AD}$ and $\overline{GKH}$ is drawn parallel to $\overline{AB}$ so as to form the smaller squares $AGKE$ and $FCHK$. If the area of $AGKE$ is 36, find the area of $FCHK$.

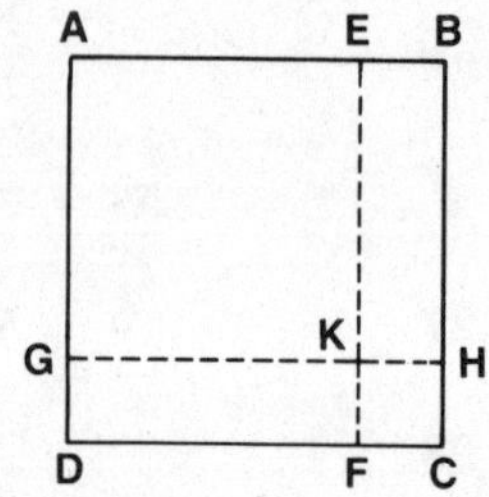

13. The area of a rectangle is represented by $(x^2 + 2x)$ and the length by $(x + 2)$. Express the width of the rectangle in terms of x.
14. If each side of a rectangle is doubled, then its area always
(1) remains the same (3) is multiplied by 8
(2) is increased by 2 (4) is multiplied by 4

In 15–20, only an algebraic solution will be accepted.

15. The sides of a rectangle are x and $x + 6$. The area of the rectangle is 55. Find the lengths of the sides.
16. The area of a rectangle is 40. If the length is 6 more than the width, find the dimensions of the rectangle.
17. The ratio of two adjacent sides of a rectangle is 3:4. If the area of this rectangle is 192 square feet, find the number of feet in the length and the width.
18. The side of a certain square is 3 feet longer than that of another square. The sum of their areas is 117 square feet. Find the length of a side of the smaller square.
19. The width of a rectangle is 2 less than a side of a square, and the length is 1 more than a side of the square. If the area of the rectangle is 28, find the length of a side of the square.
20. A square and a rectangle have the same area. The length of the rectangle is 8 more than a side of the square, and the width of the rectangle is 4 less than a side of the square. Find a side of the square.

21. The base of a parallelogram is $12\frac{1}{2}$ inches and the altitude to that base is 8 inches. Find the number of square inches in the area of the parallelogram.
22. If the base of a parallelogram is represented by $3x - 2$ and the altitude to that base is represented by $4x$, then the area of the parallelogram is represented by
(1) $14x - 4$ (2) $12x^2 - 2$ (3) $12x^2 - 8$ (4) $12x^2 - 8x$
23. The measure of the base of a parallelogram is 4 meters greater than the measure of the altitude to that base. If the area of the parallelo-

gram is 32 square meters, find the number of meters in the measures of the base and altitude.

24. If the base of a triangle is represented by $2x$ and the altitude to that base is equal to 10, express the area of the triangle in terms of x.

25. The area of a triangle is 48. If the base of this triangle is 12, what is the length of the altitude to this base?

26. The area of a triangle is $\frac{1}{2}bh$. If the base of the triangle is doubled and the height of the triangle is doubled, then the area is
(1) doubled (3) multiplied by 8
(2) multiplied by 4 (4) increased by 4

27. If the altitude of a trapezoid is 10 and the bases are 12 and 14, the area of the trapezoid is (1) 260 (2) 130 (3) 200 (4) 230

28. If the area of a trapezoid is 48 and the bases are 5 and 7, then the altitude is (1) 8 (2) 4 (3) 12 (4) 6

29. Figure I is a rectangle, II is a square, III is a right triangle, and IV is an acute scalene triangle. The lengths of particular segments of the figures are indicated. Which figure has the greatest area?
(1) I (2) II (3) III (4) IV

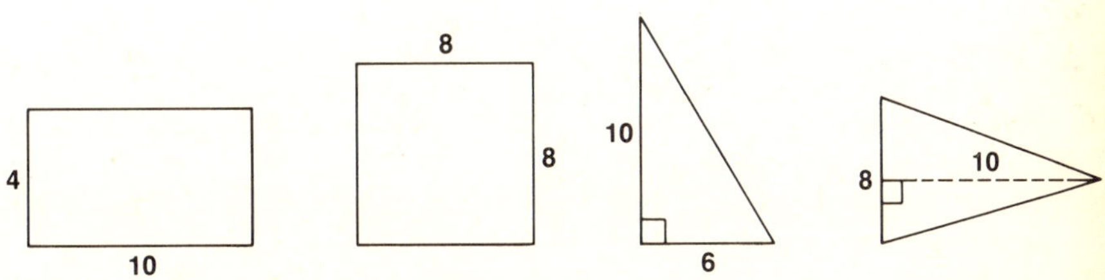

I II III IV

30. Figure I is a rectangle, II is a right triangle, and III is a trapezoid. The lengths of particular segments of the figures are indicated. Which figures have equal areas?
(1) I and II only
(2) I and III only
(3) II and III only
(4) I, II, and III

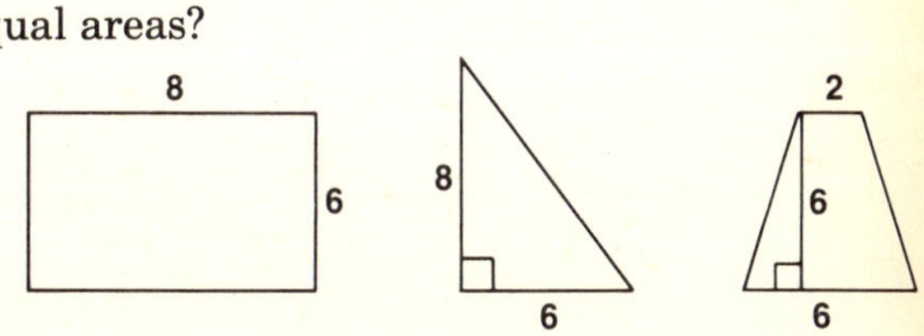

I II III

31. In the accompanying diagram, triangle ABC is a right triangle with the right angle at C. Segment $\overline{DE}$ is perpendicular to $\overline{AC}$ at E, $BC = 8$, $AC = 6$, and $DE:BC = 1:2$.

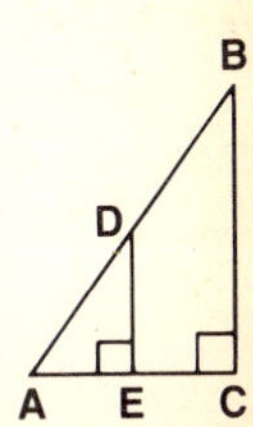

a. Find DE.
b. Find AB.
c. Find AE.
d. Find the area of $\triangle ABC$.
e. Find the area of trapezoid $ECBD$.

9.3 THE PYTHAGOREAN THEOREM; APPLICATIONS TO PERIMETER AND AREA PROBLEMS

The name of the ancient Greek mathematician Pythagoras is given to the important relation that exists among the three sides of a right triangle. This relation is called the ***Pythagorean Theorem.*** In words, the relation states:

◆ The square of the measure of the hypotenuse of a right triangle is equal to the sum of the squares of the measures of the legs.

Algebraically, this relation may be written:

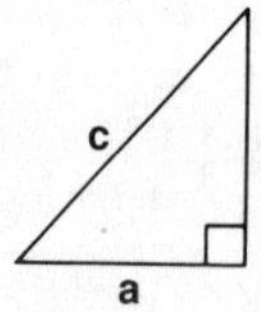

$$c^2 = a^2 + b^2$$

In this formula, c represents the length of the hypotenuse of a right triangle, and a and b represent the lengths of the legs.

Geometrically, this relation means that if squares are built on the hypotenuse and on each leg of a right triangle, then the area of the square on the hypotenuse is equal to the sum of the areas of the squares on the legs:

◆ The number of square units in the square on the hypotenuse is equal to the sum of the numbers of square units in the squares on both legs.

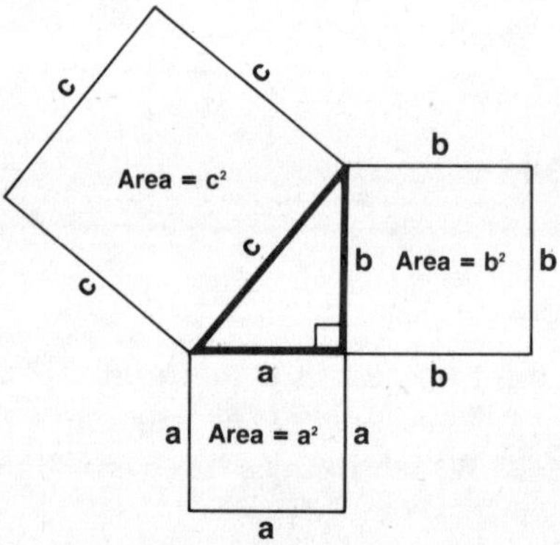

The Pythagorean Theorem is useful in many respects. First, if two of the three sides of a right triangle are known, we can use the Pythagorean Theorem to find the unknown third side.

Examples

1. Find the length of the hypotenuse of a right triangle whose legs are 4 and 7.

Solution:

Let $a = 4$, $b = 7$. Find c.

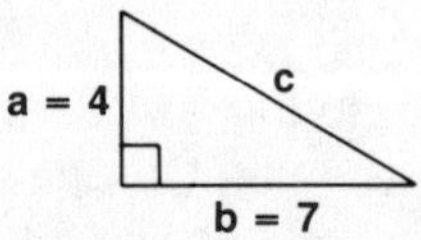

$$c^2 = a^2 + b^2$$
$$c^2 = 4^2 + 7^2$$
$$c^2 = 16 + 49$$
$$c^2 = 65$$
$c = \pm\sqrt{65}$ Reject $-\sqrt{65}$, since a length must be positive.

Answer: The length of the hypotenuse is $\sqrt{65}$ units.

2. The length of the hypotenuse of a right triangle is 17. If the length of one leg is 15, find the length of the other leg.

Solution:

Let $c = 17$, $a = 15$. Find b.

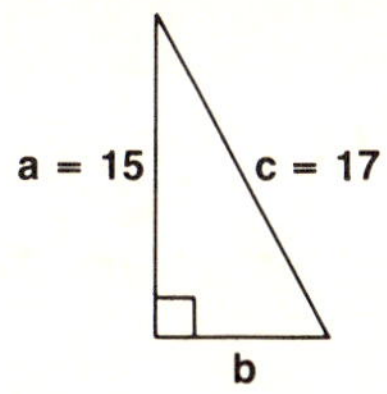

$$\begin{aligned} c^2 &= a^2 + b^2 \\ 17^2 &= 15^2 + b^2 \\ 289 &= 225 + b^2 \\ 64 &= b^2 \\ \pm\sqrt{64} &= b \\ \pm 8 &= b \quad \text{Reject } -8. \end{aligned}$$

Answer: The length of the other leg is 8 units.

The two examples show how the Pythagorean Theorem is used to find one side of a right triangle if two of the sides are known. Most frequently, the answers will be irrational numbers, since only three perfect squares that satisfy the Pythagorean relation will result in rational numbers. In fact, the only commonly used numbers that "work nicely"—that is, yield rational numbers—are:

3, 4, 5   5, 12, 13   8, 15, 17

and multiples of these ***triplets.***

Example

Find the length of the hypotenuse of a right triangle whose legs are 6 and 8.

Solution:

Let $a = 6$, $b = 8$. Find c.

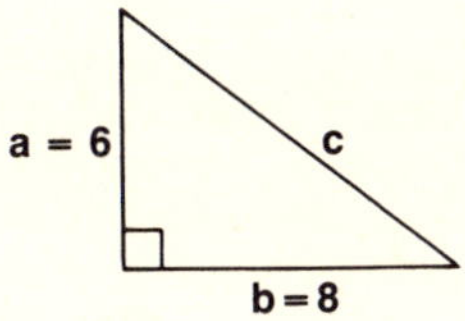

$$\begin{aligned} c^2 &= a^2 + b^2 \\ c^2 &= 6^2 + 8^2 \\ c^2 &= 36 + 64 \\ c^2 &= 100 \\ c &= \pm\sqrt{100} \\ c &= \pm 10 \quad \text{Reject } -10. \end{aligned}$$

Answer: The length of the hypotenuse is 10 units.

In the example, the numbers "worked nicely" because the triplet 6, 8, 10 is a multiple of the triplet 3, 4, 5 (that is, $6 = 3 \cdot 2$, $8 = 4 \cdot 2$, and $10 = 5 \cdot 2$).

As you go on in your study of mathematics, you will find it useful to remember the triplets and to look for their multiples before you resort to working out the problem by substituting in the formula. Although the formula will always work, remembering the triplets eliminates the need to do the calculations.

Examples

1. The hypotenuse of a right triangle is 15 and one leg is 9. Find the other leg.

Solution 1 (using a triplet):

3, 4, 5

9 is 3 · <u>3</u> 15 is 3 · <u>5</u>

Therefore, the missing leg must be 3 · <u>4</u> or 12. *Ans.*

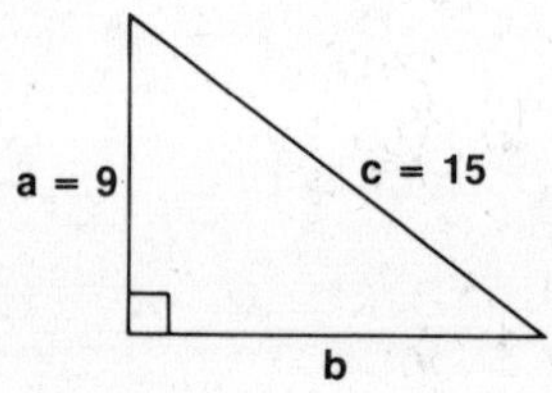

Solution 2 (using the formula):

Let $c = 15$, $a = 9$. Find b.

$$c^2 = a^2 + b^2$$
$$15^2 = 9^2 + b^2$$
$$225 = 81 + b^2$$
$$144 = b^2$$
$$\pm\sqrt{144} = b$$
$$\pm 12 = b \quad \text{(Reject } -12.\text{)}$$

Answer: The other leg is 12 units.

2. Find the length of the hypotenuse of the right triangle whose legs are 10 and 24.

Solution 1 (using a triplet):

5, 12, 13

10 is 2 · <u>5</u> 24 is 2 · <u>12</u> Therefore, the missing hypotenuse must be 2 · <u>13</u> or 26. *Ans.*

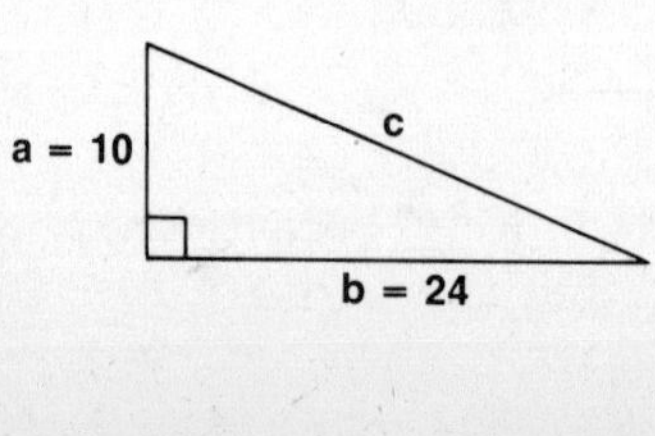

Solution 2 (using the formula):

Let $a = 10$, $b = 24$. Find c.

$$c^2 = a^2 + b^2$$
$$c^2 = 10^2 + 24^2$$
$$c^2 = 100 + 576$$
$$c^2 = 676$$
$$c = \pm\sqrt{676}$$
$$c = \pm 26 \quad \text{(Reject } -26.\text{)}$$

Answer: The length of the hypotenuse is 26 units.

3. Find the sides of a right triangle if their lengths are consecutive even integers.

Solution 1 (using triplets):

The only triplet that involves consecutive integers is 3, 4, 5. If we double this triplet, we see that the triplet 6, 8, 10 satisfies the condition that the sides are to be consecutive even integers.

Answer: The lengths of the sides are 6, 8, and 10.

Solution 2 (using the formula):

Let the sides of the triangle be represented as

$$\begin{aligned} x &= a \text{ (one leg)} \\ x + 2 &= b \text{ (the other leg)} \\ x + 4 &= c \text{ (the hypotenuse)} \end{aligned}$$

(*Note:* x, $x + 2$, and $x + 4$ represent the consecutive even integers.)

$$\begin{aligned} c^2 &= a^2 + b^2 \\ (x + 4)^2 &= (x)^2 + (x + 2)^2 \\ x^2 + 8x + 16 &= x^2 + x^2 + 4x + 4 && \text{Multiply to remove ( ).} \\ x^2 + 8x + 16 &= 2x^2 + 4x + 4 && \text{Combine } x^2 \text{ terms on right.} \\ 8x + 16 &= x^2 + 4x + 4 && \text{Subtract } x^2. \\ 16 &= x^2 - 4x + 4 && \text{Subtract } 8x. \\ 0 &= x^2 - 4x - 12 && \text{Subtract 16.} \\ 0 &= (x - 6)(x + 2) && \text{Factor.} \end{aligned}$$

(The three subtraction steps: Put the quadratic equation in standard form.)

$$\begin{array}{r|l} x - 6 = 0 & x + 2 = 0 \\ x = 6 & x = -2 \text{ (Reject } x = -2.) \\ x + 2 = 8 & \\ x + 4 = 10 & \end{array}$$

Answer: The lengths of the sides are 6, 8, and 10.

(*Note:* Although the solution using the triplets is obviously easier, a formal examination may call for an algebraic solution. In this situation, the only type of solution that is considered acceptable is one like Solution 2.)

The Pythagorean Theorem is also useful in more complicated geometric problems involving perimeter and area. In these cases, the Pythagorean Theorem is a tool that enables us to find a missing part needed to solve the actual problem.

Examples

1. Find the area of a square whose diagonal is 6 inches.

Solution: $A_{\text{square}} = s^2$

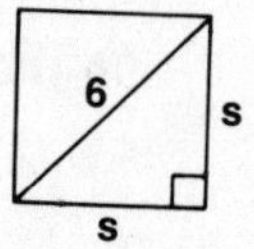

Use the given value of the diagonal in the Pythagorean Theorem to find s.

$$\begin{aligned} c^2 &= a^2 + b^2 \\ 6^2 &= s^2 + s^2 \quad (c = 6, a = s, b = s) \\ 36 &= 2s^2 \\ 18 &= s^2 \end{aligned}$$

We can stop here since we are interested in the area, s^2, of the square.

Answer: The area of the square is 18 square inches.

2. Find the area of the parallelogram shown.

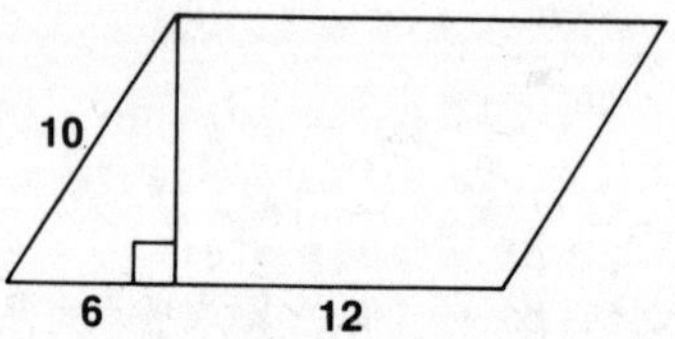

Solution: $A_{\text{parallelogram}} = b \cdot h$

To calculate the area of the parallelogram, we require two pieces of information: the length of the base and the length of the altitude. The length of the base is 18, the sum of 6 and 12. The length of the altitude may be determined by the Pythagorean Theorem:

Let $c = 10, a = 6$. Find b.
(b is h of the parallelogram)

$$\begin{aligned} c^2 &= a^2 + b^2 \\ 10^2 &= 6^2 + h^2 \\ 100 &= 36 + h^2 \\ 64 &= h^2 \\ \pm\sqrt{64} &= h \\ \pm 8 &= h \quad \text{(Reject } -8.) \end{aligned}$$

The altitude of the parallelogram is 8. (*Note:* The use of the triplet 3, 4, 5 multiplied by 2 would have eliminated the calculation.)

Substitute $b = 18$ and $h = 8$ in

$$\begin{aligned} A_{\text{parallelogram}} &= b \cdot h \\ &= 18 \cdot 8 = 144 \text{ square units} \quad \textit{Ans.} \end{aligned}$$

3. Find the perimeter of the isosceles trapezoid shown.

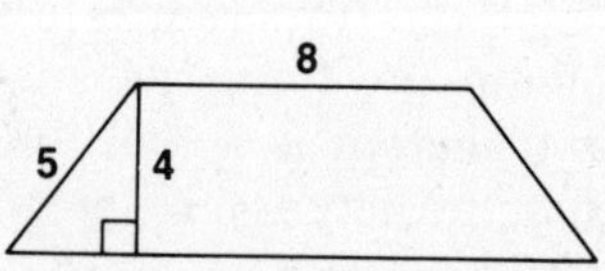

Solution:

The perimeter of the isosceles trapezoid is the sum of the four sides. Two of the sides are given: one leg is 5 and the shorter base is 8. Thus, a third side, the other leg, must be 5 because in an isosceles trapezoid the legs are equal in length.

The only information missing now is the length of the longer base. We can use the Pythagorean Theorem to determine a piece, namely 3, of the longer base. The remainder of the longer base can be determined if another altitude is drawn, thus breaking the isosceles trapezoid into a rectangle and two congruent right triangles. From the rectangle, another piece of the longer base, namely 8, is known. From the right triangle, the last piece of the longer base, namely 3, is known. Thus, the length of the longer base is 3 + 8 + 3, or 14.

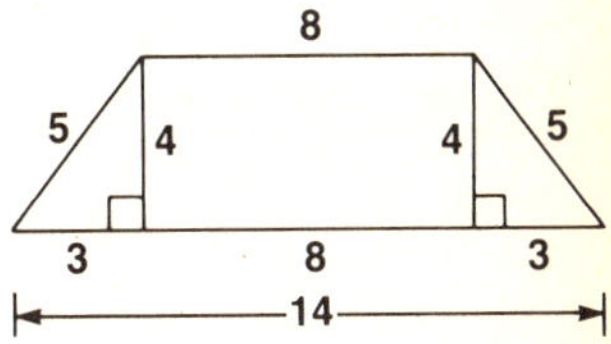

$$\begin{aligned} \text{Therefore, } P_{\text{trapezoid}} &= \text{leg} + \text{shorter base} + \text{leg} + \text{longer base} \\ &= 5 + 8 + 5 + 14 \\ &= 32 \end{aligned}$$

Answer: The perimeter of the isosceles trapezoid is 32 units.

4. As shown in the accompanying figure, $ABCD$ is a rectangle. P is a point on $\overline{CD}$, $CP = 6$, $BP = 10$, and $AB = 30$. Find the:

a. length of $\overline{BC}$
b. perimeter of trapezoid $ABPD$
c. area of $\triangle BCP$
d. area of rectangle $ABCD$
e. area of trapezoid $ABPD$

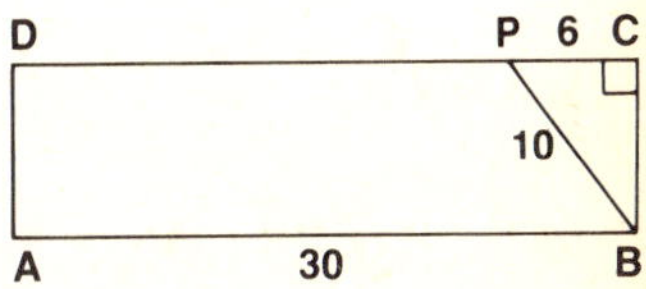

Solution:

a. In right triangle BCP, use the Pythagorean Theorem to find BC.

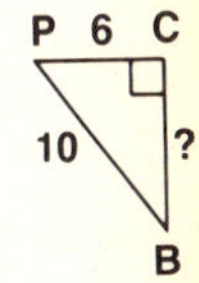

$$\begin{aligned} c^2 &= a^2 + b^2 \\ 10^2 &= 6^2 + (BC)^2 \quad (c = 10, a = 6, b = BC) \\ 100 &= 36 + (BC)^2 \\ 64 &= (BC)^2 \\ \pm\sqrt{64} &= BC \\ 8 &= BC \quad \textit{Ans.} \qquad \text{(Reject } -8.) \end{aligned}$$

(*Note:* Using the triplet 3, 4, 5, multiplied by 2, we would have obtained the same answer.)

Since $\overline{AD} \cong \overline{BC}$ and $BC = 8$, $AD = 8$. Also, $PD = 30 - 6$ or 24.

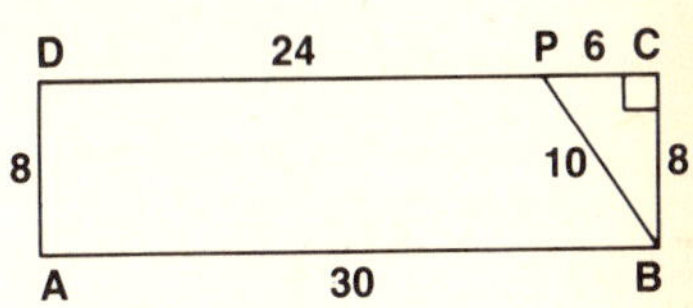

b.

$$\begin{aligned} P_{\text{trapezoid } ABPD} &= AB + BP + PD + AD \\ &= 30 + 10 + 24 + 8 \\ P_{\text{trapezoid } ABPD} &= 72 \text{ units} \quad \textit{Ans.} \end{aligned}$$

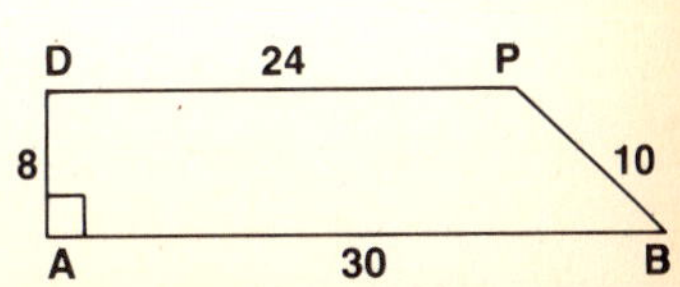

c. $A_{\triangle BCP} = \frac{1}{2}b \cdot h$
$= \frac{1}{2}(CP)(BC)$
$= \frac{1}{2}(6)(8)$
$A_{\triangle BCP} = 24$ square units *Ans.*

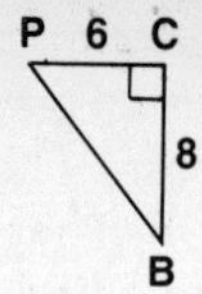

d. $A_{\text{rectangle } ABCD} = b \cdot h$
$= (AB)(BC)$
$= (30)(8)$
$A_{\text{rectangle } ABCD} = 240$ sq. units *Ans.*

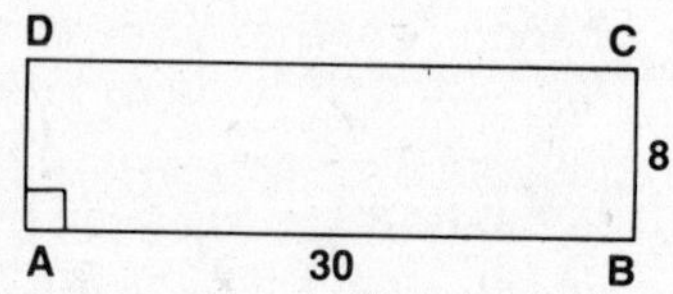

e. $A_{\text{trapezoid } ABPD} = A_{\text{rectangle } ABCD} - A_{\triangle BCP}$
$= 240 - 24$
$A_{\text{trapezoid } ABPD} = 216$ square units *Ans.*

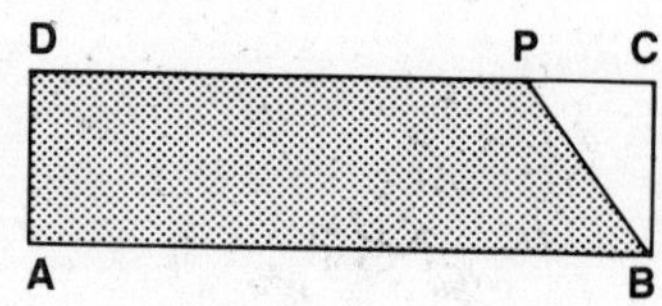

Alternate Solution:

$A_{\text{trapezoid } ABPD} = \frac{1}{2}h(b_1 + b_2)$
$= \frac{1}{2}(AD)(DP + AB)$
$= \frac{1}{2}(8)(24 + 30)$
$= 4(54)$
$A_{\text{trapezoid } ABPD} = 216$ square units *Ans.*

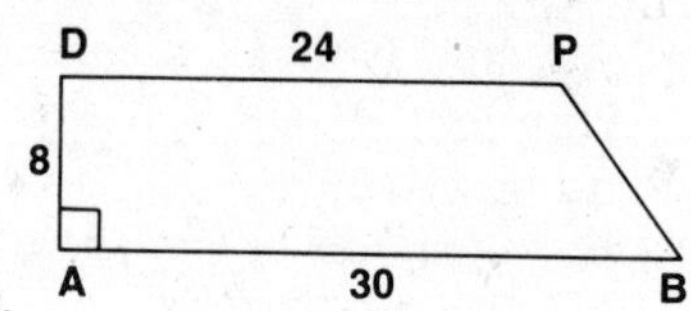

leg $\overline{AD}$ = altitude

5. In the accompanying figure, $\overline{BE}$ is a diagonal of rectangle $ABCE$ and of trapezoid $ABDE$. $AE = 10$, $BE = 26$, $ED = 16$.

a. Find AB.
b. Find the area of rectangle $ABCE$.
c. Find the area of trapezoid $ABDE$.
d. Find the area of triangle BCD.

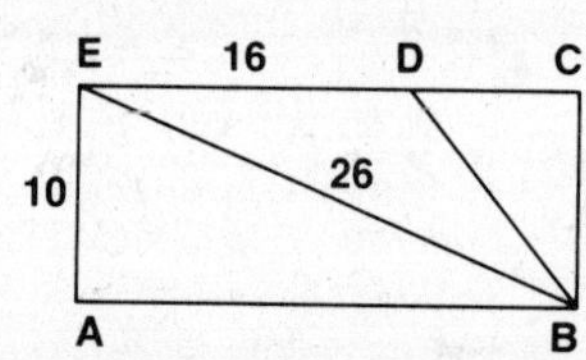

Solution:

a. In right triangle ABE, use the Pythagorean Theorem to find AB.

$c = BE = 26$, $a = AE = 10$. Find $b = AB$.

$c^2 = a^2 + b^2$
$(26)^2 = (10)^2 + (AB)^2$
$676 = 100 + (AB)^2$
$576 = (AB)^2$
$\pm\sqrt{576} = AB$
$24 = AB$ *Ans.* (Reject -24.)

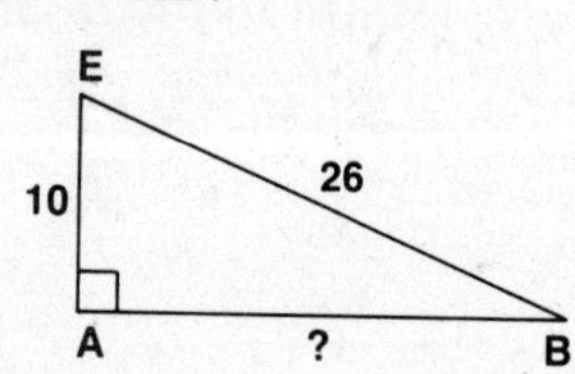

(*Note:* Using the triplet 5, 12, 13 multiplied by 2 would have resulted in the same answer.)

Since $\overline{EC} \cong \overline{AB}$ and $AB = 24$, $EC = 24$. Also, $DC = 24 - 16$ or 8.

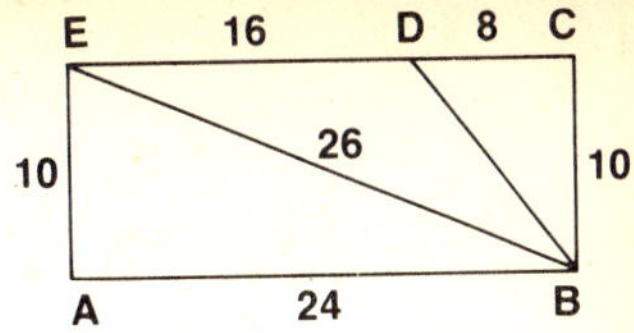

b. $$\begin{aligned} A_{\text{rectangle } ABCE} &= b \cdot h \\ &= (AB)(AE) \\ &= (24)(10) \\ A_{\text{rectangle } ABCE} &= 240 \quad \textit{Ans.} \end{aligned}$$

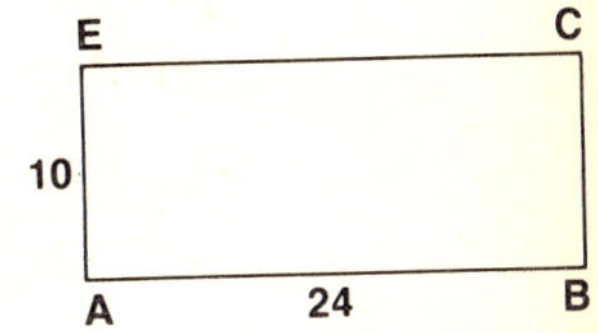

c. $$\begin{aligned} A_{\text{trapezoid } ABDE} &= \tfrac{1}{2}h(b_1 + b_2) \\ &= \tfrac{1}{2}(AE)(ED + AB) \\ &= \tfrac{1}{2}(10)(16 + 24) \\ &= 5(40) \\ A_{\text{trapezoid } ABDE} &= 200 \quad \textit{Ans.} \end{aligned}$$

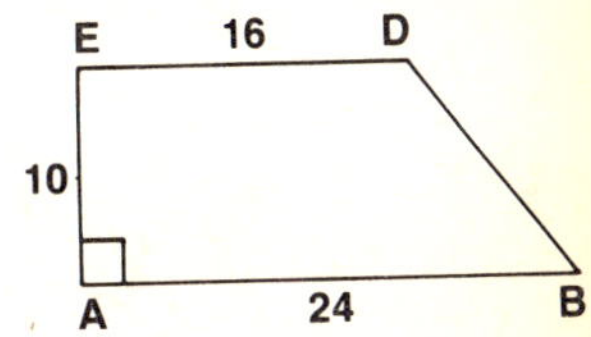

leg $\overline{AE}$ = altitude

d. $$\begin{aligned} A_{\triangle BCD} &= A_{\text{rectangle } ABCE} - A_{\text{trapezoid } ABDE} \\ &= 240 - 200 \\ A_{\triangle BCD} &= 40 \quad \textit{Ans.} \end{aligned}$$

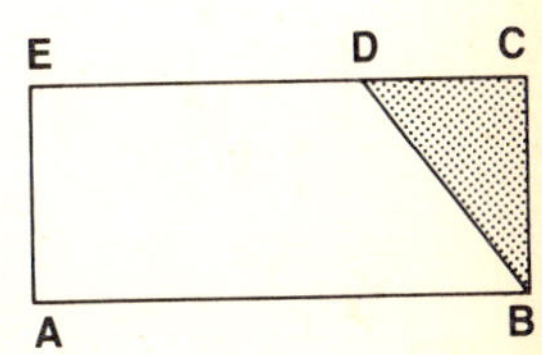

Alternate Solution:

$$\begin{aligned} A_{\triangle BCD} &= \tfrac{1}{2}b \cdot h \\ &= \tfrac{1}{2}(CD)(BC) \\ &= \tfrac{1}{2}(8)(10) \\ &= 4(10) \\ A_{\triangle BCD} &= 40 \quad \textit{Ans.} \end{aligned}$$

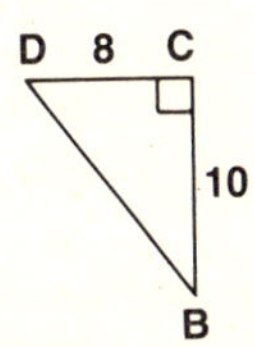

leg $\overline{BC}$ = altitude

In the work of this section, we have concentrated upon the fact that if a triangle is a right triangle, it must satisfy the Pythagorean Theorem. The converse of this statement is also true. That is, if a triangle satisfies the Pythagorean Theorem, it must be a right triangle. This information becomes important in future work when we will want to know if, given certain information, a triangle is indeed a right triangle.

EXERCISES

1. The lengths of two legs of a right triangle are 5 and 12. Find the length of the hypotenuse.
2. If one of the legs of a right triangle is 3 and the other leg is 5, then the length of the hypotenuse is (1) $\sqrt{34}$ (2) 34 (3) $\sqrt{8}$ (4) 8

3. The legs of a right triangle have lengths of 2 and 7. Express, in radical form, the length of the hypotenuse.

4. The lengths of the legs of a right triangle are 2 and 5. Find, in radical form, the length of the hypotenuse.

5. The hypotenuse of a right triangle is 10. If the length of one leg is 8, find the length of the other leg.

6. If the length of the hypotenuse of a right triangle is 13 and the length of one leg is 12, find the length of the other leg.

7. The length of the hypotenuse of a right triangle is 8 and the length of one leg is 5. The length of the other leg is
(1) $\sqrt{39}$ (2) $\sqrt{89}$ (3) 3 (4) 13

8. The length of a rectangle is 8 centimeters and its width is 6 centimeters. Find the number of centimeters in the length of the diagonal.

9. Express in radical form the length of the diagonal of a square whose sides are each 2.

10. The length and width of a rectangle are 15 and 8, respectively. Find the length of a diagonal.

11. Find the diagonal of a rectangle whose sides are 5 and 12.

12. The length of one leg of a right triangle is 8. The lengths of the other leg and the hypotenuse are consecutive odd integers. Find the length of the hypotenuse. [Only an algebraic solution will be accepted.]

13. As shown in the accompanying figure, $ABCD$ is a rectangle, E is a point on $\overline{AB}$, $DE = 13$, $AE = 5$, and $DC = 20$.

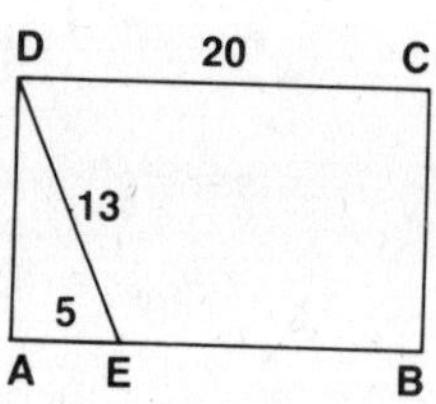

a. Find AD.
b. Find the perimeter of trapezoid $EBCD$.
c. Find the area of $\triangle AED$.
d. Find the area of rectangle $ABCD$.
e. Find the area of trapezoid $EBCD$.

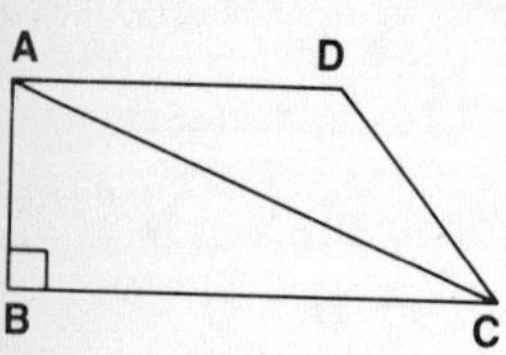

14. In the accompanying figure, $\overline{AB}$ is the height in trapezoid $ABCD$, $AB = 5$, $BC = 12$, and $AD = 8$. Diagonal $\overline{AC}$ is drawn.
a. Find AC.
b. Find the area of trapezoid $ABCD$.
c. Find the area of triangle ABC.
d. Find the area of triangle ACD.

15. As shown in the accompanying diagram, $ABCD$ is a rectangle and a line segment drawn from B intersects $\overline{CD}$ at E.

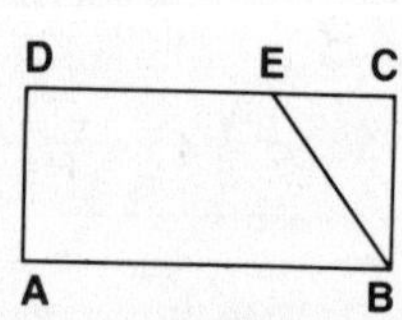

a. If $AB = 9$ and the area of the rectangle is 36, find AD.
b. The point E separates $\overline{DC}$ into two segments such that $DE:EC = 2:1$. Find DE and EC.
c. Find BE.
d. Find the area of trapezoid $ABED$.

16. In the diagram, $\angle BDE$ and $\angle BCA$ are right angles, $\triangle ABC$ is similar to $\triangle EBD$, $BC = 12$, $AC = 9$, and $BD = 8$.
   a. Find DE.
   b. Find BE.
   c. Find the area of $\triangle ABC$.
   d. Find the area of $\triangle BDE$.
   e. Find the area of trapezoid $CAED$.

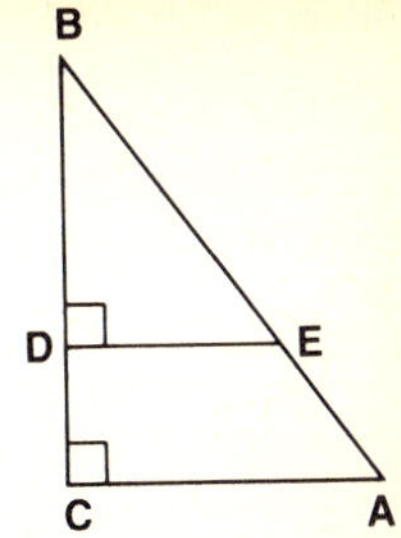

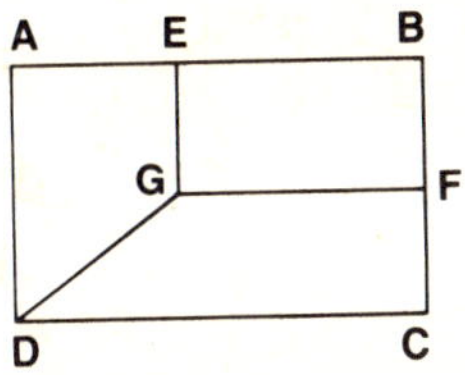

17. In the accompanying figure, $ABCD$ and $EBFG$ are rectangles such that $\overline{AEB}$ and $\overline{BFC}$ are straight line segments. If $AB = 10$, $EB = 6$, $BC = 6$, and $BF = 3$, find the length of $\overline{DG}$. (*Hint:* Extend $\overline{EG}$.)

9.4 CIRCUMFERENCE AND AREA OF A CIRCLE

The distance around a circle is called the ***circumference.*** The circumference, C, of a circle is a little more than three times the length of its diameter, d. The number π, approximately equal to 3.14 or to $\frac{22}{7}$, is the number by which the diameter must be multiplied to produce the circumference. That is:

$$C = \pi d$$

(*Note:* Since the number π cannot be expressed exactly as a ratio of two integers, π is an irrational number.)

Since the diameter of a circle is equal in length to two radii (the length of one radius is r), another formula for the circumference of a circle is:

$$C = 2\pi r$$

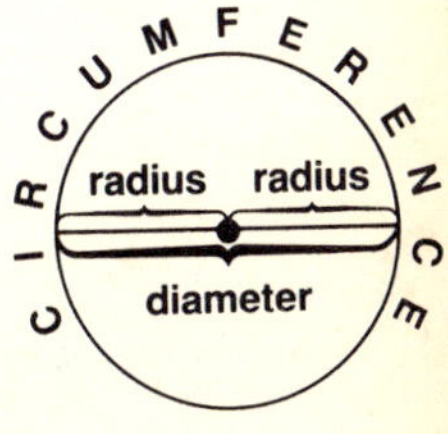

Examples

1. Find the length of the circumference of a circle if the length of a radius is 3 inches.

   Solution:

$$C = 2\pi r$$
$$C = 2\pi(3)$$
$$C = 6\pi \text{ inches} \quad \textit{Ans.}$$

2. What is the length of a diameter of the circle whose circumference is 10π meters?

Solution:

$$C = \pi d$$
$$10\pi = \pi d$$
Divide by π. $\quad 10 = d$ meters *Ans.*

3. If a radius of a circle is tripled, then the circumference of the circle
(1) is increased by 3
(2) is multiplied by 3
(3) is multiplied by 6
(4) is multiplied by 9

Solution:

$$C = 2\pi r$$
Replace r by $3r$: $\quad C = 2\pi(3r)$
$$C = 6\pi r$$

The new circumference, $6\pi r$, is three times the original circumference, $2\pi r$.

Answer: (2)

4. A racetrack for cars is in the form of a rectangle and two semicircles joined as shown.

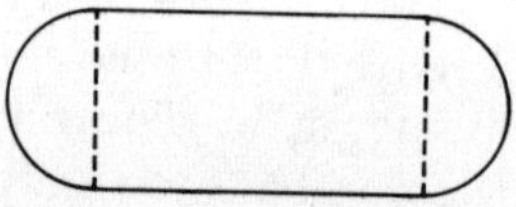

If the length of the base of the rectangle is 1 mile and the radius of each semicircle is $\frac{1}{4}$ mile, find the total mileage a car must go if it is to race the entire track. Express the answer to the nearest tenth of a mile.

Solution:

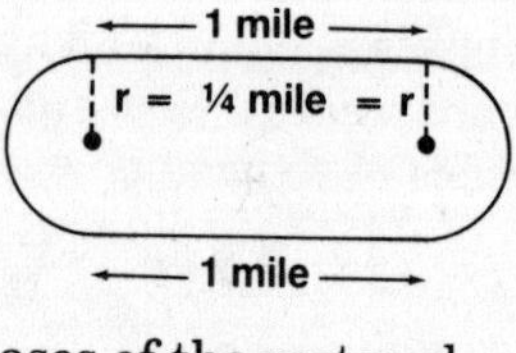

The entire track perimeter = 2 bases of the rectangle + 2 semicircles

perimeter = 2(1 mile) + 2($\frac{1}{2}$ circumference)

$$P = 2 \text{ miles} + 1 \text{ circumference}$$
$$P = 2 \text{ miles} + 2\pi r$$
$$P \approx 2 \text{ miles} + 2(3.14)(\tfrac{1}{4} \text{ mile})$$
$$P \approx 2 \text{ miles} + (6.28)(\tfrac{1}{4}) \text{ miles}$$
$$P \approx 2 \text{ miles} + 1.57 \text{ miles}$$
$$P \approx 3.57 \text{ miles}$$
$$P \approx 3.6 \text{ miles}$$

Answer: The length of the entire track is approximately 3.6 miles.

The area of a circle represents the measure of the portion of the plane enclosed by the circle. The area of a circle is calculated by the formula:

$$A = \pi r^2$$

Examples

1. Find the area of a circle whose diameter is 12 inches. [Answer may be left in terms of π.]

Solution: Since $d = 12$, $r = 6$.

$$A = \pi r^2$$
$$A = \pi(6)^2$$
$$A = 36\pi \text{ square inches} \quad \textit{Ans.}$$

2. Find a diameter of a circle whose area is 81π square centimeters. [Answer may be left in terms of π.]

Solution:

$$A = \pi r^2$$
$$81\pi = \pi r^2$$
$$81 = r^2 \quad \text{Divide by } \pi.$$
$$\pm\sqrt{81} = r$$
$$\pm 9 = r \quad (\text{Reject } -9.)$$

Diameter $= 2r = 2 \cdot 9 = 18$ centimeters *Ans.*

3. If a radius of a circle is multiplied by 4, then the area of the circle
   (1) is increased by 4
   (2) is multiplied by 4
   (3) is multiplied by 8
   (4) is multiplied by 16

Solution: $A = \pi r^2$

Replace r by $4r$: $A = \pi(4r)^2$

$$A = \pi \cdot 16r^2$$
$$A = 16\pi r^2$$

The new area, $16\pi r^2$, is 16 times the original area, πr^2.

Answer: (4)

A circle is ***circumscribed about*** a polygon when all the vertices of the polygon are on the circumference of the circle. The polygon is ***inscribed in*** the circle.

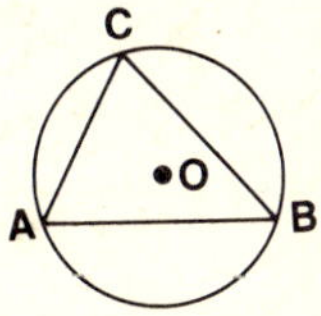

Circle O is circumscribed about $\triangle ABC$. $\triangle ABC$ is inscribed in circle O.

When a square is inscribed in a circle (the circle is circumscribed about the square), the diagonals of the square are diameters of the circle.

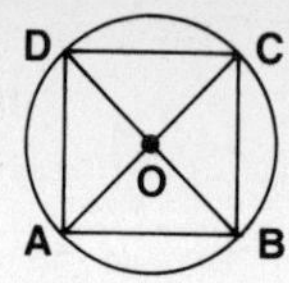

When a circle is inscribed in a square (the square is circumscribed about the circle), the sides of the square have the same length as a diameter of the circle.

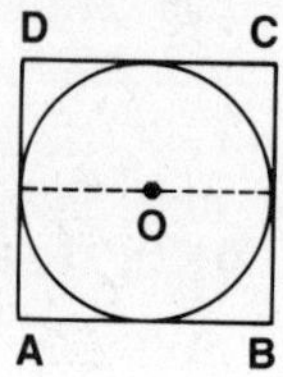

Examples

Square $ABCD$ is inscribed in circle O as shown in the diagram. If a side of the square = 10, find:

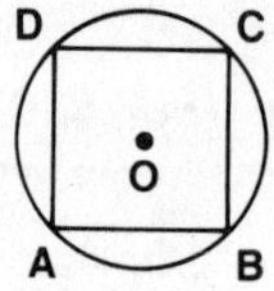

1. The length of a diagonal of the square. [Answer may be left in radical form.]
2. The area of the circle. [Answer may be left in terms of π.]

Solutions:

1. Apply the Pythagorean Theorem to right triangle ABC.

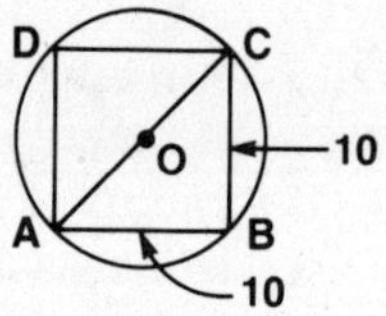

$$\begin{aligned}(AC)^2 &= (AB)^2 + (BC)^2\\ (AC)^2 &= 10^2 + 10^2\\ (AC)^2 &= 100 + 100\\ (AC)^2 &= 200\\ AC &= \pm\sqrt{200} \quad (\text{Reject } -\sqrt{200}.)\\ AC &= \sqrt{200} \text{ or } \sqrt{100}\cdot\sqrt{2} = 10\sqrt{2} \quad \textit{Ans.}\end{aligned}$$

2. Each diagonal of the square is also a diameter of the circle. Thus, diameter = $10\sqrt{2}$ and radius = $\frac{1}{2}(10\sqrt{2})$ or $5\sqrt{2}$.

$$\begin{aligned}\text{Area of circle} &= \pi r^2\\ &= \pi(5\sqrt{2})^2\\ &= \pi\cdot 25\cdot 2\\ &= 50\pi \text{ square units} \quad \textit{Ans.}\end{aligned}$$

EXERCISES

1. Find the circumference of a circle whose radius is 6. [Answer may be left in terms of π.]
2. If the circumference of a circle is 20π, what is a radius of the circle?
3. If a radius of a circle is multiplied by 5, then the circumference
   (1) remains the same
   (2) is increased by 5
   (3) is multiplied by 5
   (4) is multiplied by 25

4. If $C = 2\pi r$ and $d = 2r$, the ratio $\frac{C}{d}$ is
(1) 1 (2) 2 (3) $\sqrt{2}$ (4) π

5. A design is formed from two equilateral triangles, two semicircles, and a square joined as shown in the figure. The length of each side of the equilateral triangles and of the square is 10 inches. If the outer perimeter of the figure is to be trimmed in tape, how many inches of tape are required? [Answer may be left in terms of π.]

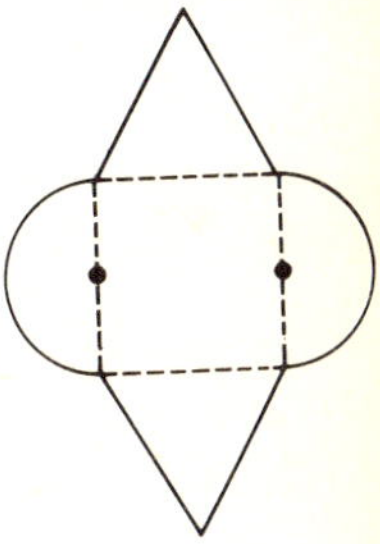

6. What is the area of a circle whose radius is 5?
(1) 100π (2) 25π (3) 10π (4) 5π
7. If a radius of a circle is 10, the area of the circle is
(1) 10π (2) 20π (3) 100π (4) 100
8. If a diameter of a circle is 14, find the area of the circle in terms of π.
9. If the area of a circle is 64π, what is a radius of the circle?
10. Find a radius of a circle whose area is 25π.
11. If the area of a circle is 16π, what is the length of a radius?
12. Find a radius of a circle if its area is 49π.
13. If a radius of a circle is tripled, then the area of the circle is multiplied by (1) 27 (2) 9 (3) 3 (4) 6

14. In the accompanying figure, $\overline{AC}$ is a diameter of circle O. The measure of arc AB is 80°.

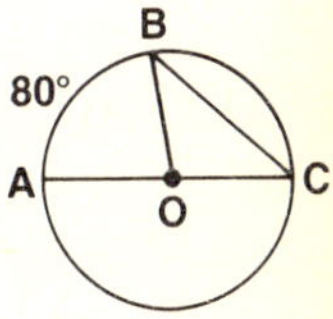

a. Find, in degrees, the measure of $\angle AOB$.
b. Find, in degrees, the measure of $\angle ACB$.
c. If the area of the circle is 49π, find the length of a radius.
d. If the area of the circle is 49π, find the circumference of the circle in terms of π.
e. Triangle COB is
(1) scalene (2) isosceles (3) equilateral (4) right

15. Square $ABCD$ is inscribed in a circle as shown in the diagram. If a radius of the circle is 6, find:

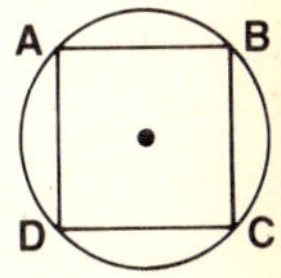

a. the length of a diagonal of the square
b. the length of a side of the square [Answer may be left in radical form.]
c. the area of the square

9.5 MORE DIFFICULT AREA PROBLEMS

Some area problems combine circles with other figures and require addition or subtraction of areas.

Examples

1. Two unequal semicircles, a square, and a rectangle are joined as shown. If the dimensions of the rectangle are 6 by 12, find the area of the entire region. [Answer may be left in terms of π.]

Solution:

Mark the diagram as follows:

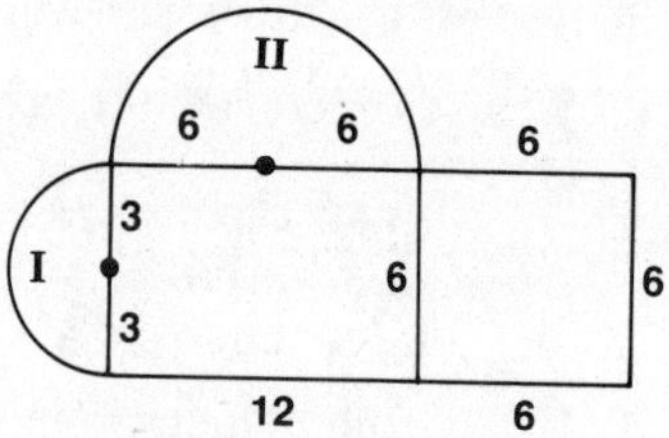

$$
\begin{aligned}
\text{Required Area} &= A_{\text{semicircle I}} + A_{\text{semicircle II}} + A_{\text{square}} + A_{\text{rectangle}}\\
\text{Required Area} &= \tfrac{1}{2}(\pi r^2) + \tfrac{1}{2}(\pi R^2) + s^2 + b \cdot h\\
&= \tfrac{1}{2}(\pi \cdot 3^2) + \tfrac{1}{2}(\pi \cdot 6^2) + 6^2 + 6 \cdot 12\\
&= \tfrac{1}{2}(9\pi) + \tfrac{1}{2}(36\pi) + 36 + 72\\
&= \tfrac{9}{2}\pi + \tfrac{36}{2}\pi + 36 + 72\\
&= (\tfrac{45}{2}\pi + 108) \text{ square units} \quad \textit{Ans.}
\end{aligned}
$$

2. As shown in the figure, one circle is cut out of a larger circle. Find the area of the shaded portion if the radii are 3 and 8, respectively. [Answer may be left in terms of π.]

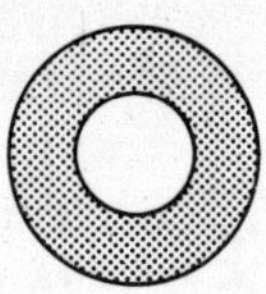

Solution:

$$
\begin{aligned}
\text{Required Area} &= A_{\text{large circle}} - A_{\text{small circle}}\\
&= \pi R^2 - \pi r^2\\
&= \pi(8)^2 - \pi(3)^2\\
&= 64\pi - 9\pi\\
&= 55\pi \text{ square units} \quad \textit{Ans.}
\end{aligned}
$$

3. As shown in the accompanying figure, four congruent quarter-circles are cut out of a square of side 10. Find the area of the shaded region. [Answer may be left in terms of π.]

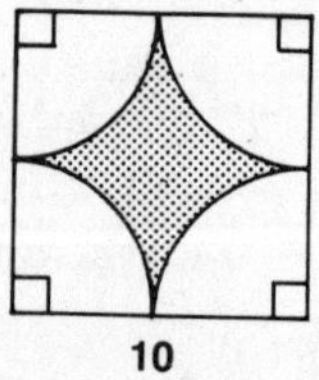

Solution:

Since four congruent quarter-circles = 1 circle:

$$
\begin{aligned}
\text{Required Area} &= A_{\text{square}} - A_{\text{circle}}\\
&= s^2 - \pi r^2\\
&= 10^2 - \pi(5)^2 \quad \text{Radius of circle} = \tfrac{1}{2} \text{ side of square.}\\
&= (100 - 25\pi) \text{ square units} \quad \textit{Ans.}
\end{aligned}
$$

EXERCISES

1. As shown in the accompanying diagram, a square with side s is inscribed in a circle with radius r. Which expression represents the area of the shaded region?
(1) $s^2 - \pi r^2$ (3) $\pi r^2 - 4s$
(2) $\pi r^2 - s^2$ (4) $4s - \pi r^2$

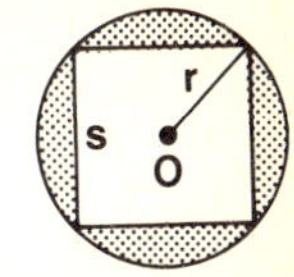

In 2–10, answers may be left in terms of π or in radical form.

2. In the accompanying figure, square $ABCD$ is circumscribed about circle O. A side of the square and a diameter of the circle are both 10. Find the:
   a. circumference of the circle
   b. perimeter of the square
   c. area of the square
   d. area of the circle
   e. area of the shaded portion

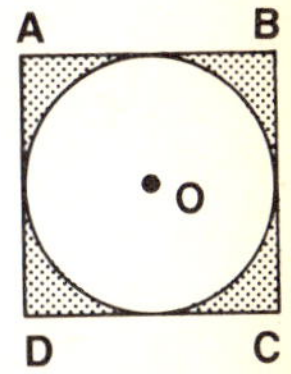

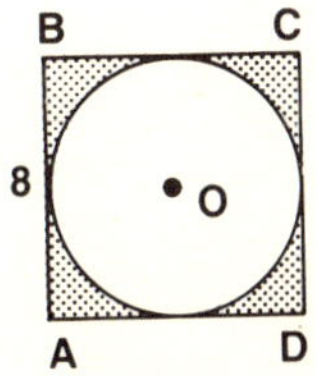

3. In the accompanying figure, square $ABCD$ is circumscribed about circle O. The sides of the square have the same length as a diameter of the circle and $AB = 8$. Find the:
   a. circumference of the circle
   b. area of the circle
   c. area of the square
   d. area of the shaded portion
   e. length of $\overline{AC}$

4. Given: Rectangle $ABDC$ with two circles removed from the rectangle. The length of the rectangle is 35 and its width is 14. A diameter of each circle is 12. Find the:
   a. perimeter of the rectangle
   b. circumference of either circle
   c. area of the rectangle
   d. area of either circle
   e. area of the shaded portion

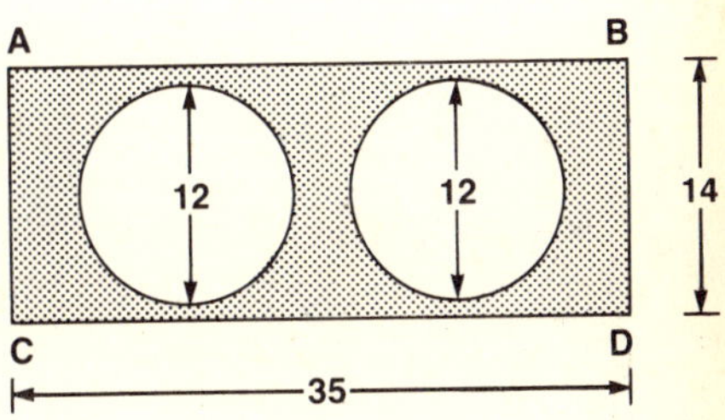

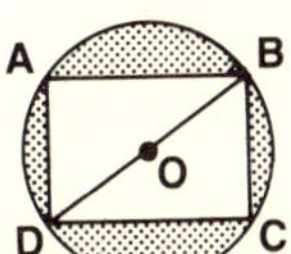

5. In the accompanying figure, rectangle $ABCD$ is inscribed in circle O and $\overline{DB}$ is a diameter. A radius of the circle is 5 and $CD = 8$. Find the:
   a. area of the circle
   b. length of $\overline{BC}$
   c. area of $\triangle BCD$
   d. area of rectangle $ABCD$
   e. area of the shaded portion

6. In the diagram below, arcs $\overset{\frown}{AB}$, $\overset{\frown}{BC}$, and $\overset{\frown}{CD}$ are semicircles with diameters $\overline{AB}$, $\overline{BC}$, and $\overline{CD}$, respectively. $ABCD$ is a rectangle, $BC = 28$, and $AB = 14$. Find the area of:

a. rectangle $ABCD$
b. the region enclosed by diameter $\overline{BC}$ and semicircle $\overset{\frown}{BC}$
c. the region enclosed by diameter $\overline{AB}$ and semicircle $\overset{\frown}{AB}$
d. the entire region

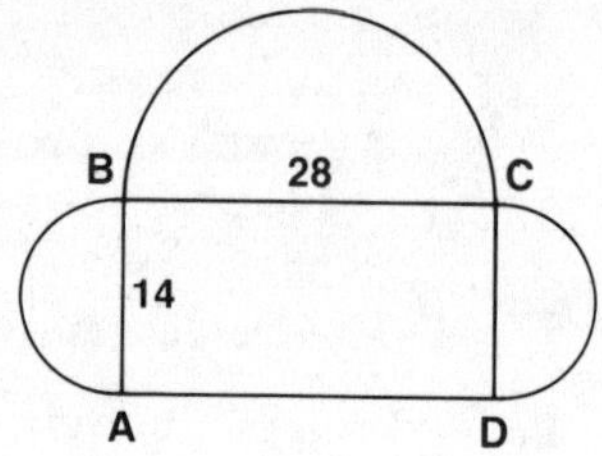

7. As shown in the accompanying figure, circle O is inscribed in square $ABFG$, $ACEG$ is a rectangle, $ADEG$ is a trapezoid, $AG = 6$, $FE = 4$, and $ED = 10$. Find the:

a. area of circle O
b. area of rectangle $ACEG$
c. length of $\overline{CD}$
d. area of triangle CDE
e. area of trapezoid $ADEG$

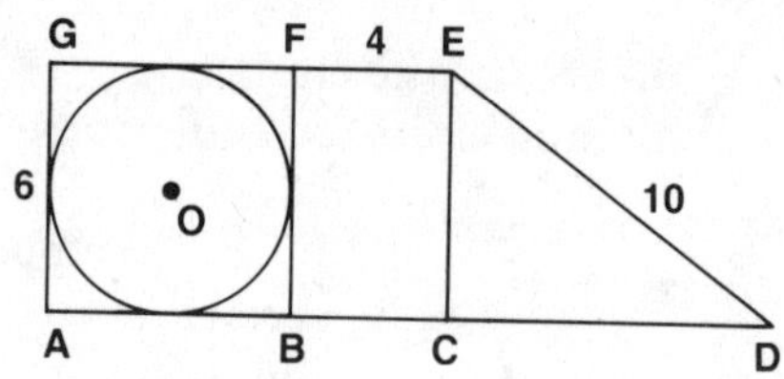

8. In the accompanying figure, isosceles right triangle ABC is inscribed in a semicircle of diameter 20. Find the area of the shaded region.

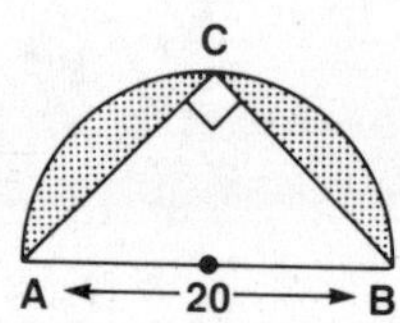

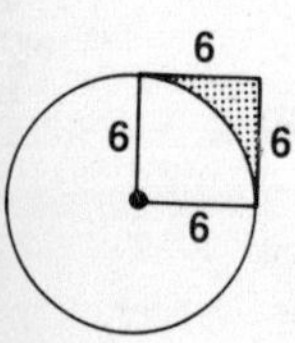

9. In the accompanying figure, a square of side 6 and a circle of radius 6 are joined as shown. Find the area of the shaded region.

10. A square joins the centers of four congruent circles as shown. If a side of the square is 8, find the area of the shaded region.

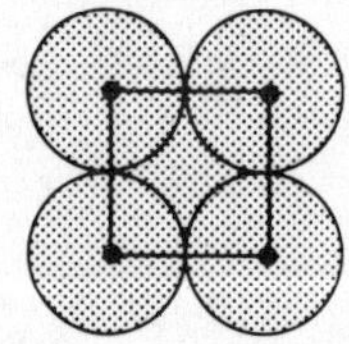

9.6 VOLUME OF SOME SOLID FIGURES

A *plane* figure has *two* dimensions; a *solid* figure has *three* dimensions.

Plane Figure

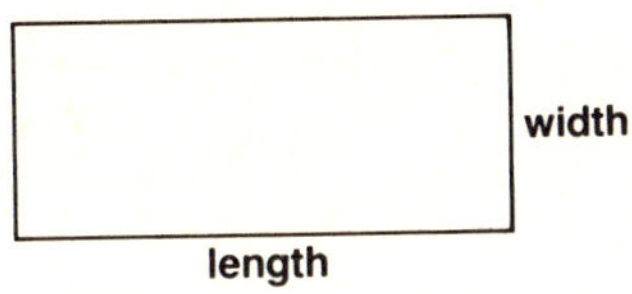

Solid Figure

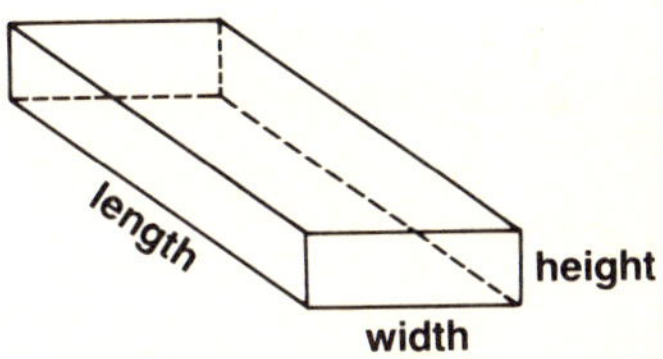

Just as *area* is the *number of one square units* contained in a *plane* figure, so ***volume*** is the number of *one cubic units* contained in a *solid* figure.

Area in a Plane Figure

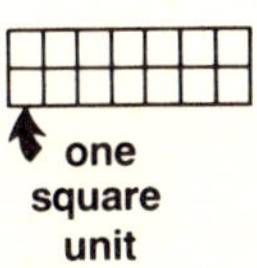

Volume in a Solid Figure

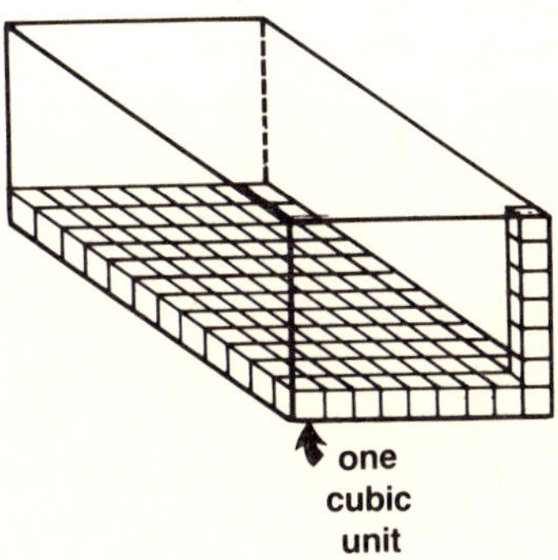

Area of a Square

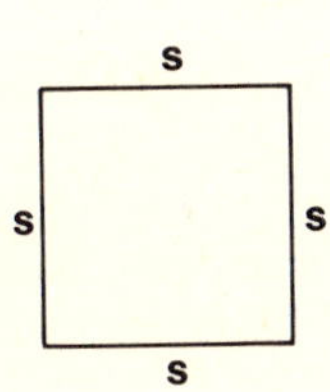

There are 4 equal sides to a square.

$$A_{\text{square}} = s^2$$

Volume of a Cube

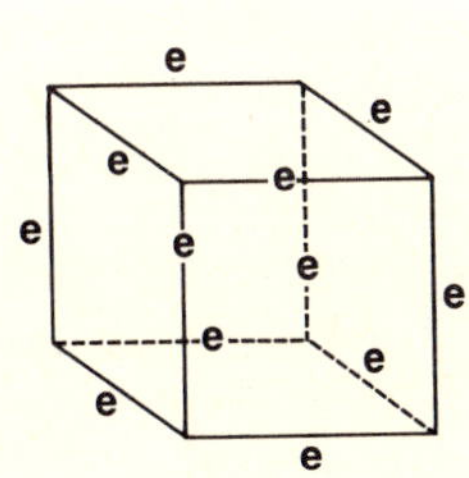

There are 12 equal *edges* to a cube and 6 equal *faces*. Each face is a square of side *e*.

$$V_{\text{cube}} = e^3$$

Area of a Rectangle

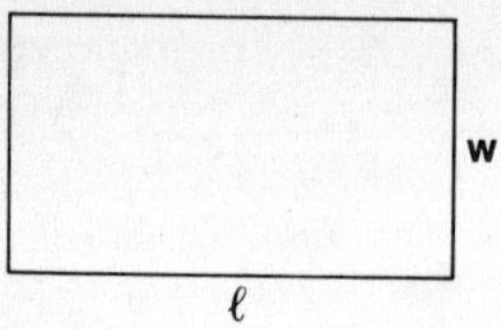

$A_{\text{rectangle}} = \ell \cdot w$

Volume of a Rectangular Solid (Prism)

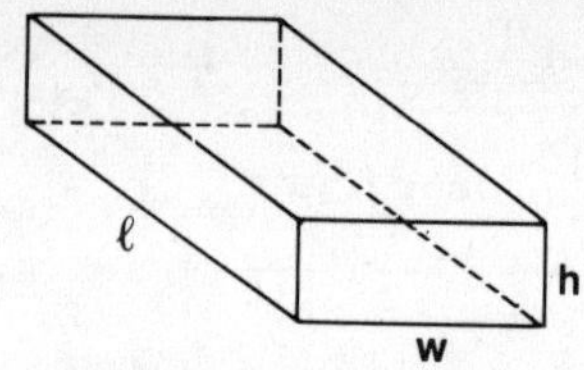

$V_{\text{rectangular solid}} = \ell \cdot w \cdot h$

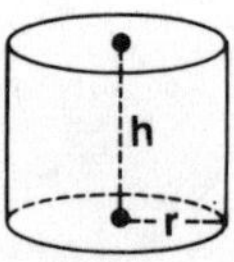

Another common solid figure is a ***cylinder*** (think of a closed tin can). The volume of this type of cylinder, a right circular cylinder, is found by multiplying the area of the circular base by the height of the cylinder.

$V_{\text{cylinder}} = \pi r^2 h$ where r = radius of the base and h = height of cylinder

Two other important solid figures are the cone (right circular cone) and the sphere. The volume of each of these figures is found by the following formulas:

Cone (Right Circular Cone)

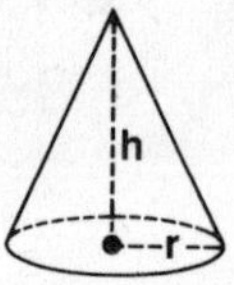

$V_{\text{cone}} = \frac{1}{3}\pi r^2 h$

(*Note:* The volume of a right circular cone is one-third the volume of a right circular cylinder.)

Sphere

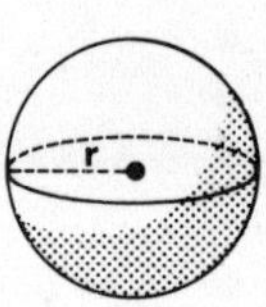

$V_{\text{sphere}} = \frac{4}{3}\pi r^3$

As in area problems, volume problems are solved by beginning with the appropriate formula. Substitute the known values in the proper places and solve for the unknown value.

Examples

1. Find the volume of a cube of edge 4 inches.

Solution:

$$\begin{aligned} V_{\text{cube}} &= e^3 \\ &= 4^3 \\ &= 64 \text{ cubic inches} \quad \textit{Ans.} \end{aligned}$$

2. Find the length of an edge of a cube whose volume is 27 cubic centimeters.

Solution:

$$V_{\text{cube}} = e^3$$
$$27 = e^3$$
$$\sqrt[3]{27} = e$$
$$e = 3 \text{ centimeters} \quad \textit{Ans.}$$

3. Find the area of a face of a cube whose volume is 8 cubic inches.

Solution:

Each of the six faces of a cube is a square whose side, s, is an edge, e, of the cube.

The area of a square face is s^2 or e^2. Thus, we must first find the length of an edge of the cube.

$$V_{\text{cube}} = e^3$$
$$8 = e^3$$
$$\sqrt[3]{8} = e$$
$$e = 2$$

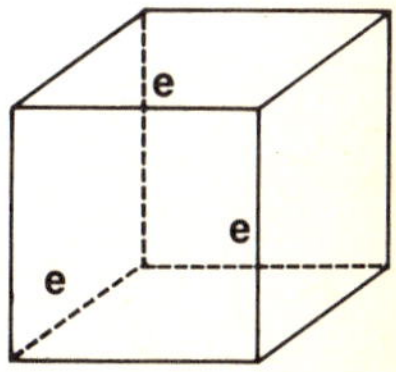

Use the length of the edge, $e = 2$ inches, to find the area of the square face.

$$A_{\text{square}} = s^2 \text{ or } e^2$$
$$= 2^2 = 4 \text{ square inches} \quad \textit{Ans.}$$

4. Find the volume of a prism (rectangular solid) whose length is 4 feet, width is 2 feet, and height is 3.5 feet.

Solution:

$$V_{\text{prism}} = \ell \cdot w \cdot h$$
$$= (4)(2)(3.5)$$
$$= 8(3.5)$$
$$= 28 \text{ cubic feet} \quad \textit{Ans.}$$

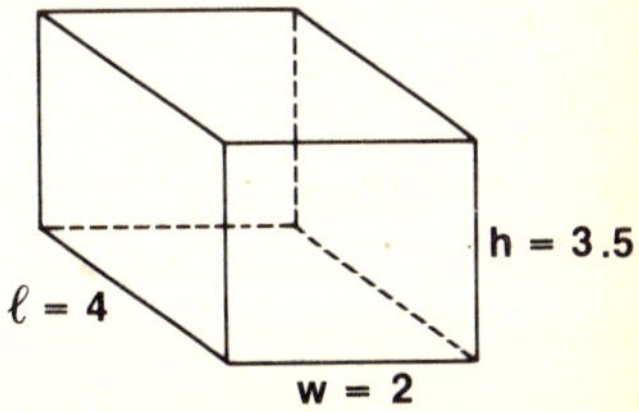

5. The volume of a rectangular solid is 48 cm^3. If the height is 1 cm and the length exceeds the width by 13 cm, find the width.

Solution:

Let x = the width of the rectangular solid.
Then $x + 13$ = the length of the rectangular solid.

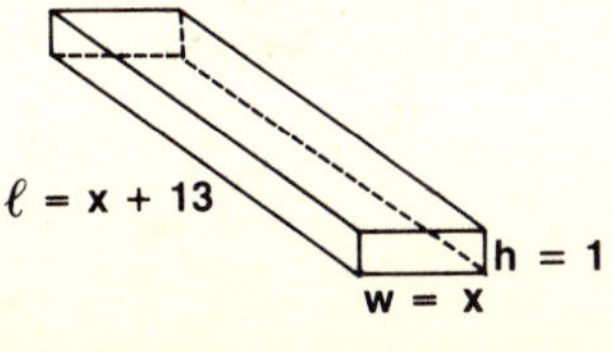

$$
\begin{aligned}
V_{\text{rectangular solid}} &= \ell \cdot w \cdot h \\
48 &= (x + 13)(x)(1) \\
48 &= x^2 + 13x \\
x^2 + 13x - 48 &= 0 \\
(x + 16)(x - 3) &= 0
\end{aligned}
$$

$x + 16 = 0$	$x - 3 = 0$
$x = -16$	$x = 3$ width
Reject.	$x + 13 = 16$ length

Check:

$$
\begin{aligned}
V &= \ell \cdot w \cdot h \\
48 &\stackrel{?}{=} (16)(3)(1) \\
48 &= 48 \checkmark
\end{aligned}
$$

Answer: The width is 3 cm.

6. Find the volume of a cylinder if the radius of the base is 6 inches and the height of the cylinder is 4 inches. [Answer may be left in terms of π.]

Solution:

$$
\begin{aligned}
V_{\text{cylinder}} &= \pi r^2 h \\
&= \pi(6)^2 4 \\
&= \pi(36)4 \\
&= \pi(144) \\
&= 144\pi
\end{aligned}
$$

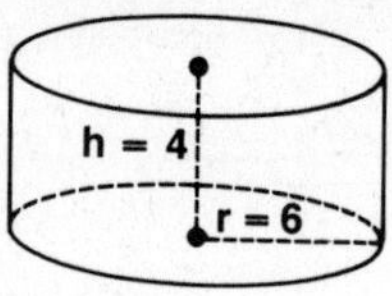

Answer: The volume of the cylinder is 144π cubic inches.

7. If a radius of the base of a cylinder is doubled, then the volume
(1) is multiplied by 2
(2) is multiplied by 4
(3) is multiplied by 8
(4) is multiplied by 16

Solution:

$$V_{\text{cylinder}} = \pi r^2 h$$

Replace r by $2r$: $V_{\text{cylinder}} = \pi(2r)^2 h = \pi 4r^2 h = 4\pi r^2 h$

Thus, the new volume, $4\pi r^2 h$, is four times the original volume, $\pi r^2 h$.

Answer: (2)

8. Find a radius of a cone whose volume is 100π cubic inches and whose height is 12 inches.

Solution:

$$
\begin{aligned}
V_{\text{cone}} &= \tfrac{1}{3}\pi r^2 h \\
100\pi &= \tfrac{1}{3}\pi r^2 (12) \\
100\pi &= 4\pi r^2 && \tfrac{1}{3} \cdot 12 = 4 \\
25 &= r^2 && \text{Divide by } 4\pi. \\
\pm\sqrt{25} &= r \\
\pm 5 &= r && \text{Reject } r = -5.
\end{aligned}
$$

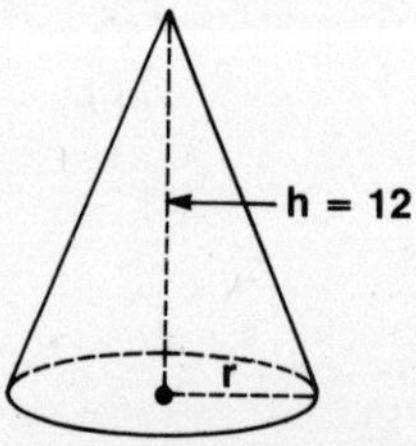

Answer: A radius of the cone is 5 inches.

9. Find the volume of a right circular cone whose radius is 7 feet and whose altitude is 3 feet. [Answer may be left in terms of π.]

Solution:

$$\begin{aligned} V_{\text{cone}} &= \tfrac{1}{3}\pi r^2 h \\ &= \tfrac{1}{3}\pi(7)^2(3) \\ &= \pi(7)^2 \quad \text{Cancel 3's.} \\ &= 49\pi \end{aligned}$$

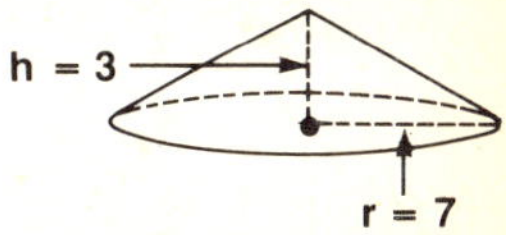

Answer: The volume of the cone is 49π cubic feet.

10. If a radius and the altitude of a right circular cone are doubled, the volume is multiplied by (1) 2 (2) 4 (3) 6 (4) 8

Solution: $V_{\text{cone}} = \tfrac{1}{3}\pi r^2 h$

Replace r by $2r$ and replace h by $2h$:

$$\begin{aligned} V &= \tfrac{1}{3}\pi(2r)^2(2h) \\ &= \tfrac{1}{3}\pi(4r^2)(2h) \\ &= \tfrac{1}{3}\pi(8r^2h) \\ &= 8(\tfrac{1}{3}\pi r^2 h) \end{aligned}$$

Thus, the new volume, $8(\tfrac{1}{3}\pi r^2 h)$, is eight times the original volume, $\tfrac{1}{3}\pi r^2 h$.

Answer: (4)

11. Find, to the nearest tenth of a cubic inch, the volume of a sphere whose radius is 2.1 inches.

Solution:

$$\begin{aligned} V_{\text{sphere}} &= \tfrac{4}{3}\pi r^3 \\ &\approx \tfrac{4}{3}(3.14)(2.1)^3 \quad \pi \approx 3.14 \\ &\approx (4.18)(9.26) \\ &\approx 38.7068 \\ &\approx 38.7 \text{ cubic inches} \quad \textit{Ans.} \end{aligned}$$

Calculations:

$\tfrac{4}{3}(3.14)$:

$$\begin{array}{r} 3.14 \\ \times\ 4 \\ \hline 12.56 \end{array} \qquad \begin{array}{r} 4.18 \\ 3\overline{)12.56} \end{array}$$

$(2.1)^3$:

$$\begin{array}{r} 2.1 \\ \times\ 2.1 \\ \hline 2\,1 \\ 4\,2 \\ \hline 4.4\,1 \\ \times\ 2.1 \\ \hline 4\,4\,1 \\ 8\,8\,2 \\ \hline 9.2\,6\,1 \end{array} \approx 9.26$$

12. Find a radius of a sphere whose volume is 36π cubic centimeters.

Solution:

$$\begin{aligned} A_{\text{sphere}} &= \tfrac{4}{3}\pi r^3 \\ 36\pi &= \tfrac{4}{3}\pi r^3 \\ 36 &= \tfrac{4}{3}r^3 \quad \text{Divide by } \pi. \\ \tfrac{3}{4}(36) &= \tfrac{3}{4}(\tfrac{4}{3}r^3) \quad \text{Multiply by } \tfrac{3}{4}. \\ 27 &= r^3 \\ \sqrt[3]{27} &= r \\ r &= 3 \text{ centimeters} \quad \textit{Ans.} \end{aligned}$$

EXERCISES

1. An edge of a cube has length 3. What is the volume of the cube?
2. An edge of a cube is represented by $2x$. Express, in terms of x, the volume of the cube.
3. The volume of a cube is 64. What is the length of an edge?
4. Find the area of a face of a cube whose volume is 125 cm^3.
5. The length of a rectangular solid is represented by x, the width by $2x$, and the height by $3x$. The volume of the rectangular solid is represented by
   (1) $6x$ (2) $5x^3$ (3) $6x^3$ (4) $6x^2$
6. The volume of a rectangular solid is 210 in.^{3} If the length is 6 in. and the width is 5 in., find the height.
7. The volume of a rectangular solid is 40 cubic feet. The height is 1 foot and the length exceeds the width by 3 feet. Find the width.
8. The length and width of a rectangular solid are doubled, and the height remains the same. The volume
   (1) remains the same (3) is multiplied by 4
   (2) is multiplied by 2 (4) is multiplied by 8
9. Find the volume of a cylinder whose radius is 8 inches and whose height is 10 inches. [Answer may be left in terms of π.]
10. The volume of a cylinder is found by using the formula $V = \pi r^2 h$. If h is doubled, then the volume is
    (1) doubled (3) increased by 4
    (2) increased by 2 (4) multiplied by 4
11. Find the volume of a right circular cone whose radius is 4 centimeters and whose altitude is 6 centimeters. [Answer may be left in terms of π.]
12. Find the altitude of a cone whose volume is 27π cubic inches and whose radius is 3 inches.
13. A radius of a cone is doubled and the altitude remains the same. The volume is multiplied by
    (1) 2 (2) 4 (3) 6 (4) 8
14. Find, to the nearest cubic foot, the volume of a sphere whose radius is 2 feet.
15. Find a radius of a sphere whose volume is $\frac{256}{3}\pi$ cubic centimeters.
16. If a radius of a sphere is doubled, the volume is multiplied by
    (1) 2 (2) 4 (3) 6 (4) 8
17. If $C = 2\pi r$, solve for r in terms of C and π.
18. If $V = \ell wh$, solve for h in terms of V, ℓ, and w.

9.7 SUMMARY EXERCISES

1. If the lengths of the sides of a triangle are represented by $3x - 1$, $4x + 2$, and $x + 5$, express the perimeter of the triangle in terms of x.
2. If the perimeter of a square is represented by $16x - 8$, express the length of one side of the square in terms of x.
3. If each side of regular octagon $ABCDEFGH$ is represented by $2x - 1$, which expression represents the perimeter of the octagon?
(1) $10x - 7$ (2) $16x - 1$ (3) $16x - 8$ (4) $8x$
4. If the length and width of a rectangle are tripled, then its perimeter
(1) is tripled (3) is multiplied by 9
(2) is multiplied by 6 (4) is increased by 9
5. The length of a rectangle is 2 inches less than three times the width. If the perimeter of the rectangle is 28 inches, find the dimensions of the rectangle.
6. The length of a rectangle is equal to the length of a side of an equilateral triangle, and the width of the rectangle is 2 centimeters less than the side of the triangle. If the perimeter of the rectangle exceeds the perimeter of the triangle by 10 centimeters, find the length of a side of the triangle.
7. In the accompanying diagram, $\angle A \cong \angle P$, $\angle B \cong \angle Q$. If $AB = 8$, $BC = 6$, $AC = 3$, and $PR = 1\frac{1}{2}$, what is the perimeter of $\triangle PQR$?

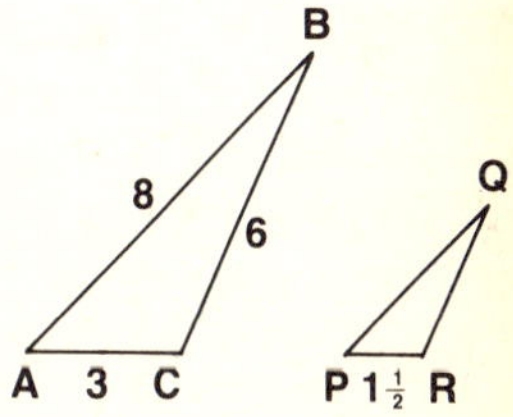

8. If two rectangles have equal perimeters, which statement must be true?
(1) The rectangles have equal areas.
(2) The rectangles are squares.
(3) The lengths of the rectangles are equal.
(4) The sum of the length and width of one rectangle is equal to the sum of the length and width of the other.
9. The length of a rectangle exceeds the width by 2. If the perimeter of the rectangle is greater than 68, what are the smallest possible integral dimensions of the rectangle?
10. The length of a rectangle is 3 less than the width. If the perimeter of the rectangle is less than 26, what are the largest possible integral dimensions of the rectangle?
11. Find the area of a square whose perimeter is 32.
12. The area of a square whose perimeter is $8k$ is
(1) $8k^2$ (2) $4k\sqrt{2}$ (3) $4k^2\sqrt{2}$ (4) $4k^2$
13. Two squares have areas of 9 square centimeters and 16 square centimeters. What is the ratio of the perimeter of the smaller square to the perimeter of the larger square?
(1) 81:256 (2) 9:16 (3) 3:4 (4) $\sqrt{3}:2$

14. If the length of a rectangle is represented by $3x - 2$ and its width is represented by $2x$, what is the area of the rectangle in terms of x?
(1) $10x - 4$ (2) $10x - 2$ (3) $6x^2 - 4x$ (4) $6x^2 - 2$

15. If the area of a rectangle is represented by $x^2 - 25$ and the length is represented by $x + 5$, which of the following represents the width?
(1) $x + 5$ (2) $x - 5$ (3) x (4) $5x$

16. The width of a rectangle is 1 less than a side of a square. The length of the rectangle is 3 more than a side of the square. If the area of the rectangle exceeds the perimeter of the square by 45, find the length of a side of the square.

17. A hexagon *ABCDEF* is composed of five congruent squares, as shown in the accompanying diagram. If the perimeter of *ABCDEF* is 36, find the area of *ABCDEF*.

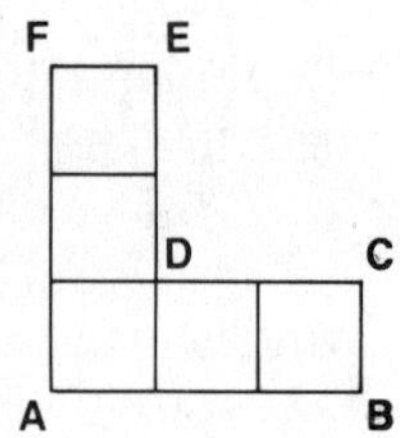

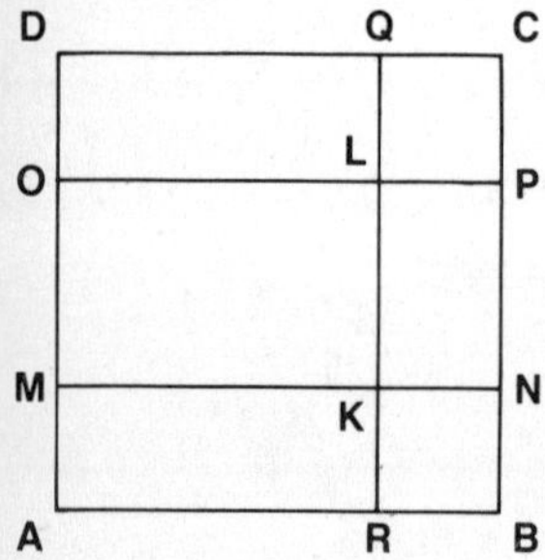

18. $\overline{MN}$, $\overline{OP}$, and $\overline{QR}$ are drawn in square *ABCD* so that a collection of rectangles and squares is formed within *ABCD*, as shown in the accompanying diagram. If the area of square *ABCD* is 144 and the area of each of the two congruent squares *QCPL* and *KNBR* is 9, find the area of rectangle *ARKM*.

19. The lengths of the sides of two unequal squares are in the ratio 1:3. If the sum of the areas of the two squares is 1,000, find the perimeter of the smaller square.

20. The width of a rectangle is 1 less than a side of a square. The length of the rectangle is 5 more than a side of the square. If the area of the rectangle exceeds the area of the square by 23, find the length of a side of the square.

21. Find the area of the parallelogram whose dimensions are shown in the accompanying diagram.

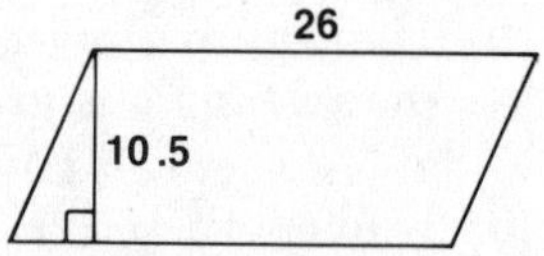

22. If the area of a parallelogram is 48 and the length of a base is 12, what is the length of an altitude drawn to that base?

23. The measure of the base of a parallelogram is 5 meters more than the measure of the altitude to that base. If the area of the parallelogram is 36 square meters, find the number of meters in the measures of the base and altitude.

24. If, in a parallelogram, both the base and an altitude drawn to that base are doubled, the area of the parallelogram (1) is doubled (2) is increased by 2 (3) is multiplied by 4 (4) is increased by 4

25. In the accompanying diagram, $ABCD$ is a trapezoid with $\overline{BC} \parallel \overline{AD}$ and altitude $\overline{BE}$. If $BC = 7$, $BE = 4$, and $AD = 12$, what is the area of $ABCD$?
(1) 23 (2) 38 (3) 76 (4) 336

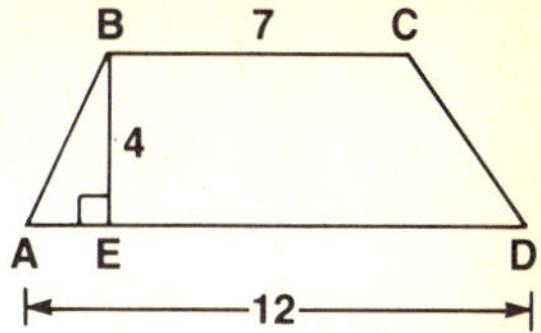

26. The area of parallelogram $ABCD$ is equal to the area of a square whose side is 6. If $AB = 12$, find the length of an altitude drawn to side $\overline{AB}$.

27. The lengths of the bases of a trapezoid are 11 centimeters and 15 centimeters. If the area of the trapezoid is 78 square centimeters, find the number of centimeters in the length of an altitude.

28. The area of a trapezoid is 100 square centimeters, an altitude is 10 centimeters, and the length of one base is 5 centimeters. What is the number of centimeters in the length of the other base?
(1) 5 (2) 10 (3) 15 (4) 20

29. In the accompanying diagram, $ABCD$ is a trapezoid in which $\overline{AB}$ and $\overline{CD}$ are the bases. If $m\angle B = 90$, $AB = 13$, $BC = 4$, $CD = 10$, and $DA = 5$, find the area of $ABCD$.

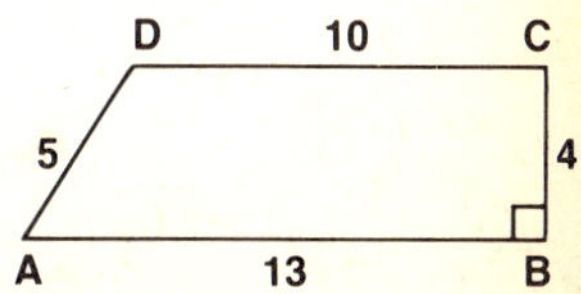

30. The area of a triangle is 150 square inches. If the length of a base of the triangle is 50 inches, find the number of inches in the altitude drawn to that base.

31. Find the area of a right triangle whose legs have lengths 5 and 10.

32. The length of the base $\overline{AB}$ of $\triangle ABC$ and the length of a side of a square are each 8. If the area of the triangle is equal to the area of the square, find the length of an altitude drawn to $\overline{AB}$.

33. In the accompanying figure, quadrilateral $ABCD$ is a trapezoid with $\overline{AB} \parallel \overline{DC}$, $\overline{AD} \perp \overline{AB}$, and $\overline{DB} \perp \overline{BC}$; $AB = AD = 4$, and $DC = 8$.

a. Find DB in radical form.
b. Find the area of $\triangle ABD$.
c. Find the area of trapezoid $ABCD$.
d. Find the area of $\triangle DBC$.

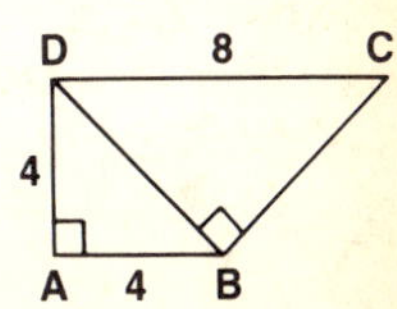

34. Figure I is a rectangle, II is a right triangle, and III is a trapezoid. The lengths of particular segments of the figures are indicated. Which figures have equal area?
(1) I and II only
(2) I and III only
(3) II and III only
(4) I, II, and III

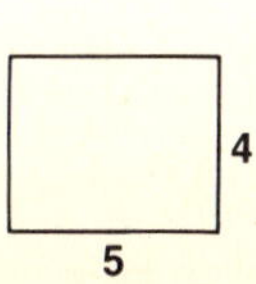

I

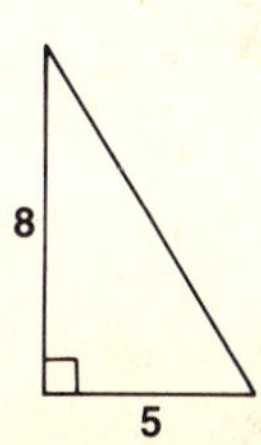

II

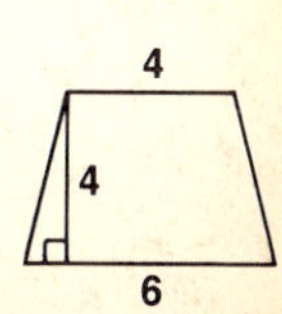

III

35. In the accompanying diagram, $\triangle ABC$ and $\triangle AED$ are right triangles drawn as shown. If $CB = 6$, $AC = 15$, and $DE:CB = 3:1$, find:
   a. DE
   b. AD
   c. the area of $\triangle ABC$
   d. the area of trapezoid $EBCD$

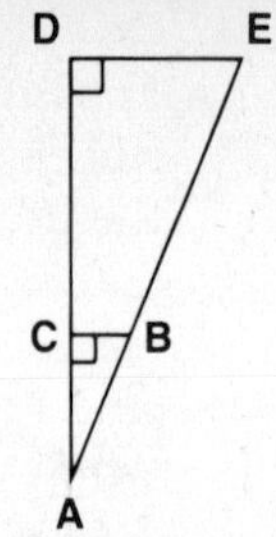

36. The lengths of the two legs of a right triangle are 9 and 12. Find the length of the hypotenuse.

37. The length of the hypotenuse of a right triangle is 26 and the length of the longer leg is 24. Find the length of the shorter leg.

38. In the accompanying diagram of rectangle $ABCD$, $AB = 10$ and $BC = 4$. Find, in radical form, the length of diagonal $\overline{AC}$.

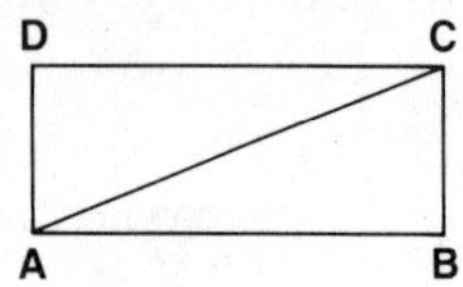

39. A rectangle has a diagonal of length 10 and one side of length 6. What is the perimeter of the rectangle?
   (1) 14 (2) 21 (3) 28 (4) 48

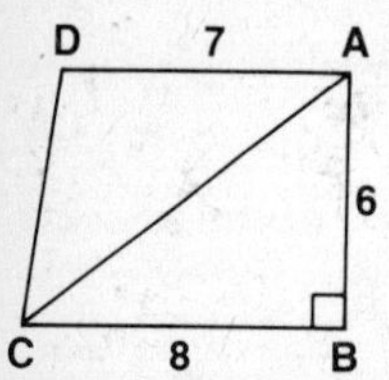

40. In the accompanying figure, $\overline{AB}$ is the height in trapezoid $ABCD$, $AB = 6$, $BC = 8$, and $AD = 7$. Diagonal $\overline{AC}$ is drawn.
   a. Find AC.
   b. Find the area of trapezoid $ABCD$.
   c. Find the area of triangle ABC.
   d. Find the area of triangle ADC.

41. As shown in the accompanying diagram, $ABCD$ is a rectangle, E is a point on $\overline{CD}$, $AB = 24$, $AD = 12$, and $AE = 13$.
   a. Find ED.
   b. Find the area of $\triangle AED$.
   c. Find the area of rectangle $ABCD$.
   d. Find the area of trapezoid $ABCE$.

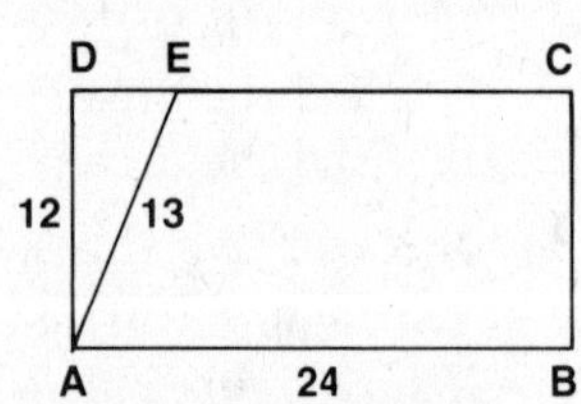

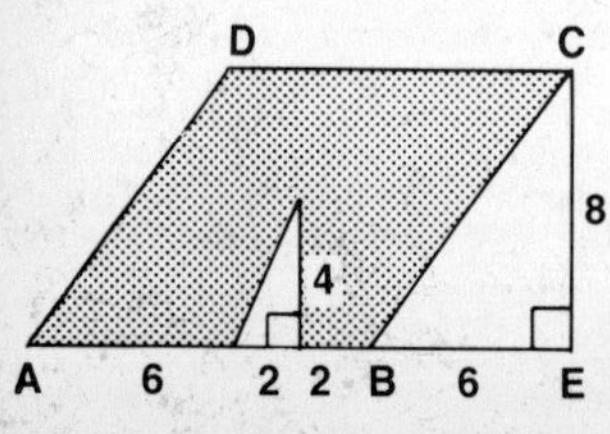

42. As shown in the accompanying diagram, two right triangles are cut out of trapezoid $AECD$. If $ABCD$ is a parallelogram, use the dimensions indicated and find the area of the shaded portion.

43. In the accompanying diagram, triangle BCE is inscribed in square $ABCD$. If E is the midpoint of $\overline{AD}$ and the length of side $\overline{BC}$ = 4 centimeters, what is the area in square centimeters of the shaded portion of the diagram?

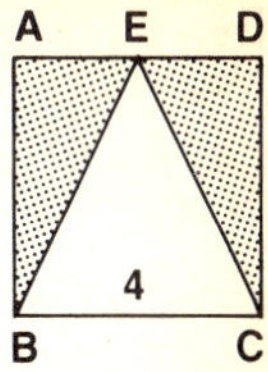

44. Find the circumference of a circle whose diameter is 12. [Answer may be left in terms of π.]

45. Find the area of a circle whose circumference is 20π. [Answer may be left in terms of π.]

46. If a radius of a circle is doubled, then the circumference
(1) is increased by 2
(2) is doubled
(3) is increased by 4
(4) is multiplied by 4

47. The value of π is
(1) rational and equal to 3.14
(2) irrational and equal to 3.14
(3) rational and between 3.14 and 3.15
(4) irrational and between 3.14 and 3.15

48. As shown in the accompanying diagram, a small circle lies in the interior of a larger circle. If the lengths of radii of the two circles are 6 and 9, respectively, find the area of the shaded region. [Answer may be left in terms of π.]

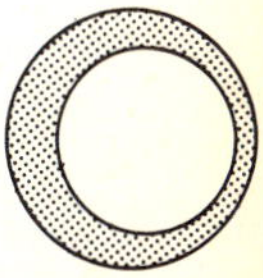

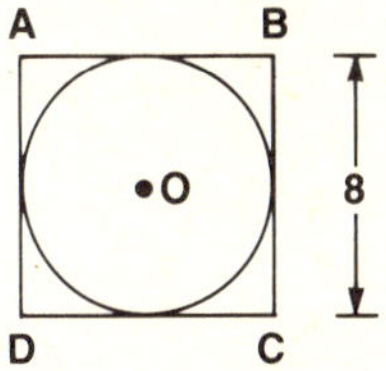

49. In the accompanying diagram, circle O is inscribed in square $ABCD$. If BC = 8, what is the area of the circle?
(1) 16π (2) 8π (3) 16 (4) 8

50. As shown in the accompanying diagram, rectangle $ABCD$ is inscribed in circle O. The length of diameter $\overline{DOB}$ of the circle is 13 and BC = 5. Find the length of $\overline{AB}$.

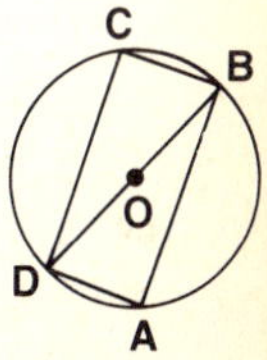

51. Two unequal semicircles, a rectangle, and an isosceles right triangle are joined as shown in the accompanying diagram. If the dimensions of the rectangle are 4 by 12, find the area of the entire region. [Answer may be left in terms of π.]

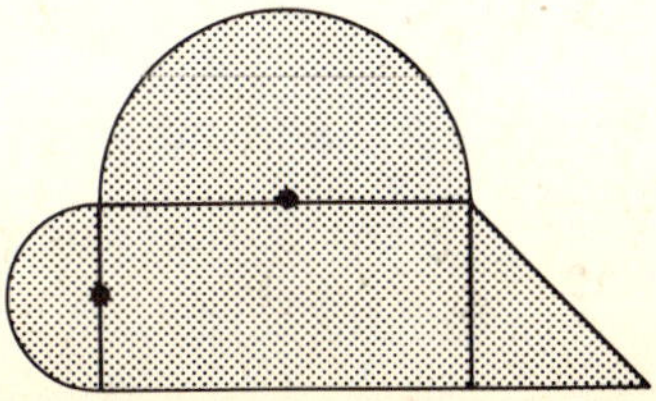

CHAPTER 10

Probability

10.1 DEFINITIONS AND SIMPLE PROBABILITY

The ***probability*** that an event will occur is the ratio of the number of ways that the event can occur to the total number of possible outcomes. The definition says, in essence, that *probability is a ratio.* Many types of notation can be used to express this concept.

$n(E)$ represents the number of ways that the event can occur.

$$P(E) = \frac{n(E)}{n(S)}$$

$P(E)$ represents the probability that the event will occur.

$n(S)$ represents the total number of possible outcomes, or the number of elements in the ***sample space.***

(*Note:* The formula is read: "*P* of *E* equals *n* of *E* over *n* of *S*.")

The following information is useful for solving the types of problems associated with probability:

A *die* (plural, *dice*), unless otherwise specified, has 6 *faces,* numbered 1, 2, 3, 4, 5, 6. A *fair die* means that any one of the faces can turn up in a given roll of the die. The number of elements in the sample space for a die is 6, or the total number of possible outcomes when the die is rolled is 6.

A coin has 2 sides, designated *heads* and *tails*. In a given toss of a *fair coin,* it is equally likely that either side will land face up. The number of elements in the sample space for a coin is 2, or the total number of possible outcomes when the coin is tossed is 2.

A *standard deck* of playing cards has 52 cards. There are 4 *suits:* hearts, diamonds, spades, and clubs. Two of the suits, spades and clubs, are black and 2 of the suits, hearts and diamonds, are red. Each suit contains 13 cards: ace, 2, 3, 4, 5, 6, 7, 8, 9, 10, jack, queen, king. The picture cards are the jacks, queens, and kings. The number of elements in the sample space for the entire deck of cards is 52, and the number of elements in the sample space for each suit is 13.

The English alphabet contains 26 letters: a, b, c, d, e, f, g, h, i, j, k, l, m, n, o, p, q, r, s, t, u, v, w, x, y, z. Five of these letters are called *vowels:* a, e, i, o, u. (We will not consider "y" a vowel.) The remainder of the letters are called *consonants*. The number of elements in the sample space for the entire alphabet is 26; for the set of vowels, 5; and for the set of consonants, 21.

Random selection means choosing "without looking." *At random* means all outcomes are equally likely.

Examples

1. What is the probability of rolling an even number in one roll of a fair die?

Solution:

$$P(E) = \frac{n(E)}{n(S)}$$

$P(\text{even}) = \frac{3}{6}$ 3 → There are 3 even possibilities: 2, 4, 6. 6 → The number of elements in the sample space is 6, the total number of ways the die can fall.

Answer: $\frac{3}{6}$ or $\frac{1}{2}$

2. What is the probability of drawing a red ace from a standard deck of cards if one card is drawn at random?

Solution:

$$P(E) = \frac{n(E)}{n(S)}$$

$P(\text{red ace}) = \frac{2}{52}$ 2 → There are 4 aces, but only 2 of them are red. 52 → The number of elements in the sample space is 52, the total number of cards.

Answer: $\frac{2}{52}$ or $\frac{1}{26}$

3. What is the probability of selecting an integer divisible by 4 if an integer is chosen at random from the first 6 positive even integers?

Solution:

Write out the sample space, which consists of the first 6 positive even integers: 2, 4, 6, 8, 10, 12. Choose those divisible by 4: 4, 8, 12.

$$P(E) = \frac{n(E)}{n(S)}$$

$P(\text{divisible by 4}) = \frac{3}{6}$ 3 → There are 3 integers divisible by 4. 6 → The number of elements in the sample space is 6, the first 6 positive even integers.

Answer: $\frac{3}{6}$ or $\frac{1}{2}$

4. A cookie jar contains 8 oatmeal cookies, 5 lemon cookies, and 3 raisin cookies. What is the probability of selecting a lemon cookie in one random draw from the cookie jar?

Solution:

$$P(E) = \frac{n(E)}{n(S)}$$

$$P(\text{lemon}) = \frac{5}{8 + 5 + 3} = \frac{5}{16}$$

5 ⟶ There are 5 lemon cookies.
16 ⟶ The number of elements in the sample space is 16, the total number of cookies.

Answer: $\frac{5}{16}$

EXERCISES

In 1–15, find the probability of:

1. tossing a "head" in one toss of a fair coin.
2. rolling a "5" in one roll of a fair die.
3. rolling an odd number in one roll of a fair die.
4. drawing a black queen from a standard deck of cards if one card is drawn at random.
5. selecting an integer divisible by 3 if an integer is chosen at random from the first 10 positive integers.
6. drawing a "7" from a standard deck of cards if one card is drawn at random.
7. rolling an even number greater than 4 in one roll of a fair die.
8. rolling an odd number less than 5 in one roll of a fair die.
9. rolling an odd number greater than 3 in one roll of a fair die.
10. choosing the letter "e" if one letter is chosen at random from the set of vowels.
11. choosing the letter "q" if one letter is chosen at random from the set of consonants.
12. choosing the letter "m" if one letter is chosen at random from the alphabet.
13. selecting a day whose name begins with the letter "S" if a day is selected at random from the set {names of the days of the week}.
14. selecting a prime number if one number is chosen at random from the set {2, 3, 4, 5}.
15. selecting a perfect square if one number is chosen at random from the set {1, 2, 3, 4, 5, 6, 7, 8, 9, 10}.

In 16–21, the name of a month is selected at random from the set {names of the months of the year}. What is the probability of selecting the name of a month that:

16. begins with the letter "J"?
17. has 31 days?

18. has more than 29 days?
19. ends with the letter "y"?
20. ends with the letter "r"?
21. begins with a vowel?

22. A letter is chosen at random from the letters of the word "HUNDRED." What is the probability that the letter chosen is an "E"?
23. What is the probability of selecting a vowel if a letter is chosen at random from the word "SQUARE"?
24. What is the probability of selecting a consonant if a letter is chosen at random from the word "TRIANGLE"?
25. In the accompanying figure, the spinner has 5 equal sections numbered 1 through 5. If the arrow is equally likely to land on any of the sections, what is the probability that it will land on an even number on the next spin?

26. The digits of the number 1977 are in a bag. What is the probability that a 7 will be drawn on a single random draw?
27. An advertising display has 20 red lights, 15 blue lights, 15 green lights, and 10 yellow lights. What is the probability that the first light to burn out is red?
28. The mail consists of 3 bills, 2 advertisements, and 1 letter. If the mail is opened randomly, what is the probability that an advertisement is opened first?
29. A cookie jar has 4 vanilla cookies, 3 chocolate cookies, and 2 lemon cookies. If a cookie is drawn at random, what is the probability that it will be chocolate?
30. A jar contains 5 white marbles, 3 blue marbles, and 2 green marbles. If one marble is drawn at random, what is the probability that it will be blue?

10.2 THE PROBABILITY OF CERTAINTY, IMPOSSIBILITY, AND THE NEGATION OF AN EVENT

When every one of the possible outcomes in the sample space is also a way in which the given event can occur, the event is ***certain*** to happen.

For example, what is the probability of rolling a number less than 10 in one roll of a fair die? Since all of the numbers on the die are less than 10, the event—rolling a number less than 10—is certain to happen. In this case:

$$P(E) = \frac{n(E)}{n(S)}$$

$$P(\text{number} < 10) = \frac{6}{6} \begin{array}{l}\longrightarrow \text{The number of ways that the event can occur.} \\ \longrightarrow \text{The number of elements in the sample space.}\end{array}$$

◆ Thus, ***P(certainty) = 1,*** since the numerator and denominator of the probability ratio are equal.

When none of the possible outcomes in the sample space is a way in which the given event can occur, it is ***impossible*** for the event to occur.

For example, what is the probability of rolling a number greater than 10 in one roll of a fair die? Since none of the numbers on the die is greater than 10, it is impossible for the event—rolling a number greater than 10—to occur. In this case:

$$P(E) = \frac{n(E)}{n(S)}$$

$$P(\text{number} > 10) = \frac{0}{6}\begin{array}{l}\longrightarrow \text{The number of ways that the event can occur.}\\ \longrightarrow \text{The number of elements in the sample space.}\end{array}$$

◆ Thus, ***P(impossibility) = 0,*** since the numerator of the probability ratio is 0.

The number 1 represents the maximum probability that an event will occur, namely, the probability that it will *always* occur.

The number 0 represents the minimum probability that an event will occur, namely, the probability that it will *never* occur.

Thus, the probability that an event that is neither impossible nor certain can occur is between 0 and 1.

For each event, A, there is an opposite event, $\sim A$ (the negation of A). For example, if A represents the event that a baseball team will win a particular game, then $\sim A$ represents the event that the team will not win the game. If A represents the event that a particular day will not be sunny, then $\sim A$ represents the event that the day will be sunny.

The sum of the probabilities that an event will occur and will not occur is equal to 1. That is:

$$P(A) + P(\sim A) = 1 \text{ and } P(\sim A) = 1 - P(A)$$

◆ $P(\sim A) = 1 - P(A)$ means: If the probability that an event will occur is known to be a, then the probability that the event will *not* occur is $1 - a$. (*Note:* The negation of an event is related to the negation of a statement in logic and the complement of a set.)

Examples

1. If the probability that an event will happen is $\frac{1}{7}$, what is the probability that the event will not happen?

   Solution:

   The probability that the event will happen is $\frac{1}{7}$.
   The probability that the event will not happen is $1 - \frac{1}{7}$ or $\frac{6}{7}$. *Ans.*

2. What is the probability of not rolling a "5" on a single roll of a fair die?

   Solution:

   The probability of rolling a "5" is $\frac{1}{6}$.
   The probability of not rolling a "5" is $1 - \frac{1}{6}$ or $\frac{5}{6}$. *Ans.*

3. A jar contains 3 red marbles, 4 white marbles, and 5 blue marbles. If one marble is selected at random, what is the probability that the marble selected is: **a.** not red? **b.** not black?

Solution:

a. The probability of selecting a red marble is $\frac{3}{12}$. (The number of red marbles is 3. The total number of marbles is 12.) The probability of selecting a marble that is not red is $1 - \frac{3}{12}$ or $\frac{9}{12}$ or $\frac{3}{4}$. *Ans.*

b. There are no black marbles in the jar. Thus, the probability that the marble selected is black is 0 (impossible). Since it is impossible to select a black marble, it is certain that a marble that is not black is selected. The probability that the marble selected is not black is 1. *Ans.*

EXERCISES

1. The digits of the number 333 are in a bag. What is the probability that a "3" will be drawn on a single random draw?
2. The digits of the number 357 are in a bag. What is the probability that an even integer will be drawn on a single random draw?
3. A single fair die is rolled once. What is the probability of obtaining a number greater than 6?
4. An integer is chosen at random from the first 5 positive odd integers. What is the probability of selecting an integer not exactly divisible by 2?
5. An urn (jar) contains 4 red marbles and 6 white marbles. If one marble is drawn at random, what is the probability that it will be blue?
6. A name is drawn at random from the set {names of the days of the week}. What is the probability of selecting a name that ends in the letter "y"?
7. In a leap year, a name is drawn at random from the set {names of the months of the year}. What is the probability that the name drawn is that of a month with exactly 28 days?
8. A letter is selected at random from the word "DEED." What is the probability that the letter selected appears in the word exactly twice?
9. A letter is selected at random from the word "SQUARE." What is the probability that the letter selected appears in the word more than once?
10. A number is chosen at random from the set {12, 1,212, 111,222}. What is the probability that the number selected has digits that appear with equal frequency?
11. If the probability of an event's happening is $\frac{2}{5}$, what is the probability of the event's *not* happening?
12. The probability that we will win the baseball game is $\frac{3}{5}$. What is the probability that we will *not* win the game?

13. If the probability that Jones will win the election is 0.6, what is the probability that Jones will *not* win the election?
14. If q represents the probability that event Q will occur, what is the probability that event $\sim Q$ will occur?
15. If r represents the probability that event R will *not* occur, what is the probability that event R will occur?
16. If t represents the probability that event T will occur, what is the probability that event $\sim T$ will *not* occur?

In 17 and 18, given the event that a letter is chosen at random from the word "PROBABILITY." What is the probability that the letter chosen:

17. is *not* a "B"?
18. is *not* a vowel?

In 19–23, given the event that a number is chosen at random from the set {2, 3, 4, 5, 6, 7, 8, 9}. What is the probability that the number chosen:

19. is *not* even?
20. is *not* a prime?
21. is *not* greater than 5?
22. is *not* less than 3?
23. is less than 10?

24. An advertising display has 20 red lights, 15 blue lights, 15 green lights, and 10 yellow lights. What is the probability that the first light to burn out is *not* red?
25. On a test, the probability of getting the correct answer to a certain question is represented by $\frac{x}{7}$. Which *cannot* be a value of x?

(1) 1 (2) -1 (3) 0 (4) 7

10.3 THE PROBABILITY OF CONNECTIVES

Consider the probability of drawing a red ace in one random drawing of a card from a standard deck. This problem poses two conditions: (1) that the card is red ***and*** (2) that the card is an ace. One method of solution was given earlier:

$$P(E) = \frac{n(E)}{n(S)}$$

$$P(\text{red ace}) = \frac{2}{52}$$

2 ⟶ There are 4 aces, but only 2 of them are red.
52 ⟶ The number of elements in the sample space is 52, the total number of cards.

This same problem can be thought of in terms of sets. The number of ways in which the desired card—a red ace—can be selected is the *intersection* of the set of red cards and the set of aces. (The intersection of two sets is the set containing those elements common to the two sets.)

The set of red cards contains 26 elements. The set of aces contains 4 elements. The intersection of the two sets contains 2 elements.

The probability that the event specifying the two conditions will occur is then determined by the ratio of the number of ways the event can occur to the total number of possible outcomes (the number of elements in the sample space).

◆ In general, the probability that the event $A \wedge B$ can occur is:

$$P(A \wedge B) = \frac{n(A \cap B)}{n(S)}$$

$n(A \cap B)$ → The number of elements in the intersection of the two sets.
$n(S)$ → The number of elements in the sample space.

The probability of drawing a red card ***or*** an ace is more challenging. There are 26 red cards and 4 aces, but two cards—the red aces—would be counted twice if 26 and 4 were simply added. These two cards must be subtracted from the sum of the two sets, obtaining the *union* of the two sets. (The union of two sets is the sum of the two sets decreased by the intersection of the two sets; that is, $A \cup B = (A) + (B) - (A \cap B$.) Therefore, $26 + 4 - 2$, or 28, is the number of ways in which a red card *or* an ace can be drawn. The probability that the event can occur is, therefore, $\frac{28}{52}$.

◆ In general, the probability that the event $A \vee B$ can occur is:

$$P(A \vee B) = \frac{n(A \cup B)}{n(S)}$$

$n(A \cup B)$ → The number of elements in the union of the two sets.
$n(S)$ → The number of elements in the sample space.

Examples

1. In one roll of a fair die, what is the probability of rolling:

   a. an even number less than 5?
   b. a number that is even or is less than 5?

Solution:

Let A represent the set of even numbers on a die. $A = \{2, 4, 6\}$
Let B represent the set of numbers < 5 on a die. $B = \{1, 2, 3, 4\}$

a. There are two elements in the intersection of the two sets:

$$A \cap B = \{2, 4\} \text{ and } n(A \cap B) = 2$$

$$P(A \wedge B) = \frac{n(A \cap B)}{n(S)} = \frac{2}{6}$$

2 → The number of ways the event can occur (the number of elements in the intersection of the sets).
6 → The number of possible outcomes (the number of elements in the sample space).

Answer: $\frac{2}{6}$ or $\frac{1}{3}$

b. There are 5 elements in the union of the sets:

$$A \cup B = \{1, 2, 3, 4, 6\} \text{ and } n(A \cup B) = 5$$

$$P(A \vee B) = \frac{n(A \cup B)}{n(S)} = \frac{5}{6}$$

5 → The number of ways the event can occur (the number of elements in the union of the sets).

6 → The number of possible outcomes (the number of elements in the sample space).

Answer: $\frac{5}{6}$

2. A cookie jar contains 6 vanilla cookies, 3 chocolate cookies, and 8 lemon cookies. If one cookie is randomly picked from the jar, what is the probability that it will be either vanilla or lemon?

Solution:

Since there are no intersecting sets here, the union of any two sets will be a simple addition. That is, a cookie is either vanilla or lemon; it cannot be both. Thus, $n(\text{vanilla} \cup \text{lemon}) = 6 + 8$, or 14.

$$P(\text{vanilla} \vee \text{lemon}) = \frac{n(\text{vanilla} \cup \text{lemon})}{n(S)}$$

$$= \frac{14}{6 + 8 + 3}$$

14 → The number of ways the event can occur (the number of elements in the union of the sets).

6 + 8 + 3 → The number of possible outcomes (the number of elements in the sample space). The total number of cookies.

Answer: $\frac{14}{17}$

EXERCISES

In 1–7, the event is "a fair single die is thrown once." What is the probability of obtaining each of the following?

1. an odd number less than 4 **2.** a "1" or a "2"
3. an even number greater than 1 **4.** a "3" and a "7"
5. an even number that is not a prime
6. an odd number that is a prime
7. a number that is neither even nor odd

In 8–11, the event is "drawing a name at random from the set {names of the months}." What is the probability of drawing a name that:

8. begins with "J" and ends with "y"?
9. begins with a vowel and ends with a consonant?
10. begins with "J" or ends with "y"?
11. ends with "r" or begins with a vowel?

In 12–18, the event is "drawing a card at random from a standard deck of playing cards." In each case, one of the given choices will answer the question "What is the probability of drawing:"

12. a black king? (1) $\frac{2}{52}$ (2) $\frac{4}{52}$ (3) $\frac{28}{52}$ (4) $\frac{30}{52}$
13. either a black card or a king? (1) $\frac{2}{52}$ (2) $\frac{4}{52}$ (3) $\frac{28}{52}$ (4) $\frac{30}{52}$
14. either an ace or a red king? (1) $\frac{8}{52}$ (2) $\frac{6}{52}$ (3) $\frac{6}{26}$ (4) $\frac{3}{14}$
15. either a king or an ace? (1) $\frac{8}{52}$ (2) $\frac{6}{52}$ (3) $\frac{4}{52}$ (4) $\frac{2}{52}$
16. either a spade or a club? (1) $\frac{8}{52}$ (2) $\frac{2}{52}$ (3) $\frac{13}{52}$ (4) $\frac{26}{52}$
17. either a black card or a red card? (1) $\frac{13}{52}$ (2) $\frac{26}{52}$ (3) 0 (4) 1
18. a black heart? (1) 0 (2) $\frac{1}{3}$ (3) $\frac{1}{2}$ (4) 1

In 19–22, the event is "a single marble is randomly picked from an urn containing 2 red marbles, 4 blue marbles, and 3 yellow marbles." What is the probability that the marble selected:

19. is either red or blue?
20. is neither red nor blue?
21. is not yellow?
22. is not red and is blue?

In 23 and 24, the event is "a single piece of mail is randomly selected from a stack of mail containing 4 letters, 3 bills, and 3 advertisements." What is the probability that the piece of mail selected:

23. is either a letter or a bill? **24.** is neither a letter nor a bill?

10.4 THE COUNTING PRINCIPLE

So far we have looked at problems that involved a *single* activity, such as choosing one item from a given set of items. In these cases, we could easily count the number of elements in the sample space.

When there are two or more activities to be performed, it is more difficult to determine the number of elements in the sample space. We will study the methods of determining the number of elements in a sample space for two or more activities by listing or diagramming the actual elements of the sample space. First, however, let us state a shortcut for telling just the number of elements in the sample space.

◆ ***The Counting Principle:***

The number of ways in which one activity can be performed.	×	The number of ways in which the second activity can be performed.	=	The total number of ways in which both activities can be performed.

The Counting Principle is a short way of determining the *number of elements* in a sample space. This principle can be extended for more than two activities.

Examples

1. There are 5 roads from town Q to town R and 10 roads from town R to town S. How many different routes can be taken from town Q to town S through town R?

Solution:

The first activity, choosing a road from town Q to town R, can be performed in 5 ways.

The second activity, choosing a road from town R to town S, can be performed in 10 ways.

According to the Counting Principle, the number of ways in which the two activities can be performed is $5 \cdot 10$ or 50.

Answer: There are 50 possible routes.

2. An outfit consists of a jacket, a blouse, and a skirt. How many different outfits can be made if there are 4 jackets, 3 blouses, and 3 skirts?

Solution:

The first activity, choosing a jacket, can be performed in 4 ways.
The second activity, choosing a blouse, can be performed in 3 ways.
The third activity, choosing a skirt, can be performed in 3 ways.
According to the Counting Principle, the number of ways in which the three activities can be performed is $4 \cdot 3 \cdot 3$ or 36.

Answer: There are 36 different ways in which an outfit can be made.

EXERCISES

1. There are 3 roads from town A to town B and 5 roads from town B to town C. How many different routes could be driven from town A to town C through town B?
(1) 5 (2) 8 (3) 15 (4) 18
2. There are 6 roads between Plattsburgh and Lake Placid. In how many different ways can a person travel from Plattsburgh to Lake Placid and back to Plattsburgh?
3. If a girl has 6 shirts and 4 pairs of slacks, how many different outfits of a shirt and slacks does she have?
(1) 24 (2) 10 (3) 6 (4) 4
4. A woman has 10 blouses, 6 sweaters, and 4 skirts. How many possible outfits consisting of a blouse, a sweater, and a skirt can she choose?
5. The menu for dinner lists 2 soups, 3 meats, and 2 desserts. How many different meals that have one soup, one meat, and one dessert are possible?
(1) 7 (2) 8 (3) 12 (4) 24

6. In a school, there are 5 math teachers, 5 science teachers, 6 English teachers, and 3 history teachers. If Alice takes these 4 subjects, in how many different ways can her teachers be chosen?
7. From 5 violinists, 3 pianists, and 6 drummers, how many trios of different instruments can be formed?

10.5 THE COUNTING PRINCIPLE AND PROBABILITY

The Counting Principle as applied to probability states that if the probability of one event's occurring is m and the probability of a second event's occurring is n, then the probability of both events' occurring is $m \cdot n$. (*Note:* This principle can be extended for more than two events.)

Examples

1. There are 4 black marbles and 5 green marbles in an urn. One marble is drawn at random, its color is noted, and it is returned. A second marble is drawn at random and its color is noted. What is the probability that:
   a. both marbles are black?
   b. the first marble is black and the second marble is green?

Solution:

a. The probability that the first marble is black is:

$\frac{4}{9}$ 4 ⟶ There are 4 black marbles.
9 ⟶ There are 9 marbles in all.

The probability that the second marble is black is:

$\frac{4}{9}$ 4 ⟶ There are still 4 black marbles, since the first marble chosen, even if it was black, was returned.
9 ⟶ There are still 9 marbles in all, since the first marble chosen was returned.

According to the Counting Principle, the probability that both events can occur is the product of the individual probabilities. Thus, the probability that both marbles are black is $\frac{4}{9} \cdot \frac{4}{9}$ or $\frac{16}{81}$. *Ans.*

b. The probability that the first marble is black is:

$\frac{4}{9}$ 4 ⟶ There are 4 black marbles.
9 ⟶ There are 9 marbles in all.

The probability that the second marble is green is:

$\frac{5}{9}$ 5 ⟶ There are still 5 green marbles, since the first marble chosen, even if it was green, was returned.
9 ⟶ There are still 9 marbles in all, since the first marble chosen was returned.

According to the Counting Principle, the probability that both events can occur is the product of the individual probabilities. Thus, the probability that the first marble is black and the second marble is green is $\frac{4}{9} \cdot \frac{5}{9}$ or $\frac{20}{81}$. *Ans.*

2. A single die is rolled twice. What is the probability of obtaining a "5" on the first roll and then a "3" on the second roll?

Solution:

The probability of obtaining a "5" on the first roll is:

$\frac{1}{6}$ 1 ⟶ The number of ways to obtain a "5."
6 ⟶ The total number of ways the die can fall.

The probability of obtaining a "3" on the second roll is:

$\frac{1}{6}$ 1 ⟶ The number of ways to obtain a "3."
6 ⟶ The total number of ways the die can fall.

According to the Counting Principle, the probability that both events can occur is the product of the individual probabilities. Thus, the probability of obtaining a "5" on the first roll and then a "3" on the second roll is $\frac{1}{6} \cdot \frac{1}{6}$ or $\frac{1}{36}$. *Ans.*

3. A single card is drawn at random from a standard deck. The card is identified and then returned to the deck. Another card is drawn at random. What is the probability that:

a. both cards are black?
b. the first card is an ace and the second card is a "7"?
c. the first card is a black "5" and the second card is a jack?

Solution:

a. The probability that the first card is black is:

$\frac{26}{52}$ 26 ⟶ The number of black cards.
52 ⟶ The total number of cards.

The probability that the second card is black is:

$\frac{26}{52}$ 26 ⟶ There are still 26 black cards, since the first card, even if it was black, was returned.
52 ⟶ There are still 52 cards in all, since the first card drawn was returned.

According to the Counting Principle, the probability that both events can occur is the product of the individual probabilities. Thus, the probability that both cards are black is $\frac{26}{52} \cdot \frac{26}{52}$ or $\frac{1}{2} \cdot \frac{1}{2}$ or $\frac{1}{4}$. *Ans.*

b. The probability that the first card is an ace is:

$\frac{4}{52}$ 4 ⟶ There are 4 aces.
52 ⟶ There are 52 cards.

The probability that the second card is a "7" is:

4 ⟶ There are still four 7's, since the first card, even if it was a "7," was returned.
52 ⟶ There are still 52 cards, since the first card was returned.

By the Counting Principle, the probability that the first card is an ace and the second card is a "7" is $\frac{4}{52} \cdot \frac{4}{52}$ or $\frac{1}{13} \cdot \frac{1}{13}$ or $\frac{1}{169}$. *Ans.*

c. The probability that the first card is a black "5" is:

2 ⟶ There are 2 black 5's.
52 ⟶ There are 52 cards.

The probability that the second card is a jack is:

4 ⟶ There are still 4 jacks, since the first card, even if it was a jack, was returned.
52 ⟶ There are still 52 cards, since the first card was returned.

By the Counting Principle, the probability that the first card is a black "5" and the second card is a jack is $\frac{2}{52} \cdot \frac{4}{52}$ or $\frac{1}{26} \cdot \frac{1}{13}$ or $\frac{1}{338}$. *Ans.*

EXERCISES

1. The probability that the Bears will beat the Eagles is $\frac{1}{2}$. The probability that the Bears will beat the Cubs is $\frac{3}{5}$. What is the probability that the Bears will win both games?
2. A single die is rolled twice. What is the probability of obtaining a "3" on the first roll and then a "4" on the second roll?
3. A bag contains 5 red and 4 white marbles. One marble is drawn at random, its color is noted, and the marble is returned. A second marble is drawn at random and its color is noted. What is the probability that both marbles are red?
(1) $\frac{25}{72}$ (2) $\frac{20}{72}$ (3) $\frac{25}{81}$ (4) $\frac{16}{81}$

In 4–10, there are 10 black marbles, 4 red marbles, 8 white marbles, and 6 green marbles in an urn. One marble is drawn at random, its color is noted, and the marble is returned. A second marble is drawn at random and its color is noted. What is the probability that:

4. both marbles are red?
5. both marbles are green?
6. the first marble is red and the second marble is green?
7. neither marble is red?
8. neither marble is green?
9. the first marble is red and the second marble is not white?
10. the first marble is not red and the second marble is not green?

In 11–25, a single card is drawn at random from a standard deck. The card is identified and then returned to the deck. Another card is drawn at random. What is the probability that:

11. both cards are red?
12. the first card is red and the second card is black?
13. both cards are kings?
14. neither card is a king?
15. the first card is a king and the second card is not a king?
16. the first card is a king and the second card is an ace?
17. both cards are black kings?
18. neither card is a black king?
19. the first card is a black king and the second card is not a black king?
20. the first card is a black king and the second card is a jack?
21. the first card is a black king and the second card is a red jack?
22. both cards are diamonds?
23. neither card is a diamond?
24. the first card is a diamond and the second card is not a diamond?
25. the first card is a diamond and the second card is black?

10.6 PROBABILITY WITHOUT REPLACEMENT

In the previous section, we saw problems in which two selections were made. When the first selection was made, the item was replaced, and then the second selection was made. In those problems, the second selection was not affected by the first selection.

Now let us consider some situations in which the first item selected is *not* returned. The second selection has, therefore, fewer items from which to be made. Also, the first selection might or might not have another effect on the second selection, as we shall see in the following examples.

Examples

1. An urn contains 10 red marbles and 6 yellow marbles. One marble is drawn at random from the urn and that marble is put aside. A second marble is then drawn at random from the urn. What is the probability that:

a. both marbles are red?
b. both marbles are yellow?
c. the first marble is red and the second marble is yellow?

Solution:

a. The probability that the first marble drawn is red is:

$\frac{10}{16}$ 10⟶There are 10 red marbles.
16⟶There are 16 marbles in all.

Since one red marble has been removed from the urn, there are only 9 red marbles remaining in the urn, and there are only 15 marbles in all. Thus, the probability that the second marble drawn is red is $\frac{9}{15}$.

By the Counting Principle, the probability that both marbles are red is $\frac{10}{16} \cdot \frac{9}{15}$ or $\frac{3}{8}$. *Ans.*

b. The probability that the first marble drawn is yellow is:

$\frac{6}{16}$ 6 ⟶ There are 6 yellow marbles.
16 ⟶ There are 16 marbles in all.

Since one yellow marble has been removed from the urn, there are only 5 yellow marbles remaining in the urn, and there are only 15 marbles in all. Thus, the probability that the second marble drawn is yellow is $\frac{5}{15}$.

By the Counting Principle, the probability that both marbles are yellow is $\frac{6}{16} \cdot \frac{5}{15}$ or $\frac{1}{8}$. *Ans.*

c. The probability that the first marble drawn is red is:

10 ⟶ There are 10 red marbles.
16 ⟶ There are 16 marbles in all.

Since one red marble has been removed from the urn, there are only 9 red marbles remaining in the urn. There are still 6 yellow marbles in the urn. There are 15 marbles in all. Thus, the probability that the second marble drawn is yellow is $\frac{6}{15}$.

By the Counting Principle, the probability that the first marble is red and the second marble is yellow is $\frac{10}{16} \cdot \frac{6}{15}$ or $\frac{1}{4}$. *Ans.*

2. One card is drawn at random from a standard deck and it is put aside. A second card is then drawn at random from that deck. What is the probability that:

a. both cards are aces?
b. neither card is an ace?
c. both cards are red?
d. the first card is a diamond and the second card is black?
e. the first card is a heart and the second card is green?

Solution:

a. The probability that the first card is an ace is:

4 ⟶ There are 4 aces.
52 ⟶ There are 52 cards in all.

Since one ace has been removed from the deck, there are 3 aces remaining, and there are 51 cards remaining. Thus, the probability that the second card is an ace is $\frac{3}{51}$.

By the Counting Principle, the probability that both cards are aces is $\frac{4}{52} \cdot \frac{3}{51}$ or $\frac{1}{221}$. *Ans.*

b. The probability that the first card is not an ace is:

$\frac{48}{52}$ 48 → Since there are 4 aces, there are 52 − 4 or 48 cards that are not aces.
52 → There are 52 cards in all.

Since one non-ace has been removed, there are 47 cards that are not aces remaining, and there are 51 cards remaining. Thus, the probability that the second card is not an ace is $\frac{47}{51}$.

By the Counting Principle, the probability that neither card is an ace is $\frac{48}{52} \cdot \frac{47}{51}$ or $\frac{188}{221}$. *Ans.*

c. The probability that the first card is red is:

$\frac{26}{52}$ 26 → There are 26 red cards.
52 → There are 52 cards in all.

Since one red card has been removed, there are 25 red cards remaining, and there are 51 cards remaining. Thus, the probability that the second card is red is $\frac{25}{51}$.

By the Counting Principle, the probability that both cards are red is $\frac{26}{52} \cdot \frac{25}{51}$ or $\frac{25}{102}$. *Ans.*

d. The probability that the first card is a diamond is:

$\frac{13}{52}$ 13 → There are 13 diamonds.
52 → There are 52 cards in all.

Since one diamond has been removed, there are 51 cards remaining. Of these 51 cards, 26 are still black. (The removal of the diamond, which is red, has no effect on the number of black cards.) Thus, the probability that the second card is black is $\frac{26}{51}$.

By the Counting Principle, the probability that the first card is a diamond and the second card is black is $\frac{13}{52} \cdot \frac{26}{51}$ or $\frac{13}{102}$. *Ans.*

e. The probability that the first card is a heart is:

$\frac{13}{52}$ 13 → There are 13 hearts.
52 → There are 52 cards in all.

There are no green cards in a standard deck. Thus, the probability that the second card is green is 0.

By the Counting Principle, the probability that the first card is a heart and the second card is green is $\frac{13}{52} \cdot 0$ or 0. *Ans.*

Note: In problems of this type, when the probability of one of the events is 0, there is no need for any calculation. The probability of the combined events must be 0.

There are situations regarding nonreplacement that do not ask for the probability of combined events but ask, simply, in how many ways the combined events can occur (the number of elements in the sample space).

Example

There are 5 doors to a house. In how many ways can a person enter by one door and leave by another?

Solution:

There are two activities to be performed: (1) choosing a door for entry and (2) choosing a door for exit that is different from the entry door.

The first activity, choosing a door for entry, can be performed in 5 ways.

The second activity, choosing a door for exit that is different from the entry door, can be performed in only 4 ways (one door has been used for entry and can not be used for exit).

According to the Counting Principle, the number of ways in which the two activities can be performed is $5 \cdot 4$ or 20. *Ans.*

EXERCISES

In 1–6, an urn contains 15 red marbles, 12 white marbles, and 10 blue marbles. One marble is drawn at random from the urn and put aside. A second marble is then drawn at random from that urn. What is the probability that:

1. both marbles are blue?
2. neither marble is blue?
3. the first marble is red and the second marble is blue?
4. the first marble is blue and the second marble is red?
5. both marbles are green?
6. both marbles are not white?

In 7–14, a card is drawn at random from a standard deck and put aside. A second card is then drawn at random from that deck. What is the probability that:

7. both cards are black?
8. both cards are kings?
9. the first card is a red king and the second card is a black king?
10. neither card is a king?
11. both cards are clubs?
12. the first card is a diamond and the second card is a heart?
13. the first card is a diamond and the second card is red?
14. both cards are yellow?

In 15–19, a stack of mail consists of 8 bills, 10 letters, and 6 advertisements. One piece of mail is drawn at random from the stack and put aside. Then a second piece of mail is drawn at random from the stack. What is the probability that:

15. both pieces of mail are letters?
16. both pieces of mail are bills?
17. the first piece of mail is a letter and the second piece of mail is a bill?
18. neither piece of mail is a bill?
19. the first piece of mail is either a bill or a letter and the second piece of mail is an advertisement?

20. A box contains 4 nickels, 3 dimes, and 2 quarters. One coin is drawn at random, put aside, and then another coin is drawn at random. What is the probability that the two coins total 9¢.

21. A school has 12 doors. In how many different ways could a student enter by one door and leave by another?

10.7 ILLUSTRATING PROBABILITY

Among the ways that probability can be illustrated are a *listing*, a *graph*, and a *tree diagram*.

Consider rolling a pair of fair 4-sided dice, each with faces numbered 1 through 4. On each roll, the side facing down is the one counted. What is the probability of rolling a sum of 5 or higher? We will illustrate this probability in each of the three ways just mentioned.

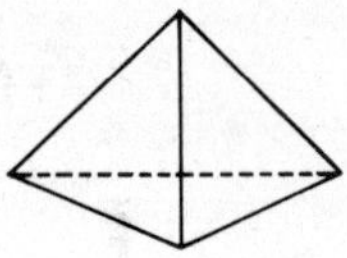

ILLUSTRATION 1: a **listing** of ordered pairs

There are 16 possibilities in the sample space:

(1, 1)	(2, 1)	(3, 1)	(4, 1)*
(1, 2)	(2, 2)	(3, 2)*	(4, 2)*
(1, 3)	(2, 3)*	(3, 3)*	(4, 3)*
(1, 4)*	(2, 4)*	(3, 4)*	(4, 4)*

Each ordered pair tells the numeral on the down face of the first 4-sided die followed by the numeral on the down face of the second 4-sided die.

The ordered pairs marked * represent the ways in which the sum of the numerals on the down faces of the two 4-sided dice is 5 or higher. There are 10 such ordered pairs.

Thus, the probability of rolling a 5 or higher in one roll of a pair of 4-sided dice is:

$\frac{10}{16}$

10 → The number of ways in which the event—rolling a sum of 5 or higher—can occur.

16 → The number of elements in the sample space, or the total number of ways in which the pair of 4-sided dice can fall.

ILLUSTRATION 2: a **graph**

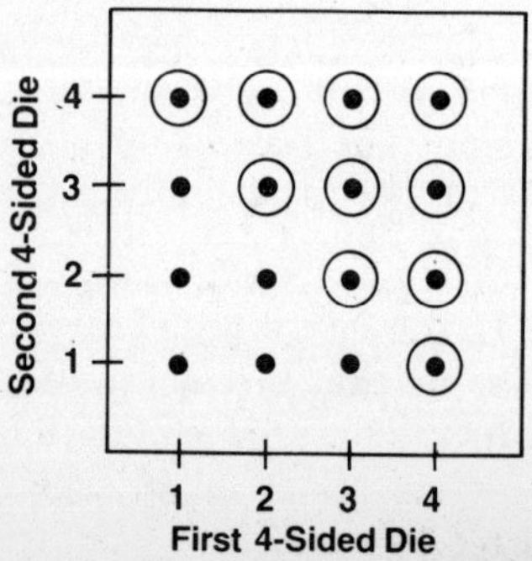

Each point on the graph tells the numeral on the down face of the first 4-sided die and the numeral on the down face of the second 4-sided die. The points are ordered pairs.

There are 16 possibilities in the sample space. The circled points represent the ways in which the event—a sum of 5 or more—can occur. There are 10 such points.

Thus, the probability of rolling a sum of 5 or higher in one roll of a pair of 4-sided dice is, again, $\frac{10}{16}$.

ILLUSTRATION 3: a **tree diagram**

The first set of branches represents the 4 possibilities for the first 4-sided die. Each of these branches has 4 branches, the possibilities for the second 4-sided die.

There are 16 possibilities in the sample space, and 10 of these possibilities, marked *, are ways in which the event can occur. The probability of rolling a sum of 5 or higher in one roll of a pair of 4-sided dice is $\frac{10}{16}$.

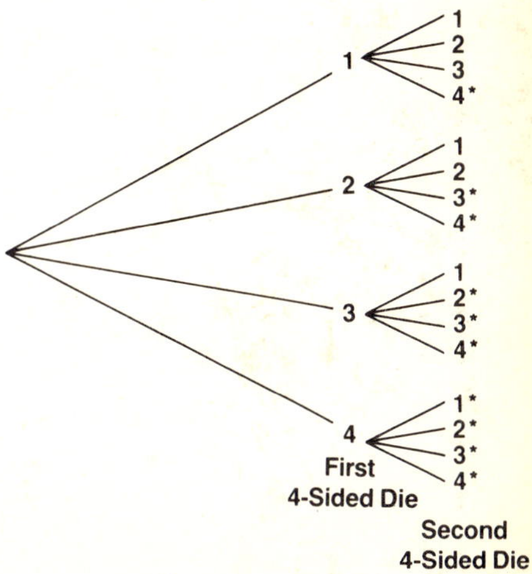

Examples

1. A coin is tossed at the same time that a 6-sided die is rolled. What is the probability of getting a "head" on the coin and a "4" or higher on the die?

Solution 1: *a listing of ordered pairs*

There are 12 possibilities in the sample space, (coin, die):

$(H, 1)$	$(T, 1)$
$(H, 2)$	$(T, 2)$
$(H, 3)$	$(T, 3)$
$(H, 4)$*	$(T, 4)$
$(H, 5)$*	$(T, 5)$
$(H, 6)$*	$(T, 6)$

The ordered pairs marked * represent the ways in which there is a "head" on the coin and a "4" or higher on the die. There are 3 such ordered pairs.

Thus, the probability of getting a "head" on the coin and a "4" or higher on the die is:

$\frac{3}{12}$
3 → The number of ways in which the event (getting a "head" on the coin and a "4" or higher on the die) can occur.
12 → The number of elements in the sample space, or the total number of ways in which the coin and the die can fall.

Answer: $\frac{3}{12}$ or $\frac{1}{4}$

Solution 2: a graph

There are 12 points on the graph. The circled points represent the desired outcome. There are 3 such points.

Thus, the probability of getting a "head" on the coin and a "4" or higher on the die is $\frac{3}{12}$ or $\frac{1}{4}$. *Ans.*

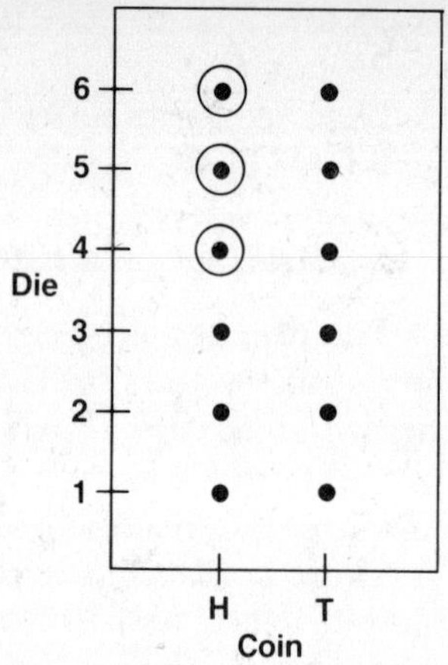

Solution 3: a tree diagram

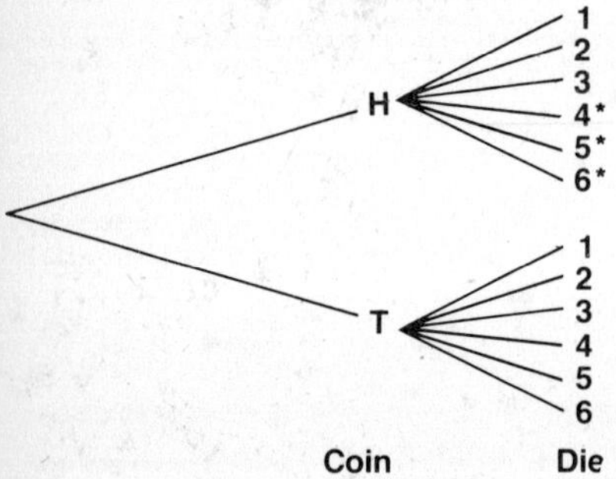

The first set of branches represents the 2 possibilities for the coin. Each of these branches has 6 branches, the possibilities for the die. There are 12 possibilities in the sample space. Three of the possibilities, marked *, are ways in which the event can occur. The probability of getting a "head" on the coin and a "4" or higher on the die is $\frac{3}{12}$ or $\frac{1}{4}$. *Ans.*

2. Walter has 6 books in his locker: math, French, chemistry, physics, history, and English. Without looking, Walter pulls out one book and then, without replacing the first book, he pulls out a second book.

a. Using the letters *M, F, C, P, H,* and *E,* write the sample space or draw a tree diagram for pulling 2 books from the locker.

b. What is the probability that Walter pulled first the chemistry book and then the math book?

c. What is the probability that he pulled out the same kind of book both times?

Solution:

a. Sample space for pulling 2 books from the locker (a listing of ordered pairs):

(M, F)	(F, M)	(C, M)	(P, M)	(H, M)	(E, M)
(M, C)	(F, C)	(C, F)	(P, F)	(H, F)	(E, F)
(M, P)	(F, P)	(C, P)	(P, C)	(H, C)	(E, C)
(M, H)	(F, H)	(C, H)	(P, H)	(H, P)	(E, P)
(M, E)	(F, E)	(C, E)	(P, E)	(H, E)	(E, H)

There are 30 elements in the sample space.

Tree diagram for pulling 2 books from the locker.

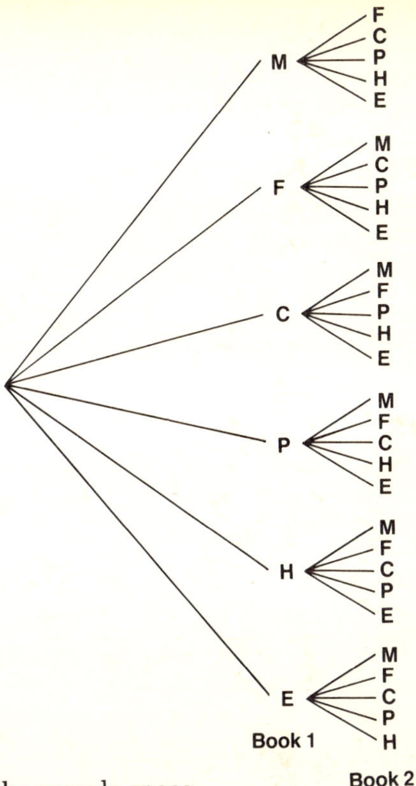

There are 30 elements in the sample space.

b. $P(C \text{ then } M) = \frac{1}{6} \cdot \frac{1}{5}$

1 → There is only 1 C. 6 → There are 6 books in all.

1 → There is only 1 M. 5 → There are 5 books remaining after the first book has been pulled.

$= \frac{1}{30}$ *Ans.*

c. *Answer:* 0 (There are no books of the same kind.)

EXERCISES

1. If a 6-sided die and a coin are tossed simultaneously, how many pairs (outcomes) are there in this sample space?
2. In a single toss of 2 coins, what is the probability of coming up with one "head" and one "tail"?
3. If two 6-sided dice are rolled, what is the probability of getting a sum of 3? (1) $\frac{1}{36}$ (2) $\frac{2}{36}$ (3) $\frac{3}{36}$ (4) $\frac{2}{6}$
4. A 6-sided die is rolled at the same time that a coin is tossed. What is the probability of getting a "tail" on the coin and an even number on the die? (1) 1 (2) $\frac{1}{2}$ (3) $\frac{1}{4}$ (4) 2
5. Two fair dice are tossed. Each die has 6 faces numbered 1 to 6. What is the probability that each die shows a "5"?

6. A 4-sided die is rolled at the same time that a coin is tossed. What is the probability of getting an odd number on the die and a "head" on the coin?

In 7–15, a 4-sided die is rolled at the same time that a 6-sided die is rolled.

7. Draw a graph to represent the sample space.
8. List the sample space as ordered pairs.
9. Draw a tree diagram to represent the sample space.

What is the probability of obtaining:

10. a "2" on both dice?
11. an even number on both dice?
12. a sum of 7?
13. a sum greater than 5?
14. a sum less than 8?
15. a sum greater than 10?

16. A silver dollar, a half-dollar, a quarter, a dime, and a nickel are in a box. One coin is drawn and not replaced. A second coin is drawn.
    a. Draw a tree diagram or list the sample space of all possible pairs that are outcomes for this experiment.
    b. What is the probability that the value of the first coin drawn is greater than the value of the second coin drawn?
    c. What is the probability that the value of the 2 coins drawn is greater than $1.00?
    d. What is the probability that the sum of the values of the 2 coins drawn is exactly $1.00?
17. An urn contains 3 orange marbles numbered 1, 2, and 3, respectively, and 2 blue marbles numbered 1 and 2, respectively. One marble is removed, its color is noted, and the marble is not replaced. A second marble is removed and its color is noted.
    a. Draw a tree diagram or list the sample space of all possible pairs of outcomes.
    b. What is the probability that:
        (1) both marbles drawn are orange?
        (2) neither marble drawn is orange?
        (3) one of the marbles drawn is green?
18. A contest offers a first prize of $10,000 or a car. The second prize is $500 or a television set or a set of encyclopedias.
    a. Draw a tree diagram or write the sample space of all possible pairs of first and second prizes.
    b. If each prize is equally likely to be chosen, find the probability that:
        (1) both prizes are money.
        (2) neither prize is money.
        (3) one prize is money but the other is not.
19. A student has the following 5 books in his locker: math, history, Latin, science, and English. Without looking at the books, he pulls out one book and then, without replacing the first book, pulls out a second book.
    a. Using the letters *M, H, L, S,* and *E,* draw a tree diagram or write the sample space for pulling 2 books from the locker.

b. What is the probability that he first pulled the math book and then the English book?
c. What is the probability that he pulled out 2 different books?

20. Three coins are tossed simultaneously.
a. Draw a tree diagram or list the sample space for all possible outcomes.
b. What is the probability that all 3 coins will come up "heads"?
c. In how many ways can a "head" and 2 "tails" appear?
d. What is the probability of obtaining a "head" and 2 "tails"?

21. A certain game requires spinning a marker and then tossing a single die. The marker must stop over one of 3 equally likely colors (red, green, blue), and the die is a standard fair 6-sided type.
a. Draw a tree diagram or list the sample space of all possible pairs of outcomes for spinning the marker and tossing the die.
b. What is the probability of obtaining the result (red, 4)?
c. What is the probability of obtaining the result (green, 8)?
d. What is the probability of obtaining the result (blue, even number)?

22. In a certain class, there are 4 students in the first row: 3 girls, Ann, Barbara, and Cathy, and one boy, David. The teacher called one of these students to the board to solve a problem. When the problem was done, the teacher called one of the remaining students in the first row to do a second problem at the board.
a. Draw a tree diagram or list the sample space of all possible pairs of names for calling two students to the board.
b. Find the probability that the teacher called Ann first and Barbara second.
c. Find the probability that the teacher called 2 girls to the board.
d. Find the probability that David was one of the 2 students called.

23. A die and a coin are tossed simultaneously. The die is fair and has 6 faces.
a. Draw a tree diagram or list the sample space of all possible pairs of outcomes.
b. What is the probability of obtaining a "6" on the die and a "head" on the coin?
c. What is the probability of obtaining an odd number on the die and a "tail" on the coin?
d. What is the probability of obtaining a "head" on the coin?

24. The first step of an experiment is to pick one number from the set {1, 2, 3}. The second step of the experiment is to pick one number from the set {1, 4, 9}.
a. Draw a tree diagram or list the sample space of all possible pairs of outcomes.
b. Determine the probability that:
(1) both numbers are the same.
(2) both numbers are odd.
(3) the second number is the square of the first.

10.8 PERMUTATIONS

Combinatorics involves computing the number of ways in which r things can be chosen from a total of n things.

The original n things can be all different or not.

The arrangement, or order, of the chosen r things can matter or not.

If the *arrangement is important,* then *xyz, yxz,* and *zxy* are all different choices. Each such choice is called a ***permutation.***

If the *arrangement is not important,* then *xyz, yxz,* and *zxy* are all the same choice. Each such choice is called a ***combination.***

In this course, the major concern is with *permutations* (arrangements are important) of things that are *all different.*

The Counting Principle applies to permutations. That is, if one activity can occur in any of m ways and a second activity can occur in any of n ways, then the total number of ways both activities can occur is $m \cdot n$. (*Note:* This principle can also be extended for more than two activities.)

We can solve permutation problems of things that are different more easily by using various illustrations. Among these illustrations are the three we used for probability: a listing, a graph, and a tree diagram. Another handy device for illustrating permutation problems is a series of "slots" to represent each of the activities.

Consider the number of permutations possible in arranging the letters x, y, and z if all 3 letters are to be used each time and no letter is to be used more than once in an arrangement.

ILLUSTRATION 1: a **listing** of ordered triples

(x, y, z) (y, x, z) (z, x, y)
(x, z, y) (y, z, x) (z, y, x)

There are 6 possible arrangements.

ILLUSTRATION 2: a **tree diagram**

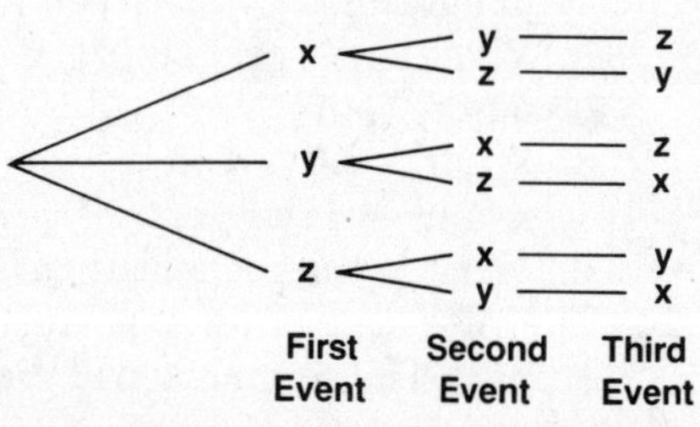

There are 6 possible arrangements.

ILLUSTRATION 3: filling **slots**

Let _____ _____ _____ represent the 3 slots to be filled.
first event, second event, third event

The first slot can be filled in any of 3 ways.

The second slot can be filled in either of 2 ways. (One way has been used for the first slot and cannot be used again.)

The third slot can be filled in only 1 way. (One way has been used for the first slot and a second way has been used for the second slot; neither of these 2 ways can be used again.)

Fill in the slots and multiply: $\underset{\text{first event}}{\underline{\;3\;}} \cdot \underset{\text{second event}}{\underline{\;2\;}} \cdot \underset{\text{third event}}{\underline{\;1\;}} = 6$

There are 6 possible arrangements.

(*Note:* A graph-like solution is not offered for this problem since it would require 3 axes instead of 2.)

Consider the number of possible outcomes if a fair coin and a fair 6-sided die are simultaneously tossed once. In Example 1 on pages 305 and 306, we saw the listing of ordered pairs, the graph, and the tree diagram that could be used to show the number of possible outcomes, 12, for these conditions, simultaneously tossing a fair coin and a fair 6-sided die. Let us now add:

filling slots

$$\underset{\text{possibilities for the coin}}{\underline{\;2\;}} \bullet \underset{\text{possibilities for the die}}{\underline{\;6\;}} = 12 \text{ possible outcomes}$$

Factorial

In work with permutations, the result is often a series of factors, each being one less than the preceding factor (for example, $6 \cdot 5 \cdot 4 \cdot 3 \cdot 2 \cdot 1$). Such a series of factors is written 6! and is read "***factorial*** 6" or "6 factorial." $6! = 720$.

◆ In general, $n! = n(n - 1)(n - 2)(n - 3) \ldots 3 \cdot 2 \cdot 1$ where n is a natural number. (*Note:* $0! = 1$ and $1! = 1$.)

The notation used for the number of possible permutations P of r things chosen from n things is ${}_nP_r$. The numerical value of this notation is obtained from the following formula:

$${}_nP_r = n(n - 1)(n - 2)(n - 3) \ldots (n - r + 1)$$

For example, ${}_7P_3$ means the number of possible permutations of 3 things chosen from 7 things, or the number of possible ways to arrange 7 things taken 3 at a time. In this example, $n = 7$, $r = 3$, and $n - r + 1 = 5$. Thus, the numerical value of ${}_7P_3 = 7 \cdot 6 \cdot 5$, or 210.

We can add this formula as another method of solution for permutation problems. In the following examples, several solutions are offered for the first two problems. The methods are chosen from: listing of ordered pairs (triples, etc.), graph, tree diagram, filling slots, and formula.

Examples

1. How many different 2-digit numbers can be made from the digits 3, 4, 5, and 6 if a digit can appear just once in a number?

Solution 1: *filling slots*

There are 2 slots to be filled to form 2-digit numbers. Once a slot has been filled, the digit that was used to fill the slot cannot be used again. Thus, there are 4 ways to fill the first slot and 3 ways to fill the second slot.

$$\underset{\text{first digit}}{\underline{4}} \cdot \underset{\text{second digit}}{\underline{3}} = 12$$

Answer: 12 2-digit numbers

Solution 2: *tree diagram*

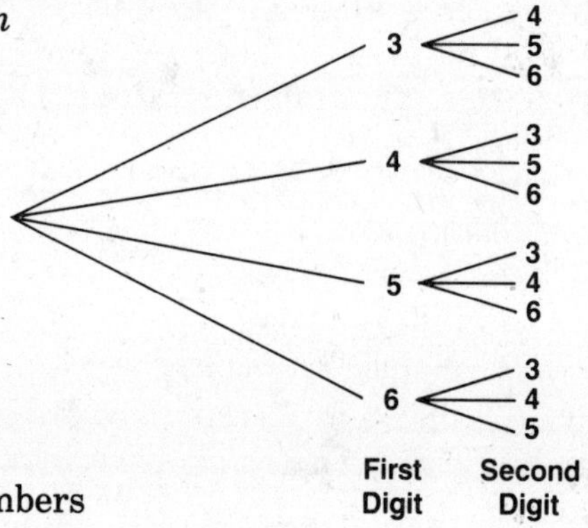

Answer: 12 2-digit numbers

Solution 3: *listing of ordered pairs*

(3, 4)	(4, 3)	(5, 3)	(6, 3)
(3, 5)	(4, 5)	(5, 4)	(6, 4)
(3, 6)	(4, 6)	(5, 6)	(6, 5)

Answer: 12 2-digit numbers

Solution 4: *formula*

$$_nP_r = n(n-1)(n-2)\ldots(n-r+1)$$
$$n = 4, r = 2, n - r + 1 = 3$$
$$_4P_2 = 4 \cdot 3 = 12$$

Answer: 12 2-digit numbers

2. How many 4-digit numbers can be formed from the digits 5, 6, 7, and 8 if no repetition is allowed?

Solution 1: *filling slots*

There are 4 slots to be filled. Once a slot has been filled, the digit that was used to fill the slot cannot be used again. Thus, the first slot can be filled in any of 4 ways.

The second slot can be filled in any of 3 ways.
The third slot can be filled in either of 2 ways.
The fourth slot can be filled in only 1 way.

$$\underset{\substack{\text{1st}\\\text{digit}}}{\underline{\ 4\ }} \cdot \underset{\substack{\text{2nd}\\\text{digit}}}{\underline{\ 3\ }} \cdot \underset{\substack{\text{3rd}\\\text{digit}}}{\underline{\ 2\ }} \cdot \underset{\substack{\text{4th}\\\text{digit}}}{\underline{\ 1\ }} = 24$$

Answer: 24 4-digit numbers

Solution 2: *formula*

$$_nP_r = n(n-1)(n-2)\ldots(n-r+1)$$
$$n = 4, r = 4, n - r + 1 = 1$$
$$_4P_4 = 4 \cdot 3 \cdot 2 \cdot 1 = 24, \text{ or } 4!$$

Answer: 24 4-digit numbers

3. In how many ways can 5 people be seated on a bench that:
a. seats 2? **b.** seats 5?

a. *Solution* 1: *filling slots*

There are 2 slots to be filled (representing the number of available seats).

$$\underset{\substack{1^{st}\\\text{seat}}}{\underline{\ 5\ }} \cdot \underset{\substack{2^{nd}\\\text{seat}}}{\underline{\ 4\ }} = 20$$

Answer: There are 20 ways to seat 5 people on a bench that seats 2.

Solution 2: *formula*

$$_nP_r = n(n-1)(n-2)\ldots(n-r+1)$$
$$n = 5, r = 2, n - r + 1 = 4$$
$$_5P_2 = 5 \cdot 4 = 20$$

Answer: There are 20 ways to seat 5 people on a bench that seats 2.

b. *Solution* 1: *filling slots*

There are 5 slots to be filled (representing the number of available seats).

$$\underset{\substack{1^{st}\\\text{seat}}}{\underline{\ 5\ }} \cdot \underset{\substack{2^{nd}\\\text{seat}}}{\underline{\ 4\ }} \cdot \underset{\substack{3^{rd}\\\text{seat}}}{\underline{\ 3\ }} \cdot \underset{\substack{4^{th}\\\text{seat}}}{\underline{\ 2\ }} \cdot \underset{\substack{5^{th}\\\text{seat}}}{\underline{\ 1\ }} = 120$$

Answer: There are 120 ways to seat 5 people on a bench that seats 5.

Solution 2: *formula*

$$_nP_r = n(n-1)(n-2)\ldots(n-r+1)$$
$$n = 5, r = 5, n - r + 1 = 1$$
$$_5P_5 = 5 \cdot 4 \cdot 3 \cdot 2 \cdot 1 = 120 \text{ or } 5!$$

Answer: There are 120 ways to seat 5 people on a bench that seats 5.

4. Evaluate $\frac{5!}{3!}$.

Solution: $\frac{5!}{3!} = \frac{5 \cdot 4 \cdot \not{3} \cdot \not{2} \cdot \not{1}}{\not{3} \cdot \not{2} \cdot \not{1}} = 5 \cdot 4 = 20$ *Ans.*

5. Show that $(6 - 2)! \neq 6! - 2!$

Solution:

$(6 - 2)!$	$6! \quad - 2!$
$4!$	$6 \cdot 5 \cdot 4 \cdot 3 \cdot 2 \cdot 1 - 2 \cdot 1$
$4 \cdot 3 \cdot 2 \cdot 1$	$720 \quad - 2$

$$24 \neq 718$$

EXERCISES

1. How many 3-digit numbers can be formed from the digits, 3, 4, and 5 if no repetition is allowed?
2. How many 4-digit numbers can be formed from the digits 3, 4, 5, and 6 if no repetition is allowed?
3. How many different 2-digit numbers can be made from the digits of 345 if a digit can appear just once in a number?
4. What is the number of possible 4-letter arrangements of letters F, O, N, and Z if each letter is used only once in each arrangement?
5. What is the total number of possible 5-letter arrangements of the letters D, I, S, C, O if each letter is used only once in each arrangement?
6. How many arrangements of 2 letters can be formed from the letters O, L, Y, M, P, I, C, S if each letter is used only once in each arrangement?
7. How many arrangements of 3 letters can be formed from the letters C, A, P, S, U, L, E if each letter is used only once in each arrangement?
8. In how many different ways can 4 students be arranged in a row?
9. In how many ways can 6 students be seated on a bench that seats only 4?
10. Four people enter a room in which there are 6 empty chairs. In how many ways can they take seats?
11. The value of 3! is (1) 6 (2) 5 (3) 3 (4) -3
12. The value of 5! is (1) $\frac{1}{5}$ (2) 15 (3) 120 (4) 5

In 13–20, find the value of the given expression.

13. $\frac{7!}{4!}$ **15.** $3 \cdot (5 - 2)!$ **17.** $0!$ **19.** ${}_5P_5$

14. $3! \cdot 4!$ **16.** $3 \cdot \frac{(7 - 3)!}{3!}$ **18.** ${}_5P_3$ **20.** ${}_5P_1$

10.9 SUMMARY EXERCISES

In 1–5, a fair die is rolled once. What is the probability of:

1. rolling a number greater than 3?
2. rolling a "7"?
3. rolling a prime number greater than 1?
4. rolling a "3" or a "4"?
5. rolling a "2" and a "6"?

In 6–10, what is the probability of:

6. choosing the letter "i" if one letter is drawn at random from the set of vowels?
7. choosing a day beginning with the letter "T" if one day is chosen at random from the set {names of the days of the week}?
8. choosing a prime number if one number is chosen at random from the set {2, 3, 5, 7, 9, 10}?
9. choosing a month that begins with the letter "A" if one month is chosen at random from the set {names of the months that begin with a vowel}?
10. choosing a month that either begins with the letter "M" or ends with the letter "y" if one month is chosen at random from the set {names of the months of the year}?

In 11–20, one marble is selected at random from an urn containing 8 yellow marbles, 5 red marbles, and 7 green marbles. The color of the marble is noted, and the marble is returned to the urn. A second marble is then drawn at random from that urn. What is the probability that:

11. both marbles are green?
12. neither marble is yellow?
13. the first marble is red and the second marble is green?
14. the first marble is green and the second marble is not yellow?
15. the first marble is yellow and the second marble is white?
16. the first marble is either red or yellow and the second marble is green?
17. the first marble is either red or blue and the second marble is green?
18. the first marble is not red and the second marble is not green?
19. the first marble is either red or yellow and the second marble is not red?
20. the first marble is either red or blue and the second marble is not blue?

In 21–30, one card is drawn at random from a standard deck and that card is put aside. A second card is then drawn at random from that deck. What is the probability that:

21. both cards are red?
22. neither card is red?
23. the first card is red and the second card is black?

24. both cards are red aces?
25. neither card is a black queen?
26. the first card is a "4" and the second card is a red "3"?
27. the first card is red and the second card is not red?
28. the first card is red and the second card is blue?
29. the first card is either a queen or a king and the second card is a "7"?
30. the first card is a black "5" and the second card is either a jack or a queen?

In 31–35, the probability that the Lions will beat the Tigers is $\frac{2}{5}$. The probability that the Lions will beat the Leopards is $\frac{3}{4}$. What is the probability that:

31. the Lions will not beat the Tigers?
32. the Lions will win both games?
33. the Lions will lose both games?
34. the Lions will win the game against the Tigers and lose the game against the Leopards?
35. the Lions will lose the game against the Tigers and win the game against the Leopards?

36. If a fair 4-sided die and a fair 6-sided die are rolled simultaneously, how many pairs (outcomes) are there in the sample space?
37. In a single toss of 2 fair coins, what is the probability of coming up with 2 "heads"?
38. In a single toss of 2 fair coins, what is the probability of coming up with one "head" and one "tail"?
39. A fair 4-sided die is rolled at the same time that a fair coin is tossed. What is the probability of getting a "head" on the coin and an odd number on the die?
40. A fair 6-sided die is rolled at the same time that a fair coin is tossed. What is the probability of getting a "tail" on the coin and an odd number on the die?

In 41–50, a fair 4-sided die and a fair 5-sided die are rolled simultaneously.

41. List the sample space for all possible outcomes.
42. Draw a graph to represent the sample space.
43. Draw a tree diagram to represent the sample space.
44. How many ways are there to get an even number on both dice?
45. How many ways are there to get a sum greater than 4?
46. What is the probability of getting a sum greater than 9?
47. What is the probability of getting a sum less than 12?
48. What is the probability of getting a sum of 7?
49. What is the probability of getting a sum of 6?
50. What is the probability of getting a sum that is not 4?

In 51–55, 2 fair coins are tossed simultaneously with the spinning of the pointer on a marker of equal sections labeled 1, 2, and 3.

51. List the sample space for all possible outcomes.

52. Draw a tree diagram to represent the sample space.

53. How many ways are there to get a "head" on both coins and an odd number on the spinner?

54. In how many ways can one "head," one "tail," and an even number appear?

55. What is the probability of getting the same face on each coin and an odd number on the spinner?

56. Jim bought 3 packets of vegetable seeds: beans, carrots, and radishes; and 2 packets of flower seeds: marigolds and petunias. Finding that his garden was too small, he decided to give away one packet of vegetable seeds and one packet of flower seeds, each selected randomly.

a. Draw a tree diagram or list all possible pairs of packets in the sample space that he might select to give away.

b. Find the probability that he did *not* give away the packet of bean seeds.

c. Find the probability that he gave away two packets of vegetable seeds.

d. Find the probability that he gave away the packet of petunia seeds.

57. In how many ways can 7 people be seated on a bench that seats 7?

58. In how many ways can 7 people be seated on a bench that seats 5?

59. In how many ways can 7 people be seated on a bench that seats 2?

60. How many arrangements of 3 letters can be formed from the letters T, R, A, P, E, Z, O, I, D if each letter is used only once in each arrangement?

61. A snack bar serves 3 kinds of soup, 10 kinds of sandwiches, and 5 kinds of beverages. If a meal consists of a soup, a sandwich, and a beverage, how many different meals are possible?

62. There are 8 roads between City 1 and City 2 and 6 roads between City 2 and City 3. How many different routes could be taken from City 1 to City 3 through City 2?

63. Mr. Dagwood has 7 kinds of sliced bread. In how many ways can he make a sandwich so that the bottom slice of bread is a kind different from the top slice of bread?

64. Find the value of $5 \cdot 5!$

65. The value of $(7 - 3)!$ is

(1) $7! - 3!$ (2) $7! - 3$ (3) $7 - 3!$ (4) $4!$

66. Find the value of ${}_{10}P_3$.

CHAPTER 11

Statistics

11.1 THE FREQUENCY HISTOGRAM

Statistics is the branch of mathematics that concerns itself with the organization, presentation, and interpretation of data. A kind of graph called a ***frequency histogram*** is useful for displaying data. We can learn important information about the data from the histogram.

In order to draw a frequency histogram, we first construct a table in which the data are grouped into intervals of equal length. (*Note:* The intervals can be of any length: 1, 3, 5, 10, etc.)

Examples

1. The table shows the number of pushups done by 40 women in an exercise class during a timed period of 3 minutes. Construct a frequency histogram for the grouped data.

Interval	Number (Frequency)
16–18	5
13–15	10
10–12	10
7–9	12
4–6	2
1–3	1

Solution:

Show the intervals on a horizontal scale marked off in units of equal length and labeled from the interval column of the table. Give this scale a title, telling what the numbers represent.

Show the frequencies on a vertical scale marked off in units convenient to fit the given numbers. The scale starts at 0 and increases to include the highest frequency. In this case, the frequencies range from 1 through 12. Thus, a breakdown by ones, labeled by fives, is convenient.

When drawing the vertical bars, do not leave gaps between the intervals.

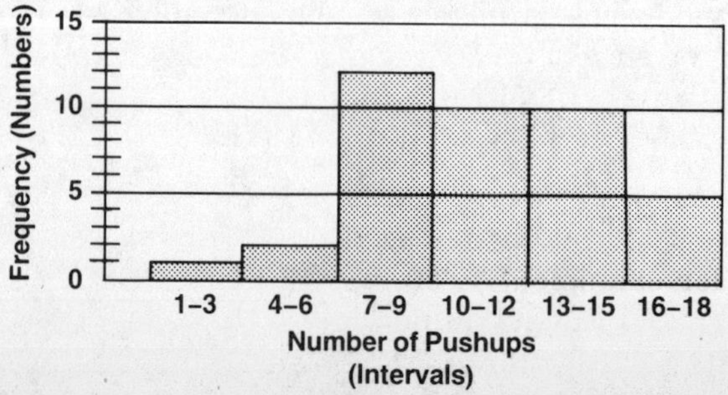

2. Answer the following questions for the data considered in Example 1:

   a. Which interval contains the highest frequency?
   b. Which interval contains the lowest frequency?
   c. How many women were able to do 10 or more pushups in the allotted time?
   d. What percent of the women were able to do more than 15 pushups in the allotted time?
   e. How many of the women were not able to do at least 10 pushups in the allotted time?

Answers:

a. 7–9
b. 1–3
c. 25 (10 + 10 + 5)
d. $12\frac{1}{2}\%$ ($\frac{5}{40} = \frac{1}{8} = 8\overline{)1.000}$ = .125)
e. 15 (12 + 2 + 1)

Often, we must first ***tally*** the given data in order to determine the frequencies. See the following example for such a case.

Example

The weights, in pounds, of 20 students in an elementary school class are: 72, 64, 56, 60, 66, 72, 48, 66, 58, 60, 60, 50, 68, 72, 68, 62, 72, 58, 60, 68.

a. Construct a frequency histogram based on the data. (Use intervals of length 5.)
b. Determine the range of the weights.
c. Which interval contains the highest frequency? the lowest frequency?
d. How many of the 20 students weigh less than 60 pounds?
e. What percent of the students weigh at least 65 pounds?

Solution:

a. Set up a table of 3 columns: Interval, Tallies, Number (Frequency). The given instruction is to use 5-pound intervals. This choice will accommodate the highest and the lowest weights. (Other interval lengths could have been chosen. Also, the beginnings and endings of the intervals could be taken differently, for example, 46–50, 51–55, 56–61, etc.) Enter the individual weights by tally marks.

Interval	Tallies	Number (Frequency)
70–74	||||	4
65–69	~~||||~~	5
60–64	~~||||~~ |	6
55–59	|||	3
50–54	|	1
45–49	|	1

Draw and label horizontal and vertical scales. Use a vertical scale marked off and labeled in ones. Draw the bars, leaving no gaps between intervals.

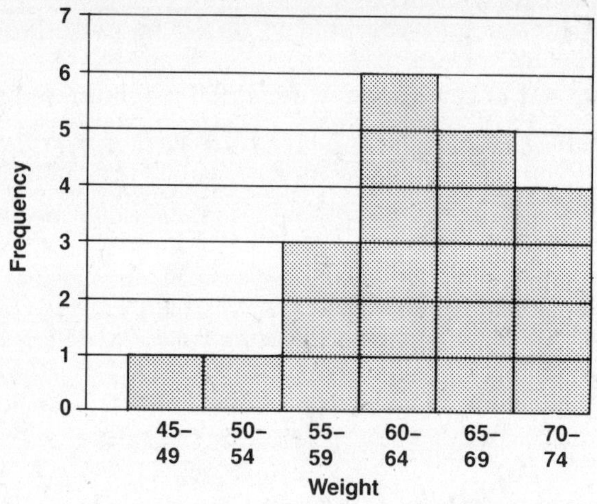

b. The ***range*** of the weights is the difference between the highest weight and the lowest weight. Thus, the range of the weights is 72–48, or 24.

c. The highest frequency lies in the interval 60–64. The lowest frequency appears twice: in the interval 45–49 and in the interval 50–54.

d. The number of students weighing less than 60 pounds is 5. (3 + 1 + 1, or 5)

e. The percent of students weighing at least 65 pounds is 45%. $\left(\frac{4+5}{20} = \frac{9}{20} = 20\overline{)9.00}^{\,.45}\right)$

EXERCISES

1. The table shows the gas mileage (in miles per gallon) attained by 26 cars in a consumer study.

Interval	Number (Frequency)
28–31	2
24–27	7
20–23	9
16–19	8

a. Construct a frequency histogram.

b. Which interval contains the highest frequency? the lowest frequency?

c. How many of the cars were not able to get at least 20 miles to the gallon?

d. What percent of the cars were able to get at least 24 miles to the gallon? (Write the answer to the nearest tenth of a percent.)

2. The heights, in inches, of 40 students in a class are: 60, 62, 65, 58, 61, 57, 63, 62, 68, 56, 62, 64, 59, 67, 60, 66, 65, 63, 59, 61, 60, 58, 57, 64, 60, 60, 61, 64, 62, 62, 63, 61, 62, 63, 54, 52, 61, 63, 61, 62.

a. What is the range of the heights?

b. Construct a frequency histogram. (Use intervals of length 3. Begin with the interval 51–53.)
c. Which interval contains the highest frequency?
d. Which interval contains the lowest frequency?
e. If the interval 60–62 is considered "average height" for this group of students, how many students are above "average height"? below "average height"?

3. The noon temperatures, in degrees Fahrenheit, for the 30 days of April as recorded in Central Park were: 46, 50, 47, 49, 52, 48, 53, 47, 48, 50, 49, 51, 53, 52, 51, 48, 50, 51, 50, 52, 54, 52, 51, 53, 54, 52, 50, 51, 51, 49.
   a. What is the range of the temperatures?
   b. Construct a frequency histogram. (Use intervals of length 3. Begin with the interval 44–46.)
   c. Which interval contains the highest frequency? the lowest frequency?
   d. If the interval 50–52 is considered "average noon temperature" for this month, how many days were above "average noon temperature"?
   e. What percent of the days were below "average noon temperature"?

4. On a test consisting of 10 questions, 30 students received scores as shown in the table.
   a. Construct a frequency histogram.
   b. Which interval contains the highest frequency?
   c. If 6 is considered the passing score, how many of the 30 students passed?
   d. What percent of the students failed? (Write the answer to the nearest tenth of a percent.)

Grade	Number (Frequency)
10	2
9	4
8	3
7	8
6	6
5	1
4	3
3	0
2	1
1	1
0	1

5. The weights, in pounds, of 40 people entering a contest are:

   161, 193, 163, 194, 126, 170, 171, 166, 210, 163,
   172, 195, 119, 157, 219, 190, 162, 141, 198, 150,
   102, 126, 127, 126, 111, 148, 175, 183, 200, 181,
   204, 200, 220, 200, 201, 203, 140, 221, 199, 148

   a. Determine the range of the weights.
   b. Construct a frequency histogram. (Use intervals of length 25. Begin with the interval 100–124.)
   c. Construct a frequency histogram. (Use intervals of length 10. Begin with the interval 100–109.)

In 6–13, refer to the following histogram.

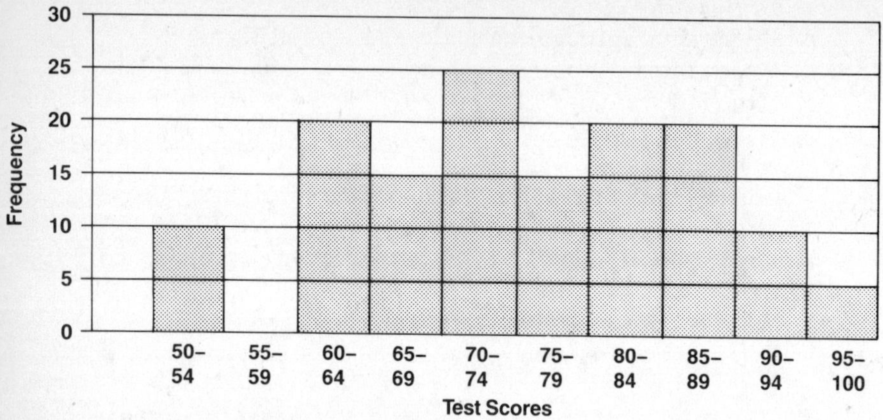

6. What is the length of each interval?
7. How many students took the test?
8. Which interval contains the highest frequency?
9. Which interval contains the lowest frequency?
10. If 65 is the passing grade, how many students passed?
11. What percent of the students failed?
12. How many students had grades of 80 or better?
13. What percent of the students had grades of 90 or better?

11.2 THE CUMULATIVE FREQUENCY HISTOGRAM

The bars of a frequency histogram represent the frequency per interval. In a ***cumulative frequency histogram,*** the bars represent the *sum of the frequencies* up through a given interval.

To construct a cumulative frequency histogram, we follow much the same procedure as for the frequency histogram, but the graph itself is based on a new column in the table. That column is entitled "Cumulative Frequency," and its entries are arrived at by a series of additions. The following example will illustrate the differences between the two types of histograms.

Example

The weights, in pounds, of 36 male wrestlers are:

155, 162, 168, 147, 152, 173, 178, 181, 180, 158, 163, 177,
171, 168, 183, 154, 180, 158, 157, 160, 171, 164, 171, 175,
183, 174, 180, 169, 175, 166, 149, 150, 170, 163, 158, 162

Construct a frequency histogram and a cumulative frequency histogram for the given data. (Use intervals of length 10.)

Solution:

Prepare a table as follows:

Interval	Tallies	Number (Frequency)	Cumulative Frequency
180–189	𝍸 |	6	36
170–179	𝍸 𝍸	10	30
160–169	𝍸 𝍸	10	20
150–159	𝍸 |||	8	10
140–149	||	2	2

Since the intervals are arranged from the lowest up to the highest, the cumulative frequencies are obtained by additions from the bottom up.

The corresponding frequency histogram is:

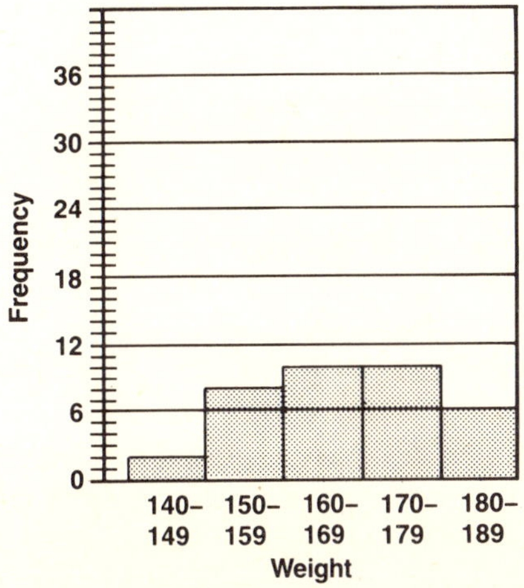

The corresponding cumulative frequency histogram is:

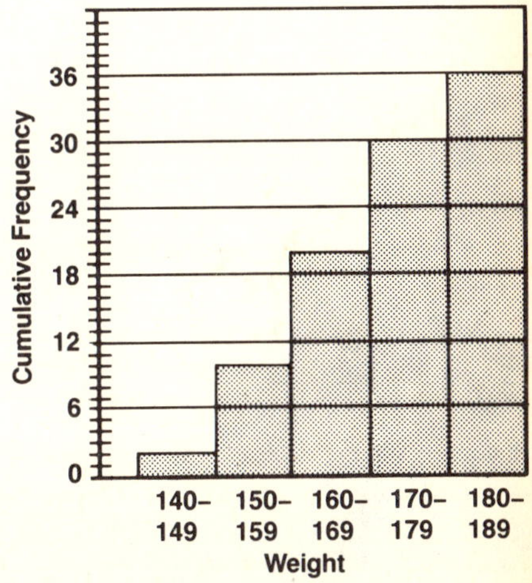

Note 1: Since the bars of the frequency histogram represent the frequency per interval, the tallest bar tells the highest frequency and the shortest bar tells the lowest frequency.

Note 2: Since the bars of the cumulative frequency histogram represent sums of frequencies, the bars keep on growing taller, and the last bar tells the total frequency.

EXERCISES

In 1–4, refer to the cumulative frequency histogram shown.

1. How many children were weighed?
2. How many children weighed less than 60 pounds?
3. What percent of the children weighed 70 pounds or more?
4. In which interval is the highest frequency?

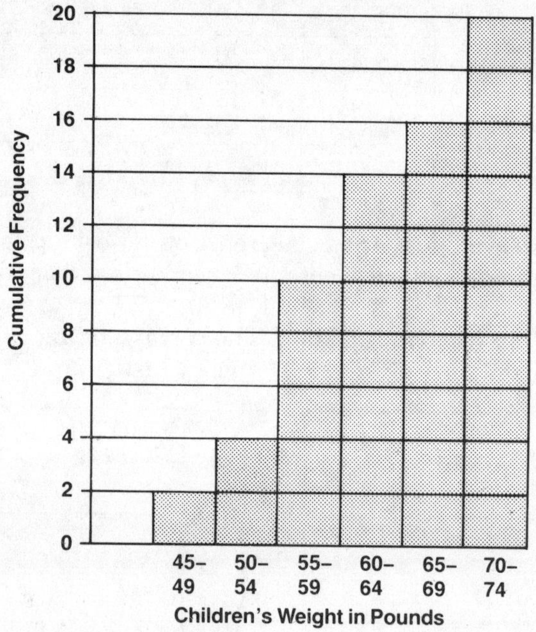

In 5–9, use the sets of data given for Exercises 1–5 in section 11.1 (pages 320–321), and construct a cumulative frequency histogram for each set.

11.3 PERCENTILES

Certain statistically useful measures can be read from the cumulative frequency histogram. One such measure is known as a ***percentile.*** The nth percentile is the measure at or below which $n\%$ of the measures fall. Thus, for example, the 40th percentile is the measure at or below which 40%, or $\frac{2}{5}$, of all the measures fall. The 50th percentile is the measure at or below which 50%, or $\frac{1}{2}$, of all the measures fall.

Another term referred to in percentile measure is ***quartile.*** Quartile means that the total of 100% is broken into *four* equal parts: 25%, 50%, 75%, 100%. The 25th percentile is called the ***lower quartile,*** the 75th percentile is called the ***upper quartile,*** and the 50th percentile is called the ***median.***

Reconsider the table from the example in the previous section that grouped the 36 given weights. The quartile information for this group of data is:

Interval	Number (Frequency)	Cumulative Frequency
180–189	6	36
170–179	10	30
160–169	10	20
150–159	8	10
140–149	2	2

Since 25% (or $\frac{1}{4}$) of 36 is 9, the lower quartile falls in the interval 150–159.

Since 50% (or $\frac{1}{2}$) of 36 is 18, the median falls in the interval 160–169.

Since 75% (or $\frac{3}{4}$) of 36 is 27, the upper quartile falls in the interval 170–179.

If more exact information about the quartiles is desired from the graph, a ***cumulative frequency polygon*** is drawn on the cumulative frequency histogram. This is done by connecting a series of points to form the cumulative frequency polygon. (Note that this special polygon is not a closed figure.) The points to be connected are the upper right corners of the bars of the histogram. Begin at the lower left corner of the first bar.

If the quartile information discussed above is entered on the graph as dotted horizontal lines that are extended until these dotted horizontal lines intersect the cumulative frequency polygon, the intersections will tell how to read within the intervals. That is, by dropping dotted vertical lines from the intersections of the dotted horizontal lines and the polygon to the interval scale, we can read values within the intervals.

Reading from the graph:

The lower quartile (the measure at or below which 25% of the weights fall) appears to be 158. (The actual value is 158.)

The median (the measure at or below which 50% of the weights fall) appears to be 168. (The actual value is 168.)

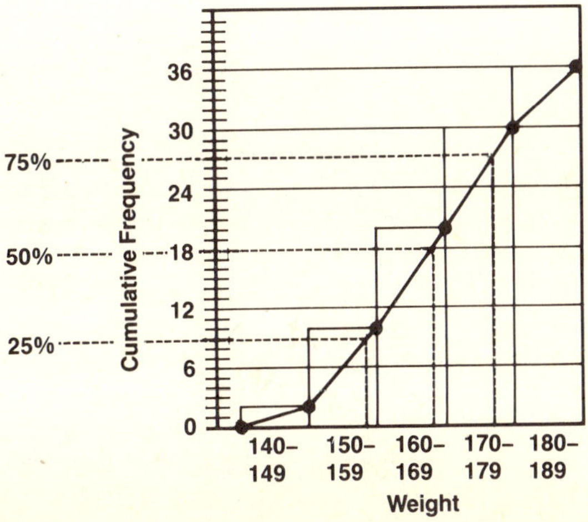

The upper quartile (the measure at or below which 75% of the weights fall) appears to be 177. (The actual value is 175.)

Note 1: There are ways to determine the exact values, but these methods require more calculation. Unless the exactness of the value is most important, the error introduced by approximating from the graph is small.

Note 2: In the example used, the interval lengths were rather long (10 units). Thus, the approximations may be less accurate than they would have been with shorter intervals. However, the use of shorter intervals in this case would have made the graph itself more cumbersome to draw.

EXERCISES

1. The graph shows the distribution of scores on a mathematics test. How many students took the mathematics test?

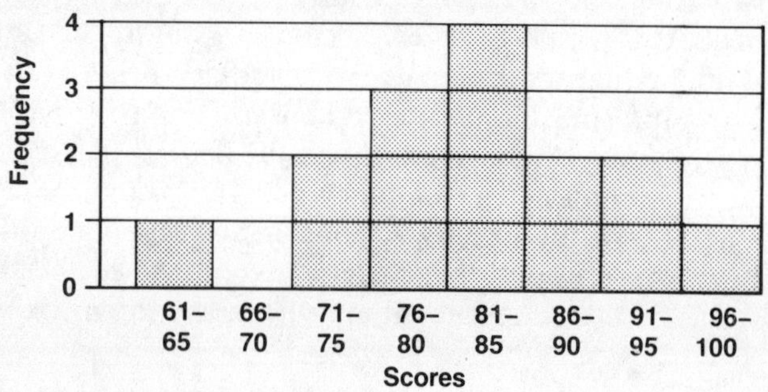

2. The following data are heights (in centimeters) of a group of 15 students: 165, 160, 173, 150, 188, 150, 173, 155, 163, 152, 175, 183, 151, 163, 178.

a. Copy and complete the table.

b. On graph paper, construct a frequency histogram based on the grouped data.

c. In what interval is the median for the grouped data?

Interval	Number (Frequency)
180–189	
170–179	
160–169	
150–159	

3. The following data represent the examination marks of 10 students in a certain class: 87, 71, 80, 84, 83, 81, 86, 90, 99, 89.

a. Copy and complete the table.

b. On graph paper, construct a frequency histogram based on the data.

c. Which interval contains the median?

Interval	Number (Frequency)
69–76	
77–84	
85–92	
93–100	

4. The following data are test scores for a class of 16 students: 96, 83, 91, 77, 58, 88, 80, 62, 89, 100, 87, 93, 64, 98, 88, 86.
   a. Copy and complete the table.
   b. On graph paper, construct a frequency histogram based on the data.
   c. Which interval contains the median?
   d. Which interval contains the lower quartile?

Interval	Number (Frequency)
91–100	
81–90	
71–80	
61–70	
51–60	

5. The following data represent the heights of 14 students in a certain class: 65, 63, 68, 59, 74, 59, 68, 61, 64, 60, 69, 72, 55, 64.
   a. On your answer paper, copy and complete the table.
   b. On graph paper, construct a frequency histogram based on the data.
   c. Which interval contains the median?

Interval	Number (Frequency)
55–58	
59–62	
63–66	
67–70	
71–74	

6. Given the data in the table representing test scores for a class of 20 students:
   a. On graph paper, construct a frequency histogram based on the data.
   b. Which interval contains the median?
   c. Which interval contains the lower quartile?

Interval	Number (Frequency)
51–60	1
61–70	3
71–80	3
81–90	7
91–100	6

In 7–11, draw cumulative frequency polygons for the frequency histograms in exercises 2–6.

12. The following diagram is a cumulative frequency histogram of raw scores on a mathematics examination.

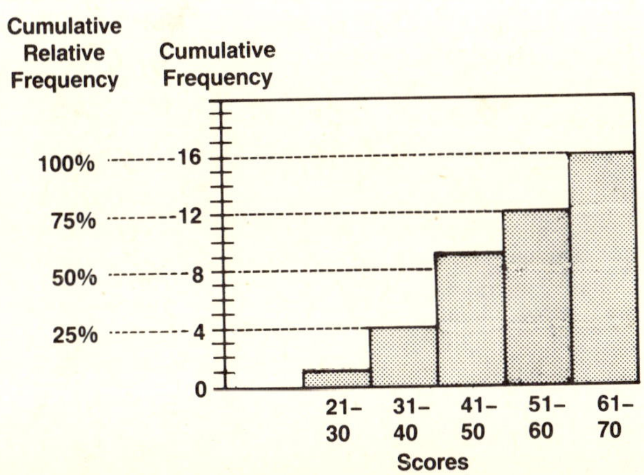

   a. How many students took the examination?
   b. How many students had a score less than or equal to 60?

c. What percent of the students had a score less than or equal to 60?
d. Which interval contains the median?
e. Which interval contains the lower quartile?

13. The table represents the ages of the teachers at a school.
a. In what interval is the median?
b. A teacher is chosen at random from this school. What is the probability that the teacher's age is in the interval 33–37?
c. What is the probability that the age of a teacher from this school is less than 38?
d. What is the probability that a teacher from this school is older than 57?
e. What percent of the teachers are in the interval 43–47?

Interval	Number (Frequency)
53–57	4
48–52	8
43–47	6
38–42	4
33–37	2
28–32	4
23–27	2

14. The test scores of 15 students are: 86, 57, 69, 82, 91, 87, 75, 84, 68, 65, 91, 88, 81, 62, 72.
a. Copy and complete the table.
b. On graph paper, draw a frequency histogram based on the data.
c. In which interval does the median lie?
d. What is the probability that a student, selected at random, scored above 89 on this test?

Interval	Number (Frequency)
90–99	
80–89	
70–79	
60–69	
50–59	

11.4 MEASURES OF CENTRAL TENDENCY

The ***median,*** previously discussed as a quartile value, is also called a ***measure of central tendency,*** since it represents a value referring to the middle (center) of the data.

The median of a small set of values for which no graph or table is prepared can be found by arranging the values in order, either ascending or descending, and choosing the one in the middle.

Examples

1. Determine the median value of 4, 3, 1, 4, 4, 6, 3.

Solution:

Arrange the values in ascending order and choose the middle value:

1, 3, 3, 4, 4, 4, 6
↑
median

Answer: The median value is 4.

2. Determine the median value of 3, 4, 1, 1, 4, 6.

Solution:

Arrange the values in ascending order. Since there is an even number of values, there is no middle value. Take the number halfway between the two values surrounding the exact middle:

1, 1, 3, ↑ 4, 4, 6
median

Answer: The median is 3.5.

3. Determine the median of 1, 3, 1, 3, 6, 4.

Solution: 1, 1, 3, ↑ 3, 4, 6
median

Answer: The median is 3.

Another measure of central tendency is the ***arithmetic mean,*** or ***average,*** of a set of values. The arithmetic mean is calculated by adding all of the values and dividing that sum by the number of values.

Examples

1. Determine the arithmetic mean for the set of values: 30, 41, 32, 46, 37.

Solution:

$$\text{arithmetic mean} = \frac{30 + 41 + 32 + 46 + 37}{5} = \frac{186}{5} = 37\frac{1}{5}$$

Answer: The arithmetic mean is $37\frac{1}{5}$.

Note that in this case the arithmetic mean, or simply mean, is not one of the original values.

2. Determine the mean for the set of values: 2, 1, 6, 3, 500, 3, 1.

Solution:

$$\text{mean} = \frac{2 + 1 + 6 + 3 + 500 + 3 + 1}{6} = \frac{516}{6} = 86$$

Answer: The arithmetic mean is 86.

Note that in this case the mean is not a good representative of the central tendency of the values. A better representative of the middle of these values would be 3, the median.

A third measure of central tendency is called the ***mode***. The mode is the measure that occurs most frequently, and it is easily read from a frequency histogram: The tallest bar of the histogram represents the mode. (*Note:* The tallest bar of a cumulative frequency histogram—that is, the last bar—represents the total frequency, and is not to be confused with the mode.) Sometimes, a set of values has more than one mode.

Examples

1. Determine the mode from the following histogram:

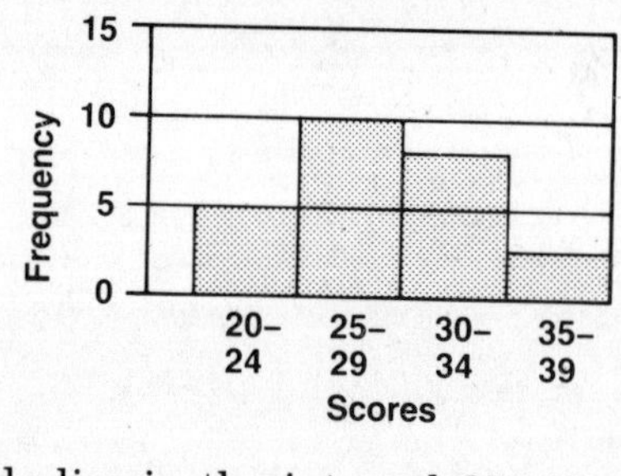

Answer: The mode lies in the interval 25–29. (The tallest bar of a frequency histogram represents the mode.)

2. Determine the mode for the following set of values: 3, 1, 1, 3, 1, 4.

Answer: The mode is 1. (The value occurring most frequently is the mode.)

3. Determine the mode for the following set of values: 4, 3, 1, 1, 3, 1, 3, 4.

Answer: The modes are 1 and 3. (Two values occur with the same highest frequency.) This set of values is said to be ***bimodal;*** that is, there are two modes.

4. For which set of data do the mean, median, and mode all have the same value?

(1) 2, 6, 6, 6, 10 (3) 2, 2, 2, 4, 10
(2) 2, 2, 4, 10, 12 (4) 2, 2, 6, 10, 20

Answer: (1) (The mean, median, and mode for this set of values are all 6.)

5. A student received test scores of 73, 83, and 86. What score must that student receive on a fourth test so that the average of the 4 scores will be exactly 82?

Solution:

Let x = the score on the fourth test.

$$\text{average} = \frac{73 + 83 + 86 + x}{4} = 82$$

$$\frac{242 + x}{4} = 82$$

$$242 + x = 328 \quad \text{Multiply by 4.}$$

$$x = 86 \quad \text{Subtract 242.}$$

Answer: The student must receive 86 on the fourth test for an average of 82.

6. Express the mean (average) of $x - 1$ and $5x + 5$ as a binomial.

Solution:

$$\text{average} = \frac{(x - 1) + (5x + 5)}{2}$$

$$= \frac{6x + 4}{2}$$

$$= 3x + 2 \quad \textit{Ans.}$$

7. Express the mean (average) of $x - 1$, $x + 2$, $2x - 3$, and $4x + 6$ as a binomial.

Solution:

$$\text{average} = \frac{(x - 1) + (x + 2) + (2x - 3) + (4x + 6)}{4}$$

$$= \frac{8x + 4}{4} = 2x + 1 \quad \textit{Ans.}$$

EXERCISES

1. What is the mean score for the test results of 60, 60, 70, 75, and 80?
(1) 60 (2) 69 (3) 70 (4) 75
2. Jane's test grades were 75, 83, 87, 90, and 90. What is the median of these grades?
3. The high temperatures during 5 days were 82°, 86°, 91°, 79°, and 91°. Find the mode for these temperatures.

4. If student heights are 176 cm, 172 cm, 160 cm, and 160 cm, what is the mean height of these students?
5. What is the median score for the test results of 60, 60, 70, 75, and 80?
(1) 60 (2) 69 (3) 70 (4) 75
6. Find the mode for the following data: 4, 5, 3, 4, 5, 3, 5.
7. For which set of data do the mean, median, and mode all have the same value?
(1) 1, 3, 3, 3, 5 (2) 1, 1, 2, 5, 6 (3) 1, 1, 1, 2, 5 (4) 1, 1, 3, 5, 10
8. A student received test scores of 72, 84, and 86. What score must the student receive on a fourth test so that the mean of these scores will be 85?
9. If the heights, in centimeters, of 5 students are 176, 172, 160, 158, and 158, find the median height for this group of 5 students.
10. Which set of data has more than one mode?
(1) 2, 2, 4, 6, 7, 9 (3) 2, 2, 2, 6, 7, 9
(2) 2, 2, 4, 6, 9, 9 (4) 2, 3, 4, 6, 9, 9
11. Express the mean (average) of $x + 1$ and $3x - 3$ as a binomial.
12. Express the mean (average) of $3x - 1$, $2x$, and $x + 7$ as a binomial.
13. A student received test scores of 82, 94, and 96. What must she receive as a fourth score so that the mean of these 4 scores will be exactly 90?
14. The heights of the starting 5 players on a high school basketball team are 77″, 72″, 69″, 74″, and 77″. What is the median height of these players?
15. For which set of numbers will the mean, median, and mode all be equal? (1) 2, 2, 5 (2) 2, 5, 5 (3) 2, 3, 3, 4 (4) 2, 2, 5, 5
16. If a group of data consists of the numbers 2, 2, 5, 6, and 15, which is true?
(1) median > mean (3) mode < median
(2) mean = mode (4) median = mode
17. Five girls in a club reported on the number of boxes of cookies that they sold: 20, 20, 40, 50, and 70. Which is true?
(1) The median is 20. (3) The median is equal to the mean.
(2) The mean is 20. (4) The median is equal to the mode.
18. If a group of data consists of the numbers 5, 5.5, 7, and 7.5, which is true?
(1) The median and the mean are each 6.5.
(2) The median and the mode are each 6.5.
(3) The median and the mean are each 6.25.
(4) The median and the mode are each 6.25.
19. If a group of data consists of the numbers 6, 6, 6.5, 8, and 8.5, which is true?
(1) median < mean (3) median < mode
(2) median = mean (4) median = mode
20. The following data are numbers of minutes a student spent on homework: 25, 35, 30, 50, and 38. What is the median of the data?

21. The scores on a test were 75, 75, 85, 90, and 100. Which statement about these scores is true?
(1) The mean and the median are the same.
(2) The mode is greater than the median.
(3) The mode is greater than the mean.
(4) The mean is less than the median.

22. The accompanying histogram shows the distribution of student ages in a ninth grade class. Which age is the mode?

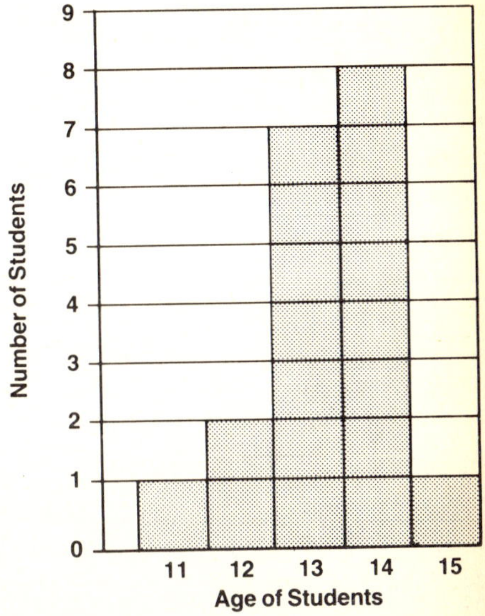

23. The data represent the distribution of test grades of students on a mathematics test.
a. Find the total frequency.
b. Find the mode.
c. Find the median.
d. Find the mean.

Grade	Number (Frequency)
60	1
65	1
70	1
75	2
80	3
85	4
90	2
95	1
100	0

24. The points scored by Rosa in twenty basketball games are 35, 33, 27, 35, 29, 37, 32, 35, 35, 32, 23, 37, 32, 29, 26, 30, 28, 31, 29, 35.

a. Find the mode.
b. Copy and complete the table.
c. Construct a frequency histogram based on the table completed in part **b.**
d. In what interval does the median lie?

Interval	Tally	Frequency
35–37		
32–34		
29–31		
26–28		
23–25		

Grades	Frequency	Cumulative Frequency
12		
13		
14		
15		
16		
17		
18		
19		

25. On a test, 15 students received the following grades: 17, 14, 16, 18, 17, 19, 15, 15, 16, 13, 17, 12, 18, 16, 17.

a. Copy and complete the table.
b. Find the median.
c. Find the mode.
d. Find the 75th percentile.

26. A class record, shown in the accompanying table, gave the number of misspelled words in each of 25 essays.

a. On graph paper, construct a frequency histogram based on the data.
b. Find the mean number of misspelled words.
c. Find the median number of misspelled words.
d. Find the mode number of misspelled words.

Misspelled Words	Frequency (Number of Essays)
0	1
1	0
2	3
3	5
4	4
5	9
6	3

11.5 SUMMARY EXERCISES

In 1–11, given the following heights, in inches, of 27 basketball players: 75, 86, 74, 69, 83, 75, 81, 70, 72, 84, 84, 83, 74, 73, 74, 71, 70, 70, 77, 69, 75, 76, 75, 73, 79, 69, 71.

1. What is the range, in inches, of the heights?
2. Construct a frequency histogram for the data. (Use intervals of length 3. Begin with the interval 67–69.)
3. Which interval contains the highest frequency?
4. Which interval contains the lowest frequency?
5. How many of the 27 players are taller than 6 feet?
6. What percent of the 27 players are 6 feet or less in height?
7. Construct a cumulative frequency histogram for the data.
8. In which interval does the lower quartile fall?
9. In which interval does the median fall?
10. In which interval does the 75th percentile fall?
11. Construct a cumulative frequency polygon for the data.

In 12–18, the data in the table represent the scores achieved by the students in a class on a quiz.

12. Find the total frequency.
13. Find the mode.
14. Find the median.
15. Find the mean.
16. What is the probability that a student chosen at random from this class achieved a score greater than the mode?
17. What is the probability that a student chosen at random from this class achieved a score equal to the median?
18. What is the probability that a student chosen at random from this class achieved a score less than the mean?

Score	Number (Frequency)
10	3
9	6
8	7
7	8
6	5
5	4
4	2
3	2
2	1
1	1
0	1

In 19–25, the table shows the distribution of scores of 30 students on a test.

Scores	Frequency	Cumulative Frequency
91–100	3	
81–90	11	
71–80	8	
61–70	6	
51–60	1	
41–50	1	

19. Using the data in the Frequency column of the table, draw a frequency histogram.
20. Copy the table and complete the column for Cumulative Frequency.
21. Using the data in the Cumulative Frequency column of the table, draw a cumulative frequency histogram.
22. In which interval does the mode fall?
23. In which interval does the median fall?
24. What is the probability that a student chosen at random from this group achieved a score less than the median?
25. What is the probability that a student chosen at random from this group achieved a score greater than the mode?

In 26–33, use the set of numbers {10.7, 10.3, 10, 10.1, 10}.

26. Find the mean.
27. Find the median.
28. Find the mode.
29. Which of the following is true?
(1) median > mean (3) median < mode
(2) median = mean (4) median < mean
30. What is the probability that a number selected at random from the set is equal to the mode?
31. What is the probability that a number selected at random from the set is not equal to the median?
32. What is the probability that a number selected at random from the set is not equal to the mean?
33. What is the probability that a number selected at random from the set is equal to either the median or the mode?

34. Express the mean (average) of $2x + 1$ and $2x + 3$ as a binomial.
35. Express the mean (average) of $x + 2$, $2x - 4$, and $3x + 5$ as a binomial.

CHAPTER 12

The Rectangular Coordinate System

12.1 DEFINITIONS

The French mathematician René Descartes introduced a method of linking numbers with points in a plane. This method evolved into a system that bears his name, the ***Cartesian Coordinate System.*** This system is also called the ***Rectangular Coordinate System.***

The rectangular coordinate system begins with a pair of perpendicular lines called ***axes*** (plural of ***axis***). The *horizontal axis,* or ***x-axis,*** tells directions right or left; the *vertical axis,* or ***y-axis,*** tells directions up or down. All four of these directions are taken with reference to the ***origin,*** the point at which the two axes intersect. On the x-axis, positive numbers are used for the direction "right of the origin" and negative numbers are used for the direction "left of the origin." Similarly, on the y-axis, positive numbers represent the direction "up from the origin" and negative numbers represent the direction "down from the origin."

The origin is a point and is labeled by the letter O. Since the axes are lines and thus are infinitely long, they are indicated as such by arrowheads. Some numbers are shown on each part of each axis to tell the directions and scales used. Since each axis is a number line, each contains the number 0. The origin, the point at which the axes intersect, therefore contains the number 0 twice. This is indicated as (0, 0).

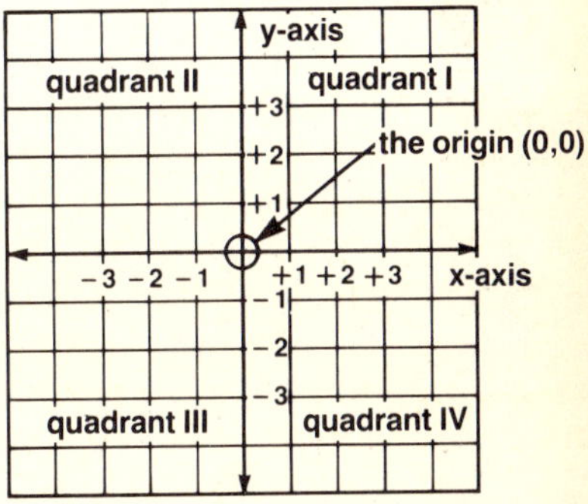

The axes divide the plane into four equal parts called ***quadrants.*** Starting from the upper right, these quadrants are numbered in a counterclockwise direction.

Each point in the plane is described by a set of two numbers whose order tells first the horizontal direction of the point (right or left) from the origin, and second the vertical direction of the point (up or down) from the origin. The set of numbers for each point is, therefore, an ***ordered pair*** and is called the ***coordinates*** of the point. The first coordinate is called the ***abscissa*** and represents the x-value, or horizontal distance of the point from the origin. The second coordinate is called the ***ordinate*** and represents the y-value, or vertical distance of the point from the origin. A set of coordinates for a point is written as (x, y).

◆ Every point in the ***coordinate plane*** is thus accounted for as follows:

1. All points in quadrant I have a positive abscissa and a positive ordinate $(+, +)$. These coordinates show how to move into quadrant I from the origin: go right and then go up.
2. All points in quadrant II have a negative abscissa and a positive ordinate $(-, +)$. These coordinates show how to move into quadrant II from the origin: go left and then go up.
3. All points in quadrant III have a negative abscissa and a negative ordinate $(-, -)$. These coordinates show how to move into quadrant III from the origin: go left and then go down.
4. All points in quadrant IV have a positive abscissa and a negative ordinate $(+, -)$. These coordinates show how to move into quadrant IV from the origin: go right and then go down.
5. All points on the x-axis have an ordinate of 0 $(+ \text{ or } -, 0)$, or $(x, 0)$. These coordinates show that after going right or left, no up or down direction is taken.
6. All points on the y-axis have an abscissa of 0 $(0, + \text{ or } -)$, or $(0, y)$. These coordinates show that no direction of right or left is taken first, but a direction of up or down is taken second.
7. The origin has both an abscissa and an ordinate of 0 $(0, 0)$.

Examples

Tell the locations of the given points.

	Answers:
1. $(3, 4)$	**1.** quadrant I $(+, +)$
2. $(-3, 4)$	**2.** quadrant II $(-, +)$
3. $(-3, -4)$	**3.** quadrant III $(-, -)$
4. $(3, -4)$	**4.** quadrant IV $(+, -)$
5. $(3, 0)$	**5.** positive portion of x-axis $(+, 0)$
6. $(-3, 0)$	**6.** negative portion of x-axis $(-, 0)$
7. $(0, 4)$	**7.** positive portion of y-axis $(0, +)$
8. $(0, -4)$	**8.** negative portion of y-axis $(0, -)$
9. $(0, 0)$	**9.** origin

EXERCISES

In 1–10, tell the locations of the given points.

1. $(2, 3.4)$	**4.** $(-3.4, -2)$	**7.** $(-2, 0)$	**9.** $(0, 2)$
2. $(3.4, 2)$	**5.** $(-2, 3.4)$	**8.** $(0, -2)$	**10.** $(2, 0)$
3. $(-2, -3.4)$	**6.** $(-3.4, 2)$		

11. Tell why (2, 3) and (3, 2) do not describe the same point.

In 12–15, fill in the blanks using the word "positive" or "negative."

12. A point in quadrant II has a ____ abscissa and a ____ ordinate.
13. A point in quadrant IV has a ____ abscissa and a ____ ordinate.
14. A point in quadrant I has an abscissa and an ordinate that are both ____.
15. A point in quadrant III has an abscissa and an ordinate that are both ____.

16. In the accompanying diagram that shows line segment $\overline{PQR}$, $\overline{PQ} \cong \overline{QR}$, the coordinates of P are $(-2, 3)$ and the coordinates of Q are $(0, 0)$. What are the coordinates of R?

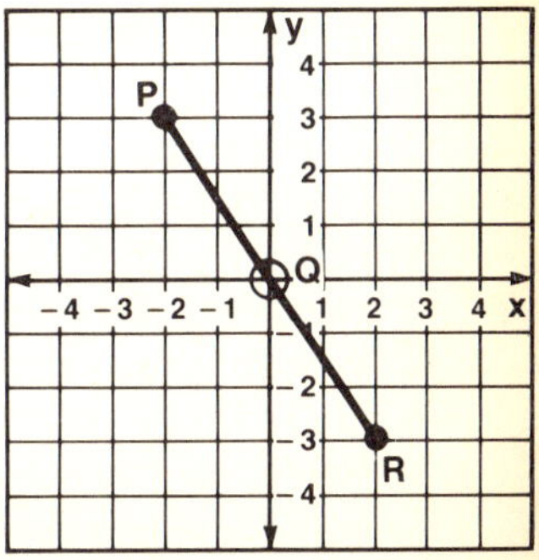

12.2 SIMPLE AREAS IN THE COORDINATE PLANE

Recall the following area formulas:

$$\text{Area}_{\text{rectangle}} = b \cdot h \qquad \text{Area}_{\text{square}} = s^2 \qquad \text{Area}_{\text{triangle}} = \tfrac{1}{2}b \cdot h$$

$$\text{Area}_{\text{parallelogram}} = b \cdot h \qquad \text{Area}_{\text{trapezoid}} = \tfrac{1}{2}h(b_1 + b_2)$$

If these figures are placed in the coordinate plane, their areas can be computed. The lengths of the parts of the figures needed for the area formulas can sometimes be determined by simply counting boxes on the graph, as in the following examples and exercises.

Examples

1. Find the area of the rectangle.

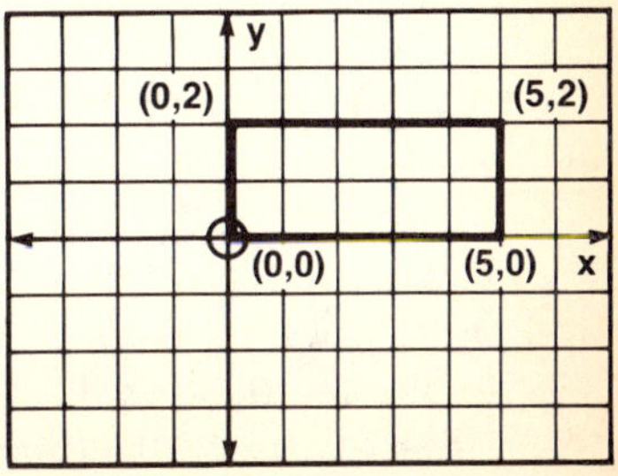

Solution:

The base of the rectangle is 5 units long, and the height is 2 units long.

$$\begin{aligned} A_{\text{rectangle}} &= b \cdot h \\ &= 5 \cdot 2 \\ &= 10 \text{ square units} \quad \textit{Ans.} \end{aligned}$$

2. Find the area of the triangle whose vertices are (0, 0), (5, 0), and (0, 8).

Solution:

Place the triangle on a graph.

The length of b is 5.
The length of h is 8.

$$\begin{aligned} A_{\text{triangle}} &= \tfrac{1}{2}b \cdot h \\ &= \tfrac{1}{2}(5) \cdot (8) \\ &= \tfrac{1}{2}(40) \\ &= 20 \text{ square units} \quad \textit{Ans.} \end{aligned}$$

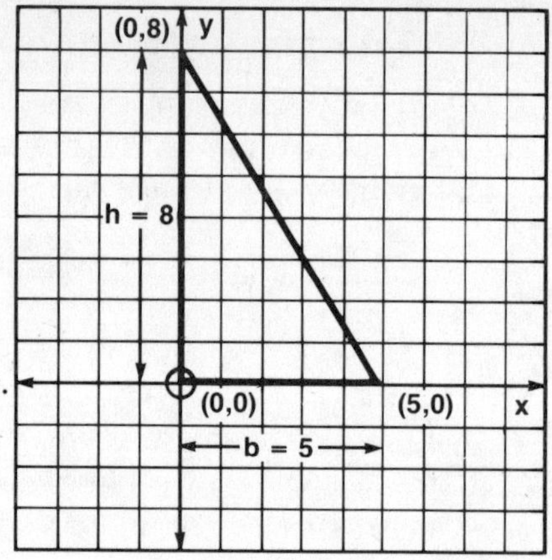

3. Find the area of the parallelogram whose vertices are (0, 0), (10, 0), (14, 5), and (4, 5).

Solution:
Place the parallelogram on a graph.

The length of b is 10.
The length of h is 5.

$$\begin{aligned} A_{\text{parallelogram}} &= b \cdot h \\ &= (10)(5) \\ &= 50 \end{aligned}$$

Ans. 50 square units

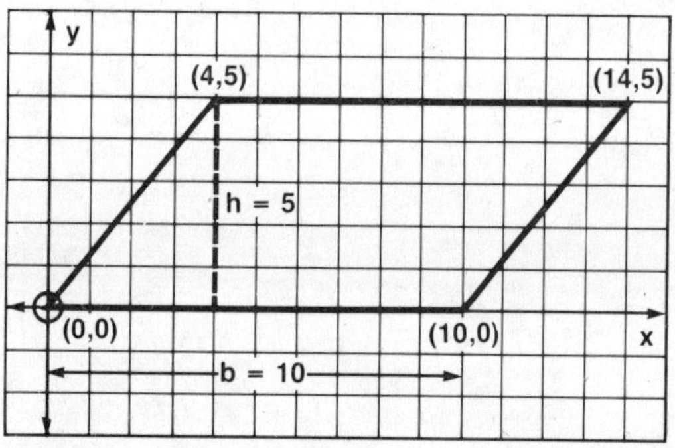

EXERCISES

In 1–12, find the area of the given figure whose vertices have the indicated coordinates.

1. rectangle: (0, 0), (3, 0), (0, 2), (3, 2)
2. rectangle: (0, 0), (3, 0), (0, 10), (3, 10)
3. rectangle: (−2, 0), (2, 0), (−2, 5), (2, 5)
4. square: (−5, 0), (5, 0), (−5, 5), (5, 5)
5. triangle: (0, 0), (8, 0), (2, 6)
6. triangle: (−6, 0), (6, 0), (0, 3)
7. triangle: (−2, 0), (7, 0), (0, 4)
8. parallelogram: (0, 0), (8, 0), (2, 3), (10, 3)
9. parallelogram: (−5, 0), (5, 0), (−2, 6), (8, 6)
10. trapezoid: (0, 0), (8, 0), (2, 5), (5, 5)
11. trapezoid: (−4, 0), (10, 0), (−2, 4), (3, 4)
12. trapezoid: (−2, 5), (3, 5), (−5, −7), (5, −7)

13. As shown in the diagram, the dimensions of each congruent rectangle are 4 by 3. What are the coordinates of point Q?

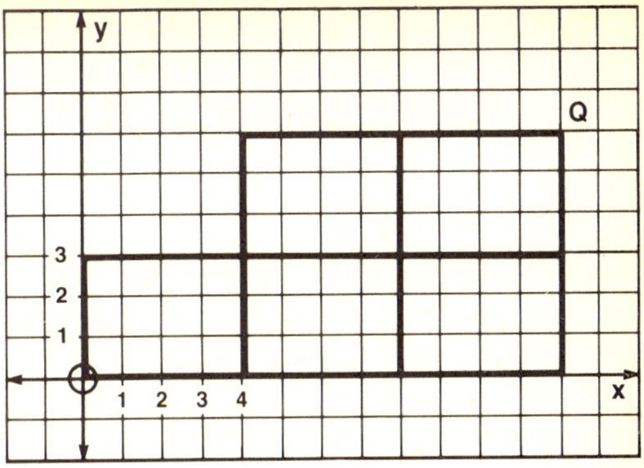

12.3 THE SLOPE OF A LINE

A line is an infinite set of points. When drawn in the coordinate plane, a straight line is one of the following:

horizontal, parallel to the x-axis
vertical, parallel to the y-axis
oblique, not parallel to either axis

Each oblique line has a slant, or ***slope,*** that is unique to it. The slope of an oblique line is defined as a ratio: $\frac{\text{the change in } y}{\text{the change in } x}$. These changes can be determined by working with two points on the line. If two points on a line are represented by (x_1, y_1) and (x_2, y_2), then

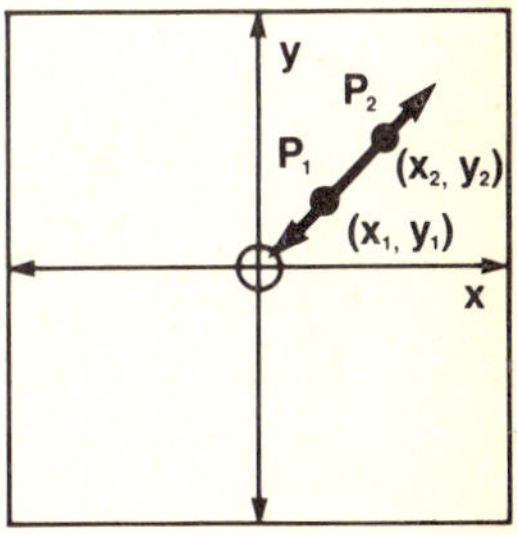

$$\text{slope} = \frac{y_2 - y_1}{x_2 - x_1}$$

Example

Determine the slope of the line joining the points (7, 4) and (−3, 2).

Solution:

Let $(x_2, y_2) = (7, 4)$
$(x_1, y_1) = (-3, 2)$

$$\text{slope} = \frac{y_2 - y_1}{x_2 - x_1}$$
$$= \frac{(4) - (2)}{(7) - (-3)} = \frac{2}{7 + 3}$$
$$= \frac{2}{10}, \text{ or } \frac{1}{5} \quad \textit{Ans.}$$

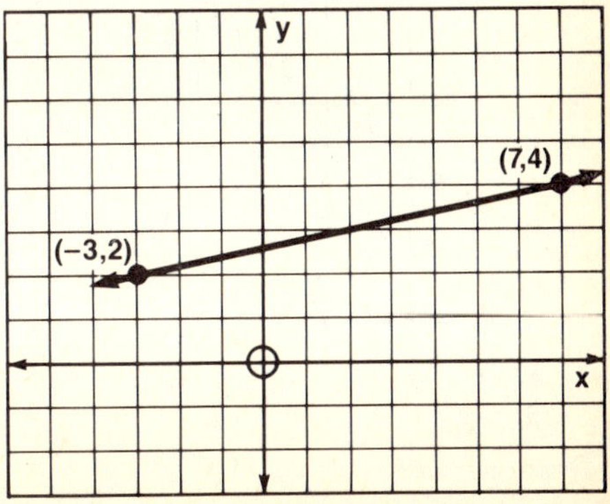

In the example, $\frac{1}{5}$ was found to be the slope of the line in question. This means several things, among them:

1. To move from one point to another on this line, we can go "up 1 and right 5." Using the following diagram, start at $A(-3, 2)$ and trace these directions to move to points B, C, and D.
2. Since $\frac{-1}{-5} = \frac{1}{5}$, we can also follow the directions "down 1 and left 5" to move from one point to another on this line. Return to $A(-3, 2)$ and trace these directions to move to points P and Q.
3. Any directions that are equivalent to the ratio $\frac{1}{5}$ will lead from one point to another point on this line. For example, since $\frac{2}{10} = \frac{1}{5}$, the directions to move from $A(-3, 2)$ directly to $C(7, 4)$ are go "up 2 and right 10." Also, since $\frac{4}{20} = \frac{1}{5}$, the directions to move from $Q(-13, 0)$ to $C(7, 4)$ are go "up 4 and right 20."
4. Any directions that are not equivalent to the ratio $\frac{1}{5}$ will lead off the line.

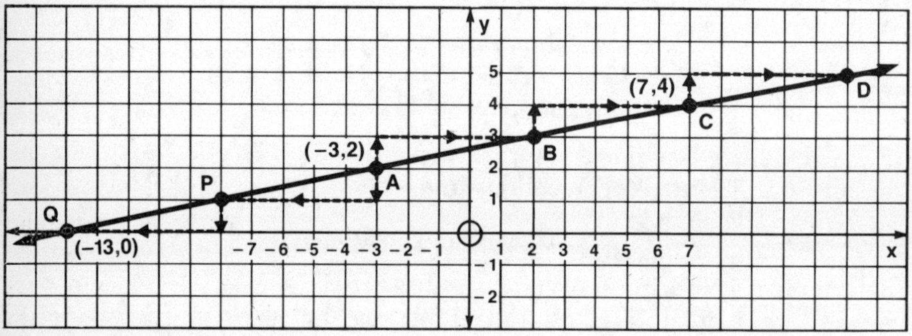

Consider the following lines and their slopes as determined from the given points:

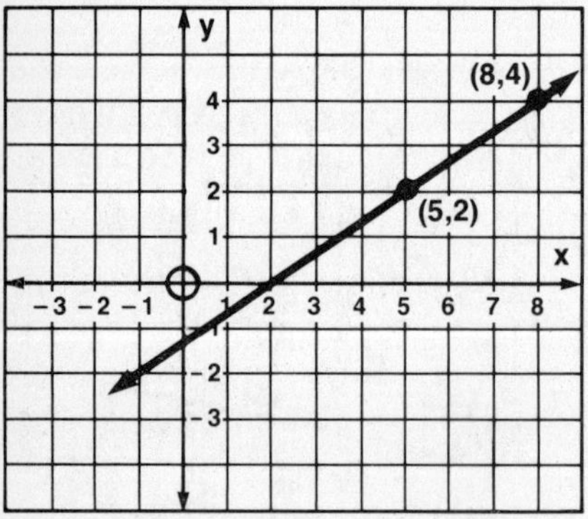

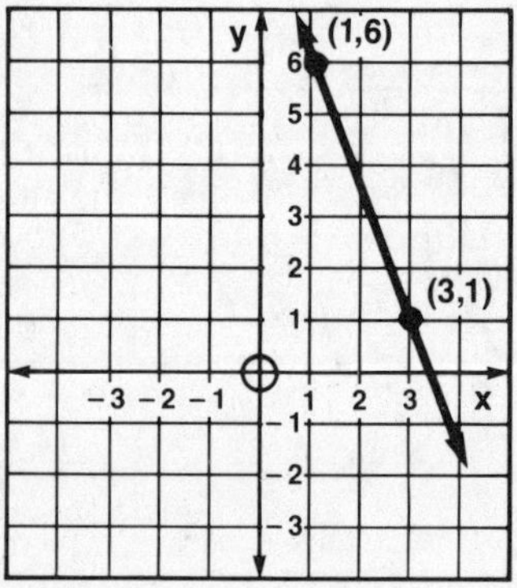

$$\text{slope} = \frac{y_2 - y_1}{x_2 - x_1} = \frac{(4) - (2)}{(8) - (5)} = \frac{2}{3}$$

$$\text{slope} = \frac{y_2 - y_1}{x_2 - x_1} = \frac{(6) - (1)}{(1) - (3)} = \frac{5}{-2} = -\frac{5}{2}$$

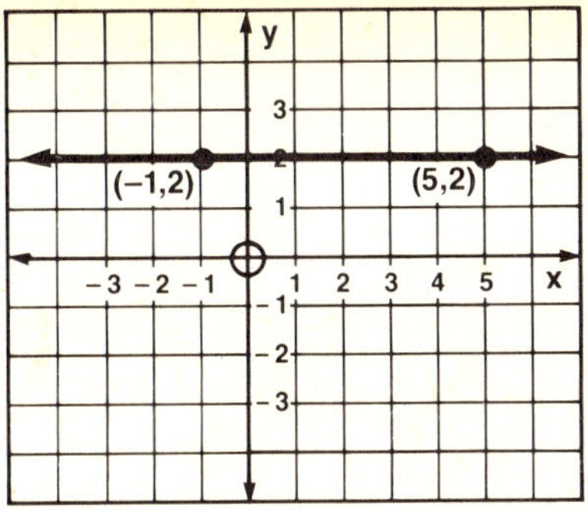

$$\text{slope} = \frac{y_2 - y_1}{x_2 - x_1} = \frac{(2) - (2)}{(5) - (-1)}$$
$$= \frac{0}{6} = 0$$

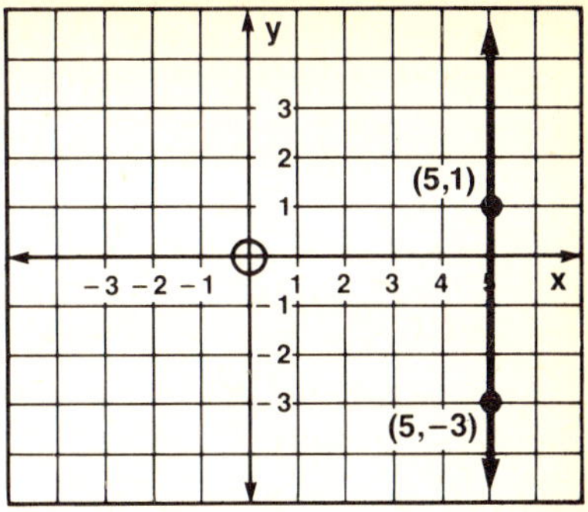

$$\text{slope} = \frac{y_2 - y_1}{x_2 - x_1} = \frac{(1) - (-3)}{(5) - (5)}$$
$$= \frac{4}{0}, \text{ undefined}$$

(We cannot divide by zero.)

◆ The following generalizations are true:

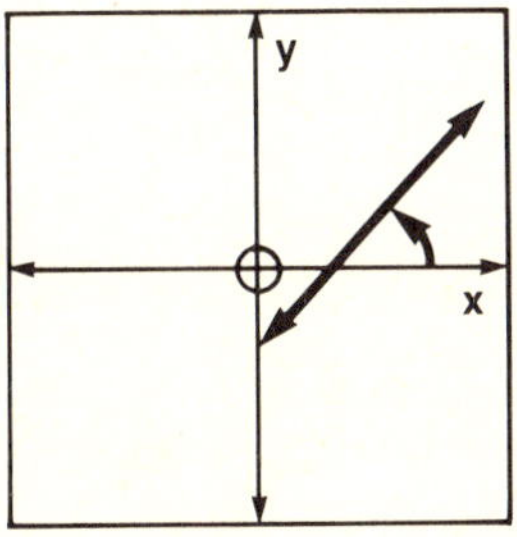

An oblique line making an acute angle with the x-axis has a positive slope.

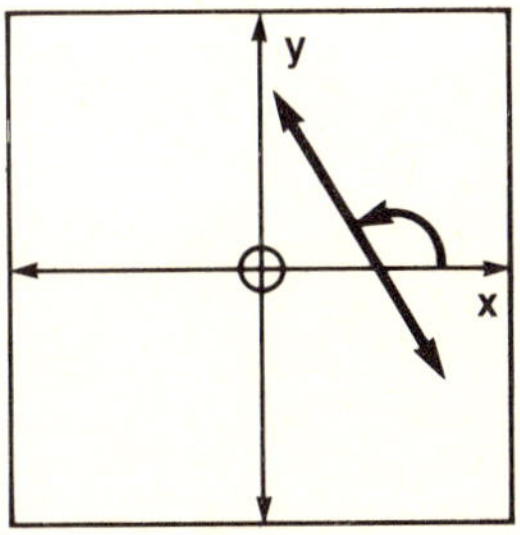

An oblique line making an obtuse angle with the x-axis has a negative slope.

(*Note:* We are referring always to the angle whose rotation is measured from the positive portion of the x-axis and going in a counterclockwise direction to the oblique line. This angle is called the ***angle of inclination.***)

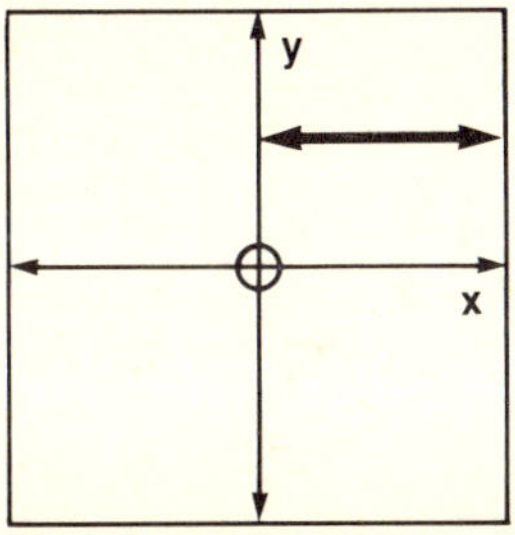

A horizontal line has a slope of 0.

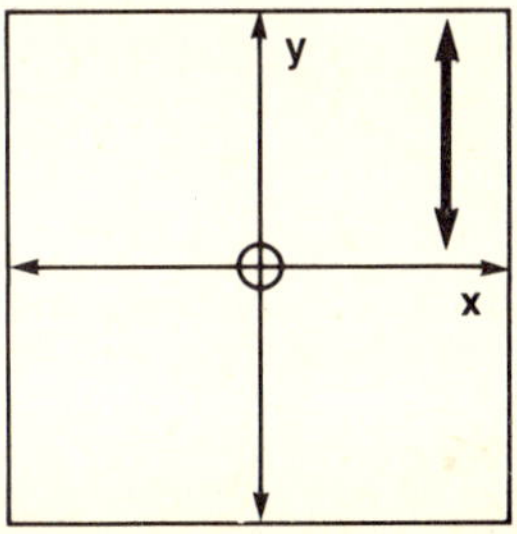

The slope of a vertical line is undefined.

Consider the following lines and their slopes as determined from the given points.

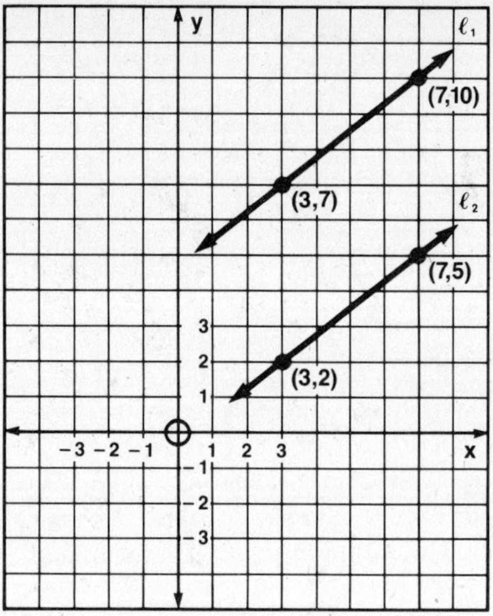

slope of $\ell_1 = \frac{(10) - (7)}{(7) - (3)} = \frac{3}{4}$

slope of $\ell_2 = \frac{(5) - (2)}{(7) - (3)} = \frac{3}{4}$

◆ The following two generalizations are true:

1. If two lines have equal slopes, then they are parallel.
2. If two lines are parallel, then they have equal slopes.

(*Note:* The two statements are equivalent to the biconditional: "two lines are parallel ↔ the slopes are equal.")

Examples

1. Determine by slopes if the line joining (−3, 2) and (5, 4) is parallel to the line joining (−2, 4) and (14, 8).

Solution:

The slope of the line joining (−3, 2) and (5, 4) is $\frac{(4) - (2)}{(5) - (-3)} = \frac{2}{8} = \frac{1}{4}$.

The slope of the line joining (−2, 4) and (14, 8) is $\frac{(8) - (4)}{(14) - (-2)} = \frac{4}{16} = \frac{1}{4}$.

Answer: Since the lines have equal slopes, they are parallel.

2. Find the value of k so that the slope of the line joining the points $(k, -3)$ and $(4, 2)$ is $\frac{1}{2}$.

Solution:

The slope of the line joining the points $(k, -3)$ and $(4, 2)$ is

$$\frac{2 - (-3)}{4 - k} = \frac{2 + 3}{4 - k} = \frac{5}{4 - k}.$$

We want this slope to have a value of $\frac{1}{2}$. That is:

$$\frac{5}{4-k} = \frac{1}{2}$$

$$1(4-k) = 2(5) \quad \text{Cross-multiply.}$$

$$4 - k = 10$$

$$-k = 6$$

$$k = -6$$

Answer: $k = -6$

Check: Is the slope of the line joining $(-6, -3)$ and $(4, 2)$ equal to $\frac{1}{2}$?

$$\frac{2-(-3)}{4-(-6)} \stackrel{?}{=} \frac{1}{2}$$

$$\frac{2+3}{4+6} \stackrel{?}{=} \frac{1}{2}$$

$$\frac{5}{10} \stackrel{?}{=} \frac{1}{2}$$

$$\frac{1}{2} = \frac{1}{2} \checkmark$$

3. Determine if the points $(0, -4)$, $(1, -2)$, and $(3, 2)$ lie on the same line (are ***collinear***).

Solution:

The three points are on the same line if and only if the slope between the first and the second is the same as the slope between the second and the third (or the first and the third).

Slope between $(0, -4)$ and $(1, -2)$: $\frac{-2-(-4)}{1-0} = \frac{-2+4}{1} = \frac{2}{1}$

Slope between $(1, -2)$ and $(3, 2)$: $\frac{2-(-2)}{3-1} = \frac{2+2}{2} = \frac{4}{2} = \frac{2}{1}$

Since the slopes are everywhere the same, $(0, -4)$, $(1, -2)$, and $(3, 2)$ are collinear.

Answer: $(0, -4)$, $(1, -2)$, and $(3, 2)$ are collinear.

EXERCISES

In 1–5, find the slope of the line joining each pair of points. Tell from the slope if the line is oblique, horizontal, or vertical. If the line is oblique, tell whether the angle of inclination is acute or obtuse.

1. $(5, 1)$ and $(6, -2)$
2. $(-3, 4)$ and $(-1, -3)$
3. $(4, 1)$ and $(4, 8)$
4. $(0, 3)$ and $(-2, 3)$
5. $(-8, -4)$ and $(-6, -3)$

In 6–10, tell whether the line joining the points P_1 and P_2 is parallel to the line joining the points Q_1 and Q_2.

6. $P_1(-1, -2)$, $P_2(5, 4)$ $Q_1(-1, 7)$, $Q_2(5, 5)$
7. $P_1(-1, -2)$, $P_2(4, 5)$ $Q_1(1, 6)$, $Q_2(-4, -1)$
8. $P_1(-3, 5)$, $P_2(5, 5)$ $Q_1(-3, -3)$, $Q_2(3, 3)$
9. $P_1(-2, 4)$, $P_2(4, 4)$ $Q_1(-2, -2)$, $Q_2(4, -2)$
10. $P_1(-1, -3)$, $P_2(-1, 4)$ $Q_1(2, 2)$, $Q_2(2, -5)$

11. Parallelogram $ABCD$ is shown in the accompanying figure. Which must be true?
(1) slope of $\overline{AB}$ = slope of $\overline{BC}$
(2) slope of $\overline{AB}$ = slope of $\overline{DC}$
(3) slope of $\overline{AB}$ = slope of $\overline{AD}$
(4) slope of $\overline{DC}$ = slope of $\overline{AD}$

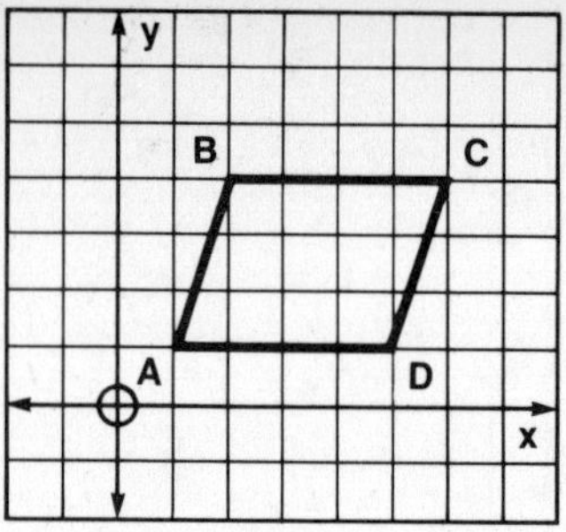

12. Find the value of k so that the slope of the line passing through the points $(2, k)$ and $(6, 8)$ is 1.
13. Find the value of k so that the line joining the points $(-2, 7)$ and $(3, k)$ is horizontal.
14. Find the value of k so that the line joining the points $(1, -3)$ and $(k, 5)$ is vertical.
15. If the line joining $P_1(3, 5)$ and $P_2(1, 2)$ is parallel to the line joining $Q_1(7, k)$ and $Q_2(5, 6)$, find the value of k.

In 16–20, determine whether the three points are collinear (lie on the same line).

16. $(2, 5), (6, 15), (4, 9)$
17. $(-1, 5), (0, 2), (1, -1)$
18. $(-7, -1), (-1, -7), (7, 1)$
19. $(2, -1), (2, 6), (2, 0)$
20. $(3, -4), (1, -4), (-1, 4)$

12.4 EQUATIONS OF LINES; INTERCEPTS

As numbers are used to name points in the rectangular coordinate system, equations are used to name lines. An equation of a line is called a ***linear equation.***

Consider the following lines and the equations by which they are represented. These diagrams are called ***graphs*** of the lines and of the equations.

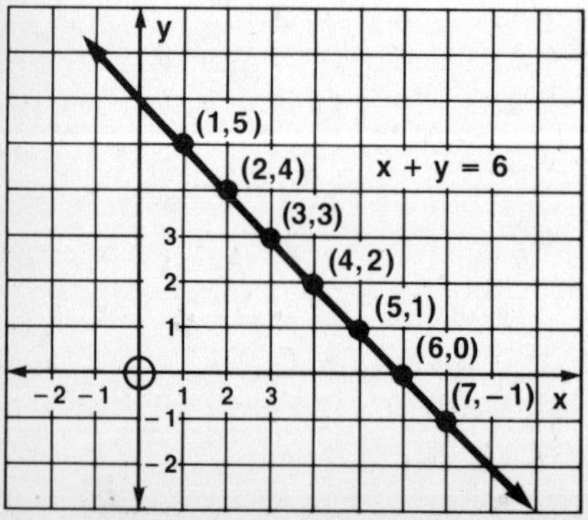

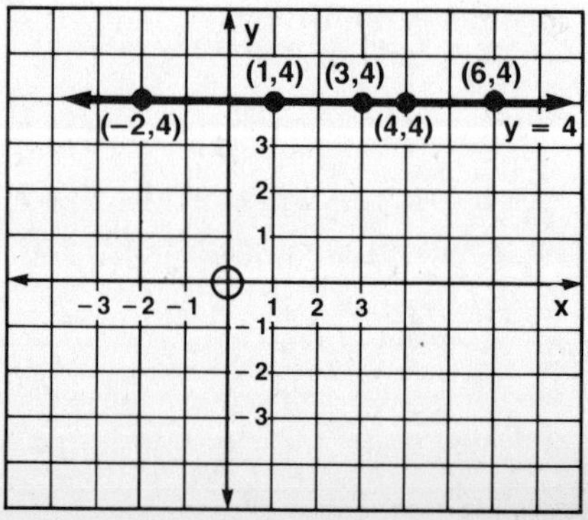

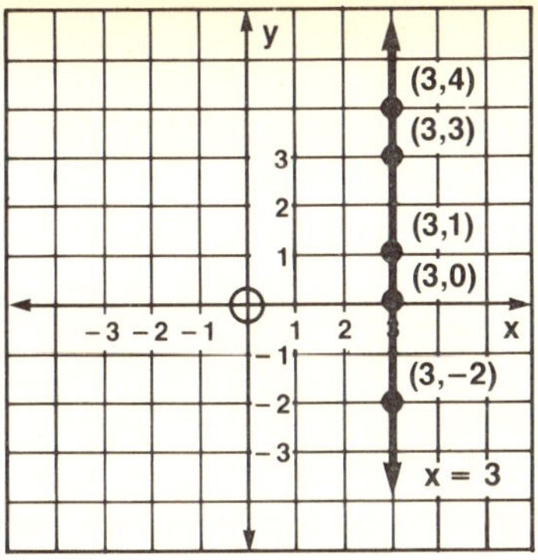

In each case, the equation tells the most important thing about the coordinates of *all* the points on the line:

1. The first case ($x + y = 6$) says that the sum of the abscissa and the ordinate is always 6.
2. The second case ($y = 4$) says that the ordinate of every point on the line is 4.
3. The third case ($x = 3$) says that the abscissa of every point on the line is 3.

◆ In general:

1. The equation of an oblique line relates the abscissa to the ordinate of every point on the line and thus contains both x and y.
2. The equation of a horizontal line tells the value of the ordinate of every point on the line and thus contains only y.
3. The equation of a vertical line tells the value of the abscissa of every point on the line and thus contains only x.

◆ One general equation for all three kinds of lines is therefore:

$$Ax + By = C$$

1. When $A \neq 0$ and $B \neq 0$, the line is oblique (example: $3x + y = 4$).
2. When $A = 0$ and $B \neq 0$, the line is horizontal (example: $2y = 6$, or $y = 3$).
3. When $A \neq 0$ and $B = 0$, the line is vertical (example: $3x = 6$, or $x = 2$).

◆ On a graph:

1. A horizontal line will always intersect only the y-axis. (*Note:* The equation of the x-axis is $y = 0$.)
2. A vertical line will always intersect only the x-axis. (*Note:* The equation of the y-axis is $x = 0$.)
3. An oblique line (extended, if necessary) will always intersect both axes. (*Note:* If an oblique line goes through the origin, it is at this one point that the line intersects both axes.)

The abscissa of the point at which a line intersects the x-axis is called the ***x-intercept,*** and the ordinate of that point is 0. The ordinate of the point at which a line intersects the y-axis is called the ***y-intercept,*** and the abscissa of that point is 0.

Examples

1. Which is true of the graph of $x = -1$?

(1) It has a slope of -1.
(2) It has a y-intercept of -1.
(3) It has an x-intercept of -1.
(4) It is parallel to the x-axis.

Solution:

The equation $x = -1$ tells that the line is a vertical line. Thus, the line $x = -1$ is parallel to the y-axis, its slope is undefined, and it intersects only the x-axis at -1 (that is, the x-intercept is -1).

Answer: (3)

2. Find the x-intercept of the line $3x + y = 6$.

Solution:

The x-intercept comes from the point whose coordinates are $(x, 0)$. Therefore, let $y = 0$, and solve for x.

$$\begin{aligned} 3x + y &= 6 \\ 3x + 0 &= 6 \\ 3x &= 6 \\ x &= 2 \end{aligned}$$

Answer: The x-intercept of the oblique line $3x + y = 6$ is 2.

3. Find the y-intercept of the line $3x + y = 6$.

Solution:

The y-intercept comes from the point whose coordinates are $(0, y)$. Therefore, let $x = 0$, and solve for y.

$$\begin{aligned} 3x + y &= 6 \\ 3(0) + y &= 6 \\ 0 + y &= 6 \\ y &= 6 \end{aligned}$$

Answer: The y-intercept of the oblique line $3x + y = 6$ is 6.

4. An equation of a line is $y = -4$. Find the x- and y-intercepts.

Solution:

$y = -4$ is an equation of a horizontal line intersecting the y-axis at -4 and never intersecting the x-axis.

Answer: The y-intercept of the line $y = -4$ is -4. There is no x-intercept.

Since there are equivalent forms for every equation, the equation of a line can be written in different forms. A variety of forms is useful because of the information available in each form.

Consider the line shown in the diagram on the following page. The equation of the line is $y = \frac{1}{2}x + 2$. (*Note:* This equation is equivalent to $x - 2y = -4$, which is of the form $Ax + By = C$.)

From the diagram, we can see that the slope is $\frac{1}{2}$ and that the y-intercept is 2. This information is reflected in the equation. The slope, $\frac{1}{2}$, is the coefficient of x, and the y-intercept, 2, is the constant term.

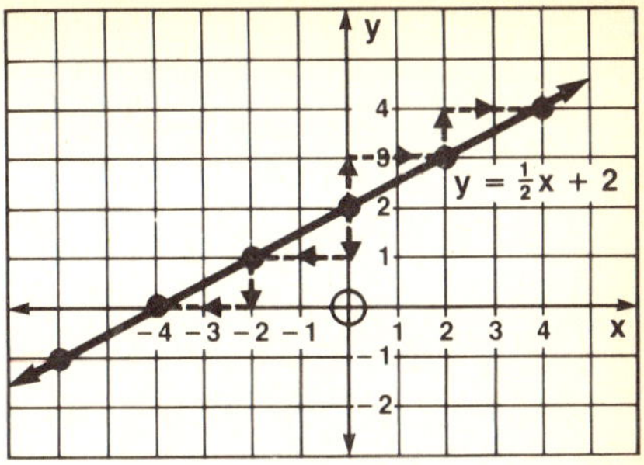

◆ In general, the equation of a line in the ***slope-intercept form*** (that is, y-intercept) is $\boldsymbol{y = mx + b}$ where m = slope and b = y-intercept.

Examples

1. Find the slope and the y-intercept of the line whose equation is $2y = 3x - 4$.

Solution:

Write the given equation.	$2y = 3x - 4$
Transform the equation into slope-intercept form. In this case, divide by 2.	$y = \frac{3}{2}x - 2$
	↕ ↕
Read the slope and y-intercept by matching the transformed equation to $y = mx + b$.	$y = mx + b$
	$m = \frac{3}{2}, b = -2$

Answer: slope $(m) = \frac{3}{2}$, y-intercept $(b) = -2$

2. Write an equation for a line whose slope is -3 and whose y-intercept is 7.

Solution:

Given: slope $= -3$, y-intercept $= 7$

The slope-intercept form of a line is:	$y = mx + b$
Substitute -3 for m and 7 for b.	$y = (-3)x + (7)$
	$y = -3x + 7$ *Ans.*

3. Is the line whose equation is $2y = x + 8$ parallel to the line whose equation is $2y = x - 6$?

Solution:

Two lines are parallel if their slopes are equal. The slope of the given lines can be determined when the lines are in $y = mx + b$ form.

$2y = x + 8$	$2y = x - 6$
$y = \frac{1}{2}x + 4$	$y = \frac{1}{2}x - 3$
$y = mx + b$	$y = mx + b$
$m = \frac{1}{2}$	$m = \frac{1}{2}$

Since the slopes of the lines are equal, the lines are parallel. *Ans.*

4. The graph of $y = 6x - 1$ is parallel to the graph of
(1) $y = x - 6$ (2) $y = 6x + 1$ (3) $y = -6x + 4$ (4) $y = 6$

Solution:

Two lines are parallel if their slopes are equal. The slope of the given line, $y = 6x - 1$, is 6, since when matched against $y = mx + b$, $m = 6$.

Examination of the choices eliminates (1) and (3), since they are in $y = mx + b$ form, and neither has a slope (m) of 6. Choice (4) may be tricky; but note that the 6 in this choice is the constant term, not the coefficient of x. Only the line in choice (2) has a slope (m) of 6, the same as the slope of the given line.

Answer: (2)

◆ The equation of a line gives information about every point on the line.

Thus:

1. If a point is on a line, its coordinates satisfy the equation of the line.

AND

2. If the coordinates of a point satisfy the equation of a line, then the point is on the line.

(*Note:* The two statements are equivalent to the biconditional: "a point is on a line ↔ the coordinates of the point satisfy the equation of the line.")

Examples

1. Given that the point $(2, k)$ is on the line $3x - 4y = 7$. Find the value of k.

Solution:

Since the point is on the line, its coordinates must satisfy the equation of the line. Substitute the x-value of the point, 2, and the y-value of the point, k, for x and y in the equation, and solve for k.

$$\begin{aligned} 3x - 4y &= 7 \\ 3(2) - 4(k) &= 7 \\ 6 - 4k &= 7 \\ -4k &= 1 \\ k &= -\tfrac{1}{4} \quad \textit{Ans.} \end{aligned}$$

2. Determine if the point $(-3, -1)$ is on the graph of the line $x - 3y = 0$.

Solution:

If the point is on the line, its coordinates satisfy the equation. Substitute the x-value of the point, -3, and the y-value of the point, -1, for x and y in the equation and see if the equation balances.

$$\begin{aligned} x - 3y &= 0 \\ (-3) - 3(-1) &\stackrel{?}{=} 0 \\ -3 + 3 &\stackrel{?}{=} 0 \\ 0 &= 0 \checkmark \end{aligned}$$

Since the coordinates of the point satisfy the equation, the point is on the graph of the line. *Ans.*

3. Which ordered pair is *not* a member of the solution set of the equation $3x - 2y = 4$?

(1) $(3, 2\frac{1}{2})$ (2) $(-2, -5)$ (3) $(4, 4)$ (4) $(-2, 0)$

Solution:

If the coordinates of a point satisfy the equation of a line, the point is on the line. To find the ordered pair that is *not* a member of the solution set, test the given ordered pairs in the equation:

$3x - 2y = 4$
test $(3, 2\frac{1}{2})$
$3(3) - 2(2\frac{1}{2}) \stackrel{?}{=} 4$
$9 - 5 \stackrel{?}{=} 4$
$4 = 4\checkmark$

Thus, $(3, 2\frac{1}{2})$ is on the line.

$3x - 2y = 4$
test $(-2, -5)$
$3(-2) - 2(-5) \stackrel{?}{=} 4$
$-6 + 10 \stackrel{?}{=} 4$
$4 = 4\checkmark$

Thus, $(-2, -5)$ is on the line.

$3x - 2y = 4$
test $(4, 4)$
$3(4) - 2(4) \stackrel{?}{=} 4$
$12 - 8 \stackrel{?}{=} 4$
$4 = 4\checkmark$

Thus, $(4, 4)$ is on the line.

$3x - 2y = 4$
test $(-2, 0)$
$3(-2) - 2(0) \stackrel{?}{=} 4$
$-6 - 0 \stackrel{?}{=} 4$
$-6 \neq 4$

Thus, $(-2, 0)$ is *not* on the line.

Answer: (4)

EXERCISES

1. Which is true of the graph of the equation $y = -2$?
(1) It has a slope of -2. (3) It passes through the point $(0, 0)$.
(2) It is parallel to the x-axis. (4) It is parallel to the y-axis.

2. An equation whose graph has the same slope as the graph of the equation $y = 5$ is
(1) $y = -5$ (2) $y = 5x$ (3) $y = x + 5$ (4) $y = x - 5$

3. The graph of the equation $y = 2$ is a line
(1) parallel to the y-axis (3) passing through the origin
(2) parallel to the x-axis (4) passing through the point $(2, 0)$

4. Which is true of the graph of $x = -3$?
(1) It has a slope of -3. (3) It is parallel to the x-axis.
(2) It passes through the origin. (4) It is parallel to the y-axis.

5. The slope of the graph shown is:
(1) 1
(2) 2
(3) 0
(4) undefined

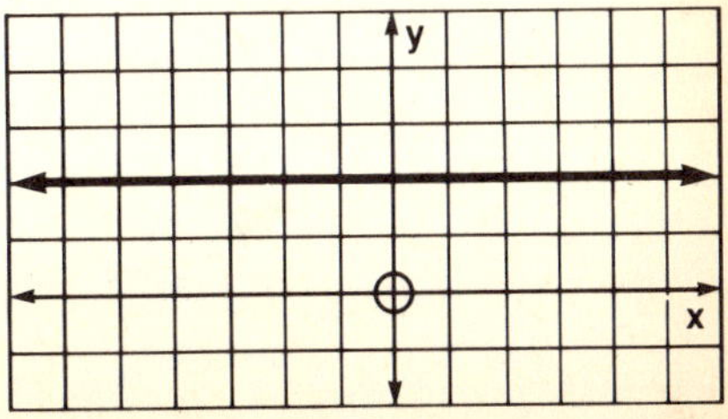

6. The graph of which equation is shown in the accompanying diagram?
(1) $x = 5$ (3) $y = 5x$
(2) $y = 5$ (4) $y = x + 5$

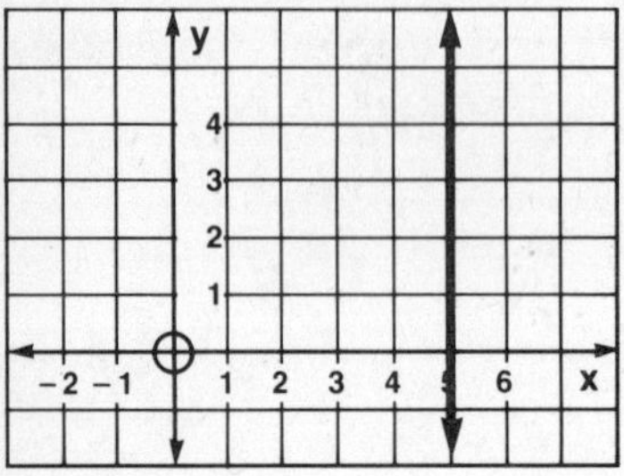

7. In the diagram, the vertical line that intersects the x-axis at -2 is the graph of which equation?
(1) $x = -2$ (3) $y = x - 2$
(2) $y = -2$ (4) $x + y = -2$

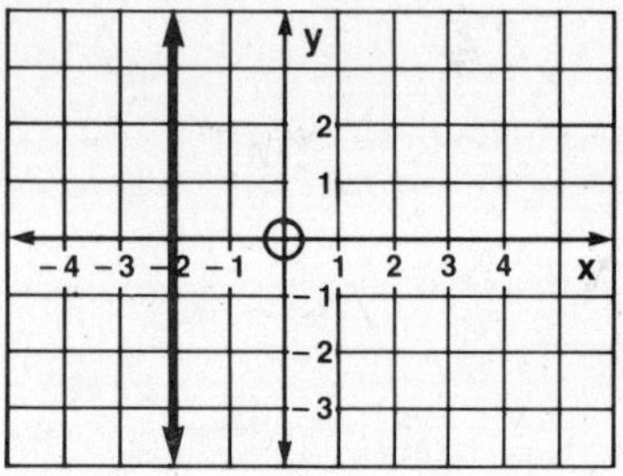

8. The equation of the graph shown is:
(1) $x = 2$
(2) $y = 2$
(3) $x = -2$
(4) $y = -2$

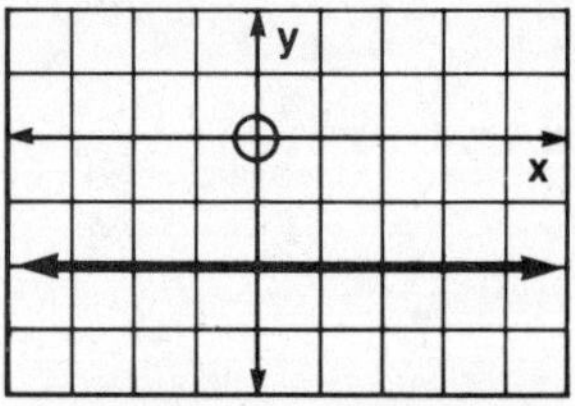

9. Which is an equation of the straight line parallel to the x-axis and 1 unit above it? (1) $y = 1$ (2) $y = -1$ (3) $x = 1$ (4) $x = -1$
10. Which is an equation of the straight line parallel to the y-axis and 1 unit to the left of it?
(1) $y = 1$ (2) $y = -1$ (3) $x = 1$ (4) $x = -1$
11. The graph of the equation $3x + 2y = 12$ intersects the y-axis at the point whose coordinates are
(1) (0, 6) (2) (6, 0) (3) (0, 4) (4) (4, 0)
12. The graph of $3x - y = 3$ intersects the x-axis at the point
(1) (1, 0) (2) (0, 1) (3) (0, 3) (4) (3, 0)
13. The graph of which equation has a slope of 3 and a y-intercept of -2?
(1) $y = 3x - 2$ (3) $y = 2x - 3$
(2) $y = 3x + 2$ (4) $y = 2x + 3$
14. What is the slope of the graph of $y = 2x + 3$?
15. What is the slope of the line whose equation is $y = 3x + 6$?
(1) $\frac{1}{2}$ (2) 2 (3) 3 (4) 6
16. What is the slope of the line whose equation is $4y = 3x + 16$?
17. The y-intercept of the graph of $y = \frac{2}{3}x + \frac{8}{3}$ is
(1) $\frac{3}{2}$ (2) 8 (3) $\frac{2}{3}$ (4) $\frac{8}{3}$
18. The y-intercept of line $y = 3x - 4$ is
(1) -3 (2) -4 (3) 3 (4) 4
19. If the y-intercept of a line is 4, which may be an equation for the line?
(1) $y = x - 4$ (2) $y = \frac{1}{2}x + 4$ (3) $4y = x$ (4) $y = 4x + 3$
20. Write an equation for a line whose slope is $\frac{1}{4}$ and whose y-intercept is 7.

21. If the slope of a line is -3 and the y-intercept is 2, which may be the equation of that line?
(1) $y - 3x + 2 = 0$ (3) $y = 2x - 3$
(2) $y + 3x = 2$ (4) $y = 3x - 2$
22. If the point $(k, 4)$ lies on the graph of $2x + y = 10$, what is the value of k?
23. A point on the graph of $y = 2x - 4$ has an x-coordinate of 3. Find the y-coordinate of this point.
24. If the point $(k, 3)$ lies on the graph of $x + 3y = 15$, what is the value of k? (1) 6 (2) 5 (3) 3 (4) 4
25. If $(r, 3)$ is in the solution set of $2x + y = 7$, then the value of r must be
(1) 1 (2) 2 (3) 3 (4) 4
26. If the point $(-1, k)$ lies on the line $2x - y = 7$, what is the value of k?
(1) 9 (2) -9 (3) 5 (4) -5
27. The graph of $3x - y = 6$ contains the point
(1) $(0, 2)$ (2) $(5, 9)$ (3) $(4, -6)$ (4) $(1, 9)$
28. Which point lies on the graph of $2x + y = 10$?
(1) $(0, 8)$ (2) $(10, 0)$ (3) $(3, 4)$ (4) $(4, 3)$
29. A point on the graph of $x + 3y = 13$ is
(1) $(4, 4)$ (2) $(-2, 3)$ (3) $(-5, 6)$ (4) $(4, -3)$
30. The graph of which equation passes through the point $(0, 0)$?
(1) $x = 3$ (2) $y = 3$ (3) $y = 3x + 3$ (4) $y = 3x$
31. Which pair of numbers represents a point that does *not* lie on the graph of $2x + 3y = 6$?
(1) $(0, 2)$ (2) $(2, 3)$ (3) $(3, 0)$ (4) $(6, -2)$
32. Which pair of numbers represents a point that does *not* lie on the graph of $2x + 3y = 6$?
(1) $(0, 2)$ (2) $(2, 3)$ (3) $(3, 0)$ (4) $(6, -2)$
33. Which ordered pair is *not* in the solution set of $x = 7 - y$?
(1) $(8, -1)$ (2) $(7, 0)$ (3) $(-7, 0)$ (4) $(0, 7)$
34. The graph of $y = 3x - 4$ is parallel to the graph of
(1) $y = 4x - 3$ (3) $y = -3x + 4$
(2) $y = 3x + 4$ (4) $y = 3$
35. Is the line $y + 4x = 7$ parallel to the line $y = 4x + 5$?
36. Is the line $2y = 6x + 5$ parallel to the line $y - 3x = 7$?
37. Write an equation of a line that is parallel to the line $2y = 6x - 1$ and has a y-intercept of -3.

38. The graph shown has a slope of:
(1) 1
(2) -1
(3) -4
(4) 4

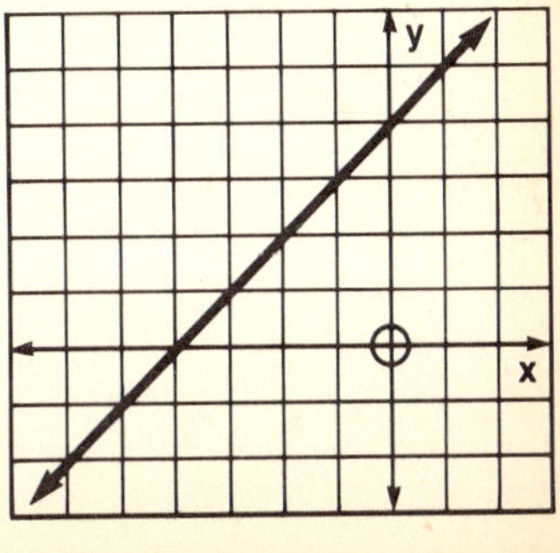

39. The graph shown has a slope of:
(1) 1 (2) -1 (3) 0 (4) $\frac{1}{0}$

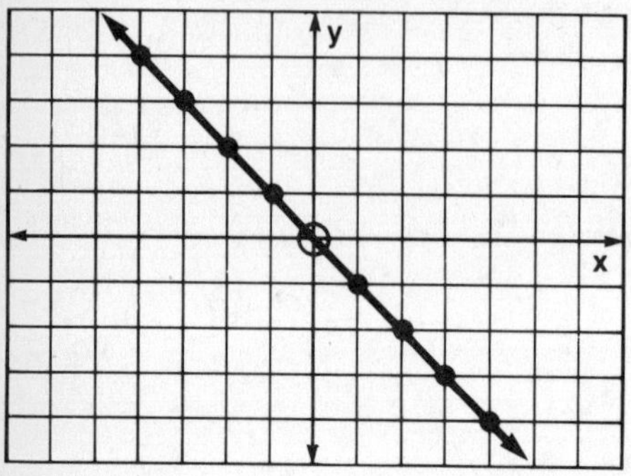

40. The lines shown both have a slope of:
(1) $\frac{1}{2}$ (2) $-\frac{1}{2}$ (3) 2 (4) -2

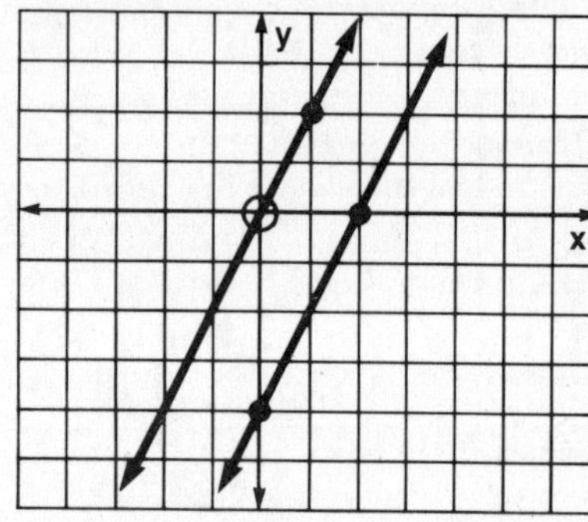

12.5 GRAPHING A LINEAR EQUATION

To draw the graph of a horizontal or vertical line is simple. We need to know only the y-intercept or the x-intercept.

Example

Draw the graph of $x = 3$.

Solution:

The graph of $x = 3$ is a vertical line intersecting the x-axis at 3, as shown in the diagram. (*Note:* All points on this line have an abscissa of 3. For example: (3, 0), (3, 1), (3, 2), (3, -1), . . .)

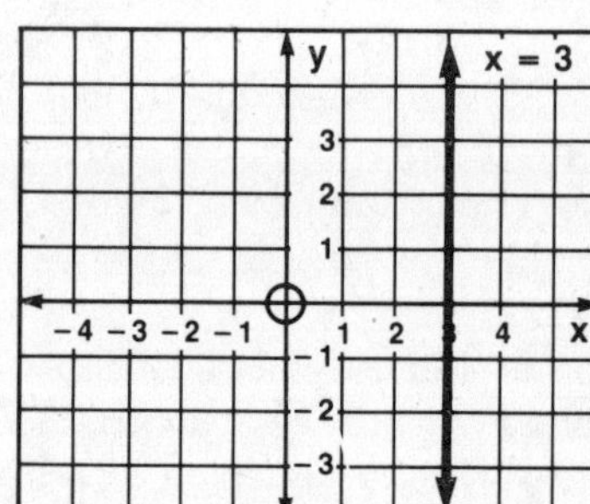

To draw the graph of an oblique line is somewhat more complicated. For example, let us graph the oblique line $y = 3x - 3$. Three methods are shown here.

Method 1: Choosing Points

Prepare a table that includes three or four x-values and their corresponding y-values. The y-values are calculated by substituting each x-value in the given equation.

x	$y = 3x - 3$	y
-2	$y = 3(-2) - 3$ $= -6 - 3$	-9
0	$y = 3(0) - 3$ $= 0 - 3$	-3
2	$y = 3(2) - 3$ $= 6 - 3$	3
3	$y = 3(3) - 3$ $= 9 - 3$	6

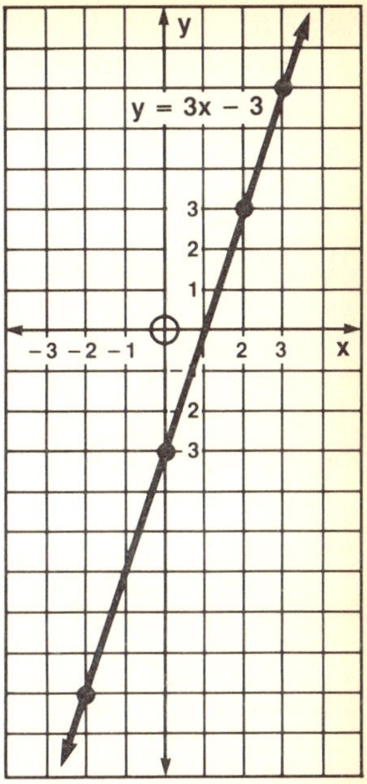

Plot the points $(-2, -9)$, $(0, -3)$, $(2, 3)$, and $(3, 6)$; and connect them with a straight line.

Note 1: Be sure to extend the line beyond the plotted points and to put arrowheads on it in order to indicate that the line is infinitely long and that the set of ordered pairs that satisfies its equation has an infinite number of elements.

Note 2: Be sure to write the equation of the line near the line.

Note 3: Be sure to choose a checkpoint to verify that the line is correct. In the example, a convenient checkpoint is (1, 0). That is, we see from the graph that (1, 0) is on the line. Substitute (1, 0) into the equation of the line and see if the equation balances.

$$(1, 0)$$
$$\begin{aligned} y &= 3x - 3 \\ 0 &\overset{?}{=} 3(1) - 3 \\ 0 &\overset{?}{=} 3 - 3 \\ 0 &= 0\checkmark \end{aligned}$$

Method 2: Slope-intercept

Determine the slope and the y-intercept of the line by matching the given equation against the slope-intercept form:

$$\begin{aligned} y &= mx + b \\ y &= 3x - 3 \end{aligned}$$

Thus, slope $(m) = 3$ and y-intercept $(b) = -3$.

Write the slope in fractional form, using equivalent negatives as well. $3 = \frac{3}{1} = \frac{+3}{+1} = \frac{-3}{-1}$. These fractions tell directions: "up 3 and right 1" or "down 3 and left 1."

From the y-intercept, plot the point $(0, -3)$.

Beginning at the point (0, −3), follow one set of directions read from the slope: "up 3 and right 1." Mark off points each time the directions are completed. Do this three or four times.

Return to the point (0, −3), and follow the alternate set of directions read from the slope: "down 3 and left 1." Mark off points each time the directions are completed. Do this three or four times.

Connect all the points with a straight line. (*Note:* If one or more of these points are not on the same line, you have made an error.) Write the equation of the line on the graph, just near the line. Be sure to include arrowheads.

Choose a checkpoint to verify that the line is correct. Read the coordinates of a point that is on the line, and substitute those coordinates into the equation of the line. Choose, for example, (2, 3).

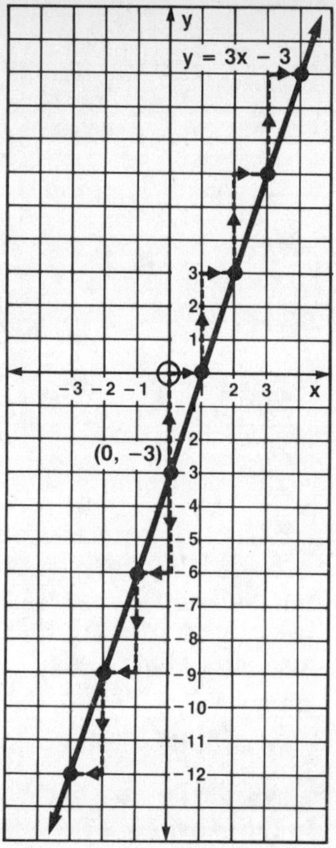

$$\begin{aligned} &(2, 3) \\ y &= 3x - 3 \\ 3 &\stackrel{?}{=} 3(2) - 3 \\ 3 &\stackrel{?}{=} 6 - 3 \\ 3 &= 3\checkmark \end{aligned}$$

Method 3: Intercepts

Only two points are needed to draw a line. Since every oblique line must cross both axes, the needed two points can be the intercepts.

We can determine the y-intercept of the line by matching the equation against the slope-intercept form:

$$\begin{aligned} y &= mx + b \\ y &= 3x - 3 \end{aligned}$$

Thus, the y-intercept $(b) = -3$.

The x-intercept is found by letting $y = 0$ and solving for x:

$$\begin{aligned} y &= 3x - 3 \\ 0 &= 3x - 3 \\ 3 &= 3x \\ 1 &= x \end{aligned}$$

(*Note:* We could also have found the y-intercept by letting $x = 0$ and solving for y.)

Using the y-intercept, plot the point (0, −3). Using the x-intercept, plot the point (1, 0). Connect these two points with a straight line. (*Note:* It is a good idea to plot a third point to insure accuracy.)

Check by choosing a point on the graph that is on the line. Substitute the coordinates of the checkpoint into the equation of the line. Choose (2, 3).

$$y = 3x - 3$$
$$3 \stackrel{?}{=} 3(2) - 3$$
$$3 \stackrel{?}{=} 6 \quad - 3$$
$$3 = 3\checkmark$$

Write the equation of the line on the graph. Include arrowheads.

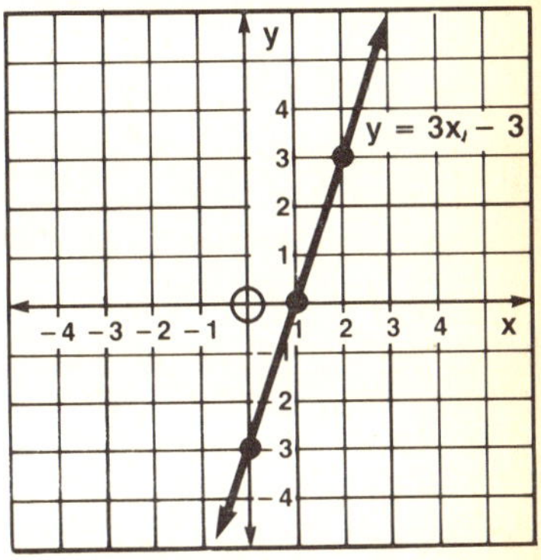

For many equations, all three methods are not as easily used as they were in the given example. Sometimes inconvenient fractions result when we try a particular method. Sometimes a line passes through the origin and, therefore, the x-intercept and y-intercept give the same point. We must exercise judgment to determine the easiest method for graphing a particular line.

Example

Draw the graph of $3x - 4y = 8$.

Solution:

In the given form of this equation, no method is as easily used as in the previous example. Since the intercepts method leads to an inconvenient fraction for the x-intercept, we will eliminate that method.

Perhaps we can use the slope-intercept method. Let us first solve the equation for y in order to put it in the slope-intercept form:

$$3x - 4y = 8$$
$$-4y = -3x + 8 \qquad \text{Subtract } 3x.$$
$$y = \frac{-3}{-4}x + \frac{8}{-4} \qquad \text{Divide by } -4.$$
$$y = \tfrac{3}{4}x - 2 \qquad \text{Adjust signs.}$$

Now we can see that the slope-intercept method is easily applied since $m = \frac{3}{4}$ and $b = -2$.

Write the slope with equivalent negatives: $\frac{3}{4} = \frac{+3}{+4} = \frac{-3}{-4}$. The directions are: "up 3 and right 4," or "down 3 and left 4."

Using the y-intercept, begin at $(0, -2)$ and follow each set of directions from the slope. Complete the graph as before. Choose a checkpoint. Write the original equation of the line on the graph. Include arrowheads.

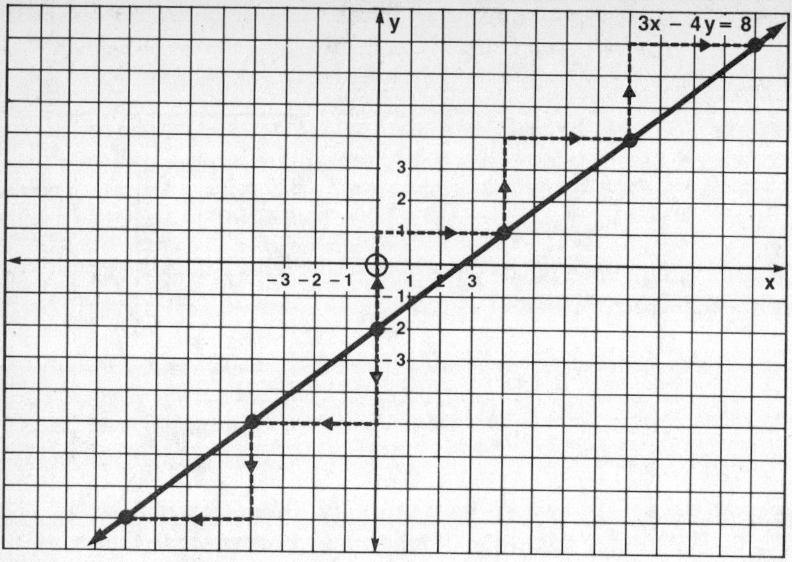

The method of choosing points can be used for this equation without first solving for y. If we do not first solve for y, we eliminate the possibility of making an error in that operation. However, we will have to do a great deal of extra arithmetic, and we will have difficulty finding appropriate x-values (those not leading to fractional values of y). Since we have already solved for y in order to apply the slope-intercept method, we know that $y = \frac{3}{4}x - 2$. Thus, to avoid fractional values of y, choose values of x that are divisible by 4.

x	$y = \frac{3}{4}x - 2$	y
−8	$y = \frac{3}{4}(-8) - 2$ $= -6 \quad - 2$	−8
0	$y = \frac{3}{4}(0) - 2$ $= 0 \quad - 2$	−2
8	$y = \frac{3}{4}(8) - 2$ $= 6 \quad - 2$	4

An appropriate table is shown here.

Plot the points $(-8, -8)$, $(0, -2)$, and $(8, 4)$. Complete the graph. Choose a checkpoint.

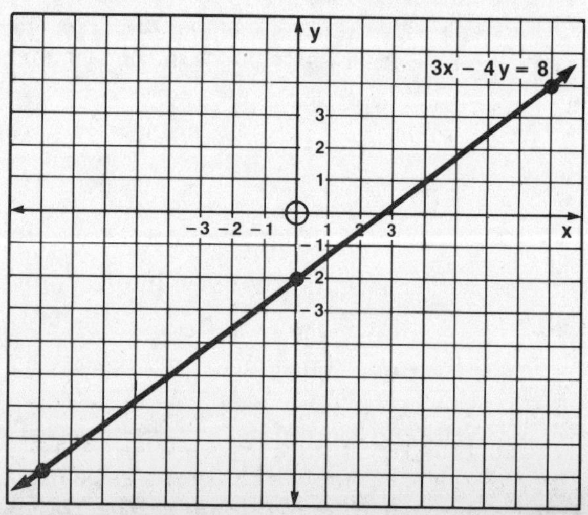

EXERCISES

In 1–5, draw the graph of each line by choosing points.

1. $y = 3x + 1$
2. $2y = 5x + 3$
3. $2y + 3x = 6$
4. $6x - 3y = 3$
5. $2x - 3y = 10$

In 6–10, draw the graph of each line by the slope-intercept method.

6. $y = \frac{2}{3}x + 4$
7. $y = -\frac{7}{5}x + 3$
8. $3y = 5x + 6$
9. $2y - x = 8$
10. $-3y + 2x = 6$

In 11–15, draw the graph of each line by the intercepts method.

11. $3y + 2x = 6$
12. $y - 3x = 9$
13. $x = 4y - 12$
14. $y = 2x + 8$
15. $7x - 2y = 14$

16. The equation of the graph shown is:
(1) $y = x$ (3) $y = 2x$
(2) $y = -x$ (4) $y = -2x$

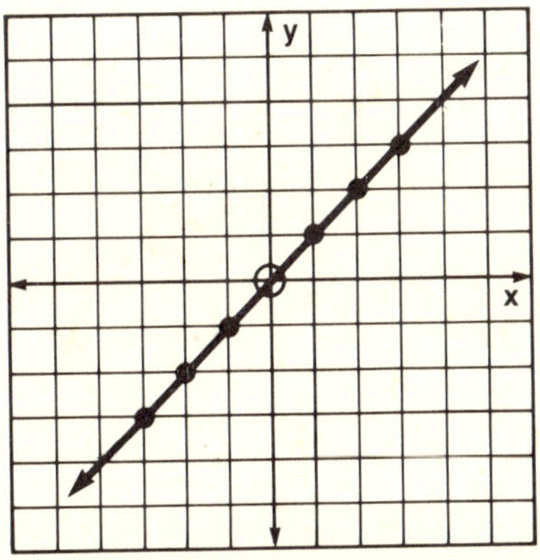

17. The equation of the graph shown is:
(1) $y - x = 0$
(2) $y + x = 0$
(3) $y - 2x = 0$
(4) $y + 2x = 0$

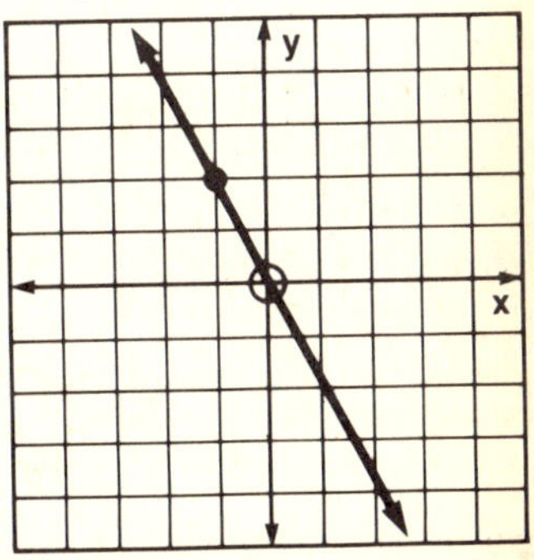

18. Which ordered pair is in the solution set of $x - y = 3$?
(1) (0, 3) (2) (3, 0) (3) (−3, 0) (4) (0, 0)

19. Which ordered pair is *not* in the solution set of $y - x = 7$?
(1) (0, 7) (2) (0, −7) (3) (1, 8) (4) (−1, 6)

20. **a.** On the same set of axes, graph the lines whose equations are:
(1) $x = -2$ (2) $x = 3$ (3) $y = 5$ (4) $y = 0$

b. Write the coordinates of the four vertices of the rectangle formed by the lines graphed in part **a.**

21. **a.** On the same set of coordinate axes, graph the lines whose equations are:
(1) $y = 2x + 1$ (2) $y = 1$ (3) $x = 2$

b. Write the coordinates of the three vertices of the triangle formed by the lines graphed in part **a.**

12.6 GRAPHING A LINEAR INEQUALITY

A linear equation produces a graph of a straight line (one of three cases: oblique, horizontal, or vertical).

What kind of graph does a linear inequality produce?

Consider the inequality $y < 3x - 3$. This inequality has the following relationship to the line whose equation is $y = 3x - 3$:

1. All points whose coordinates satisfy the equation $y = 3x - 3$ are on the line $y = 3x - 3$.
2. All points whose coordinates satisfy the inequality $y < 3x - 3$ are *not* on the line $y = 3x - 3$.

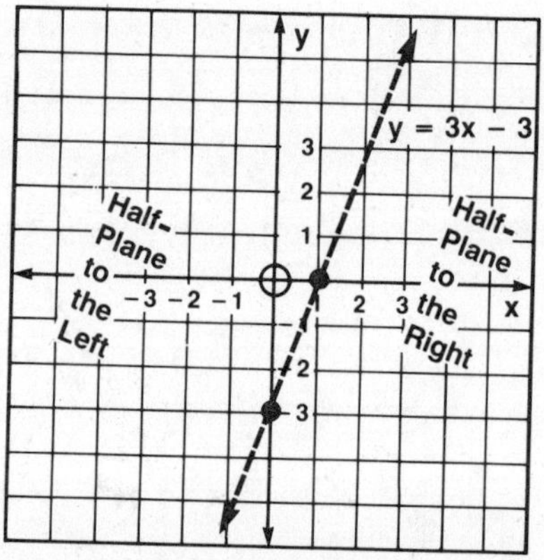

In the previous section (page 357), we saw the graph of $y = 3x - 3$. Let us reproduce that graph here, drawing the line as a dotted line since we are not interested in the points *on* the line. Notice that the line $y = 3x - 3$ divides the plane into two ***half-planes.*** One of these half-planes is to the right of the dotted line, and one of these half-planes is to the left of this line.

Choose a convenient point that is *not* on the line and is also far enough away from the line that it is obviously in one of the half-planes. For example, (4, 2) is obviously in the half-plane to the right of the line. Test this point in the inequality.

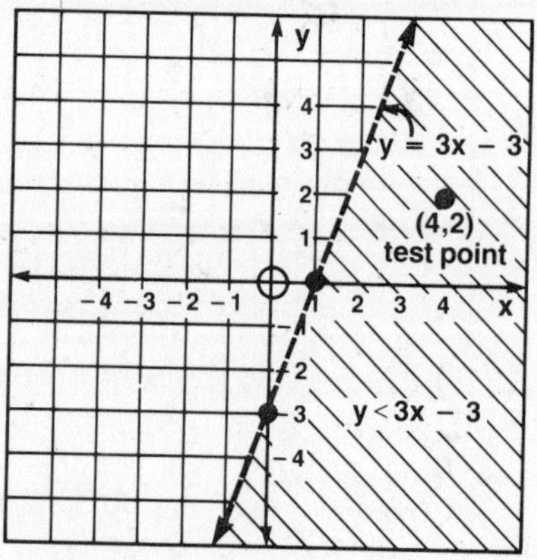

$$(4, 2)$$
$$y < 3x - 3$$
$$2 \overset{?}{<} 3(4) - 3$$
$$2 \overset{?}{<} 12 - 3$$
$$2 < 9 \quad \text{true}$$

Therefore, all points that satisfy the inequality $y < 3x - 3$ are grouped in the half-plane to the right of the line $y = 3x - 3$. The graphic solution for the inequality $y < 3x - 3$ is shown in the diagram.

◆ In general, to draw the graphic solution of a linear inequality:

1. Graph the line associated with the inequality.
   a. Draw that line as dotted if the inequality does not also contain an equals sign, for example, $y > 3x - 3$.
   b. Draw that line as solid if the inequality does also contain an equals sign, for example, $y \geq 3x - 3$. (*Note:* The graph with the solid line

indicates that the solution set for the inequality contains all the points in the appropriate half-plane plus all the points on the line.)

2. Shade the appropriate half-plane. To determine the appropriate half-plane, test a point in the inequality that is obviously not on the line. (*Note:* The origin is an easy test point since it simplifies the arithmetic. Of course, the origin can be used only if it is not already on the line.)
   a. If the test point makes the inequality "true," shade the half-plane in which it lies.
   b. If the test point makes the inequality "false," shade the half-plane in which it does *not* lie.
3. Be sure to include labels on the graph.

Examples

1. Graph the solution set of $2y \geq 5x + 6$.

Solution:

Graph the line $2y = 5x + 6$ by one of the methods previously discussed, for example, choosing points. An appropriate table is shown here.

x	$2y = 5x + 6$	y
-2	$2y = 5(-2) + 6$ $2y = -10 + 6$ $2y = -4$ $y = -2$	-2
0	$2y = 5(0) + 6$ $2y = 0 + 6$ $2y = 6$ $y = 3$	3
2	$2y = 5(2) + 6$ $2y = 10 + 6$ $2y = 16$ $y = 8$	8

Plot the points $(-2, -2)$, $(0, 3)$, and $(2, 8)$. Connect them with a solid line because the given inequality contains an equals sign.

Test a point in the inequality. Use the origin, because it is obviously not on the line.

$$2y > 5x + 6$$
$$2(0) \overset{?}{>} 5(0) + 6$$
$$0 \overset{?}{>} 0 + 6$$
$$0 > 6 \quad \text{false}$$

Thus, the required half-plane does *not* contain the origin. Shade the half-plane that does not contain the origin.

The graphic solution of $2y \geq 5x + 6$ is shown here.

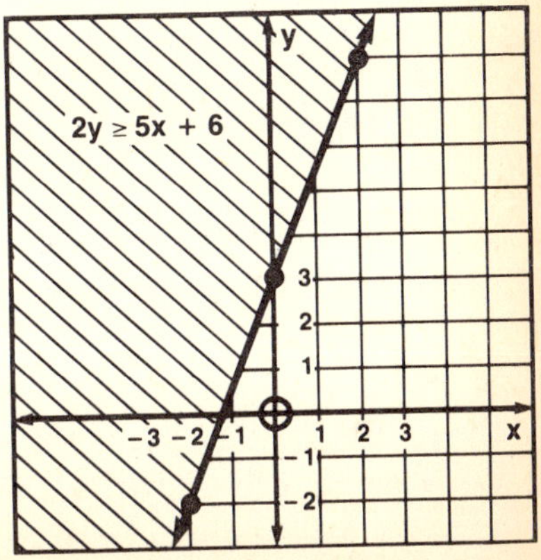

2. Draw the graph of the solution set of $y < -3$.

Solution:

Graph the horizontal line $y = -3$. Draw it as a dotted line, since the given inequality does not also contain an equals sign. The values of y that are less than -3 must lie below the line $y = -3$.

The graphic solution of $y < -3$ is shown here.

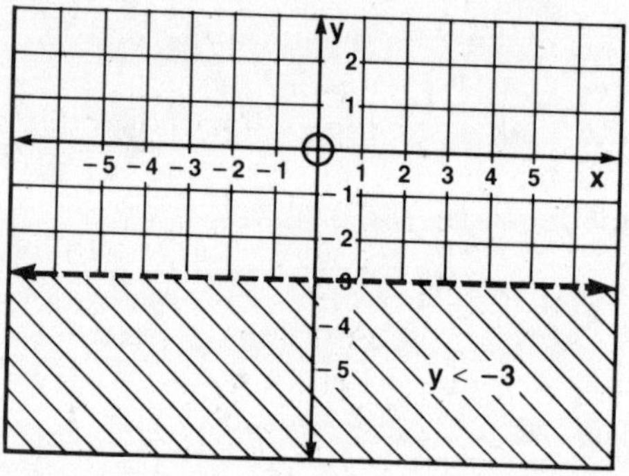

3. Which is the solution set of the given graph?

(1) $y \leq -2$ (3) $x \leq -2$
(2) $y > -2$ (4) $x > -2$

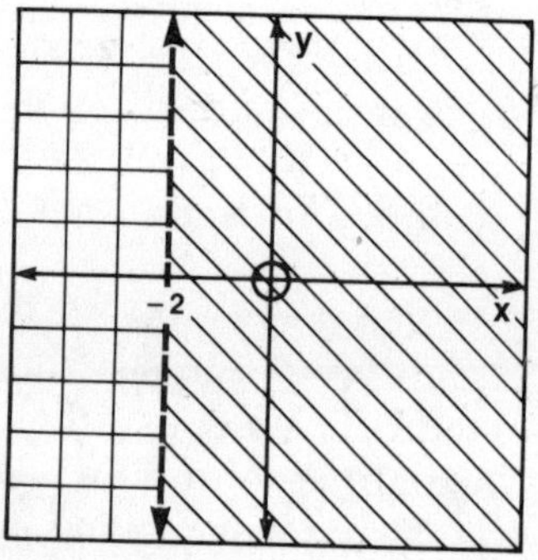

Solution:

Since the line is dotted, there is no "=" in the inequality. Thus, choices (1) and (3) are eliminated. Since the line is vertical, only x is in the inequality. Choice (2) is eliminated. The shading to the right of the vertical line indicates numbers greater than the given intercept. Thus, the graph represents the solution set of $x > -2$.

Answer: (4)

4. Which relation is shown in the graph?

(1) $\{(x, y) \mid -2 \leq x \leq 3\}$
(2) $\{(x, y) \mid -2 < x \leq 3\}$
(3) $\{(x, y) \mid -2 \leq x < 3\}$
(4) $\{(x, y) \mid -2 < x < 3\}$

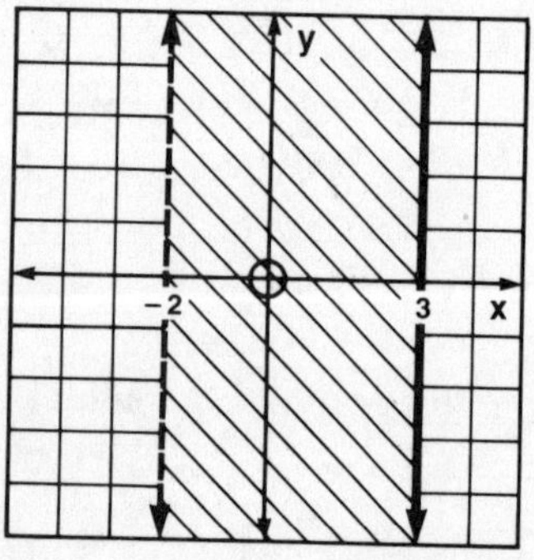

Solution:

The graph shows that x is between -2 and 3, not including -2 (because of the dotted line) but including 3 (because of the solid line). Thus, the correct solution set is $\{(x, y) \mid -2 < x \leq 3\}$.

Answer: (2)

EXERCISES

In 1–8, draw the graph of the solution set of each inequality.

1. $y \le 2x + 1$
2. $y \ge 3x - 2$
3. $y + 2x < 4$
4. $2y - x > 6$
5. $x - 3y \le 9$
6. $x \le -3$
7. $y > 4$
8. $y + 3 \le 0$

9. The graph shown at the right is the graph of:
(1) $y > x$
(2) $y < x$
(3) $y \ge x$
(4) $y \le x$

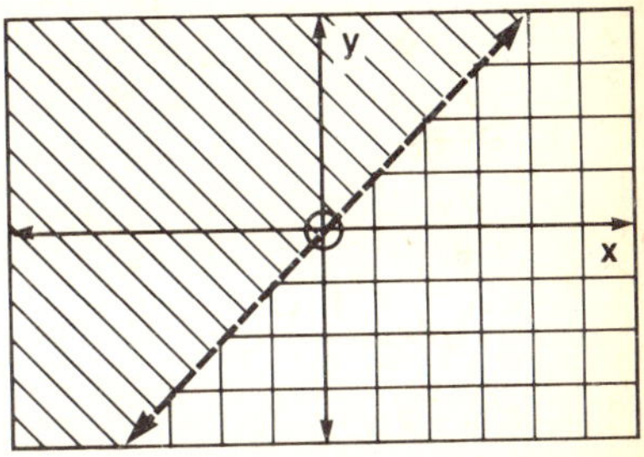

10. The graph shown below is the graph of:
(1) $y \ge x$ (3) $x \ge 0$
(2) $x \ge y$ (4) $y \ge 0$

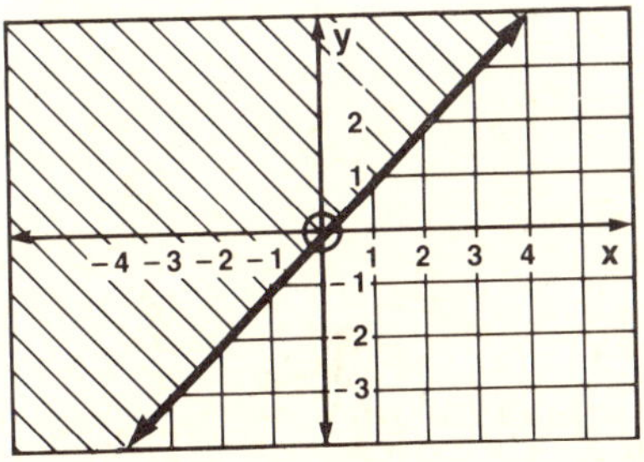

11. The graph shown below is the graph of:
(1) $y < 3$
(2) $y < x + 3$
(3) $y > x + 3$
(4) $y < 3x + 3$

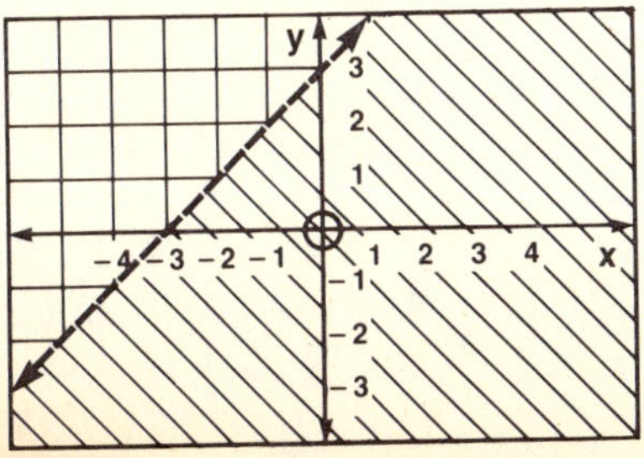

12. The graph below represents which of the following sets?
(1) $\{(x, y) \mid x < 2\}$
(2) $\{(x, y) \mid x \le 2\}$
(3) $\{(x, y) \mid y < 2\}$
(4) $\{(x, y) \mid y \le 2\}$

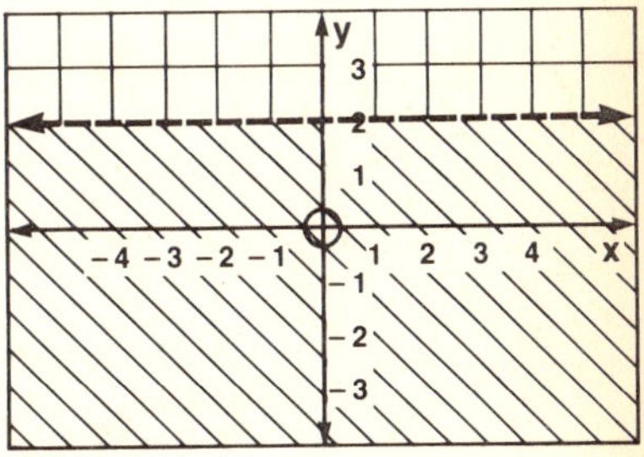

13. The graph shown below is the graph of:
(1) $y \le 0$ (3) $x \le 0$
(2) $y \ge 0$ (4) $x \ge 0$

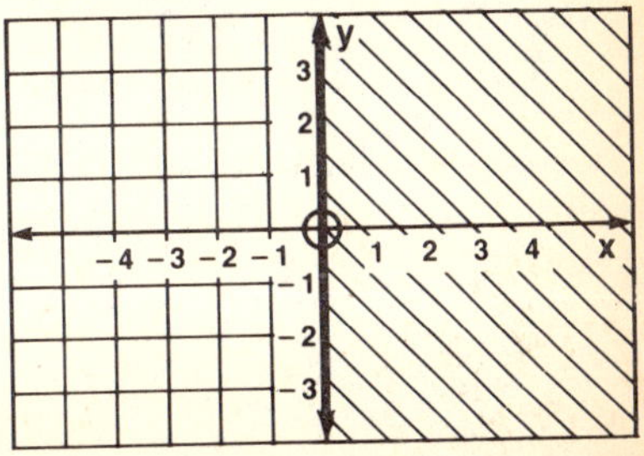

14. Which graph represents the inequality $y < 4$?

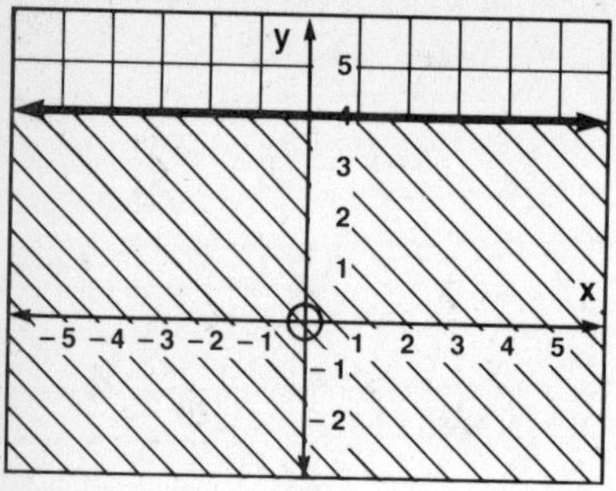

(1)

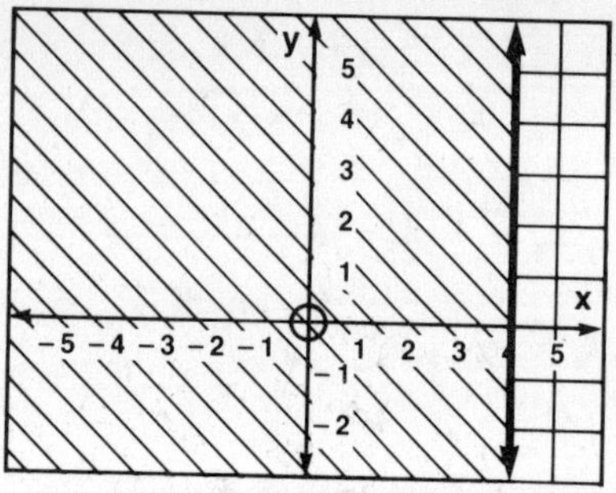

(3)

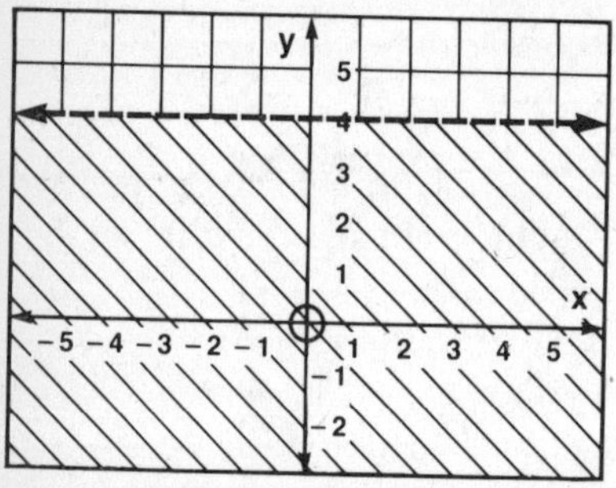

(2)

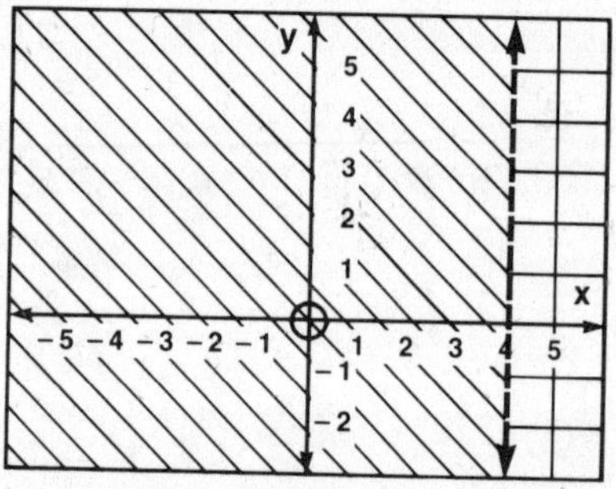

(4)

15. Which is the graph of $y \leq 2$?

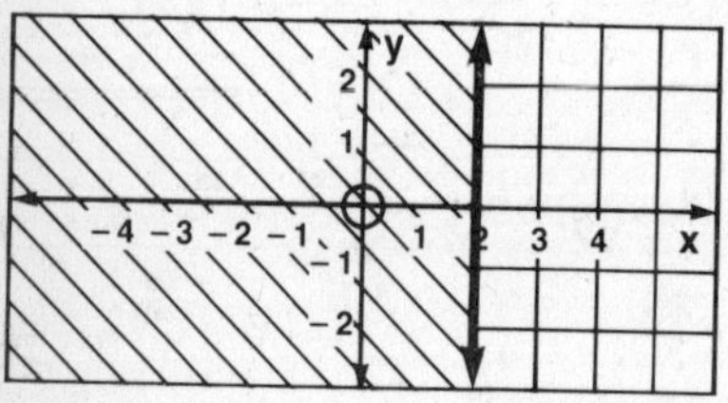

(1)

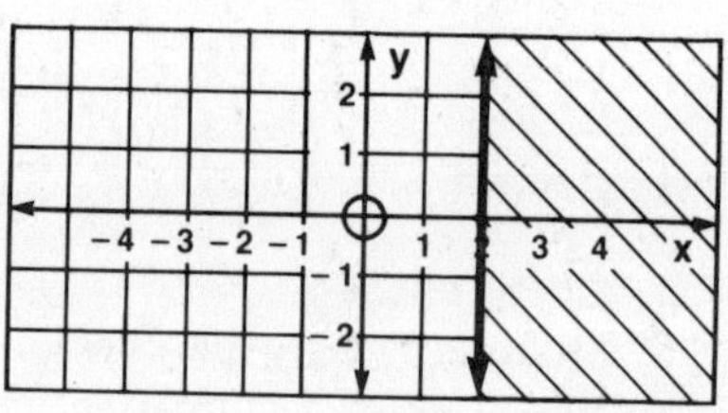

(3)

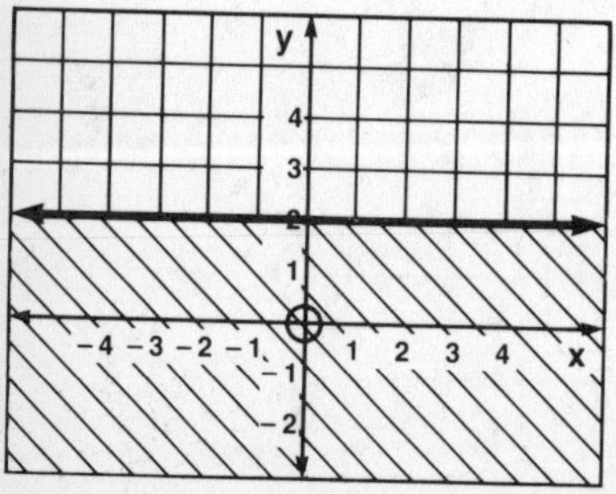

(2)

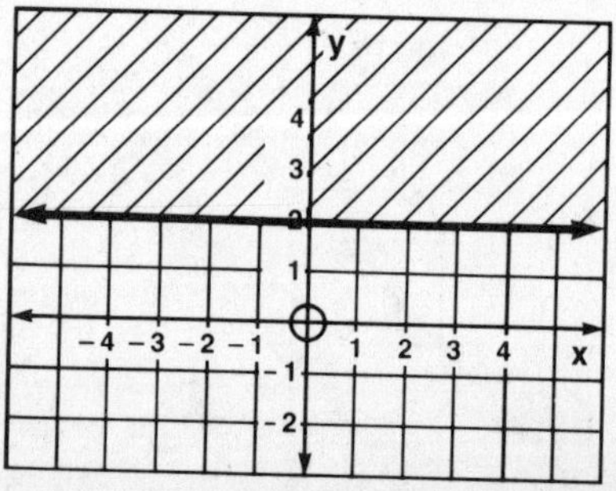

(4)

16. The graph shown is the graph of:
(1) $-1 < x \leq 4$
(2) $-1 \leq x < 4$
(3) $-1 < x < 4$
(4) $-1 \leq x \leq 4$

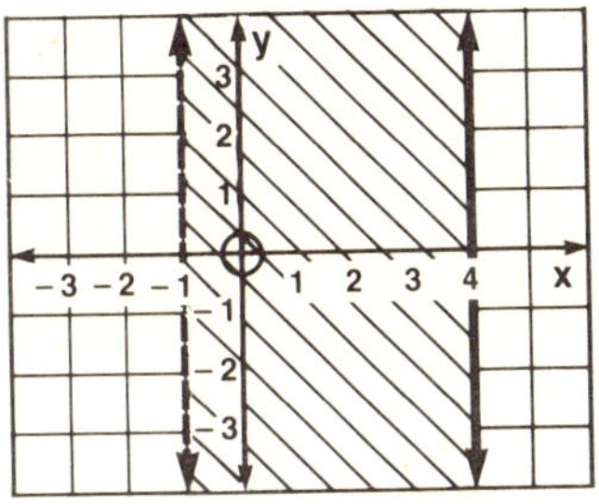

17. Which relation is shown in the graph?
(1) $\{(x, y) \mid x + y > -5\}$
(2) $\{(x, y) \mid -5 \leq x \leq 4\}$
(3) $\{(x, y) \mid -5 \leq x < 4\}$
(4) $\{(x, y) \mid -5 \leq y < 4\}$

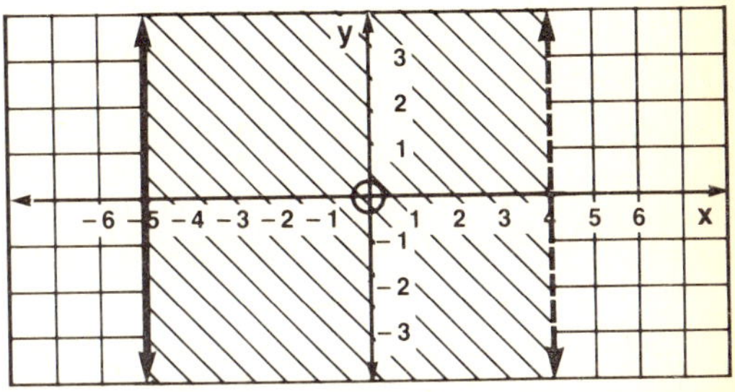

18. Which ordered pair is in the solution set of $y > 3x + 2$?
(1) (1, 5) (2) (1, 6) (3) (5, 1) (4) (0, 0)

19. Which ordered pair is *not* in the solution set of $3x - y \geq 2$?
(1) (1, 1) (2) (1, −1) (3) (−1, 1) (4) (0, −2)

12.7 SUMMARY EXERCISES

In 1–10, fill in the word "abscissa," "ordinate," "positive," "negative," or "zero."

1. The x-coordinate of a point is called the ____ and the y-coordinate is called the ____.
2. A point in quadrant I has an ____ and an ____ that are both ____.
3. A point in quadrant III has an ____ and an ____ that are both ____.
4. Both the ____ and the ____ of the origin are ____.
5. Every point on the x-axis has an ____ of ____.
6. Every point on the y-axis has an ____ of ____.
7. The slope of a horizontal line is ____.
8. The slope of an oblique line is never equal to ____.
9. The slope of a vertical line is undefined because the difference between the ____s of any two points on the line is ____, thereby leading to division by ____.
10. The slope of a line is calculated by dividing the difference between the ____s of two points on the line by the difference between the ____s of those points.

In 11–17, fill in the word "horizontal," "vertical," or "parallel."

11. A ____ line and a ____ line intersect at right angles.
12. A ____ line is ____ to the x-axis.
13. A ____ line is ____ to the y-axis.
14. If two lines are ____, then their slopes are equal.

15. If the slopes of two lines are not equal, then the lines are not ______.
16. The slope of a ______ line is zero.
17. The slope of a ______ line is undefined.

In 18–22, find the slope of the line joining the two points.

18. (1, 0) and (3, 7)
19. (−2, −3) and (−5, 0)
20. (4, −4) and (−4, 4)
21. (−3, 1) and (−3, 7)
22. (7, −2) and (5, −2)

In 23–25, determine if the three points are collinear (lie on the same line). Answer "yes" or "no."

23. (−3, 3), (3, −3), (−3, −3)
24. (7, 1), (7, 0), (7, 4)
25. (4, −2), (1, −2), (−3, −2)

In 26–30, find the value of k such that:

26. the slope of the line joining (3, k) and (6, 5) is −2.
27. the line joining (2, 2) and (−7, −7) is parallel to the line joining (2, 5) and (3, k).
28. the line joining (3, −2) and (4, k) is horizontal.
29. the line joining (−1, −1) and (k, 0) is vertical.
30. the point (k, 3) lies on the line $3x + 2y = 18$.

In 31–36, write the letter of the equation that is described by each of the following statements.

(*a*) $x + y = 5$ (*b*) $x - y = 5$ (*c*) $y = 5$ (*d*) $x = -5$

31. This line has a slope of 0.
32. This line has a slope of −1.
33. This line is parallel to the line $y + 5 = 0$.
34. This line is perpendicular to the line $x = 5$.
35. This line is parallel to the line $2y = 2x + 3$.
36. This line has a y-intercept of −5.

In 37–40, use the line whose equation is $2x - y = 6$ to answer each question.

37. What is the slope of the line?
38. What is the x-intercept?
39. What is the y-intercept?
40. Does the point (−2, −2) lie on the line?

41. If the slope of a line is −4 and the y-intercept is 1, which of the following can be the equation of that line?
(1) $y - 4x = 1$ (2) $y + 4x = 1$ (3) $x + 4y = 1$ (4) $x - 4y = 1$
42. Write an equation for the line whose slope is $-\frac{1}{3}$ and whose y-intercept is 3.
43. Which ordered pair is in the solution set of $x - y = 7$?
(1) (−7, 0) (2) (0, 7) (3) (−1, 8) (4) (−1, −8)
44. Which point is *not* on the line $x = -2$?
(1) (−2, −2) (2) (−2, 0) (3) (0, −2) (4) (−2, 2)

In 45–50, draw the graph of the solution set of each equation.

45. $2y + x = 5$
46. $4x - 3y = 12$
47. $y = \frac{1}{3}x - 2$
48. $x = -2$
49. $y = 4x + 3$
50. $y - 7 = 0$

In 51–56, draw the graph of the solution set of each inequality.

51. $2y - x \leq 6$
52. $3x + 4y > 12$
53. $y < \frac{1}{2}x - 4$
54. $x \geq -3$
55. $y - 5 \leq 0$
56. $y < 5x$

In 57–67, match each graph with a relation.

57. (1) $x \leq 4$ (2) $x \geq 4$ (3) $y \leq 4$ (4) $y \geq 4$

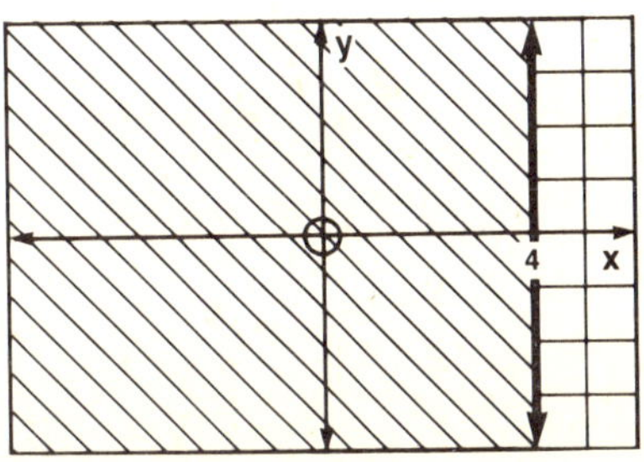

58. (1) $y \geq 2x$ (2) $y \leq 2x$ (3) $y \geq -2$ (4) $y \leq -2$

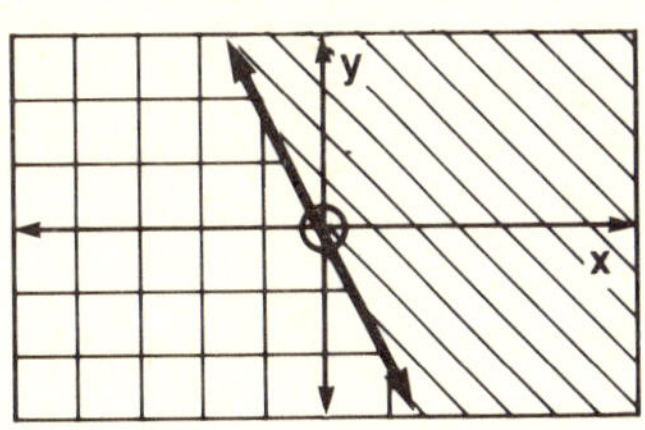

59. (1) $x + 2y < 4$
(2) $x + 2y < -4$
(3) $y + 2x < 4$
(4) $y + 2x < -4$

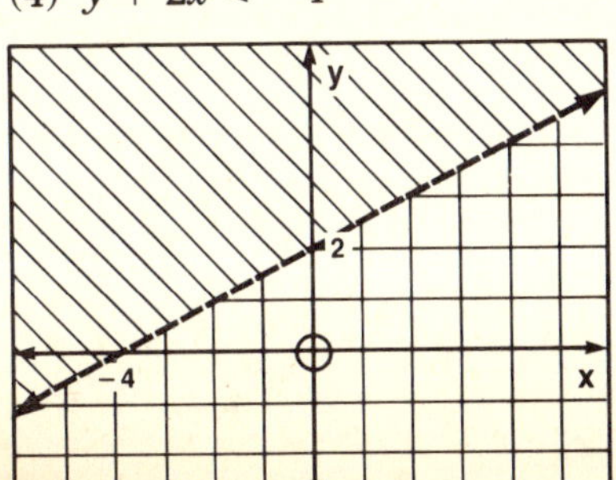

60. (1) $\{(x, y) \mid -3 < x \leq 3\}$
(2) $\{(x, y) \mid -3 \leq x < 3\}$
(3) $\{(x, y) \mid -3 < y \leq 3\}$
(4) $\{(x, y) \mid -3 \leq y < 3\}$

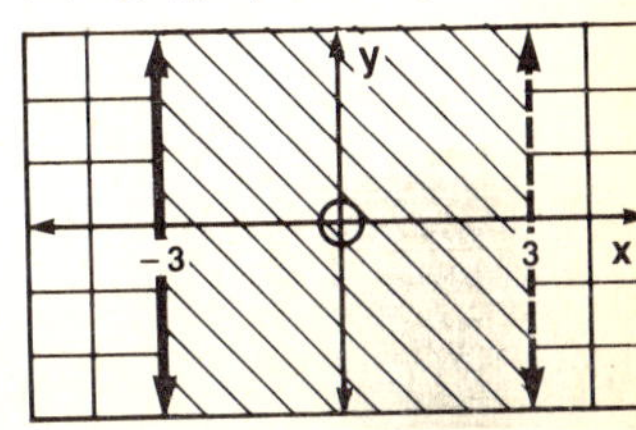

61. (1) $\{x \mid x > -3\}$
(2) $\{x \mid x < -3\}$
(3) $\{x \mid x \leq -3\}$
(4) $\{x \mid x \geq -3\}$

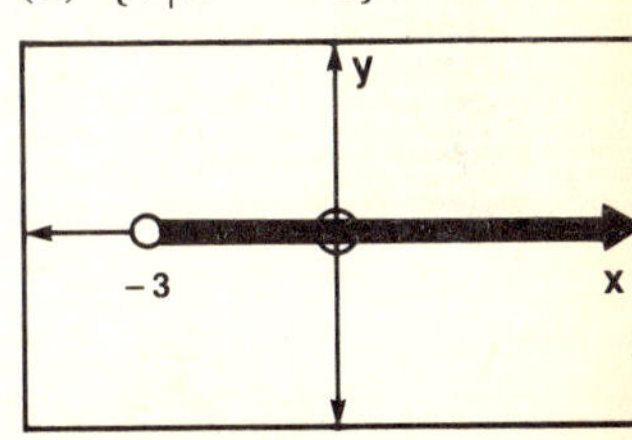

62. (1) $(x \leq 0) \wedge (x > 2)$
(2) $(x \leq 0) \vee (x < 2)$
(3) $0 \leq x < 2$
(4) $0 < x \leq 2$

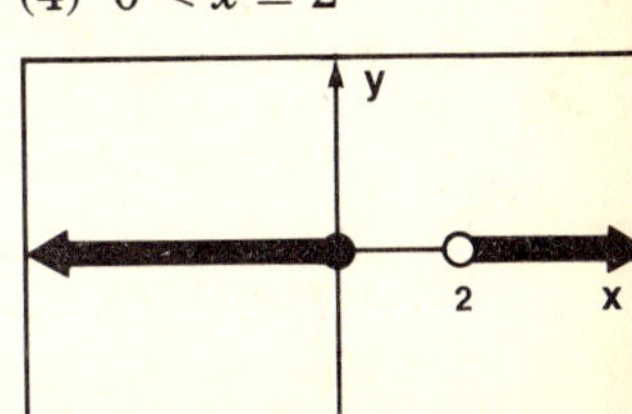

63. (1) $0 \le y < 3$
(2) $(x \ge 0) \vee (x \le 3)$
(3) $(x \le 0) \wedge (x \ge 3)$
(4) $0 \le x \le 3$

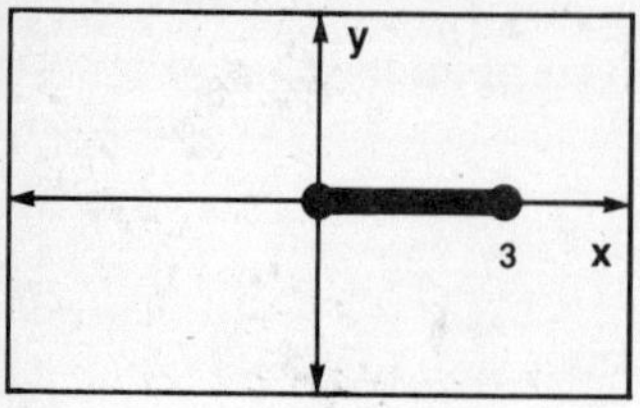

64. (1) $\{(x, y) \mid 0 \le x < 4\}$
(2) $\{(x, y) \mid 0 \le y < 4\}$
(3) $\{(x, y) \mid 0 < x \le 4\}$
(4) $\{(x, y) \mid 0 < y \le 4\}$

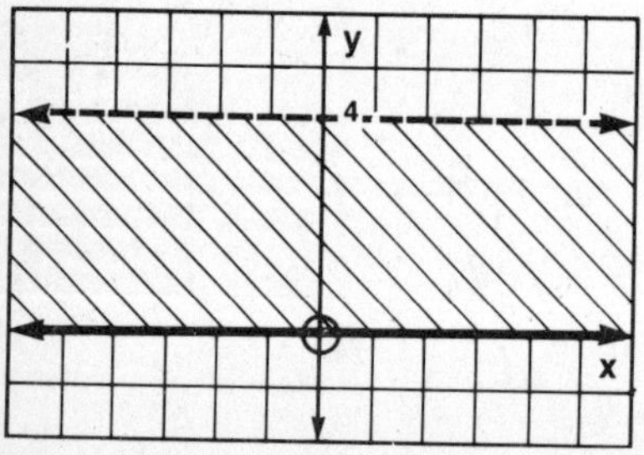

65. (1) $\{(x, y) \mid -1 \le x \le 3\}$
(2) $\{(x, y) \mid (x \le -1) \vee (x \ge 3)\}$
(3) $\{(x, y) \mid -1 \le y \le 3\}$
(4) $\{(x, y) \mid (y \le -1) \vee (y \ge 3)\}$

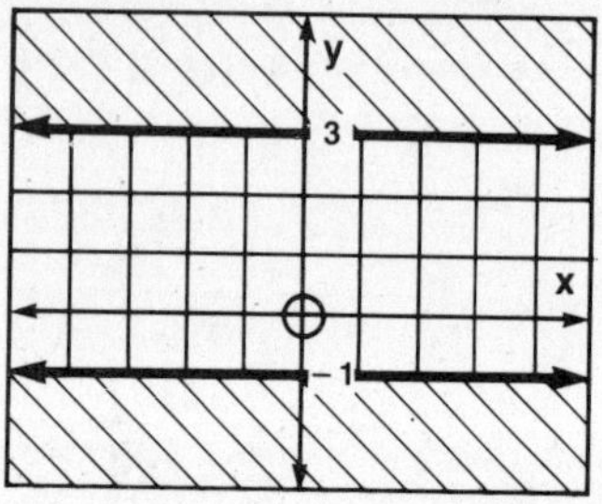

66. (1) $\{(x, y) \mid -3 < x \le 2\}$
(2) $\{(x, y) \mid -3 < y \le 2\}$
(3) $\{(x, y) \mid (x < -3) \wedge (x \ge 2)\}$
(4) $\{(x, y) \mid (x < -3) \vee (x \ge 2)\}$

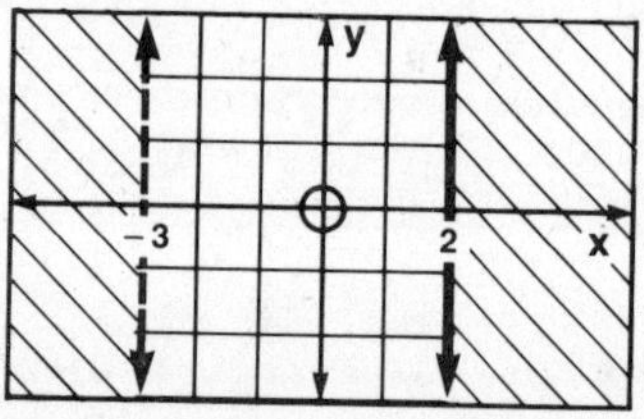

67. (1) $\{(x, y) \mid -2 \le x \le 0\}$
(2) $\{(x, y) \mid x \ge -2\}$
(3) $\{(x, y) \mid -2 \le y \le 0\}$
(4) $\{(x, y) \mid y \ge -2\}$

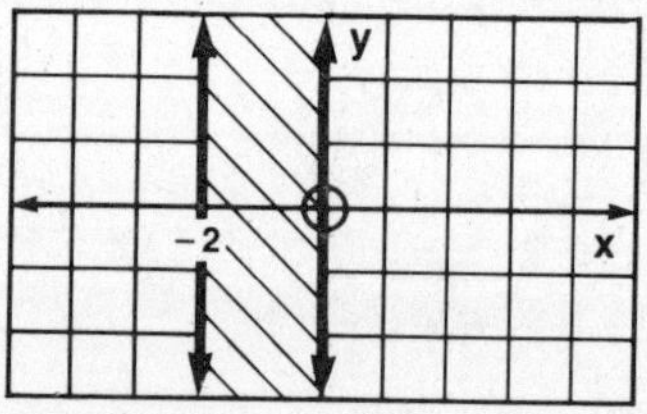

68. Which ordered pair is in the solution set of $2x - y < 8$?
(1) (4, 0) (2) (−4, 0) (3) (8, −8) (4) (0, −8)

69. Which ordered pair is in the solution set of $x - y = -1$?
(1) (0, −1) (2) (−1, 0) (3) (0, 0) (4) (1, 0)

70. Which ordered pair is *not* in the solution set of $3x + 2y \ge 6$?
(1) (2, 0) (2) (0, 6) (3) (0, −6) (4) (6, 6)

In 71–75, find the area of the given figure whose vertices have the indicated coordinates.

71. rectangle: (0, −5), (2, −5), (2, 3), (0, 3)
72. triangle: (−5, 0), (0, 0), (0, −12)
73. parallelogram: (−4, 0), (3, 0), (5, 4), (−2, 4)
74. trapezoid: (−3, −4), (4, −4), (2, 5), (−1, 5)
75. square: (−5, 0), (0, −5), (5, 0), (0, 5)

CHAPTER 13

Systems of Equations and Inequalities

13.1 GRAPHIC SOLUTION OF A SYSTEM OF LINEAR EQUATIONS

When two lines are drawn in the same plane, only one of the following three conditions is possible:

1. The two lines intersect.
2. The two lines are parallel (the two lines do not intersect).
3. The two lines coincide (the two lines are actually the same line).

If the two lines intersect, there is only one point in the entire plane, the point of intersection, that is on *both* lines. This point of intersection is called the ***solution set*** of the ***system of linear equations*** that represents the two lines. Finding the coordinates of the point of intersection of the two lines is a ***graphic solution*** of the system of linear equations that represents the two lines.

Note: In this course, when we are asked to "solve graphically," we can reasonably assume that there will be one point of intersection. If two lines are parallel, there is *no* point of intersection and the solution set for the system is empty. If two lines coincide, *every* point of one line is on the other line and the solution set for the system is infinite.

The procedure for obtaining a graphic solution of a system of linear equations is illustrated in the following example.

Example

Solve graphically and check: $x + 2y = -4$
$y = -3x + 3$

Solution:

Graph each of the equations by one of the methods previously discussed: choosing points, slope-intercept, or intercepts (x- and y-). For the line $x + 2y = -4$, suppose that we choose the intercepts method. Let $x = 0$, and solve for y.

$$x + 2y = -4$$
$$0 + 2y = -4$$
$$2y = -4$$
$$y = -2$$

Thus, the y-intercept of the line is -2.

Let $y = 0$, and solve for x:

$$\begin{aligned} x + 2y &= -4 \\ x + 2(0) &= -4 \\ x + 0 &= -4 \\ x &= -4 \end{aligned}$$

Thus, the x-intercept of the line is -4.

On a graph, plot the points associated with the x- and y-intercepts and connect them with a straight line. Include arrowheads. (*Note:* A third point is usually chosen just to insure accuracy in drawing the line.) Write the equation of the line near the line.

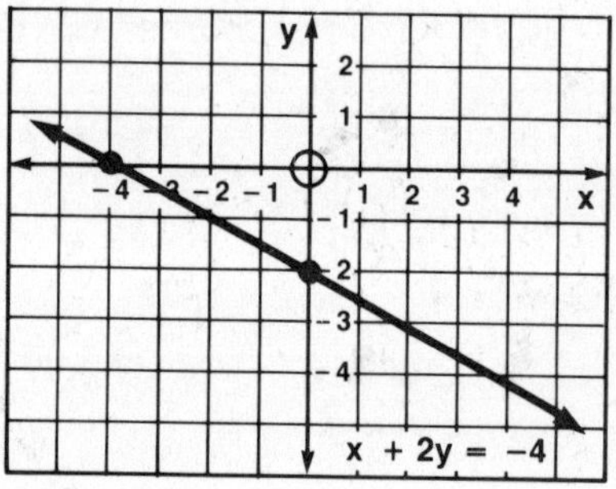

For the second line, $y = -3x + 3$, suppose that we choose some points. Prepare a table, selecting a few values for x and computing the corresponding y-values.

x	$y = -3x + 3$	y
-2	$y = -3(-2) + 3$ $= +6 \quad + 3$	9
0	$y = -3(0) + 3$ $= 0 \quad + 3$	3
3	$y = -3(3) + 3$ $= -9 \quad + 3$	-6

Plot the points $(-2, 9)$, $(0, 3)$, and $(3, -6)$ *on the same graph* used for the first line. Connect these three points with a line. Include arrowheads. Write the equation of this line near the line. Write the coordinates of the point of intersection on the graph.

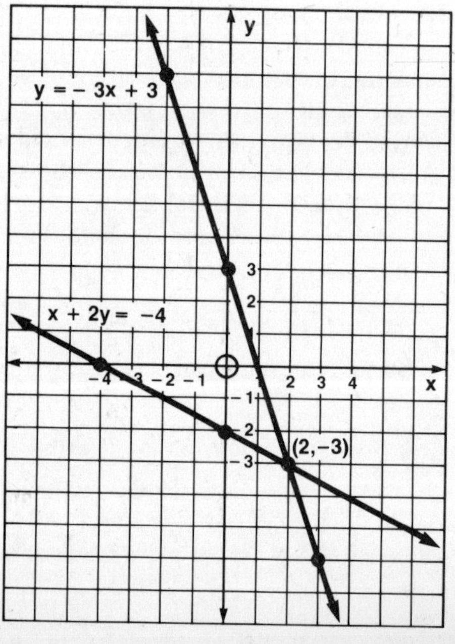

Check the coordinates of the point of intersection $(2, -3)$ in *each* equation.

$$\begin{aligned} x + 2y &= -4 \\ 2 + 2(-3) &\stackrel{?}{=} -4 \\ 2 - 6 &\stackrel{?}{=} -4 \\ -4 &= -4 \checkmark \end{aligned}$$

$$\begin{aligned} y &= -3x + 3 \\ -3 &\stackrel{?}{=} -3(2) + 3 \\ -3 &\stackrel{?}{=} -6 + 3 \\ -3 &= -3 \checkmark \end{aligned}$$

Answer: $(2, -3)$ is the solution set for the system oflinear equations.

◆ In general, the procedure for solving a system of linear equations graphically is:

1. Graph each equation, using one of the methods: choosing points, slope-intercept, or intercepts.
2. Draw the graphs of the two lines on the same set of axes. Include arrowheads on each line, and label each line with its equation.
3. Write the coordinates of the point of intersection on the graph.
4. Check the coordinates of the solution set (the point of intersection) in *each* of the two *original* equations.

Example

Solve graphically and check: $2x - y = 3$
$y = -x$

Solution:

x	$2x - y = 3$	y
-2	$2(-2) - y = 3$ $-4 - y = 3$ $-y = 7$ $y = -7$	-7
0	$2(0) - y = 3$ $0 - y = 3$ $-y = 3$ $y = -3$	-3
2	$2(2) - y = 3$ $4 - y = 3$ $-y = -1$ $y = 1$	1

x	$y = -x$	y
-2	$y = -(-2)$	2
0	$y = -(0)$	0
2	$y = -(2)$	-2

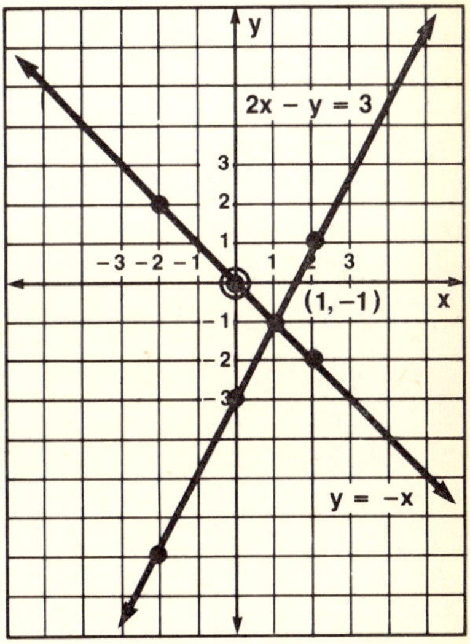

Check: (1, −1)

$$2x - y = 3$$
$$2(1) - (-1) \stackrel{?}{=} 3$$
$$2 + 1 \stackrel{?}{=} 3$$
$$3 = 3 \checkmark$$

$$y = -x$$
$$(-1) \stackrel{?}{=} -(1)$$
$$-1 = -1 \checkmark$$

Answer: (1, −1) is the solution set for the system of linear equations.

EXERCISES

1. What is the solution of the system of equations whose graph is shown at the right?
(1) $(-1, -2)$
(2) $(-2, 1)$
(3) $(-2, -1)$
(4) $(0, -2)$

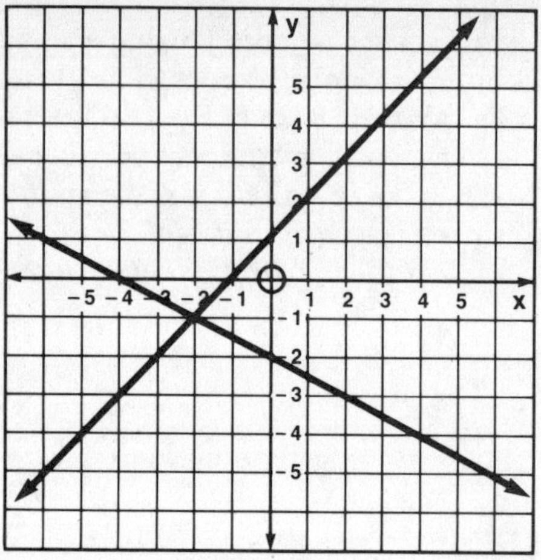

2. What is the solution of the system of equations whose graph is shown at the right?
(1) $(3, 0)$
(2) $(2, 0)$
(3) $(-3, 4)$
(4) $(4, -3)$

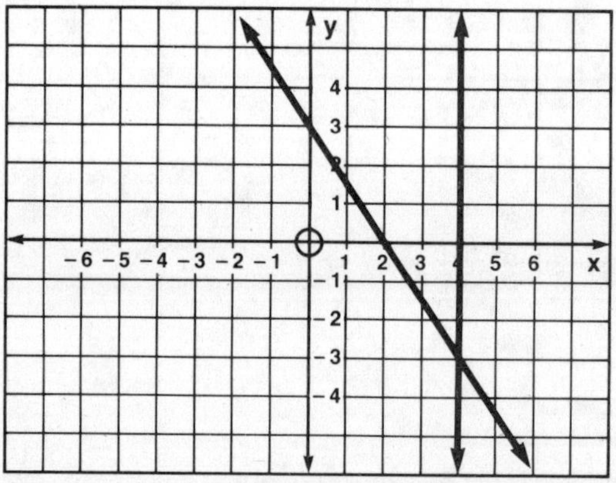

3. What is the solution of the system of equations whose graph is shown at the right?
(1) $(0, 5)$
(2) $(5, 0)$
(3) $(0, -5)$
(4) $(-5, 0)$

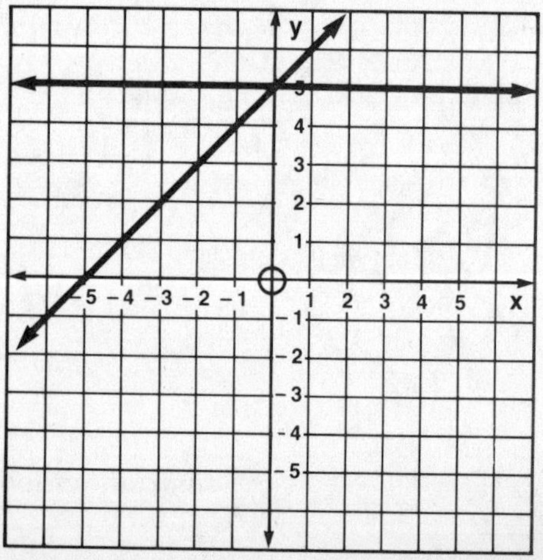

4. What is the solution of the system of equations whose graph is shown at the right?
(1) (0, −2)
(2) (−2, 0)
(3) (3, 0)
(4) (−3, 0)

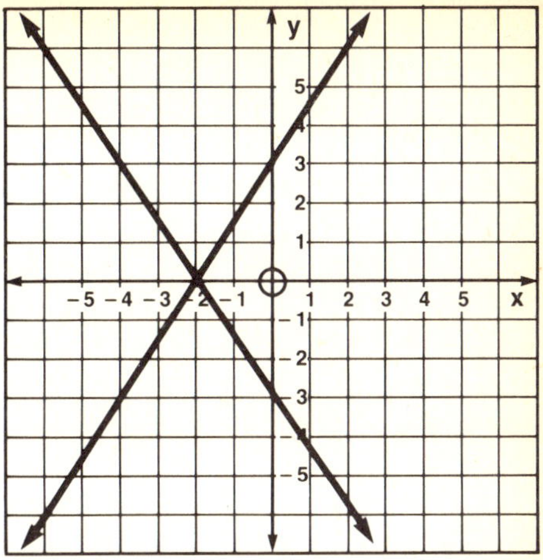

5. What is the solution of the system of equations whose graph is shown at the right?
(1) (5, 5)
(2) (−5, 5)
(3) (0, −5)
(4) (0, 0)

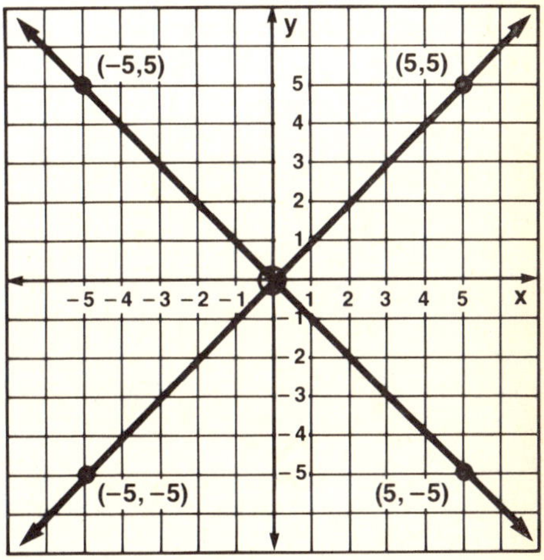

6. Which is the solution of the system of equations whose graph is shown at the right?
(1) (−5, 0)
(2) (0, 5)
(3) (−2, 3)
(4) (0, 0)

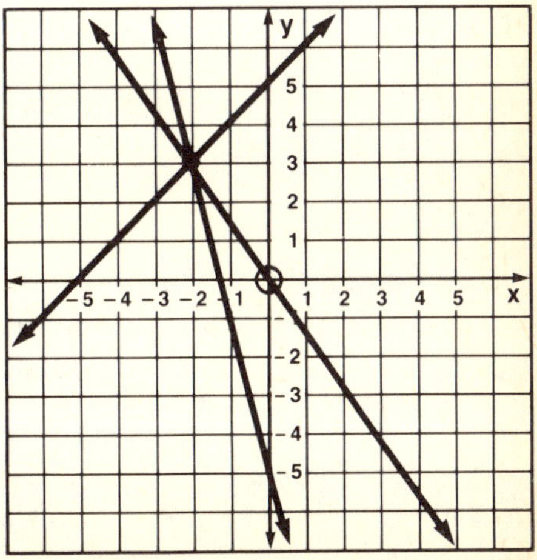

In 7–25, solve graphically and check each system of equations.

7. $2x = y + 5$
   $x + y = -2$
8. $2y = x + 6$
   $3x - y = 2$
9. $y = x + 7$
   $2x + y = 1$
10. $2x + y = 8$
    $y = 3x - 2$
11. $y = x - 3$
    $2x + y = 3$
12. $2x + y = 0$
    $y = 3x + 5$
13. $x + y = 4$
    $x - 2y = 1$
14. $x + y = 6$
    $2x - y = 0$
15. $y = 2x + 1$
    $x + 2y = 7$
16. $2x - y = 5$
    $x + y = -2$
17. $2y - x = 4$
    $y = -2x - 3$
18. $x + 2y = 7$
    $2x - y = -1$
19. $y - 2x = 5$
    $x + 2y = 0$
20. $2y = -5x$
    $y - x = 7$
21. $2y = x + 6$
    $y = 3x - 2$
22. $x + y = 8$
    $2x - y = 7$
23. $2x - y = 10$
    $x + 2y = 10$
24. $x + 2y = 7$
    $y = 2x + 1$
25. $x + 2y = 4$
    $3x - y = 5$

13.2 GRAPHIC SOLUTION OF A SYSTEM OF LINEAR INEQUALITIES

A graphic solution for a system of linear inequalities is produced in the same manner as for a system of linear equations. The one major difference is in the nature of the solution set.

While the solution set for a system of linear equalities that represented a pair of intersecting lines consists of exactly one point, the solution set for a system of linear inequalities is usually infinite. The solution set for a system of linear inequalities is comprised of all the points that satisfy *both* inequalities *simultaneously* (at the same time).

Example

Graph the solution set of the following system of inequalities and label the solution set A: $y \le x - 3$
$y > -x + 1$

Solution:

To graph the first inequality, $y \le x - 3$, we will graph the line $y = x - 3$. Since the original inequality contains an equals sign, we will draw the line solid. We will shade the appropriate half-plane. For the line $y = x - 3$, prepare a table, as shown at the right, and plot the points $(-2, -5)$, $(0, -3)$, and $(2, -1)$. Draw and label the solid line $y = x - 3$. Include arrowheads.

x	$y = x - 3$	y
-2	$y = -2 - 3$	-5
0	$y = 0 - 3$	-3
2	$y = 2 - 3$	-1

Since the origin is not on the line $y = x - 3$, use the origin as a test point in the inequality $y \leq x - 3$ for determining which half-plane is to be shaded:

$$\begin{aligned} y &\leq x - 3 \\ 0 &\overset{?}{\leq} 0 - 3 \\ 0 &\leq -3 \quad \text{false} \end{aligned}$$

Thus, the half-plane to be shaded is the one to the right of the line because it does *not* contain the origin.

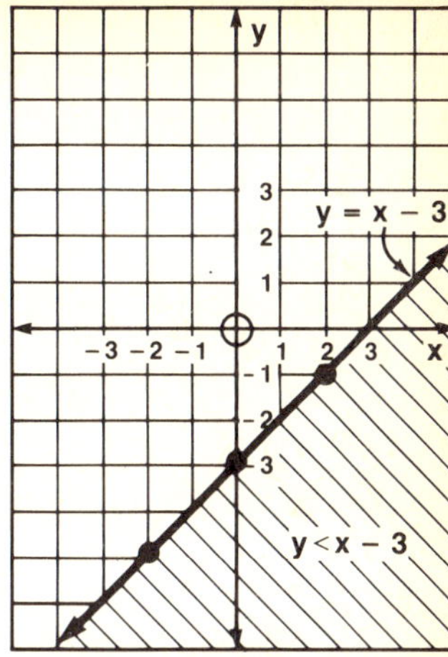

To graph the second inequality, $y > -x + 1$, we will graph the line $y = -x + 1$. Since the original inequality does not contain an equals sign, we will draw the line dotted. We will shade the appropriate half-plane for this line. This graph is to be drawn on the same set of axes as the first graph. Be very neat, or your drawing could become confusing.

For the line $y = -x + 1$, prepare a table and plot the points $(-2, 3)$, $(0, 1)$, and $(2, -1)$. Draw and label the dotted line $y = -x + 1$. Include arrowheads.

x	$y = -x + 1$	y
-2	$y = -(-2) + 1$ $= 2 \quad + 1$	3
0	$y = -(0) + 1$ $= 0 \quad + 1$	1
2	$y = -(2) + 1$ $= -2 \quad + 1$	-1

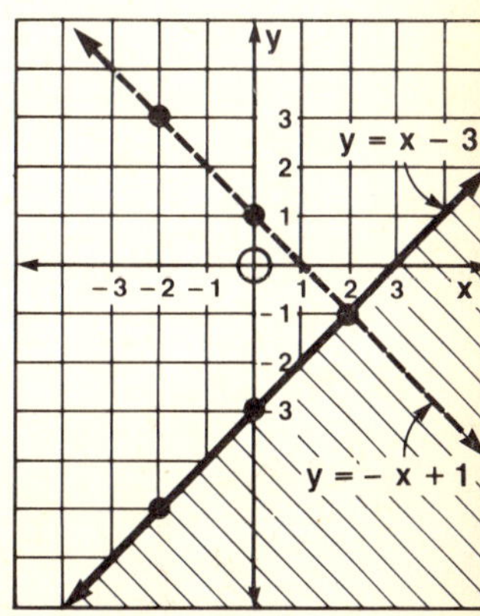

Look at the graph only with respect to the dotted line $y = -x + 1$ and see that the origin is not on this line. Thus, the origin can be used as a test point in the inequality $y > -x + 1$ to determine which half-plane is to be shaded.

$$\begin{aligned} y &> -x + 1 \\ 0 &\overset{?}{>} -(0) + 1 \\ 0 &> +1 \quad \text{false} \end{aligned}$$

Therefore, the half-plane to be shaded for this inequality is the one to the right of the line because it does *not* contain the origin. For this shading, use a direction different from the shading already on the graph so that one stands out from the other.

Notice that the two shadings *overlap* in a portion of the plane. This portion of the plane contains the points that satisfy both inequalities simultaneously. This portion of the plane is the solution set of the system of linear inequalities. (*Note:* The shading in this portion is called *cross-hatched,* that is, overlapping shadings.) The cross-hatched portion is to be labeled A according to the directions of the problem.

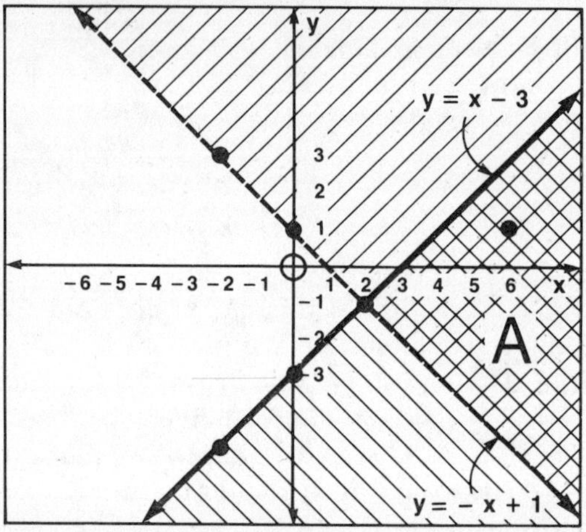

We can check to see if A is indeed the solution set of the given system of linear inequalities by choosing a point that is obviously in A and seeing if the coordinates of that chosen point satisfy both of the inequalities. Let us choose (6, 1). Substitute (6, 1) for x and y in both inequalities and see if the resulting inequalities are true.

$$\begin{array}{ll} y \leq x - 3 & y > -x + 1 \\ 1 \overset{?}{\leq} 6 - 3 & 1 \overset{?}{>} -(6) + 1 \\ 1 \leq 3 \quad \text{true} & 1 > -5 \quad \text{true} \end{array}$$

◆ In general, the procedure for solving graphically a system of linear inequalities is as follows:

1. Graph the first inequality.
   a. Graph and label the line associated with the inequality.
      (i) If the inequality also contains an equals sign, use a solid line.
      (ii) If the inequality does not contain an equals sign, use a dotted line.
   b. Shade the appropriate half-plane; use a test point in the inequality that is obviously not on the line.
      (i) If the test point makes the inequality "true," shade the half-plane in which it lies.
      (ii) If the test point makes the inequality "false," shade the half-plane in which it does *not* lie.

2. Graph the second inequality on the same set of axes, using the procedure described for graphing the first inequality. Be sure to use shading in a direction different from the first shading. The cross-hatched portion of the plane (shaded once for the first inequality and a second time for the second inequality) is the solution set for the system of inequalities. This cross-hatched portion should be labeled "solution set" or by whatever letter is given in the directions to the problem.
3. Check that the solution set is correct by choosing a point that is in the solution set. Substitute the coordinates of the checkpoint for x and y in each inequality and see if the resulting inequalities are true.

Example

Graph the solution set of the following system of linear inequalities:

$$y \geq 2x - 3$$
$$y < 2$$

Solution:

$$y \geq 2x - 3$$

Solid line for $y = 2x - 3$.

x	$y = 2x - 3$	y
-2	$y = 2(-2) - 3$ $= -4 - 3$	-7
0	$y = 2(0) - 3$ $= 0 - 3$	-3
2	$y = 2(2) - 3$ $= 4 - 3$	1

Shading: test (0, 0).

$$y \geq 2x - 3$$
$$0 \overset{?}{\geq} 2(0) - 3$$
$$0 \overset{?}{\geq} 0 - 3$$
$$0 \geq -3 \quad \text{true}$$

(Shade the half-plane in which (0, 0) lies, the half-plane to the left of the line $y = 2x - 3$.)

Check: $(-2, -2)$

$$y \overset{?}{\geq} 2x - 3 \qquad y < 2$$
$$-2 \overset{?}{\geq} 2(-2) - 3 \qquad -2 < 2 \quad \text{true}$$
$$-2 \overset{?}{\geq} -4 - 3$$
$$-2 \geq -7 \quad \text{true}$$

$$y < 2$$

Dotted line for $y = 2$.

Horizontal line intersecting the y-axis at 2.

Shading: $y < 2$ means "below 2." (Shade the half-plane below the line $y = 2$.)

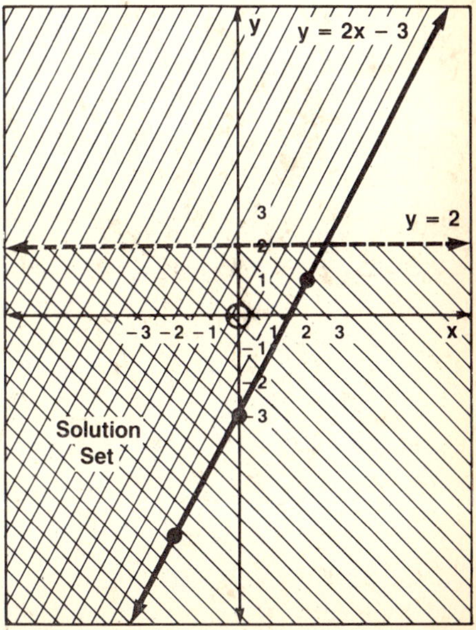

EXERCISES

In 1–15, graph each system of inequalities and label the solution set A.

1. $y \geq 3x + 1$
   $y < x + 1$
2. $y \leq 2x + 3$
   $y \geq x$
3. $y \leq 5$
   $y \leq x + 3$
4. $y > x$
   $x + y \leq 5$
5. $2y \geq x - 4$
   $y < 3x$
6. $2x - y \leq 3$
   $3x + y < 7$
7. $y \leq x - 3$
   $y > -x + 1$
8. $y > -2x + 10$
   $y \geq x + 1$
9. $y \geq 2x - 3$
   $y < 2$
10. $y < -x$
    $2x - y \leq 3$
11. $2y \geq x - 4$
    $y < 3x$
12. $x + y > 6$
    $y \leq 2x - 5$
13. $y > -2x + 4$
    $y > x + 2$
14. $2x - y \leq 3$
    $3x + y < 7$
15. $y \geq 2x - 3$
    $y < 2$

16. **a.** On the same set of coordinate axes, graph the following system of inequalities: $y < x - 2$
    $y \geq -2x + 4$

    b. Write the coordinates of a point that is *not* in the solution set of the system of inequalities graphed in part **a.**

17. **a.** On the same set of coordinate axes, graph the following system of inequalities: $y > x + 4$
    $x + y \leq 2$

    b. Which point is in the solution set of the graph drawn?
    (1) (2, 3) (2) (−5, 2) (3) (0, 6) (4) (−1, 0)

18. **a.** On the same set of axes, graph the solution set of the following system of inequalities: $y < -1$
    $y \leq 2x - 3$

    b. Name an ordered pair in the solution set of this system.

13.3 ALGEBRAIC SOLUTION OF A SYSTEM OF LINEAR EQUATIONS

In addition to solving a system of linear equations graphically, we can also solve it algebraically. By manipulating the two equations, we can eliminate one of the variables and reduce the problem to a single equation in a single variable. There are two basic methods for working with a system of linear equations.

Method 1: Elimination of One Variable by SUBSTITUTION

Consider the system of equations:
$$y = 2x - 1$$
$$3y + x = 11$$

Begin with the second equation, and replace y by $2x - 1$ from the first equation.

$$3y + x = 11$$
$$3(2x - 1) + x = 11$$

Solve this new equation for x.

$$\begin{aligned} 3(2x - 1) + x &= 11 \\ 6x - 3 + x &= 11 \\ 7x - 3 &= 11 \\ 7x &= 14 \\ \boxed{x = 2} \end{aligned}$$

Return to either of the original equations. Substitute $x = 2$, and solve for y.

$$\begin{aligned} y &= 2x - 1 \\ y &= 2(2) - 1 \\ y &= 4 - 1 \\ \boxed{y = 3} \end{aligned}$$

Check:

Check both values in *each* of the *original equations.*

$$(x, y) \stackrel{?}{=} (2, 3)$$

$$\begin{aligned} y &= 2x - 1 & 3y + x &= 11 \\ 3 &\stackrel{?}{=} 2(2) - 1 & 3(3) + 2 &\stackrel{?}{=} 11 \\ 3 &\stackrel{?}{=} 4 - 1 & 9 + 2 &\stackrel{?}{=} 11 \\ 3 &= 3\checkmark & 11 &= 11\checkmark \end{aligned}$$

Answer: $x = 2, y = 3$ or $(2, 3)$

The method of substitution may require transforming one of the given equations at the outset.

Consider the system of equations:

$$\begin{aligned} y + 3x &= 3 \\ x + 2y &= -4 \end{aligned}$$

Begin with the first equation and solve for y in terms of x.

$$\begin{aligned} y + 3x &= 3 \\ \boxed{y = -3x + 3} \end{aligned}$$

In the second equation, replace y by $-3x + 3$ from the adjusted first equation.

$$\begin{aligned} x + 2y &= -4 \\ x + 2(-3x + 3) &= -4 \end{aligned}$$

Solve this new equation for x.

$$\begin{aligned} x + 2(-3x + 3) &= -4 \\ x - 6x + 6 &= -4 \\ -5x + 6 &= -4 \\ -5x &= -10 \\ \boxed{x = 2} \end{aligned}$$

Return to any equation containing both x and y. Substitute $x = 2$, and solve for y.

$$\begin{aligned} y &= -3x + 3 \\ y &= -3(2) + 3 \\ y &= -6 + 3 \\ \boxed{y = -3} \end{aligned}$$

Check:

Check both values in *each* of the *original* equations.

$$(x, y) \stackrel{?}{=} (2, -3)$$

$$\begin{array}{rclcrcl} y + 3x &=& 3 & \quad & x + 2y &=& -4 \\ -3 + 3(2) &\stackrel{?}{=}& 3 & & 2 + 2(-3) &\stackrel{?}{=}& -4 \\ -3 + 6 &\stackrel{?}{=}& 3 & & 2 - 6 &\stackrel{?}{=}& -4 \\ 3 &=& 3\checkmark & & -4 &=& -4\checkmark \end{array}$$

Answer: $x = 2, y = -3$ or $(2, -3)$

Method 2: Elimination of One Variable by ADDITION or SUBTRACTION

Consider the system of linear equations: $x + y = 10$
$2x - y = 2$

Since y has coefficients that are equal in number and opposite in sign, y will "drop out" if the two equations are added.

Add.

$$\begin{array}{rcl} x + y &=& 10 \\ 2x - y &=& 2 \\ \hline \end{array}$$

Solve for x.

$$3x = 12$$

$$\boxed{x = 4}$$

Return to either of the original equations. Substitute $x = 4$, and solve for y.

$$x + y = 10$$
$$4 + y = 10$$
$$\boxed{y = 6}$$

Check:

Check both values in *each* of the *original* equations.

$$(x, y) \stackrel{?}{=} (4, 6)$$

$$\begin{array}{rclcrcl} x + y &=& 10 & \quad & 2x - y &=& 2 \\ 4 + 6 &\stackrel{?}{=}& 10 & & 2(4) - 6 &\stackrel{?}{=}& 2 \\ 10 &=& 10\checkmark & & 8 - 6 &\stackrel{?}{=}& 2 \\ & & & & 2 &=& 2\checkmark \end{array}$$

Answer: $x = 4, y = 6$ or $(4, 6)$

Consider the system: $2x + 10y = 11$
$2x - 4y = -3$

Since the coefficients of x are equal in number and in sign, addition will not elminate x, but subtraction will. (Remember: algebraic subtraction requires a change in sign.)

Subtract (change every sign in the subtrahend), and solve for y.

$$\begin{aligned} 2x + 10y &= 11 \\ \ominus 2x \oplus 4y &= \oplus 3 \\ \hline 14y &= 14 \\ \boxed{y = 1} \end{aligned}$$

Return to either of the original equations. Substitute $y = 1$, and solve for x.

$$\begin{aligned} 2x + 10y &= 11 \\ 2x + 10(1) &= 11 \\ 2x + 10 &= 11 \\ 2x &= 1 \\ \boxed{x = \tfrac{1}{2}} \end{aligned}$$

Check:

Check both values in *each* of the *original* equations.

$$(x, y) \stackrel{?}{=} (\tfrac{1}{2}, 1)$$

$$\begin{aligned} 2x + 10y &= 11 & 2x - 4y &= -3 \\ 2(\tfrac{1}{2}) + 10(1) &\stackrel{?}{=} 11 & 2(\tfrac{1}{2}) - 4(1) &\stackrel{?}{=} -3 \\ 1 + 10 &\stackrel{?}{=} 11 & 1 - 4 &\stackrel{?}{=} -3 \\ 11 &= 11 \checkmark & -3 &= -3 \checkmark \end{aligned}$$

Answer: $x = \frac{1}{2}$, $y = 1$ or $(\frac{1}{2}, 1)$

Sometimes the system requires a multiplier to make the coefficients of one variable equal in number.

Consider the system: $3x - y = 7$
$2x + 4y = 0$

The coefficients of y will be equal in number if the first equation is multiplied by 4.

Multiply the first equation by 4.

$$\begin{aligned} 4 \| \quad 3x - y &= 7 \\ \| \quad 2x + 4y &= 0 \end{aligned}$$

Add, and solve for x.

$$\begin{aligned} 12x - 4y &= 28 \\ 2x + 4y &= 0 \\ \hline 14x &= 28 \\ \boxed{x = 2} \end{aligned}$$

Return to any equation containing both x and y. Substitute $x = 2$, and solve for y.

$$\begin{aligned} 3x - y &= 7 \\ 3(2) - y &= 7 \\ 6 - y &= 7 \\ -y &= 1 \\ \boxed{y = -1} \end{aligned}$$

Check:

Check both values in *each* of the *original* equations.

$$(x, y) \stackrel{?}{=} (2, -1)$$

$3x - y = 7$	$2x + 4y = 0$
$3(2) - (-1) \stackrel{?}{=} 7$	$2(2) + 4(-1) \stackrel{?}{=} 0$
$6 + 1 \stackrel{?}{=} 7$	$4 - 4 \stackrel{?}{=} 0$
$7 = 7\checkmark$	$0 = 0\checkmark$

Answer: $x = 2$, $y = -1$ or $(2, -1)$

Sometimes the system requires that both equations be multiplied to make the coefficients of one variable equal in number.

Consider the system: $2x + 3y = 8$
$3x - 4y = -5$

The coefficients of x will be equal in number if the first equation is multiplied by 3 and the second equation by 2. Since the signs are the same, we then subtract to eliminate x.

(*Note:* An alternate choice would be to eliminate y. This could be done by multiplying the first equation by 4 and the second by 3. Since the signs of the coefficients are opposite, we then add to eliminate y.)

Multiply the first equation by 3. Multiply the second equation by 2.

$$3 \,\|\, 2x + 3y = 8$$
$$2 \,\|\, 3x - 4y = -5$$

Subtract (change each sign in the subtrahend), and solve for y.

$$\begin{aligned} \cancel{6x} + 9y &= 24 \\ \overset{-}{\oplus}\cancel{6x} \overset{+}{\ominus} 8y &= \overset{+}{\ominus}10 \\ \hline 17y &= 34 \end{aligned}$$

$$\boxed{y = 2}$$

Return to any equation with both x and y. Substitute $y = 2$, and solve for x.

$$\begin{aligned} 2x + 3y &= 8 \\ 2x + 3(2) &= 8 \\ 2x + 6 &= 8 \\ 2x &= 2 \end{aligned}$$

$$\boxed{x = 1}$$

Check:

Check both values in *each* of the *original* equations.

$$(x, y) \stackrel{?}{=} (1, 2)$$

$2x + 3y = 8$	$3x - 4y = -5$
$2(1) + 3(2) \stackrel{?}{=} 8$	$3(1) - 4(2) \stackrel{?}{=} -5$
$2 + 6 \stackrel{?}{=} 8$	$3 - 8 \stackrel{?}{=} -5$
$8 = 8\checkmark$	$-5 = -5\checkmark$

Answer: $x = 1$, $y = 2$ or $(1, 2)$

◆ In general, there are two procedures for obtaining an algebraic solution of a system of linear equations:

SUBSTITUTION

1. This method is especially useful if one variable is already expressed in terms of the other in one of the two equations. Use this expression in the other equation as a replacement for the first variable, thus reducing the problem to one equation in one variable.
2. Solve for that variable. Substitute this known value in either of the original equations, and solve for the second variable.
3. Check both values in *each* of the *original* equations.

ADDITION OR SUBTRACTION

1. Both methods require that the coefficients of one variable be equal in number. This may be the case from the beginning, or one or two multipliers may be needed.
2. After the coefficients of one variable are equal in number, use addition if they are opposite in sign; use subtraction if they have the same sign (remember to change each sign in the subtrahend).
3. Return to any equation involving both variables and substitute the known value. Solve for the other value.
4. Check both values in *each* of the *original* equations.

Examples

1. Solve the following system of linear equations and check:

$$y = 3x - 1$$
$$7x + 2y = 37$$

Solution:

Replace y by $3x - 1$ in the second equation, and solve for x.

$$\begin{aligned} 7x + 2y &= 37 \\ 7x + 2(3x - 1) &= 37 \\ 7x + 6x - 2 &= 37 \\ 13x - 2 &= 37 \\ 13x &= 39 \\ \boxed{x = 3} \end{aligned}$$

Substitute $x = 3$ in either of the original equations, and solve for y.

$$\begin{aligned} y &= 3x - 1 \\ y &= 3(3) - 1 \\ y &= 9 - 1 \\ \boxed{y = 8} \end{aligned}$$

Check:

Check both values in *each* of the *original* equations.

$$(x, y) \stackrel{?}{=} (3, 8)$$

$$\begin{array}{ll} y = 3x - 1 & 7x + 2y = 37 \\ 8 \stackrel{?}{=} 3(3) - 1 & 7(3) + 2(8) \stackrel{?}{=} 37 \\ 8 \stackrel{?}{=} 9 - 1 & 21 + 16 \stackrel{?}{=} 37 \\ 8 = 8\checkmark & 37 = 37\checkmark \end{array}$$

Answer: $x = 3,\ y = 8$ or $(3, 8)$

2. Solve the following system of linear equations and check:

$$\begin{aligned} 5x - 2y &= 20 \\ 2x + 3y &= 27 \end{aligned}$$

Solution 1:

To eliminate x, multiply the first equation by 2 and the second equation by 5. The coefficients of x will then be equal in number, 10, and will have the same sign. Subtract.

Multiply.

$$\begin{array}{l|l} 2 & 5x - 2y = 20 \\ 5 & 2x + 3y = 27 \end{array}$$

Subtract (change signs), and solve for y.

$$\begin{array}{rcrcr} \not{1}\not{0}x & - & 4y & = & 40 \\ \overset{-}{\not{\oplus}}\not{1}\not{0}x & \overset{-}{\oplus} & 15y & = & \overset{-}{\oplus}135 \\ \hline & - & 19y & = & -95 \end{array}$$

$$\boxed{y = 5}$$

Return to any equation with x and y. Substitute $y = 5$, and solve for x.

$$\begin{aligned} 5x - 2y &= 20 \\ 5x - 2(5) &= 20 \\ 5x - 10 &= 20 \\ 5x &= 30 \end{aligned}$$

$$\boxed{x = 6}$$

Check:

$$(x, y) \stackrel{?}{=} (6, 5)$$

$$\begin{array}{rr} 5x - 2y = 20 & 2x + 3y = 27 \\ 5(6) - 2(5) \stackrel{?}{=} 20 & 2(6) + 3(5) \stackrel{?}{=} 27 \\ 30 - 10 \stackrel{?}{=} 20 & 12 + 15 \stackrel{?}{=} 27 \\ 20 = 20\checkmark & 27 = 27\checkmark \end{array}$$

Answer: $x = 6,\ y = 5$ or $(6, 5)$

(*Note:* We could have multiplied the first equation by 2 and the second equation by -5. In this way, the coefficients of x would have been equal in number and opposite in sign. We would then have used addition to eliminate x.)

Solution 2:

To eliminate y, multiply the first equation by 3 and the second equation by 2. The coefficients of y will then be opposites (equal in number and opposite in sign). Add.

Multiply.

$$\begin{array}{c|l} 3 & 5x - 2y = 20 \\ 2 & 2x + 3y = 27 \end{array}$$

Add, and solve for y.

$$\begin{aligned} 15x - 6y &= 60 \\ 4x + 6y &= 54 \\ \hline 19x \quad &= 114 \\ \boxed{x = 6} \end{aligned}$$

Return to any equation with both x and y. Substitute $x = 6$, and solve for y.

$$\begin{aligned} 5x - 2y &= 20 \\ 5(6) - 2y &= 20 \\ 30 - 2y &= 20 \\ -2y &= -10 \\ \boxed{y = 5} \end{aligned}$$

As before, check both values in *each* of the *original* equations.

Answer: $x = 6,\ y = 5$ or $(6, 5)$

3. Solve the following system of equations and check:

$$\frac{x-1}{y} = \frac{1}{2}$$
$$2y - 3x = 6$$

Solution:

Clear of fractions in the first equation and simplify.

$$\begin{aligned} \frac{x-1}{y} &= \frac{1}{2} \\ \text{L.C.D.} &= 2y \\ 2y \cdot \frac{x-1}{y} &= \frac{1}{2} \cdot 2y \\ 2(x-1) &= 1(y) \\ 2x - 2 &= y \\ &\text{or} \\ \boxed{y = 2x - 2} \end{aligned}$$

Replace y by $2x - 2$ in the second equation, and solve for x.

$$\begin{aligned} 2y - 3x &= 6 \\ 2(2x-2) - 3x &= 6 \\ 4x - 4 - 3x &= 6 \\ x - 4 &= 6 \\ \boxed{x = 10} \end{aligned}$$

Return to any equation with both x and y. Substitute $x = 10$, and solve for y.

$$y = 2x - 2$$
$$y = 2(10) - 2$$
$$y = 20 \quad - 2$$
$$\boxed{y = 18}$$

Check:

Check both values in *each* of the *original* equations.

$$(x, y) \stackrel{?}{=} (10, 18)$$

$$\frac{x - 1}{y} = \frac{1}{2} \qquad\qquad 2y - 3x = 6$$
$$\frac{10 - 1}{18} \stackrel{?}{=} \frac{1}{2} \qquad\qquad 2(18) - 3(10) \stackrel{?}{=} 6$$
$$\frac{9}{18} \stackrel{?}{=} \frac{1}{2} \qquad\qquad 36 - 30 \stackrel{?}{=} 6$$
$$\frac{1}{2} = \frac{1}{2}\checkmark \qquad\qquad 6 = 6\checkmark$$

Answer: $x = 10,\ y = 18$ or $(10, 18)$

EXERCISES

In 1–15, solve each system of equations for x.

1. $2x + 3y = 7$
$x - 3y = 8$

2. $x + y = 7$
$2x - y = 2$

3. $3x + y = 7$
$2x - y = 8$

4. $2x + y = 7$
$x - y = 2$

5. $2x - y = 11$
$x + y = 1$

6. $x + y = 8$
$x - y = 2$

7. $3x + 2y = 19$
$x - 2y = 9$

8. $3x + y = 5$
$2x - y = 5$

9. $3x + y = 9$
$2x + y = 7$

10. $x + 9y = -6$
$2x - 9y = 15$

11. $5x - y = 20$
$y = 3x$

12. $2x - y = 5$
$y = 7 - x$

13. $2x - y = 3$
$y = 3 - x$

14. $x - y = 9$
$x + 4y = 24$

15. $x + 4y = 1$
$2x - 2y = -8$

In 16–20, solve each system of equations for x and y.

16. $x + 6y = 28$
$x - 2y = -4$

17. $x + y = 4$
$2x + y = 10$

18. $3x - y = 7$
$2x + y = 8$

19. $4x + y = 3$
$x + y = 12$

20. $x - 3y = 7$
$2x - 3y = 11$

In 21–25, select the ordered pair that is the solution set of each of the systems of equations.

21. $x + y = 5$
$y = 2$
(1) (3, 2) (2) (2, 3) (3) (5, 0) (4) (0, 5)

22. $2x + y = 7$
$x - 2y = 6$
(1) $(3, 1)$ (2) $(1, 3)$ (3) $(-1, 4)$ (4) $(4, -1)$

23. $2x + 3y = 7$
$x + y = 3$
(1) $(2, 1)$ (2) $(1, 2)$ (3) $(5, -1)$ (4) $(0, 3)$

24. $x + y = 5$
$x - y = -1$
(1) $(2, -3)$ (2) $(2, 3)$ (3) $(3, -2)$ (4) $(3, 2)$

25. $3x + 2y = 4$
$-2x + 2y = 24$
(1) $(2, -1)$ (2) $(-4, 8)$ (3) $(-4, -8)$ (4) $(2, -5)$

In 26–35, solve each system of equations for x and y and check.

26. $3x - 2y = 42$
$2x - y = 26$

27. $3x - 5y = -9$
$2x - y = 1$

28. $2x + y = -1$
$y = x + 5$

29. $4x + 3y = 27$
$y = 2x - 1$

30. $2x + y = 6$
$x = 3y + 10$

31. $2x + 3y = 6$
$3x + 5y = 9$

32. $3x - 2y = 42$
$2x - y = 26$

33. $2x + 3y = -4$
$5x + 2y = 1$

34. $3x + 2y = 1$
$2x + 3y = 9$

35. $3x - 4y = 12$
$4x - 3y = 2$

In 36–45, solve each system of equations for x and y and check.

36. $\frac{y}{2} = x + 1$
$4x - y = 6$

37. $3x - 2y = 6$
$\frac{y - 1}{x} = \frac{1}{2}$

38. $\frac{x - y}{2} = 1$
$\frac{x + y}{2} = 4$

39. $3x + y = 9$
$\frac{3x}{2} + \frac{4y}{5} = \frac{27}{5}$

40. $\frac{x}{2} + \frac{3y}{4} = 8$
$x + \frac{y}{4} = 6$

41. $\frac{3x}{2} + \frac{y}{5} = 6$
$\frac{x + 2}{4} - \frac{y - 3}{6} = 2$

42. $\frac{x}{y + 1} = \frac{2}{3}$
$x + y = 9$

43. $\frac{x}{5} - \frac{y}{10} = 1$
$\frac{x}{3} - \frac{y}{5} = 1$

44. $\frac{x}{2} + \frac{4y}{3} = 4$
$x + \frac{3y}{2} = 1$

45. $3x - 4y = 1$
$\frac{y + 2}{x} + \frac{8}{5} = \frac{20y + 4}{5x}$

13.4 USING TWO VARIABLES TO SOLVE PROBLEMS

Some types of problems require the use of two variables. Other types may be solved by using two variables or one variable. The use of two variables often makes it easier to write the necessary equations, even though the solution of the equations is lengthier than when one variable is used.

◆ In general, the procedure for solving problems by using two variables is:

1. Identify each of the two unknown quantities and represent them by letters.
2. Translate the verbal sentences relating the unknown quantities into two equations.
3. Solve the two equations simultaneously, either by the substitution method or by the addition/subtraction method.
4. Check the solution in the words of the problem.

Digit Problems

Useful information for two-digit numbers:

Let t = the tens digit.
Let u = the units digit.
Then $10t + u$ = the two-digit number,
$10u + t$ = the number with the digits reversed,
and $t + u$ = the sum of the digits.

Not all of the above information is called for in every problem. But t and u always identify the two unknown quantities.

Examples

1. The units digit of a two-digit number is one more than the tens digit. The number is five times the sum of the digits. Find the number.

Solution:

Let t = the tens digit.
Let u = the units digit.

The units digit is (one more than) (the tens digit)

$$u = t + 1$$

The number is five times the sum of the digits.

$$10t + u = 5 \cdot (t + u)$$

$$u = t + 1$$
$$10t + u = 5(t + u)$$

Substitute $u = t + 1$ in the second equation.

$$10t + u = 5(t + u)$$
$$10t + (t + 1) = 5(t + t + 1)$$

Remove ( ) by multiplication.

$$10t + t + 1 = 5t + 5t + 5$$

Collect like terms, and solve for t.

$$11t + 1 = 10t + 5$$
$$t + 1 = 5$$
$$\boxed{t = 4}$$

Substitute $t = 4$ in the first equation and solve for u.

$$u = t + 1$$
$$u = 4 + 1$$
$$\boxed{u = 5}$$

Check in the words of the problem:

5 (the units digit) is one more than 4 (the tens digit).√
45 (the number) is 5 · 9 (the sum of the digits).√

Answer: The number is 45.

2. The sum of the digits of a two-digit number is 5. The number with the digits reversed is 9 less than the original number. Find the original number.

Solution: Let t = the tens digit.
Let u = the units digit.

The sum of the digits is 5.

$$t + u = 5$$

The number reversed is (9 less than) (the original number.)

$$10u + t = 10t + u - 9$$

$$t + u = 5$$
$$10u + t = 10t + u - 9$$

Simplify the second equation to match the form of the first equation. (Subtract u, subtract $10t$, and divide by 9.)

$$9u + t = 10t - 9$$
$$9u - 9t = -9$$
$$u - t = -1$$

Write the first equation and the second equation, which has been simplified, in *matched form* (rearrange the simplified second equation). Add the two matched equations, and solve for u.

$$t + u = 5$$
$$-t + u = -1$$
$$2u = 4$$
$$\boxed{u = 2}$$

Substitute $u = 2$ in the first equation, and solve for t.

$$t + u = 5$$
$$t + 2 = 5$$
$$\boxed{t = 3}$$

Check in the words of the problem:

(3 + 2) (the sum of the digits) is 5.✓
23 (the number with the digits reversed) is 9 less than 32 (the original number).✓

Answer: The original number is 32.

Business Problems

Some business problems deal with two types of items being purchased. Either the cost of the individual items is unknown (see EXAMPLE 1) or the number of individual items purchased is unknown (see EXAMPLE 2). Both situations depend on two ideas:

1. The price of a number of the same items is calculated by multiplication (number of items × price per item).
2. The total price is a sum.

Examples

1. The manager of a bowling team bought 3 shirts and 4 caps for \$66. At another time, he bought 7 shirts and 3 caps for \$116 at the same prices as for the first purchase. What is the cost of a shirt and the cost of a cap?

   Solution:

   Let S = the cost of a shirt.
   Let C = the cost of a cap.

   The cost of 3 shirts and 4 caps: $3S + 4C = 66$
   The cost of 7 shirts and 3 caps: $7S + 3C = 116$

   Multiply the first equation by 7 and the second equation by -3 to get the coefficients of S equal in number but opposite in sign.

   $$\begin{array}{r|l} 7 & 3S + 4C = 66 \\ -3 & 7S + 3C = 116 \end{array}$$

   Add to eliminate S.

   $$\begin{array}{rcrcr} 21S & + & 28C & = & 462 \\ -21S & - & 9C & = & -348 \\ \hline & & 19C & = & 114 \end{array}$$

   $$\boxed{C = 6}$$

Substitute 6 for C in the first equation, and solve for S.

$$\begin{aligned} 3S + 4C &= 66 \\ 3S + 4(6) &= 66 \\ 3S + 24 &= 66 \\ 3S &= 42 \\ \boxed{S} &\boxed{= 14} \end{aligned}$$

Check:

Cost of 3 shirts		Cost of 4 caps	
\$14		\$ 6	
× 3		× 4	
\$42	+	\$24	= \$66✓

Cost of 7 shirts		Cost of 3 caps	
\$14		\$ 6	
× 7		× 3	
\$98	+	\$18	= \$116✓

Answer: A shirt costs \$14 and a cap costs \$6.

2. If purchased before the game, tickets to a high school baseball game cost \$1.00. If purchased at the gate, the tickets cost \$1.50. For a particular game, 500 tickets were sold and the total receipts were \$587.50. How many tickets were purchased before the game and how many tickets were purchased at the gate?

Solution:

Let B = the number of tickets purchased before the game.
Let G = the number of tickets purchased at the gate.

Total number of tickets purchased was 500: $B + G = 500$
Total amount of money received was \$587.50: $1.00B + 1.50G = 587.50$

Multiply the second equation by 100 to clear of decimals.

$$\begin{aligned} 1.00B + 1.50G &= 587.50 \\ 100B + 150G &= 58{,}750 \end{aligned}$$

Align the first equation and the simplified second equation.

$$\begin{aligned} B + G &= 500 \\ 100B + 150G &= 58{,}750 \end{aligned}$$

Multiply the first equation by 100 and the simplified second equation by -1 to obtain coefficients of B that are equal in number but opposite in sign.

$$\begin{aligned} 100 \quad & B + G = 500 \\ -1 \quad & 100B + 150G = 58{,}750 \end{aligned}$$

Add to eliminate B.
Solve for G.

$$\begin{aligned} 100B + 100G &= 50{,}000 \\ -100B - 150G &= -58{,}750 \\ \hline -50G &= -8{,}750 \\ \boxed{G = 175} & \end{aligned}$$

Substitute 175 for G in the first equation, and solve for B.

$$B + G = 500$$
$$B + 175 = 500$$
$$\boxed{B = 325}$$

Check:

Before game		At gate		
\$ 1		\$1.50		
× 325		× 175		
\$325	+	\$262.50	=	\$587.50 ✓

Answer: There were 325 tickets purchased before the game and 175 tickets purchased at the gate.

Many other types of problems can be solved by using two variables. The general procedure for using two variables, discussed at the opening of this section, should be applied.

Example

A certain fraction is equivalent to $\frac{2}{5}$. If the numerator of this fraction is decreased by 2 and its denominator is decreased by 1, the resulting fraction is equivalent to $\frac{1}{3}$. Find the original fraction.

Solution:

Let x = the numerator of the original fraction.
Let y = the denominator of the original fraction.
Then $\frac{x}{y}$ = the original fraction.

$$\underbrace{\text{The original fraction}}_{\frac{x}{y}} \; \underbrace{\text{is equivalent to}}_{=} \; \underbrace{\tfrac{2}{5}}_{\frac{2}{5}}.$$

Let $x - 2$ = the numerator of the new fraction.
Let $y - 1$ = the denominator of the new fraction.
Then $\frac{x-2}{y-1}$ = the new fraction.

$$\underbrace{\text{The new fraction}}_{\frac{x-2}{y-1}} \; \underbrace{\text{is equivalent to}}_{=} \; \underbrace{\tfrac{1}{3}}_{\frac{1}{3}}.$$

Clear of fractions in the first equation by cross-multiplication.

$$\frac{x}{y} = \frac{2}{5}$$
$$2y = 5x$$

Clear of fractions in the second equation by cross-multiplication.

$$\frac{x-2}{y-1} = \frac{1}{3}$$
$$y - 1 = 3(x - 2)$$
$$y - 1 = 3x - 6$$

Simplify.

$$y = 3x - 5$$

Substitute $3x - 5$ for y in the adjusted first equation. Solve for x.

$$2y = 5x$$
$$2(3x - 5) = 5x$$
$$6x - 10 = 5x$$
$$x = 10$$

Substitute 10 for x in any equation containing both x and y. Solve for y.

$$y = 3x - 5$$
$$y = 3(10) - 5$$
$$y = 30 - 5$$
$$y = 25$$

The original fraction is $\frac{x}{y} = \frac{10}{25}$. The new fraction is $\frac{x-2}{y-1} = \frac{8}{24}$.

Check:

The original fraction, $\frac{10}{25}$, is equivalent to $\frac{2}{5}$.

$$\frac{10}{25} \stackrel{?}{=} \frac{2}{5}$$
$$\frac{\not{5} \cdot 2}{\not{5} \cdot 5} \stackrel{?}{=} \frac{2}{5}$$
$$\frac{2}{5} = \frac{2}{5} \checkmark$$

The new fraction, $\frac{8}{24}$, is equivalent to $\frac{1}{3}$.

$$\frac{8}{24} \stackrel{?}{=} \frac{1}{3}$$
$$\frac{\not{8} \cdot 1}{\not{8} \cdot 3} \stackrel{?}{=} \frac{1}{3}$$
$$\frac{1}{3} = \frac{1}{3} \checkmark$$

Answer: The original fraction is $\frac{10}{25}$.

EXERCISES

In 1–13, only an algebraic solution is acceptable.

1. A two-digit number is 8 times the sum of the digits. Four times the units digit is 1 more than the tens digit. Find the number.
2. In a two-digit number the tens digit is 5 greater than the units digit. The number is 63 greater than the sum of its digits. Find the number.
3. The sum of the digits of a two-digit number is 9. If the digits are reversed, the new number is 9 less than the original number. Find the original number.
4. The sum of the digits of a two-digit number is 7. The number is 2 more than twice the number formed when the digits are reversed. Find the number.
5. The sum of the digits of a two-digit number is 11. The number obtained by reversing the digits is 7 more than twice the original number. Find the original number.
6. The units digit of a two-digit number is 4 more than the tens digit. If the digits of the number are reversed, a new number is formed that is 10 more than twice the original number. Find the original number.
7. The units digit of a two-digit number is 3 more than the tens digit. If the number is divided by the sum of the digits, the quotient is 4. Find the original number.

8. A woman bought 7 ears of corn and 6 oranges for \$3.66. A second woman bought 10 ears of corn and 3 oranges for \$3.39. Find the price of an ear of corn and the price of an orange.
9. At a cake sale, David bought 3 cakes and 5 cookies, and paid \$10.50. At the same prices, Lisa bought 4 cakes and 8 cookies, and paid \$14.80. What was the price of a cake and the price of a cookie?
10. The cost of a high school ring was \$75 for the large size and \$55 for the regular size. The total receipts from the sale of 120 rings were \$8,600. How many rings of each size were sold?
11. The total attendance at a school play was 850. The tickets for senior citizens were \$1.50 each, and the regular tickets were \$2.00 each. If the total receipts were \$1,650, how many tickets of each kind were sold?
12. Tickets for a high school dance cost \$2.50 each if purchased in advance of the dance, but are \$3.25 each if bought at the door. For the dance, 100 tickets were sold and \$265 was collected. How many tickets were sold at the door?
13. In a certain fraction, if 1 is subtracted from both the numerator and denominator, the resulting fraction is equivalent to $\frac{1}{2}$. If 5 is added to both the numerator and denominator of the original fraction, the resulting fraction is equivalent to $\frac{4}{5}$. Find the original fraction.

13.5 SUMMARY EXERCISES

1. Which ordered pair is the solution set of the system of equations:

$$3x - y = 9$$
$$y = -9$$

(1) (9, 0) (2) (−9, 0) (3) (0, 9) (4) (0, −9)

2. What is the solution set of the system of equations whose graph is shown?
(1) (−1, 2)
(2) (0, 2)
(3) (2, 0)
(4) (2, −1)

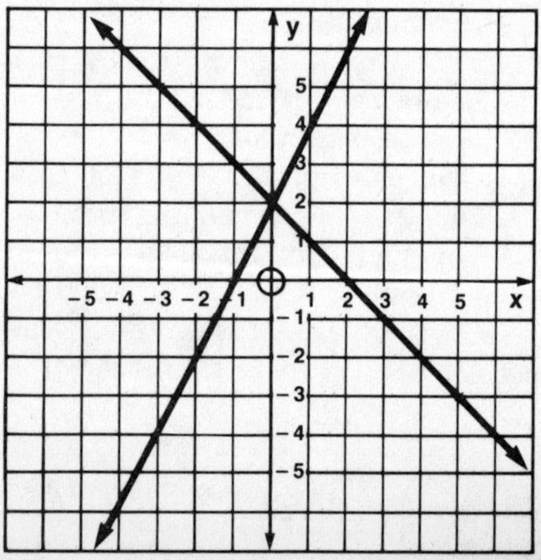

3. Which ordered pair is the solution set of the system of equations:

$$2x - y = 4$$
$$3x + y = 1$$

(1) $(-1, -2)$ (2) $(1, -2)$ (3) $(-2, 1)$ (4) $(-2, -1)$

4. What is the solution set of the system of equations whose graph is shown at the right?

(1) $(-2, 0)$ (3) $(2, 0)$
(2) $(0, -2)$ (4) $(0, 2)$

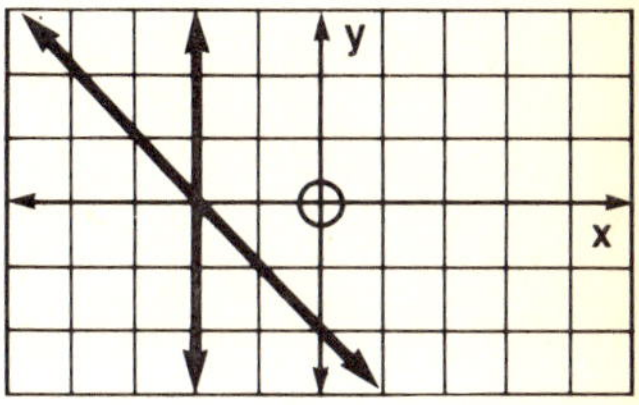

5. Which point is in the solution set of the system of inequalities whose graph is shown at the right?

(1) $(0, 0)$
(2) $(0, 2)$
(3) $(0, 3)$
(4) $(3, 0)$

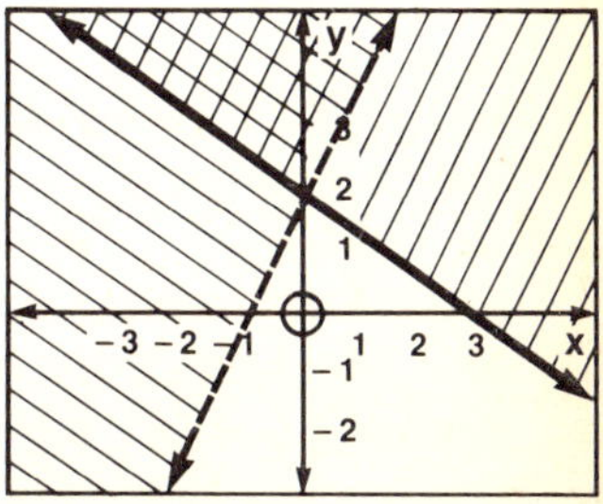

6. Which point is in the solution set of the system of inequalities whose graph is shown at the right?

(1) $(-2, 0)$
(2) $(0, -4)$
(3) $(0, -3)$
(4) $(0, 0)$

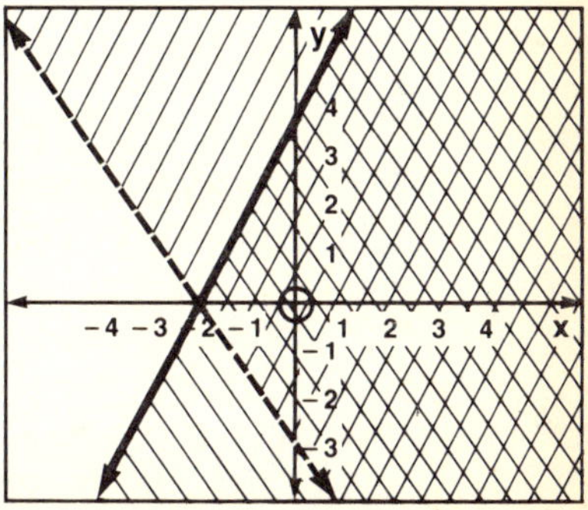

In 7–11, solve graphically and check each system of equations.

7. $3x = y + 6$
$x + y = -2$

8. $2x = y + 6$
$3x - y = 2$

9. $2x + y = 8$
$y = 5x - 6$

10. $x + y = 9$
$2x - y = 0$

11. $x + 2y = 9$
$y = 3x + 1$

In 12–16, graph each system of inequalities, and label the solution set A.

12. $y \geq 2x - 3$
$y < 4$

13. $y - 2x < 4$
$x \leq 2$

14. $x - y \leq -2$
$x + y > 2$

15. $x + y \leq 3$
$y \geq x$

16. $2x - y \geq 3$
$3x + y < 6$

In 17–20, solve each system of equations for x.

17. $3x + y = 9$
$2x - y = 6$

18. $2x - 2y = 3$
$5x + 2y = 11$

19. $6x + y = 16$
$y = 2x$

20. $4x - y = 9$
$y = x$

In 21–30, solve each system of equations for x and y and check.

21. $y = x - 2$
$3x - y = 16$

22. $x + y = 0$
$3x + 2y = 5$

23. $y = x + 4$
$3x + y = 16$

24. $x = y + 5$
$2y + x = -1$

25. $2x + 3y = 7$
$4x - 5y = 25$

26. $3x - 5y = 11$
$x + 4y = -2$

27. $2x + 3y = 3$
$3x + 4y = 3$

28. $4x + 3y = 10$
$5x + 8y = 4$

29. $\frac{x - y}{3} = 1$
$\frac{x + y}{3} = 5$

30. $x + \frac{y}{2} = 6$
$x - y = 3$

In 31–38, solve and check each problem.

31. The sum of the digits of a two-digit number is 8. If the digits are reversed, the new number is 36 less than the original number. Find the original number.

32. The tens digit of a two-digit number is 3 more than the units digit. The number is 6 less than seven times the sum of the digits. Find the number.

33. The units digit of a two-digit number is 1 more than twice the tens digit. If the digits are reversed, the new number is 1 less than twice the original number. Find the original number.

34. The tens digit of a two-digit number is 1 less than the units digit. If the number is divided by the sum of the digits, the quotient is 5. Find the original number.

35. David bought 7 shirts of equal value and 4 ties of equal value at a store and paid \$150. At the same store, Tom bought 3 of the same shirts and 5 of the same ties and paid \$84. What was the price of each shirt and each tie?

36. For a performance, 4 adult tickets and 5 senior citizen tickets cost \$27.75. For the same performance, 12 adult tickets and 8 senior citizen tickets cost \$64. Find the price of each kind of ticket.

37. The total attendance at a school play was 750. Tickets bought before the performance were \$2 each, and tickets bought at the door were \$2.75 each. How many tickets were bought at the door if the total receipts were \$1,706.25?

38. If 1 is subtracted from both the numerator and denominator of a certain fraction, the resulting fraction is equivalent to $\frac{3}{4}$. If 5 is added to both the numerator and denominator of the original fraction, the resulting fraction is equivalent to $\frac{6}{7}$. Find the original fraction.

HIGH SCHOOL MATHEMATICS: COURSE I–JUNE 1980 (1)

Part I

Answer 30 questions from this part. Each correct answer will receive 2 credits. Write your answers in the spaces provided on the separate answer sheet. Where applicable, answers may be left in terms of π or in radical form.

1. Solve for x: $3x - 5 = 16$
2. Solve for r in terms of C and π: $C = 2\pi r$
3. The length of a rectangle is 8 centimeters and its width is 6 centimeters. Find the number of centimeters in the length of the diagonal.
4. How many different two-digit numbers can be made from the digits of 345 if a digit can appear just once in a number?
5. Find the radius of a circle whose area is 25π.
6. Find the numerical value of the expression $c + xy$ when $c = 3$, $x = 4$, and $y = -5$.
7. If a 6-sided die and a coin are tossed simultaneously, how many pairs (outcomes) are there in this sample space?
8. Express the product $(x - 3)(x + 3)$ as a binomial.
9. Solve for x: $0.2x + 0.3 = 8.1$
10. Solve for c: $\frac{1}{3}c + 2 = 4$
11. Find the sum of $3x^2 + 2x - 4$ and $2x^2 - 3x + 1$.
12. If 60% of a number is 144, what is the number?
13. Find the mode for the following data: 4, 5, 3, 4, 5, 3, 5
14. Solve for the positive value of x: $x^2 - 81 = 0$
15. Factor: $x^2 - 11x + 24$
16. A farmer grew 105 bushels of grain on a 3-acre plot. How many bushels can he expect to grow under the same conditions on a 5-acre plot?
17. Solve the following system of equations for y: $$\begin{aligned} 3x + 2y &= 7 \\ -3x + y &= 8 \end{aligned}$$
18. Let p represent, "You are correct," and let q represent, "I agree." Using p and q, write in symbolic form, "You are correct and I agree."
19. In the accompanying figure, triangle ABC is congruent to triangle RST, $\overline{AB} \cong \overline{RS}$, $\overline{BC} \cong \overline{ST}$, and $\overline{AC} \cong \overline{RT}$. What angle in triangle RST is congruent to $\angle BAC$?

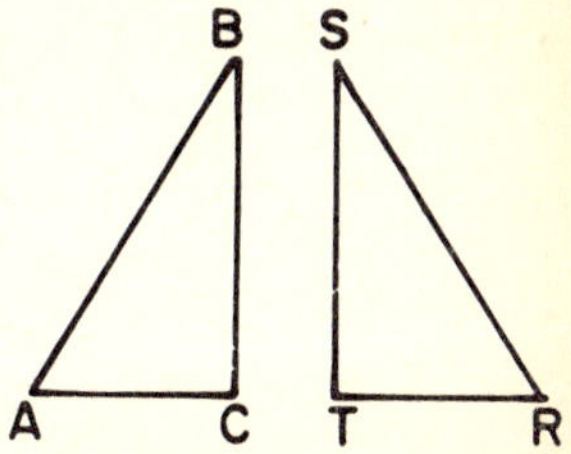

20. The area of a triangle is 40 square centimeters. If the base of the triangle is 10 centimeters, find the number of centimeters in the height.

21. In the accompanying figure, $\overleftrightarrow{AB}$ is parallel to $\overleftrightarrow{CD}$, and these lines are cut by the transversal $\overleftrightarrow{EF}$ at points E and F, respectively. If the measure of $\angle AEF$ is 70°, find the number of degrees in the measure of $\angle DFE$.

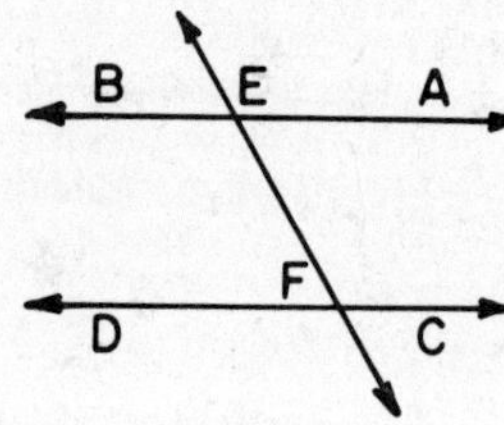

22. If the heights in centimeters of five students are 176, 172, 160, 158, and 158, find the median height for this group of five students.

23. The probability we will win the baseball game is $\frac{3}{5}$. What is the probability we will *not* win the game?

Directions (24–35): For *each* question chosen, write on the separate answer sheet the *numeral* preceding the word or expression that best completes the statement or answers the question.

24. The expression $\dfrac{-12x^6y^2}{3x^2y}$ is equivalent to

(1) $9x^3y^2$ (2) $4x^4y$ (3) $-4x^3y$ (4) $-4x^4y$

25. Two complementary angles are in the ratio 4 : 5. What is the measure in degrees of the *smaller* angle?
(1) 10 (2) 40 (3) 50 (4) 80

26. The graph of which equation is shown in the accompanying diagram?

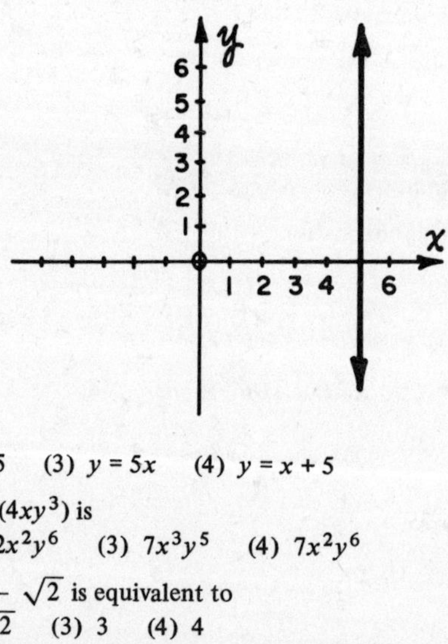

(1) $x = 5$ (2) $y = 5$ (3) $y = 5x$ (4) $y = x + 5$

27. The product $(3x^2y^2)(4xy^3)$ is
(1) $12x^3y^5$ (2) $12x^2y^6$ (3) $7x^3y^5$ (4) $7x^2y^6$

28. The expression $\sqrt{18} - \sqrt{2}$ is equivalent to
(1) $8\sqrt{2}$ (2) $2\sqrt{2}$ (3) 3 (4) 4

29. If the length of a rectangle is doubled and the width is also doubled, then the perimeter must be
(1) doubled (2) increased by 2 (3) increased by 4 (4) multiplied by 4

30. What is the inverse of the statement, "If I study, then I will pass"?
(1) If I pass, then I did study.
(2) If I do not pass, then I did not study.
(3) If I do not study, then I will not pass.
(4) If I study, then I will not pass.

31. If 2 six-sided dice are rolled, what is the probability of getting a sum of 3?
(1) $\frac{1}{36}$ (2) $\frac{2}{36}$ (3) $\frac{3}{36}$ (4) $\frac{2}{6}$

32. The graph of $3x - y = 3$ intersects the x-axis at the point
(1) (1, 0) (2) (0, 1) (3) (0, 3) (4) (3, 0)

33. Let p represent: x is odd. Let q represent: $x > 8$. When x is 6, which is true?
(1) p (2) q (3) $p \rightarrow q$ (4) $p \wedge q$

34. Given the true statement, "If I live in Lack Placid, then I live in New York State." Which statement must also be true?
(1) If I do not live in Lake Placid, then I do not live in New York State.
(2) If I do not live in New York State, then I do not live in Lake Placid.
(3) If I live in New York State, then I live in Lake Placid.
(4) I live in Lake Placid and I do not live in New York State.

35. Which ordered pair is in the solution set of $y > 3x + 2$?
(1) (1, 5) (2) (1, 6) (3) (5, 1) (4) (0, 0)

Part II

Answer four questions from this part. Show all work unless otherwise directed.

36. Solve graphically and check:

$$\begin{aligned} 2y &= x + 6 \\ 3x - y &= 2 \end{aligned} \quad [8, 2]$$

37. In triangle ABC, the measures of angles A and B are in the ratio 1:3. Angle C is 5 degrees more than angle A. Find the number of degrees in each angle. [*Only an algebraic solution will be accepted.*] [5, 5]

38. The square of a positive number is 42 more than the number itself. What is the number? [*Only an algebraic solution will be accepted.*] [5, 5]

39. In the accompanying figure, $\overline{AC}$ is a diameter of circle O. The measure of arc AB is 80°.
a. Find, in degrees, the measure of $\angle AOB$. [2]
b. Find, in degrees, the measure of $\angle ACB$. [2]
c. If the area of the circle is 49π, find the length of the radius. [2]
d. If the area of the circle is 49π, find the circumference of the circle in terms of π. [2]
e. Triangle COB is
(1) scalene (2) isosceles (3) equilateral
(4) right [2]

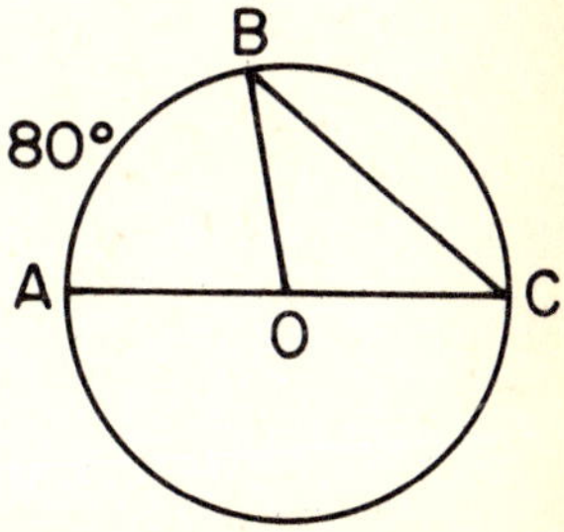

40. *a.* On your answer paper, copy and complete the truth table for the statement $[(p \to q) \wedge \sim q] \to \sim p$. [8]

p	q	$p \to q$	$\sim q$	$(p \to q) \wedge \sim q$	$\sim p$	$[(p \to q) \wedge \sim q] \to \sim p$

b. Is $[(p \to q) \wedge \sim q] \to \sim p$ a tautology? [1]
c. Justify the answer you gave in part *b.* [1]

41. In a certain class, there are 4 students in the first row: three girls, Ann, Barbara, and Cathy, and one boy, David. The teacher called one of these students to the board to solve a problem. When the problem was done, the teacher called one of the remaining students in the first row to do a second problem at the board.
a. Draw a tree diagram or list the sample space of all possible pairs of names for calling two students to the board. [4]
b. Find the probability that the teacher called Ann first and Barbara second. [2]
c. Find the probability that the teacher called two girls to the board. [2]
d. Find the probability that David was one of the two students called. [2]

42. The following data are test scores for a class of 16 students: 96, 83, 91, 77, 58, 88, 80, 62, 89, 100, 87, 93, 64, 98, 88, 86.
a. On your answer paper, copy and complete the following table. [2]

Interval	Number (frequency)
91–100	
81–90	
71–80	
61–70	
51–60	

b. On graph paper, construct a frequency histogram based on the data. [4]
c. Which interval contains the median? [2]
d. Which interval contains the lower quartile? [2]

HIGH SCHOOL MATHEMATICS: COURSE I–JANUARY 1981 (1)

Part I

Answer 30 questions from this part. Each correct answer will receive 2 credits. No partial credit will be allowed. Write your answers on a separate sheet. Where applicable, answers may be left in terms of π or in radical form.

1. Solve for c: $3(2 + c) = 12$

2. What is the value of $3x^3$ if $x = 2$?

3. Solve for z: $3z + 9 = 1 - z$

4. If a car can travel 60 kilometers on 3 liters of gasoline, how many kilometers can it travel under the same conditions on 7 liters?

5. Find the circumference of a circle whose radius is 6. [Answer may be left in terms of π.]

6. Solve for x: $0.2x + 6 = 80$

7. Express the product $(x + 2)(x - 5)$ as a trinomial.

8. Solve for r in terms of S, π, and h: $S = 2\pi rh$

9. A student received test scores of 82, 94, and 96. What must she receive as a fourth score so that the mean of these four scores will be exactly 90?

10. If p represents the statement, "I try," and q represents the statement, "I will succeed," write in symbolic form, using p and q: "If I try, then I will succeed."

11. If the three angles of a triangle are in the ratio 1:3:5, find the number of degrees in the *smallest* angle.

12. In the accompanying figure, $\angle BCD$ is an exterior angle to $\triangle ABC$ at vertex C. If $m\angle A = 50$ and $m\angle B = 60$, find the number of degrees in the measure of $\angle BCD$.

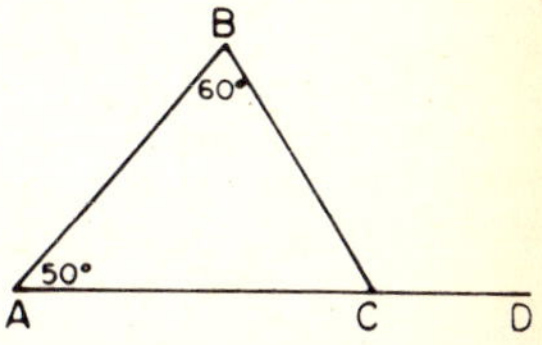

13. Factor: $x^2 + 2x - 3$

14. The lengths of the sides of a triangle are represented by $a + 2b$, $a + b$, and $2a - b$. Express the perimeter of the triangle as a binomial.

15. If the sales tax rate is 4%, what is the amount of sales tax which must be paid on a \$350 motorcycle?

16. The expression $\dfrac{4}{(x - 3)}$ is undefined for what value of x?

17. In the accompanying figure, $\triangle ABC$ is inscribed in circle O and $m\widehat{AB} = 80$. Find the number of degrees in $\angle ACB$.

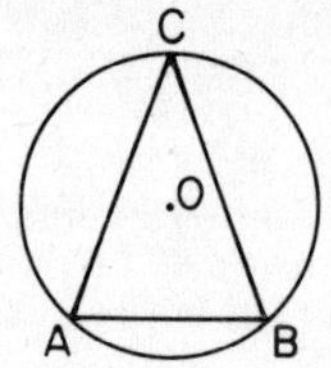

18. What is the *positive* root of the equation $x^2 - 49 = 0$?

19. A letter is chosen at random from the letters of the word "HUNDRED." What is the probability that the letter chosen is an "E"?

20. Solve the following system of equations for x: $x + y = 8$
$x - y = 2$

21. The lengths of two legs of a right triangle are 5 and 12. Find the length of the hypotenuse.

Directions (22–35): For *each* question chosen write on a separate sheet the *numeral* preceding the word or expression that best completes the statement or answers the question.

22. The symbol for n factorial is n!. The value of 3! is
(1) 6 (2) 5 (3) 3 (4) –3

23. The quotient $\dfrac{12x^3y^6}{4xy^2}$ is equivalent to
(1) $3x^2y^3$ (2) $3x^2y^4$ (3) $3x^3y^4$ (4) $3x^3y^3$

24. The sum of $\sqrt{18}$ and $15\sqrt{2}$ is
(1) $18\sqrt{2}$ (2) $24\sqrt{2}$ (3) $15\sqrt{20}$ (4) $16\sqrt{20}$

25. If a boy has 5 shirts and 4 pairs of jeans, how many different outfits of a shirt and a pair of jeans does he have?
(1) 20 (2) 9 (3) 5 (4) 4

26. What is the slope of the line whose equation is $y = 3x - 7$?
(1) –7 (2) –3 (3) 3 (4) 7

27. Which set of data has more than one mode?
(1) 2, 2, 4, 6, 7, 9 (2) 2, 2, 4, 6, 9, 9 (3) 2, 2, 2, 6, 7, 9 (4) 2, 3, 4, 6, 9, 9

28. What is the contrapositive of the statement $p \rightarrow q$?
(1) $p \longleftrightarrow q$ (2) $q \rightarrow p$ (3) $\sim p \rightarrow \sim q$ (4) $\sim q \rightarrow \sim p$

29. The graph of $3x - y = 6$ contains the point
(1) (0, 2) (2) (5, 9) (3) (4, –6) (4) (1, 9)

30. The number line below shows the solution set of which inequality?

(1) $-3 < x < 5$ (2) $-3 < x \leq 5$ (3) $-3 \leq x < 5$ (4) $-3 \leq x \leq 5$

31. The statement $r \vee s$ is false if and only if
(1) r is true and s is true (3) r is false and s is true
(2) r is true and s is false (4) r is false and s is false

32. If the length of each side of a square is represented by $2x$, which expression represents its area?
(1) $8x$ (2) $4x$ (3) $4x^2$ (4) $2x^2$

33. The volume of a cylinder is found by using the formula $V = \pi r^2 h$. If h is doubled, then the volume is
(1) doubled (2) increased by 2 (3) increased by 4 (4) multiplied by 4

34. From a standard deck of playing cards, what is the probability of randomly drawing a black card or a king on one draw?

(1) $\frac{2}{52}$ (2) $\frac{4}{52}$ (3) $\frac{28}{52}$ (4) $\frac{30}{52}$

35. Which represents the inverse of the statement, "If the base angles of a triangle are congruent, then the triangle is isosceles"?
(1) If the base angles of the triangle are not congruent, then the triangle is not isosceles.
(2) If the triangle is isosceles, then the base angles are congruent.
(3) If the triangle is not isosceles, then the base angles are not congruent.
(4) If the base angles of a triangle are not congruent, then the triangle is isosceles.

Part II

Answer four questions from this part. Show all work unless otherwise directed.

36. *a.* On the same set of coordinate axes, graph the following system of inequalities:

$$y < x - 2$$
$$y \geq -2x + 4 \quad [8]$$

b. Write the coordinates of a point which is *not* in the solution set of the system of inequalities graphed in part *a*. [2]

37. Solve for x and y and check:

$$2x + y = -1$$
$$y = x + 5 \quad [8, 2]$$

38. One number is 10 more than another number. If the product of the two numbers is –21, find both numbers. [*Only an algebraic solution will be accepted.*] [5, 5]

39. As shown in the accompanying diagram, $ABCD$ is a rectangle and a line segment drawn from B intersects $\overline{CD}$ at E.

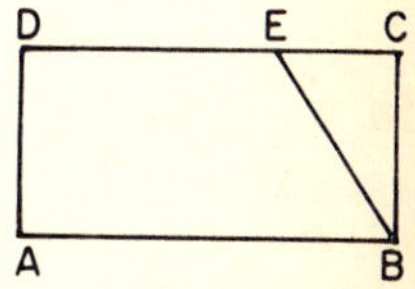

a. If the measure of $\overline{AB}$ is 9 and the area of the rectangle is 36, find AD. [2]
b. The point E separates $\overline{DC}$ into two segments such that $DE:EC = 2:1$. Find DE and EC. [2]
c. Find BE. [3]
d. Find the area of trapezoid $ABED$. [3]

40. A student has the following five books in his locker: math, history, Latin, science, and English. Without looking at the books, he pulls out one book and then, without replacing the first book, pulls out a second book.

a. Using the letters M, H, L, S, and E, draw a tree diagram or write the sample space for pulling two books from the locker. [6]

b. What is the probability that he first pulled the math book and then the English book? [2]

c. What is the probability that he pulled out two different books? [2]

41\. *a.* On your answer paper, copy and complete the truth table for the statement $(p \rightarrow q) \longleftrightarrow (\sim p \rightarrow \sim q)$. [8]

p	q	$p \rightarrow p$	$\sim p$	$\sim q$	$\sim p \rightarrow \sim q$	$(p \rightarrow q) \longleftrightarrow (\sim p \rightarrow \sim q)$

b. Which statement is true? [2]

(1) A conditional and its inverse always have the same truth value.
(2) A conditional and its contrapositive do not always have the same truth value.
(3) A conditional and its converse always have the same truth value.
(4) A conditional and its inverse do not always have the same truth value.

42\. A class record showed the following number of misspelled words in each of 25 essays.

Misspelled Words	Frequency (Number of Essays)
0	1
1	0
2	3
3	5
4	4
5	9
6	3

a. On graph paper, construct a frequency histogram based on the data. [3]

b. Find the mean number of misspelled words. [3]

c. Find the median number of misspelled words. [2]

d. Find the mode number of misspelled words. [2]

Part I

Answer 30 questions from this part. Each correct answer will receive 2 credits. No partial credit will be allowed. Write your answers on a separate sheet. Where applicable, answers may be left in terms of π or in radical form.

1. Solve for x: $4x - 3 = 41$

2. Factor: $3x + 6y$

3. Factor: $a^2 - 4a - 21$

4. What is 12.5% of 80?

5. Solve for a: $3a + 0.2 = 5$

6. What is the positive root of the equation $(x - 4)(x + 3) = 0$?

7. If the length of the hypotenuse of a right triangle is 13 and the length of one leg is 12, find the length of the other leg.

8. The lengths of the sides of $\triangle ABC$ are 5, 6, and 7. Triangle RST is similar to $\triangle ABC$. The longest side of $\triangle RST$ is 21. Find the length of the *shortest* side of $\triangle RST$.

9. The heights of the starting five players on a high school basketball team are 77″, 72″, 69″, 74″, and 77″. What is the median height of these players?

10. Two angles of a triangle are equal in measure and the measure of the third angle is 150°. Find the number of degrees in one of the two equal angles.

11. If 25 pennies weigh 42 grams, find the weight in grams of 75 pennies.

12. In the accompanying figure, the spinner has five equal sections numbered 1 through 5. If the arrow is equally likely to land on any of the sections, what is the probability that it will land on an even number on the next spin?

13. Let p represent "It is summer" and q represent "I go swimming." Using p and q, write in symbolic form: "If I go swimming, then it is summer."

14. Solve the following system of equations for x: $3x + 2y = 19$
$x - 2y = 9$

15. An urn contains 2 red marbles, 4 blue marbles, and 3 yellow marbles. If a single marble is randomly picked from the urn, what is the probability that a red or a blue marble is picked?

16. If the point $(k, 4)$ lies on the graph of $2x + y = 10$, what is the value of k?

17. As shown in the accompanying diagram, angle ABC is inscribed in circle O and the measure of arc $AC = 80°$. Find the number of degrees in the inscribed angle ABC.

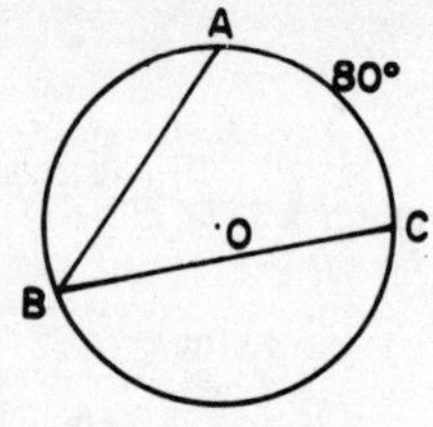

18. If the probability that Jones will win the election is 0.6, what is the probability that Jones will *not* win the election?

19. A student received test scores of 72, 84, and 86. What score must the student receive on a fourth test so that the mean of these scores will be 85?

20. The high temperatures during five days were 82°, 86°, 91°, 79°, and 91°. Find the mode for these temperatures.

Directions (21–34): For *each* question chosen, write on the separate sheet the *numeral* preceding the word or expression that best completes the statement or answers the question.

21. If the length of a rectangle is represented by $2x - 3$ and its width is represented by $3x$, what is the area of the rectangle in terms of x?
(1) $6x^2 - 9x$ (2) $6x^2 - 3$ (3) $10x - 6$ (4) $5x - 3$

22. One member of the solution set of $-5 < x \leqslant -1$ is
(1) -1 (2) -5 (3) 3 (4) -6

23. Which is a rational number?
(1) $\sqrt{6}$ (2) $\sqrt{16}$ (3) $\sqrt{18}$ (4) $\sqrt{24}$

24. If $p \leftrightarrow q$ is true, which statement must also be true?
(1) $p \wedge q$ (2) $p \vee q$ (3) $p \wedge \sim q$ (4) $p \rightarrow q$

25. In the accompanying diagram, the vertical line that intersects the x-axis at -2 is the graph of which equation?

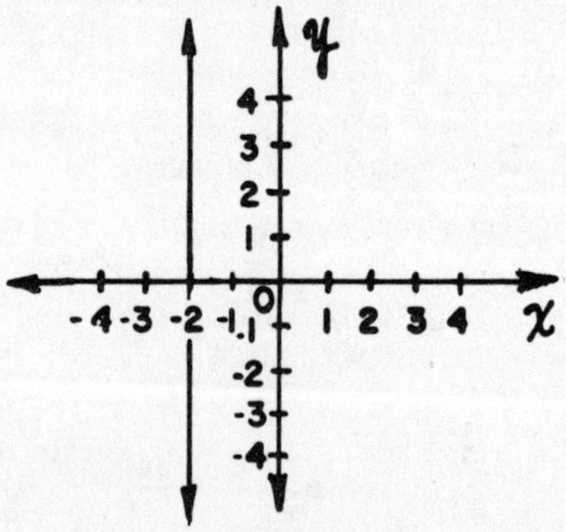

(1) $x = -2$ (2) $y = -2$ (3) $y = x - 2$ (4) $x + y = -2$

26. The product of $3x^5$ and $2x^4$ is
(1) $5x^9$ (2) $5x^{20}$ (3) $6x^9$ (4) $6x^{20}$

27. What is the inverse of $p \rightarrow \sim q$?
(1) $q \rightarrow \sim p$ (2) $\sim q \rightarrow p$ (3) $\sim p \rightarrow q$ (4) $\sim(p \rightarrow q)$

28. Which has the same truth value as the statement, "If Bill pitched, then we won the game"?
(1) If we won the game, then Bill pitched.
(2) If we did not win the game, then Bill did not pitch.
(3) If Bill pitched, then we did not win the game.
(4) If Bill did not pitch, then we did not win the game.

29. The y-intercept of the graph of $y = \frac{2}{3}x + \frac{8}{3}$ is
(1) $\frac{3}{2}$ (2) 8 (3) $\frac{2}{3}$ (4) $\frac{8}{3}$

30. If the radius of a circle is 10, the area of the circle is
(1) 10π (2) 20π (3) 100π (4) 100

31. An equation whose graph is parallel to the graph of $y = 3x + 4$ is
(1) $y = 4x - 3$ (2) $y = 3x - 4$ (3) $y = -3x + 4$ (4) $y = -4$

32. A booklet contains 30 pages. If 9 pages in the booklet have drawings, what percent of the pages in the booklet have drawings?
(1) 30% (2) 9% (3) 3% (4) $\frac{3}{10}$%

33. The value of 5! is
(1) $\frac{1}{5}$ (2) 15 (3) 120 (4) 5

34. If $p \vee q$ is false, then
(1) p is true and q is false
(2) p is false and q is true
(3) both p and q are true
(4) both p and q are false

Directions (35): Leave all construction lines on the answer sheet.

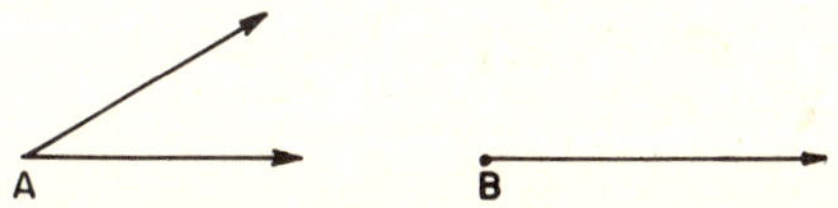

35. *On a separate sheet*, using point B as the vertex, construct an angle congruent to angle A. [Use compasses and straightedge.]

Answers to the following questions are to be written on paper provided by the school.

Part II

Answer four questions from this part. Show all work unless otherwise directed.

36. Solve graphically and check: $2x + y = 8$
$y = 3x - 2$ [8, 2]

37. The length of a rectangle is twice the length of a side of a square and the width of the rectangle is one less than the side of the square. If the perimeter of the rectangle is eight more than the perimeter of the square, find the length of a side of the square. [*Only an algebraic solution will be accepted.*] [5, 5]

38. A contest offers a first prize of \$10,000 or a car. The second prize is \$500 or a television set or a set of encyclopedias.
   a. Make a tree diagram *or* write the sample space of all possible pairs of first and second prizes. [4]
   b. If each prize is equally likely to be chosen, find the probability that:
   (1) both prizes are money [2]
   (2) neither prize is money [2]
   (3) one prize is money but the other is not [2]

39. The test scores of 15 students are 86, 57, 69, 82, 91, 87, 75, 84, 68, 65, 91, 88, 81, 62, 72.
   a. *On your answer paper*, copy and complete the table. [3]

Interval	Frequency
90–99	
80–89	
70–79	
60–69	
50–59	

   b. *On graph paper*, draw a frequency histogram based on the data. [3]
   c. In which interval does the median lie? [2]
   d. What is the probability that a student, selected at random, scored above 89 on this test? [2]

40. The length of one leg of a right triangle is 8. The lengths of the other leg and the hypotenuse are consecutive odd integers. Find the length of the hypotenuse. [*Only an algebraic solution will be accepted.*] [6, 4]

41. In the accompanying figure, rectangle $ABCD$ is inscribed in circle O and $\overline{DB}$ is a diameter. The radius of the circle is 5. [Answers may be left in terms of π.]
   a. Find the area of the circle. [2]
   b. If $CD = 8$, find BC. [3]
   c. Find the area of $\triangle BCD$. [2]
   d. Find the area of rectangle $ABCD$. [1]
   e. Find the area of the shaded portion of the figure. [2]

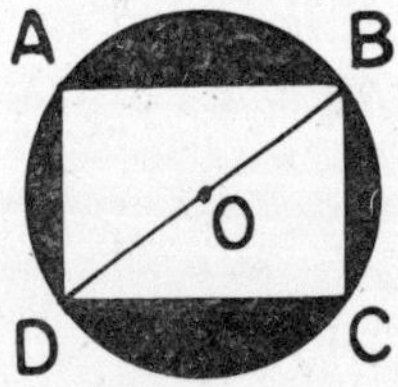

42. *a.* *On your answer paper*, copy and complete the truth table for the statement $\sim(p \rightarrow q) \leftrightarrow (p \wedge \sim q)$. [8]

p	q	$(p \rightarrow q)$	$\sim(p \rightarrow q)$	$\sim q$	$(p \wedge \sim q)$	$\sim(p \rightarrow q) \leftrightarrow (p \wedge \sim q)$
T	T					
T	F					
F	T					
F	F					

   b. Is $\sim(p \rightarrow q) \leftrightarrow (p \wedge \sim q)$ a tautology? [1]
   c. Justify the answer you gave in part *b*. [1]

HIGH SCHOOL MATHEMATICS: COURSE I–JANUARY 1982 (1)

Part I

Answer 30 questions from this part. Each correct answer will receive 2 credits. No partial credit will be allowed. Write your answers on a separate sheet. Where applicable, answers may be left in terms of π or in radical form.

1. Solve for x: $4(x - 1) - 3 = 17$

2. Factor: $x^2 - 10x + 25$

3. In a scale drawing, 2 centimeters represent 3 meters. How many centimeters will represent 36 meters?

4. Solve the following system of equations for x: $\begin{aligned} x + y &= 6 \\ 2x - y &= 0 \end{aligned}$

5. A student's scores on five tests were 86, 85, 85, 91, and 89. What is the student's median grade?

6. Express, in terms of x, the mean of $(2x - 8)$ and $(6x + 4)$.

7. What percent of 20 is 4?

8. If the probability that an event will *not* happen is $\frac{2}{3}$, what is the probability that the event will happen?

9. The lengths of the sides of $\triangle ABC$ are 16, 20, and 24. If $\triangle DEF$ is similar to $\triangle ABC$ and the shortest side of $\triangle DEF$ is 12, find the length of the longest side of $\triangle DEF$.

10. Let p represent "The temperature is 20°C," and let q represent "The room is comfortable." Using p and q, write in symbolic form: "If the room is comfortable, then the temperature is 20°C."

11. Write in symbolic form the converse of $p \rightarrow q$.

12. If the perimeter of a polygon with 5 sides of equal measure is represented by $(15x - 20)$, express the length of one side of the polygon in terms of x.

13. A purse contains 3 pennies, 2 nickels, 4 dimes, and 5 quarters. If one coin is drawn at random from the purse, what is the probability of drawing a dime?

14. In the accompanying diagram, $\angle AOC$ is a central angle of circle O and $m\angle AOC = 60$. Find the number of degrees in the measure of inscribed angle ABC.

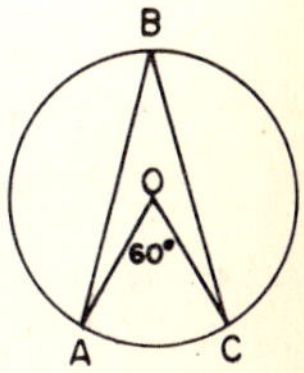

15. From $5x^2 + 3xy + 6y^2$ subtract $4x^2 + 3xy + 2y^2$.

16. The measures of the legs of a right triangle are 3 centimeters and 4 centimeters. Find the number of square centimeters in the area of the triangle.

17. The area of square $ABCD$ is 36 square centimeters. What is the number of centimeters in the perimeter of the square?

18. Solve for x in terms of a, b, and c: $ax - b = c$

Directions (19–35): For *each* question chosen, write on a separate sheet the *numeral* preceding the word or expression that best completes the statement or answers the question.

19. What is the value of x^2y when $x = 2$ and $y = -3$?
(1) -6 (2) 6 (3) -12 (4) 12

20. Which group of numbers has a mode of 5?
(1) 1, 2, 5, 6, 6 (2) 1, 5, 6, 6, 7 (3) 2, 2, 5, 7, 9 (4) 5, 5, 8, 9, 13

21. An expression equivalent to 4! is
(1) 4(3) (2) 4(4) (3) 4(3)(2)(1) (4) 4(4)(4)(4)

22. Which point is on the graph of $2x - y = 8$?
(1) $(1, -6)$ (2) $(-2, 6)$ (3) $(3, 4)$ (4) $(0, 4)$

23. Assume that there is an equal probability of a baby being born on any given day of the week and that there is also an equal probability of a baby being either male or female. What is the probability that a baby will be male *and* born on a Wednesday?
(1) $\frac{1}{14}$ (2) $\frac{9}{14}$ (3) $\frac{2}{9}$ (4) $\frac{1}{4}$

24. If the circumference of a circle is doubled, the diameter of the circle
(1) increases by 2 (2) is doubled (3) is multiplied by 4 (4) remains the same

25. An expression equivalent to $3x - 2 < 7$ is
(1) $x > \frac{5}{3}$ (2) $x < \frac{5}{3}$ (3) $x > 3$ (4) $x < 3$

26. What is the radius of a circle whose area is 100π?
(1) 100 (2) 50 (3) 10 (4) 5

27. The sum of the measures of two angles is 120°. If the measure of one of the angles is five times that of the other, which statement must be true?
(1) The angles are supplementary.
(2) The angles are complementary.
(3) Both angles are acute angles.
(4) One angle is an acute angle and the other is an obtuse angle.

28. The expression $-3x(-3x^3)$ is equivalent to
(1) $9x^4$ (2) $-9x^4$ (3) $6x^4$ (4) $9x^3$

29. Which inequality is represented by the graph?

(1) $x < 4$
(2) $-4 < x \leq 4$
(3) $-4 \leq x < 4$
(4) $-4 \leq x \leq 4$

30. If $p \vee q$ is false, then
(1) p and q are both false
(2) p and q are both true
(3) p is true and q is false
(4) p is false and q is true

31. If $x + 2y < 9$ and $(3, k)$ is an element of the solution set, then k could equal
(1) 5 (2) 2 (3) 3 (4) 6

32. Which is a root of the equation $x^2 + 6x + 8 = 0$?
(1) -8 (2) -2 (3) 6 (4) 4

33. What is the slope of the graph of the equation $y = 3$?
(1) 1 (2) -3 (3) 3 (4) 0

34. What is the total value in cents of n nickels and d dimes?
(1) $n + d$ (2) $5n + 10d$ (3) $0.05n + 0.10d$ (4) $15nd$

35. The sum of $7\sqrt{2}$ and $\sqrt{50}$ is
(1) $7\sqrt{52}$ (2) $8\sqrt{52}$ (3) $12\sqrt{2}$ (4) 24

Part II

Answer four questions from this part. Show all work unless otherwise directed.

36. Solve graphically and check: $x + y = 7$, $y = 2x - 8$ [8, 2]

37. Two positive numbers are consecutive odd integers. The square of the smaller is four more than five times the larger. Find the integers. [*Only an algebraic solution will be accepted.*] [5, 5]

38. In the accompanying figure, quadrilateral $ABCD$ is a trapezoid with $\overleftrightarrow{AB} \parallel \overleftrightarrow{DC}$, $\overleftrightarrow{AD} \perp \overleftrightarrow{AB}$, and $\overleftrightarrow{DB} \perp \overleftrightarrow{BC}$; $AB = AD = 4$, and $DC = 8$.

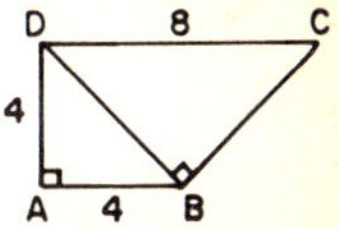

   a. Find DB in radical form. [3]
   b. Find the area of $\triangle ABD$. [2]
   c. Find the area of trapezoid $ABCD$. [3]
   d. Find the area of $\triangle DBC$. [2]

39. On a test, 15 students received the following grades: 17, 14, 16, 18, 17, 19, 15, 15, 16, 13, 17, 12, 18, 16, 17.

   a. On your answer paper, copy and complete the table below. [3]

GRADES	FREQUENCY	CUMULATIVE FREQUENCY
12		
13		
14		
15		
16		
17		
18		
19		

   b. Find the median. [2]
   c. Find the mode. [2]
   d. Find the 75th percentile. [3]

40. A fair die and a fair coin are tossed simultaneously.
   a. Show the sample space or a tree diagram of all possible outcomes. [4]
   b. What is the probability of getting a tail and an odd number? [2]
   c. What is the probability of getting a head? [2]
   d. What is the probability of getting a 6 or a head? [2]

41. In the accompanying figure, triangle ABC is an isosceles triangle, $AB = CB$, the measure of angle BCA is $2x$, and the measure of the exterior angle DAB is $3x$.

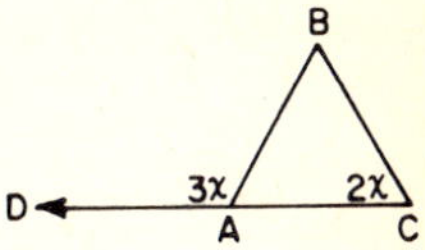

   a. Express the measure of angle BAC in terms of x. [2]
   b. What is the value of x? [3]
   c. How many degrees are in the measure of angle BAC? [1]
   d. How many degrees are in the measure of angle BAD? [1]
   e. How many degrees are in the measure of angle ABC? [3]

42. *a.* Construct a truth table for the statement $[(p \rightarrow q) \wedge p] \rightarrow q$. [9]
   b. Is this statement a tautology? [1]

HIGH SCHOOL MATHEMATICS: COURSE I–JUNE 1982 (1)

Part I

Answer 30 questions from this part. Each correct answer will receive 2 credits. Write your answers on a separate sheet.

1. Solve for y: $\frac{y}{12} = \frac{4}{3}$

2. In the accompanying figure, triangle ABC is similar to triangle DEF. If $AC = 6$, $AB = 7$, $DE = 28$, $\angle B \cong \angle E$, and $\angle C \cong \angle F$, find DF.

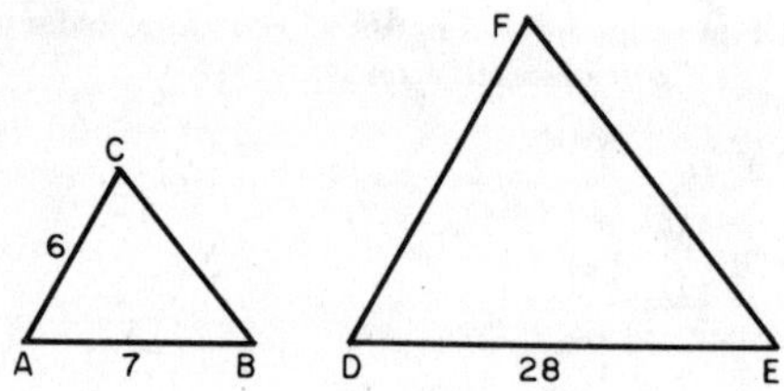

3. The graph below shows the distribution of scores on a math test. How many students took the math test?

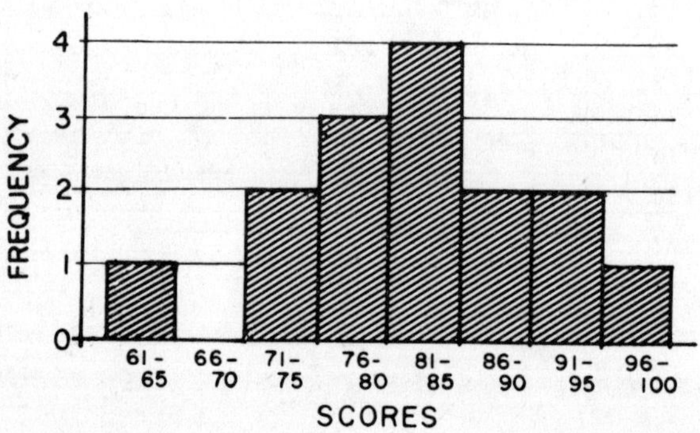

4. Solve for x: $4(2x - 1) = 20$

5. Solve for x: $0.02x = 32.6$

6. In a class of 20 students, 12 are boys and 8 are girls. If a student is randomly selected, what is the probability that the student will be a boy?

7. If p represents the perimeter of a square, represent the length of a side of the square in terms of p.

8. Express the mean of $(2x + 1)$, $(x + 1)$, and $(3x - 8)$ in terms of x.

9. The probability that an event will *not* occur is $\frac{7}{12}$. What is the probability that the event will occur?

10. The circumference of a circle is 12π. What is the radius of the circle?

11. What percent of 30 is 9?

12. Let p represent the statement "The base angles are congruent," and let q represent the statement "A triangle is isosceles." Using p and q, write in symbolic form: "A triangle is isosceles if and only if the base angles are congruent."

13. Express $\frac{2a}{3} - \frac{a}{4}$ as a single fraction.

14. If the point (2, 3) lies on the graph of the equation $2x + ky = -2$, find the value of k.

15. Factor: $25x^2 - 16$

16. How many different arrangements of four digits can be formed from the digits 2, 5, 6, and 7, if each digit is used only once in each arrangement?

17. Solve for b in terms of V and h: $V = \frac{bh}{3}$

18. The test scores for a group of students are 70, 80, 30, 80, 60, 65, and 90. What is the median score for this group of students?

19. Solve the following system of equations for x: $3x + y = 5$
$y = 5x - 3$

20. The lengths of two legs of a right triangle are 3 and 5. Find, in radical form, the length of the hypotenuse.

21. The length of a rectangle is represented by $(x + 3)$, and the width of the rectangle is represented by $(x - 8)$. Express the area of the rectangle as a trinomial in terms of x.

22. If two angles of a triangle are complementary, find the number of degrees in the third angle of the triangle.

23. In the accompanying diagram, triangle BCE is inscribed in square $ABCD$. If the length of side $\overline{BC}$ is 4 centimeters, what is the area in square centimeters of the shaded portion of the diagram?

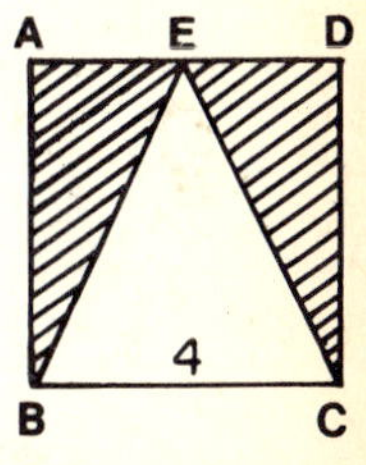

24. As shown in the accompanying diagram, angle QPR is inscribed in circle O. If arc QR measures 38°, what is the measure in degrees of angle QPR?

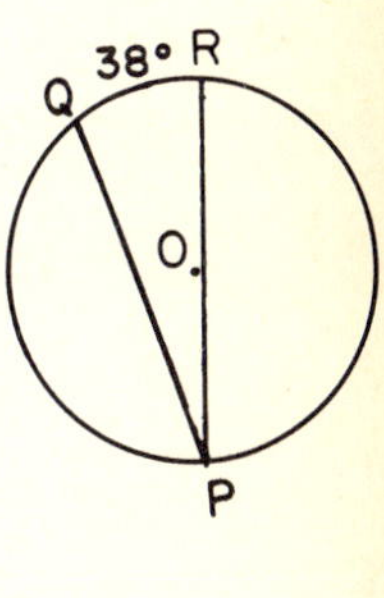

25. Factor: $x^2 - 6x + 9$

Directions (26–35): For *each* question chosen, write on a separate sheet the *numeral* preceding the word or expression that best completes the statement or answers the question.

26. The product $(-4x^2)(3x^3)$ is equal to
(1) $-12x^5$ (2) $-12x^6$ (3) $12x^5$ (4) $12x^6$

27. Which is an irrational number?
(1) $\sqrt{9}$ (2) $\sqrt{2}$ (3) $\frac{2}{3}$ (4) 0

28. The expression $6 \leq x + 4$ is equivalent to
(1) $x \geq 2$ (2) $x \leq 2$ (3) $x \leq -2$ (4) $x \geq 10$

29. Which graph represents the solution set of $-2 \leq x < 1$?

(1) 

(3) 

(2) 

(4) 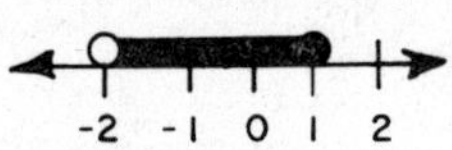

30. Which expression is undefined or meaningless when $x = 5$?

(1) $x^{\frac{0}{5}}$ (2) x^{-5} (3) $\dfrac{1}{x-5}$ (4) $\dfrac{1}{x+5}$

31. Let p represent the statement: "x is even." Let q represent the statement: "$x \leq 12$." Which is true if $x = 20$?
(1) $p \rightarrow q$ (2) $p \wedge q$ (3) $\sim p \vee q$ (4) $p \wedge \sim q$

32. Which is the equation of a line parallel to the line whose equation is $y = 3x + 1$?

(1) $y = -\frac{1}{3}x + 1$ (2) $y = 2x + 1$ (3) $y = 3x - 1$ (4) $y = -3x + 1$

33. One of the factors of $3x^3 - 6x^2$ is
(1) $3x^2$ (2) $3x^3$ (3) $-6x^2$ (4) $(x - 6x^2)$

34. Which is *always* false?
(1) $p \rightarrow q$ (2) $q \rightarrow p$ (3) $\sim p \rightarrow \sim q$ (4) $p \wedge \sim p$

35. The solution set of $x^2 - 7x + 10 = 0$ is
(1) $\{2, -5\}$ (2) $\{2, 5\}$ (3) $\{-2, 5\}$ (4) $\{-2, -5\}$

Part II

Answer four questions from this part. Show all work unless otherwise directed.

36. *a*. *On your answer paper*, copy and complete the truth table for the statement

$$\sim q \rightarrow [(p \rightarrow q) \wedge \sim p] \quad [8]$$

p	q	$p \rightarrow q$	$\sim p$	$\sim q$	$(p \rightarrow q) \wedge \sim p$	$\sim q \rightarrow [(p \rightarrow q) \wedge \sim p]$
T	T					
T	F					
F	T					
F	F					

b. Is $\sim q \rightarrow [(p \rightarrow q) \wedge \sim p]$ a tautology? [1]
c. Justify the answer you gave in part *b*. [1]

37. *a*. Solve the following system of equations algebraically:

$$\begin{aligned} x - y &= 1 \\ 3x + y &= 9 \end{aligned} \quad [4]$$

b. On a set of coordinate axes, graph the system of equations in part *a*. [6]

38. An architect wants to design a rectangular room so that its length is 8 meters more than its width, and its perimeter is greater than 56 meters. If each of the dimensions of the room must be a whole number of meters, what are the *smallest* possible measures in meters of the length and width? [*Only an algebraic solution will be accepted.*] [5, 5]

39. Mary chose one of the four numbers 1, 2, 3, and 6, at random. She then chose one of the two numbers, 1 and 5, at random.
a. Draw a tree diagram or list the sample space of all possible pairs of numbers that Mary could choose. [3]
b. Find the probability that Mary chose an even number first followed by an odd number. [2]
c. Find the probability that Mary chose *at least* one even number. [2]
d. Find the probability that both choices were the same number. [2]
e. Find the probability that Mary chose two even numbers. [1]

40. Find three consecutive odd integers such that the square of the first is equal to the second plus twice the third. [*Only an algebraic solution will be accepted.*] [5, 5]

41. In the diagram of parallelogram $ABCD$, $\overline{DE}$ is perpendicular to $\overline{AB}$, $AD = 15$, and $AE = 9$.
a. Find DE. [2]
b. If BE is 2 less than twice AE, find BE. [1]
c. Find AB. [1]
d. Find the area of triangle AED. [2]
e. Find the area of $ABCD$. [2]
f. Find the area of trapezoid $EBCD$. [2]

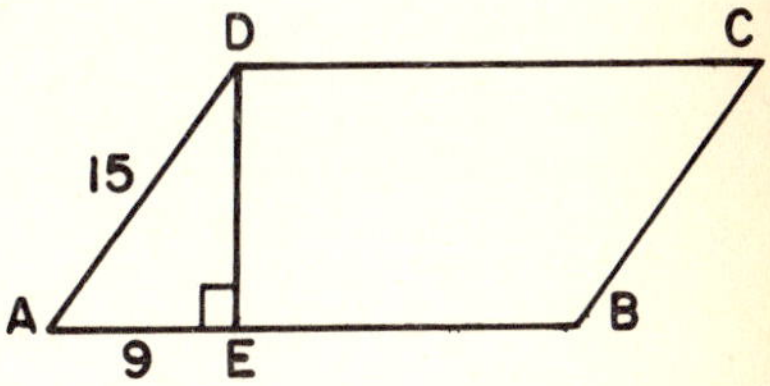

42. The following diagram is a cumulative frequency histogram of raw scores on a mathematics examination.

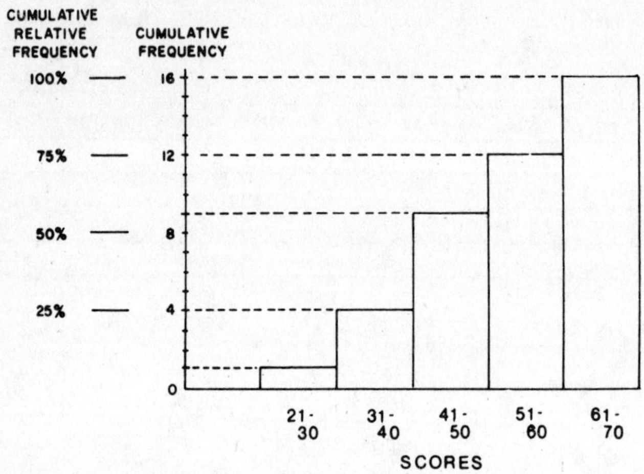

a. How many students took the examination? [2]
b. How many students had a score less than or equal to 60? [2]
c. What percent of the students had a score less than or equal to 60? [2]
d. Which interval contains the median? [2]
e. Which interval contains the lower quartile? [2]

HIGH SCHOOL MATHEMATICS: COURSE I–JANUARY 1983 (1)

Part I

Answer 30 questions from this part. Each correct answer will receive 2 credits. No partial credit will be allowed. Write your answers on a separate sheet. Where applicable, answers may be left in terms of π or in radical form.

1. The probability of rain tomorrow is 40%. What is the probability that it will *not* rain tomorrow?

2. Solve for y: $3(y-1)=9$

3. Solve for m: $\frac{m}{15}=\frac{2}{3}$

4. If $x=6$ and $y=3$, what is the sum of $\frac{4}{x}+\frac{1}{y}$?

5. A tree 10 meters in height casts a shadow 25 meters long. At the same time, a person casts a shadow 5 meters long. What is the number of meters in the height of the person?

6. Solve for y in terms of a, b, and x: $ay-bx=2$

7. Solve for x: $0.03x=36$

8. There are 3 ways of going from town A to town B and 6 ways of going from town B to town C. Find the total number of ways a person can go from town A to town B to town C.

9. Express the *sum* of $(4x-2)$ and $(3-x)$ as a binomial.

10. A bag contains 2 green marbles, 4 blue marbles, and 5 red marbles. If one marble is drawn at random from the bag, what is the probability that it will be green?

11. In the accompanying diagram, the measure of arc AB in circle O is 46°. Find the measure in degrees of angle AOB.

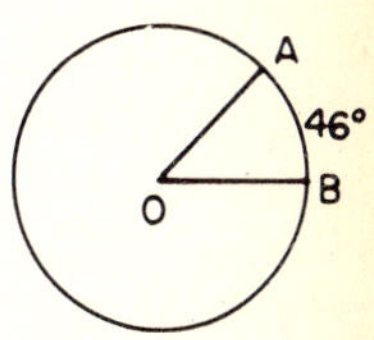

12. Solve the following system of equations for x:
$$3x+y=5$$
$$2x-y=0$$

13. Factor: $x^2-7x+10$

14. The measures of the angles of a triangle are represented by x, $2x$, and $(x+20)$. Find the number of degrees in the measure of the *smallest* angle of the triangle.

15. Express as a binomial: $3x^2(2x-5)$

16. Two angles are supplementary and congruent. How many degrees are in the measure of each angle?

17. What percent of 200 is 14?

18. The length of the edge of a cube is represented by e. Express the volume of the cube in terms of e.

19. Factor: $9x^2-1$

20. Find the length of the radius of a circle whose circumference is 18π.

21. A fair coin and a fair die are tossed simultaneously. What is the total number of possible outcomes in the sample space?

Directions (22–35): For *each* question chosen, write on the separate answer sheet the *numeral* preceding the word or expression that best completes the statement or answers the question.

22. The value of π is
    (1) rational and equal to 3.14
    (2) irrational and equal to 3.14
    (3) rational and between 3.14 and 3.15
    (4) irrational and between 3.14 and 3.15

23. The largest possible value of x in the solution set of $2x + 1 \leq 7$ is
    (1) 6 (2) 2 (3) 3 (4) 4

24. Let p represent "I am 18," and let q represent "I am going to college." Which statement is represented by $p \rightarrow q$?
    (1) I am 18, and I am not going to college.
    (2) If I am 18, then I am going to college.
    (3) I am not 18 and I am not going to college.
    (4) If I am going to college, then I am 18.

25. If p represents the statement "It is cold" and q represents the statement "It is winter," which represents "It is cold and it is not winter"?
    (1) $p \rightarrow q$ (2) $p \vee \sim q$ (3) $p \wedge q$ (4) $p \wedge \sim q$

26. The inverse of $\sim p \rightarrow q$ is
    (1) $p \rightarrow \sim q$ (2) $\sim p \rightarrow \sim q$ (3) $\sim q \rightarrow p$ (4) $q \rightarrow \sim p$

27. The graph of which equation has a slope of 2?
    (1) $y = 2x - 3$ (2) $y = -2x + 3$ (3) $y = 3x - 2$ (4) $y = -3x + 2$

28. Which point does *not* lie on the graph of $x - 2y = 10$?
    (1) (0, –5) (2) (2, –4) (3) (5, 0) (4) (6, –2)

29. If two rectangles have equal perimeters, which statement *must* be true?
    (1) The rectangles have equal areas.
    (2) The rectangles are squares.
    (3) The lengths of the rectangles are equal.
    (4) The sum of the length and width of one rectangle is equal to the sum of the length and width of the other.

30. The equation $5x + 10 = 55$ has the same solution set as the equation
    (1) $x = 45$ (2) $x + 10 = 11$ (3) $5x = 65$ (4) $5x + 15 = 60$

31. The scores on a test were 75, 75, 85, 90, and 100. Which statement about these scores is true?
    (1) The mean and the median are the same.
    (2) The mode is greater than the median.
    (3) The mode is greater than the mean.
    (4) The mean is less than the median.

32. When $3x^2$ is multiplied by $-5x^3$, the product is
    (1) $-15x^6$ (2) $-15x^5$ (3) $-8x^5$ (4) $15x^5$

33. Which inequality is represented by the accompanying graph?

    (1) $2 < x \leq -3$ (2) $-3 \leq x < 2$ (3) $-3 < x \leq 2$ (4) $-3 \leq x \leq 2$

34. The solution set of $x^2 - 2x - 8 = 0$ is
    (1) {–6, 2} (2) {4, 2} (3) {–4, 2} (4) {4, –2}

35. The sum of $\sqrt{2}$ and $3\sqrt{2}$ is
    (1) $3\sqrt{2}$ (2) $4\sqrt{2}$ (3) $3\sqrt{4}$ (4) $4\sqrt{4}$

Part II

Answer four questions from this part. Show all work unless otherwise directed.

36. On the same set of coordinate axes, graph the following system of inequalities and label the solution set S.

$$y > x + 1$$
$$x \leq -1 \qquad [8, 2]$$

37. Solve algebraically and check: $2c - d = -1$, $c + 3d = 17$ [8, 2]

38. Jim bought 3 packets of vegetable seeds: beans, carrots, and radishes; and two packets of flower seeds: marigolds and petunias. Finding that his garden was too small, he decided to give away one packet of vegetable seeds and one packet of flower seeds, each selected randomly.
   a. Draw a tree diagram or list all possible pairs of packets in the sample space that he might select to give away. [4]
   b. Find the probability that he did *not* give away the packet of bean seeds. [2]
   c. Find the probability that he gave away two packets of vegetable seeds. [2]
   d. Find the probability that he gave away the packet of petunia seeds. [2]

39. The length of a rectangle is 3 more than three times the width. The perimeter of the rectangle is 62.
   a. Find the length and width of the rectangle. [*Only an algebraic solution will be accepted.*] [4, 4]
   b. Find the area of the rectangle. [2]

40. The sum of the squares of two consecutive positive even integers is 52. Find the integers. [*Only an algebraic solution will be accepted.*] [4, 6]

41. Answer *both a and b.*
   a. Given the inequality: $8x \geq 3(x - 5)$
      (1) Solve for x. [4]
      (2) Choose one value of x from your solution in part (1), and show that it makes the inequality $8x \geq 3(x - 5)$ true. [2]
   b. On your answer paper, copy and complete the truth table for the statement $(p \rightarrow q) \leftrightarrow (p \vee q)$. [4]

p	q	$p \rightarrow q$	$p \vee q$	$(p \rightarrow q) \leftrightarrow (p \vee q)$
T	T			
T	F			
F	T			
F	F			

42. The points scored by Rosa in twenty basketball games are 35, 33, 27, 35, 29, 37, 32, 35, 35, 32, 23, 37, 32, 29, 26, 30, 28, 31, 29, 35.
   a. Find the mode. [2]
   b. On your answer paper, copy and complete the table. [2]
   c. Construct a frequency histogram based on the table completed in part *b*. [4]
   d. In what interval does the median lie? [2]

Interval	Tally	Frequency
35–37		
32–34		
29–31		
26–28		
23–25		

HIGH SCHOOL MATHEMATICS: COURSE I–JUNE 1983 (1)

Part I

Answer 30 questions from this part. Each correct answer will receive 2 credits. No partial credit will be allowed. Write your answers on a separate sheet. Where applicable, answers may be left in terms of π or in radical form.

1. Solve for x: $2(x - 3) = 8$

2. If a certain number is decreased by 9, the result is 24. What is the number?

3. If one card is drawn from a standard deck of 52 playing cards, what is the probability the card is a red seven?

4. The selling price of a radio is \$50, not including a sales tax at the rate of 7%. What is the sales tax on the purchase of the radio?

5. In the accompanying diagram, $\overleftrightarrow{AB} \parallel \overleftrightarrow{CD}$, and $\overleftrightarrow{EF}$ intersects $\overleftrightarrow{AB}$ at G and $\overleftrightarrow{CD}$ at H. If the degree measure of $\angle AGH$ is $(3x - 10)$ and the degree measure of $\angle GHD$ is 80, find the value of x.

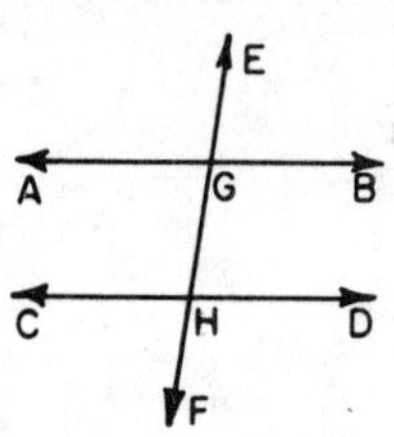

6. Express as a trinomial: $(3x - 1)(x + 2)$

7. Let p represent "I will go roller skating" and let q represent "The sun is shining." Using p and q, write in symbolic form: "I will go roller skating if and only if the sun is shining."

8. The side of a square is represented by $(x + 3)$. Express the perimeter of the square in terms of x.

9. Factor: $64 - x^2$

10. Solve for x: $0.25x + 2 = 6$

11. What is the y-intercept of the graph of the equation $y = 3x - 5$?

12. The lengths of the sides of $\triangle ABC$ are 14, 11, and 4 centimeters. The longest side of similar triangle RST is 28 centimeters. Find the length in centimeters of the shortest side of $\triangle RST$.

13. As shown in the accompanying figure, equilateral triangle ABC is inscribed in circle O. Find the measure in degrees of minor arc AB.

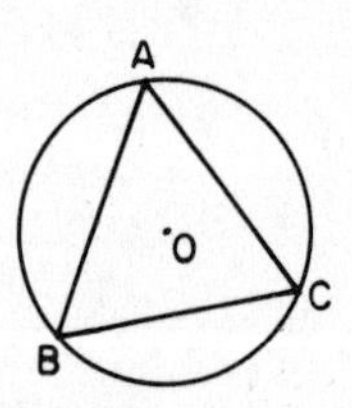

14. The degree measures of the angles of a triangle are represented by $2x$, $3x$, and $4x$. Find the number of degrees in the *smallest* angle.

15. The following data are numbers of minutes a student spent on homework: 25, 35, 30, 50, and 38. What is the median of the data?

16. Solve for E in terms of I and R: $\dfrac{E}{I} = R$

17. The lengths of the legs of a right triangle are 3 and 6. Find, in radical form, the length of the hypotenuse of the right triangle.

18. Solve for a: $\dfrac{a+2}{12} = \dfrac{5}{3}$

19. The circumference of a circle is 8π. What is the radius of the circle?

20. The accompanying histogram shows the distribution of student ages in a ninth grade class. What age is the mode?

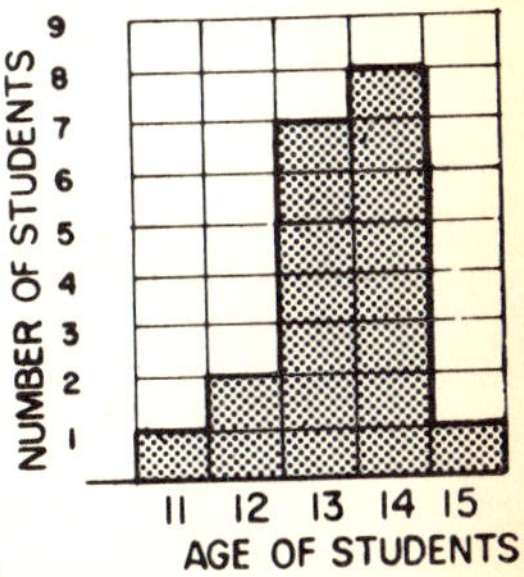

21. Express as a trinomial: $(5x^2 + 2x - 3) - (2x^2 - 3x + 7)$

Directions (22–34): For *each* question chosen, write on the separate answer sheet the *numeral* preceding the word or expression that best completes the statement or answers the question.

22. On a test the probability of getting the correct answer to a certain question is represented by $\frac{x}{7}$. Which can *not* be a value of x?
(1) 1 (2) −1 (3) 0 (4) 7

23. The expression $(-2)^2(\frac{1}{2})^3$ is equivalent to
(1) −2 (2) 2 (3) $\frac{1}{2}$ (4) $\frac{1}{32}$

24. The length of a rectangle is represented by x. If the width of the rectangle is 3 less than its length, which expression represents the area of the rectangle?
(1) $x(x-3)$ (2) $x^2 - 3$ (3) $2x - 6$ (4) $2x + 6$

25. The value of 4! is
(1) 1 (2) 12 (3) 24 (4) 4

26. The expression $18x^6 \div 3x^3$ is equivalent to
(1) $15x^2$ (2) $15x^3$ (3) $6x^2$ (4) $6x^3$

27. Which is the converse of the statement $p \rightarrow \sim q$?
(1) $p \rightarrow \sim q$ (2) $\sim p \rightarrow \sim q$ (3) $q \rightarrow \sim p$ (4) $\sim q \rightarrow p$

28. Which graph represents the solution set of $x \geq 2$?

(1) 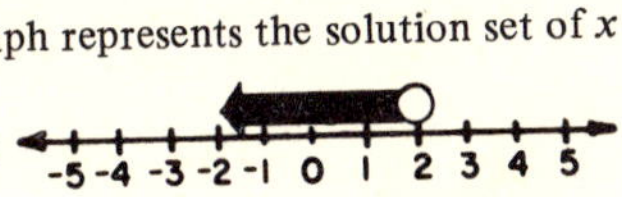

(2) 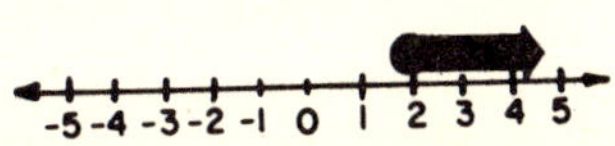

(3) 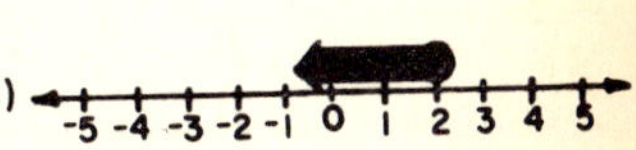

(4) 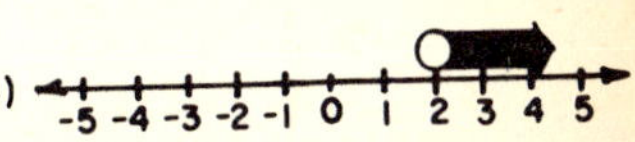

29. An equation of the line which is parallel to the x-axis and 3 units below the x-axis is
(1) $y = -3$ (2) $y = 3$ (3) $x = 3$ (4) $x = -3$

30. The area of a rectangle is represented by $x^2 + 2x - 3$. If the width of the rectangle is represented by $(x - 1)$, the length may be represented by
(1) $(x - 3)$ (2) $(x - 2)$ (3) $(x + 3)$ (4) $(x + 4)$

31. What is the solution of the system of equations whose graphs are shown?
(1) (−1, 2) (2) (0, 2) (3) (2, 0) (4) (2, −1)

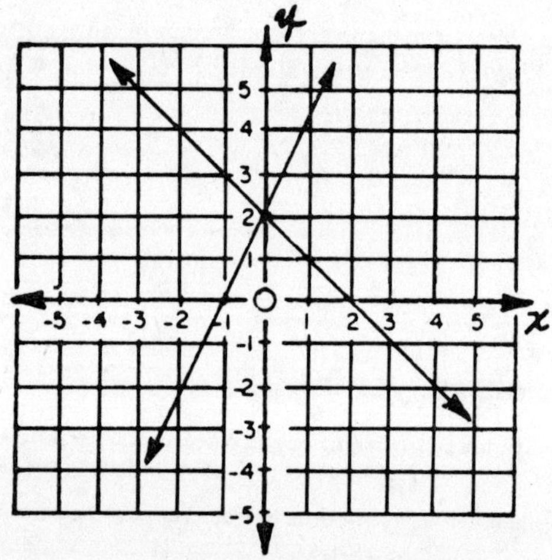

32. Which ordered pair is *not* in the solution set of $x - 2y \leq 6$?
(1) (2, −2) (2) (5, 1) (3) (0, 0) (4) (1, −6)

33. For which value of x is the expression $\frac{x - 4}{x + 3}$ undefined?
(1) −4 (2) −3 (3) 3 (4) 4

34. Which statement is false when p is false and q is false?
(1) $p \wedge q$ (2) $p \rightarrow q$ (3) $\sim p \rightarrow \sim q$ (4) $p \leftrightarrow q$

Directions (35): Use compasses and straightedge. Leave all construction lines on the answer sheet.

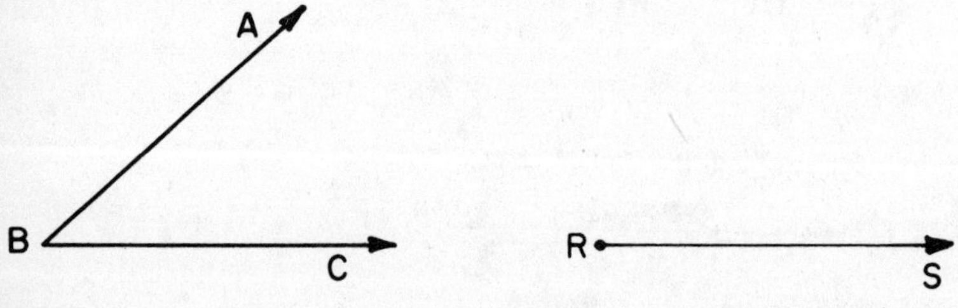

35. *On the answer sheet*, construct an angle congruent to angle ABC, using $\overrightarrow{RS}$ as one ray of the angle.

Part II

Answer four questions from this part. Show all work unless otherwise directed.

36. *a.* On the same set of coordinate axes, graph the following system of inequalities:

$$y > -2x + 7 \qquad [8]$$
$$y \leq 3x - 3$$

b. Write the coordinates of a point in the solution set of this system. [2]

37. The measure of the base of a parallelogram is 5 meters more than the measure of the altitude to that base. If the area of the parallelogram is 36 square meters, find the number of meters in the measures of the base and altitude. [*Only an algebraic solution will be accepted.*] [5, 5]

38. For a class picnic, the school cafeteria prepared a box lunch for each student, consisting of a sandwich and a cookie. The sandwiches were tuna, ham, or peanut butter and the cookies were oatmeal or chocolate chip.
a. Draw a tree diagram or list a sample space showing all possible combinations of one sandwich and one cookie that could be in the boxes. [4]
b. If each combination was equally likely to be in any one of the boxes, what is the probability that a box chosen at random contains:
(1) a peanut butter sandwich and an oatmeal cookie [2]
(2) a ham sandwich [2]
(3) a sandwich that is *not* tuna [2]

39. Solve the following system of equations algebraically and check: $3x + y = 4$, $x - 2y = 6$ [8, 2]

40. In the accompanying diagram, triangle ABC is a right triangle with the right angle at C. Segment DE is perpendicular to $\overline{AC}$ at E, $BC = 8$, $AC = 6$, and $DE : BC = 1 : 2$.
a. Find DE. [1]
b. Find AB. [2]
c. Find AE. [2]
d. Find the area of $\triangle ABC$. [2]
e. Find the area of trapezoid $ECBD$. [3]

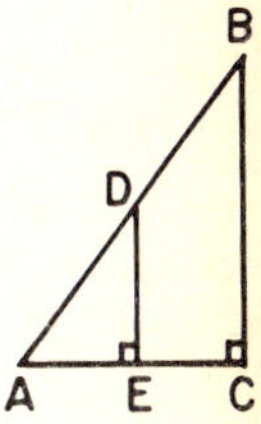

41. The following table shows the distribution of scores of 30 students on a test.

Scores	Frequency	Cumulative Frequency
91–100	3	
81–90	11	
71–80	8	
61–70	6	
51–60	1	
41–50	1	

a. Using the data in the Frequency column of the table, draw a frequency histogram. [4]
b. *On your answer paper*, copy the table and complete the column for Cumulative Frequency. [2]
c. Using the data in the Cumulative Frequency column of the table, draw a cumulative frequency histogram. [4]

42. *a.* *On your answer paper*, copy and complete the truth table for the statement $\sim(p \vee q) \leftrightarrow (\sim p \wedge \sim q)$. [8]

p	q	$p \vee q$	$\sim(p \vee q)$	$\sim p$	$\sim q$	$(\sim p \wedge \sim q)$	$\sim(p \vee q) \leftrightarrow (\sim p \wedge \sim q)$
T	T						
T	F						
F	T						
F	F						

b. Is $\sim(p \vee q) \leftrightarrow (\sim p \wedge \sim q)$ a tautology? [1]

c. Justify the answer you gave in part *b.* [1]

HIGH SCHOOL MATHEMATICS: COURSE I–JANUARY 1984 (1)

Part I

Answer 30 questions from this part. Each correct answer will receive 2 credits. No partial credit will be allowed. Write your answers in the spaces provided on the separate answer sheet. Where applicable, answers may be left in terms of π or in radical form.

1. Express the sum of $(2y + 3)$ and $(3y - 4)$ as a binomial.

2. Solve for x: $x - 0.2 = 1.8$

3. There are 3 entrances to a school and there are 2 stairways that go to the second floor. In how many different ways can a student enter the school and go to the second floor?

4. Express, in radical form, the length of the hypotenuse of a right triangle whose legs have lengths of 1 and 3.

5. Solve for x: $2(x + 3) = x + 7$

6. In the accompanying diagram of $\triangle ABC$, $m\angle A = 70$ and $m\angle B = 30$. Find the measure of exterior angle BCD.

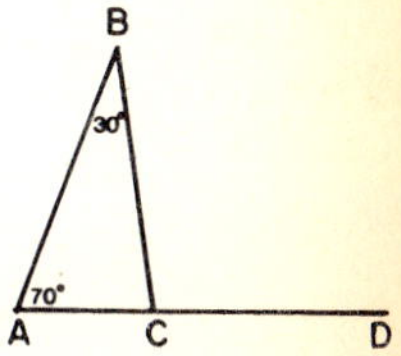

7. As shown in the accompanying diagram, $\overleftrightarrow{AB}$ and $\overleftrightarrow{CD}$ intersect at point E. If the degree measures of vertical angles AED and CEB are represented by $(3x + 20)$ and $(8x - 5)$, find the value of x.

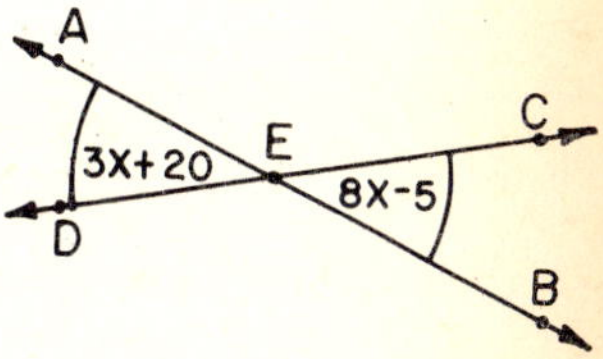

8. The measures of two complementary angles are in the ratio 2:3. Find the measure of the *larger* angle.

9. If the mean of four positive integers is exactly 10, find the sum of the four numbers.

10. Find the value of the expression $3x^3$ when $x = -2$.

11. Solve for x: $\dfrac{x+1}{8} = \dfrac{1}{2}$

12. Factor: $4x^2 - 9$

13. Find the slope of the line whose equation is $y = -2x - 4$.

14. Thirty percent of what number is 12?

15. Let p represent "I passed the test" and let q represent "I feel proud." Using p and q, write in symbolic form, "I did not pass the test and I do not feel proud."

16. There are 14 girls and 15 boys in a class. If the teacher calls on one student at random, what is the probability the student called on is a girl?

17. A rectangle has an area of 16. If the length of the rectangle is doubled and the width remains the same, what is the area of the new rectangle?

18. A box of plant food recommends adding $1\frac{1}{2}$ ounces of plant food to every 4 quarts of water. How many ounces of plant food should be added to 16 quarts of water?

19. Express as a single fraction: $\frac{x}{3} + \frac{x}{5}$

20. Express the product $(2x - 5)(4x + 5)$ as a trinomial.

21. Find the mode of the following group of numbers: 8, 8, 9, 10, 11

22. Factor: $x^2 - x - 12$

23. The inverse of a statement is $p \rightarrow \sim q$. What is the statement?

24. Solve the following system of equations for x: $x + y = 7$
$x - y = 1$

Directions (25–34): For *each* question chosen, write on the separate answer sheet the *numeral* preceding the word or expression that best completes the statement or answers the question.

25. If x represents a number, which expression represents a number which is 5 less than 3 times x?
(1) $5x - 3$ (2) $5 - 3x$ (3) $3x - 5$ (4) $3 - 5x$

26. Which is logically equivalent to $p \rightarrow q$?
(1) $\sim q \rightarrow \sim p$ (2) $\sim p \rightarrow \sim q$ (3) $q \rightarrow p$ (4) $p \wedge \sim q$

27. The length of a rectangle is three times its width. If the width is represented by x, which expression represents the perimeter of the rectangle?
(1) $6x$ (2) $8x$ (3) $3x^2$ (4) $4x$

28. The accompanying diagram shows the graph of which inequality?
(1) $y > x - 1$
(2) $y \geq x - 1$
(3) $y < x - 1$
(4) $y \leq x - 1$

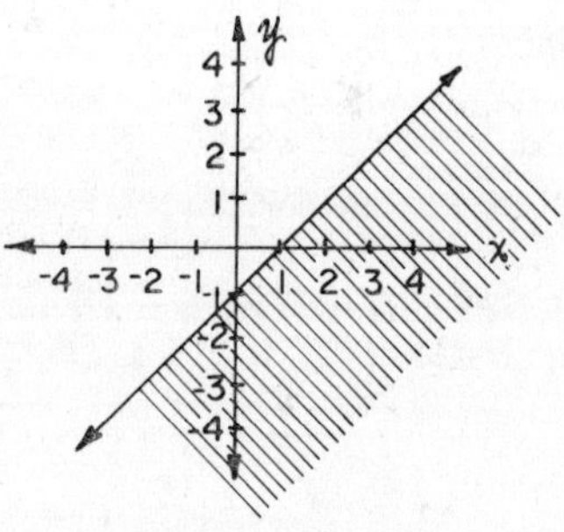

29. The graph of the equation $y = 5$ intersects the y-axis at the point whose coordinates are
(1) (–5, 5) (2) (0, 5) (3) (5, 0) (4) (5, 5)

30. If the area of a circle is 16π, the circumference of the circle is
(1) 8π (2) 8 (3) 4π (4) 4

31. If $x = 1$ and $y = -2$, which expression has a value of 3?
(1) $x + y$ (2) $x - y$ (3) xy (4) $y - x$

32. The length of an edge of a cube is represented by $5x$. Which expression represents the volume of the cube?
(1) $10x^2$ (2) $25x^2$ (3) $5x^3$ (4) $125x^3$

33. Which point does *not* lie on the graph of the equation $2x + y = 3$?
(1) $(-1, -1)$ (2) $(-1, 5)$ (3) $(0, 3)$ (4) $(\frac{1}{2}, 2)$

34. Patty needs a total of $80 to buy a bicycle. She has already saved $35. If she saves $10 a week from her earnings, what is the *least* number of weeks she must work to have enough money to buy the bicycle?
(1) 5 (2) 8 (3) 3 (4) 4

Directions (35): Use compasses and straightedge. Leave all construction lines on the answer sheet.

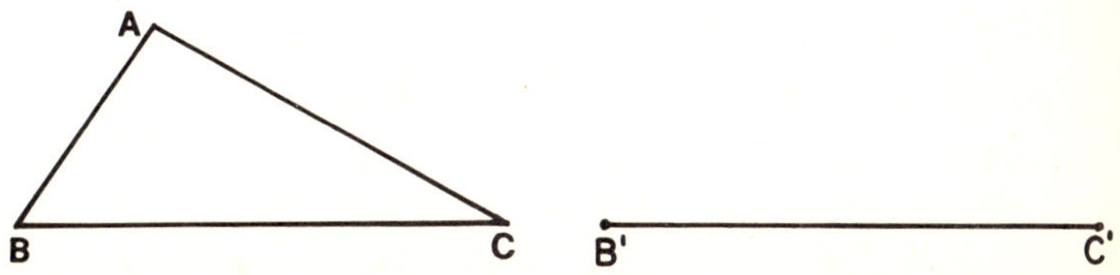

35. *On the answer sheet*, construct triangle $A'B'C'$ congruent to $\triangle ABC$, using $\overline{B'C'}$ as one side.

Part II

Answer four questions from this part. Show all work unless otherwise directed.

36. Solve the following system of equations graphically and check:

$$\begin{aligned} x + y &= -3 \\ 2x - y &= 6 \end{aligned} \quad [8, 2]$$

37. One number is 4 times another. The sum of the two numbers is less than 12. Find the largest possible values for the two numbers if both are integers. [*Only an algebraic solution will be accepted.*] [6, 4]

38. A garden is in the shape of a square. The length of one side of the garden is increased by 3 feet and the length of an adjacent side is increased by 2 feet. The garden now has an area of 72 square feet. What is the measure of a side of the original square garden? [*Only an algebraic solution will be accepted.*] [5, 5]

39. In the accompanying diagram, $\triangle ABC$ is inscribed in circle O. Triangle ABC is isosceles with $\overline{AB} \cong \overline{BC}$. Line EF, which contains point B, is parallel to line AC. The degree measure of arc BC is 150.

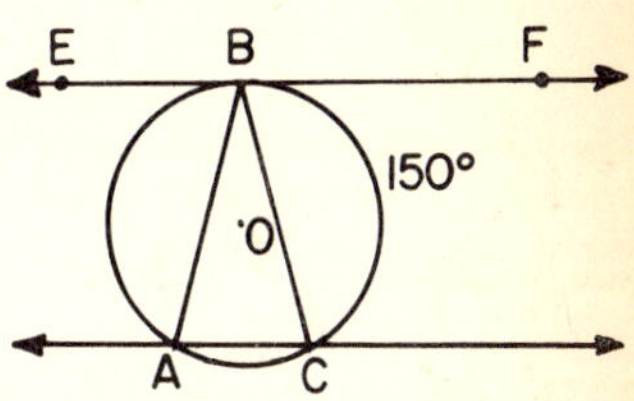

Find:
a. the measure of $\angle BAC$ [2]
b. the measure of $\angle EBA$ [2]
c. the measure of $\angle ABC$ [2]
d. the measure of arc AC [2]
e. the measure of $\angle ABF$ [2]

40. The following table gives the distribution of test scores for a class of 20 students.

Test Score Interval	Number of Students (frequency)
91–100	1
81–90	3
71–80	3
61–70	7
51–60	6

a. Draw a *frequency* histogram for the given data. [4]
b. Which interval contains the median? [2]
c. Which interval contains the lower quartile? [2]
d. What is the probability that a student selected at random scored above 90? [2]

41. The accompanying diagram represents an arrow attached to a cardboard disc. The arrow is free to spin, but cannot land on a line. The disc is divided into three regions of equal area, one of which is red and the other two blue.

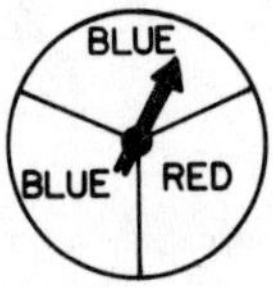

a. For any one spin, what is the probability of the arrow:
(1) landing on red [1]
(2) landing on blue [1]
b. The arrow is spun twice and each outcome is recorded. What is the probability of the arrow:
(1) landing on red on the first spin and blue on the second spin [2]
(2) landing on blue on both spins [2]
(3) *not* landing on blue on either spin [2]
(4) landing on the same color on both spins [2]

42. *a.* *On your answer paper*, copy and complete the truth table for the statement $[p \vee (p \wedge q)] \rightarrow \sim q$. [8]

p	q	$p \wedge q$	$[p \vee (p \wedge q)]$	$\sim q$	$[p \vee (p \wedge q)] \rightarrow \sim q$
T	T				
T	F				
F	T				
F	F				

b. Is $[p \vee (p \wedge q)] \rightarrow \sim q$ a tautology? [1]
c. Justify the answer you gave in part *b*. [1]

HIGH SCHOOL MATHEMATICS: COURSE I—JUNE 1984 (1)

Part I

Answer 30 questions from this part. Each correct answer will receive 2 credits. No partial credit will be allowed. Write your answers in the spaces provided on the separate answer sheet. Where applicable, answers may be left in terms of π or in radical form.

1. Solve for *m*: $\frac{m}{4} = \frac{5}{2}$

2. Solve for *x*: $5 + 3(x + 2) = 14$

3. Find the value of $3x^2y$ if $x = -2$ and $y = 3$.

4. Jean is making sandwiches for a class picnic. She is using 4 different fillings with 2 different kinds of bread. How many different kinds of sandwiches can she make using one kind of filling on one kind of bread for each sandwich?

5. If all seven of the letters of the word "REGENTS" were placed in a hat, what would be the probability of drawing an E at random on the first draw?

6. In the accompanying diagram, $\overleftrightarrow{AB} \parallel \overleftrightarrow{CD}$, $\overleftrightarrow{HG}$ intersects $\overleftrightarrow{AB}$ at *E* and $\overleftrightarrow{CD}$ at *F*. If $m\angle CFE = 95$ and $m\angle BEF = 3x + 5$, find *x*.

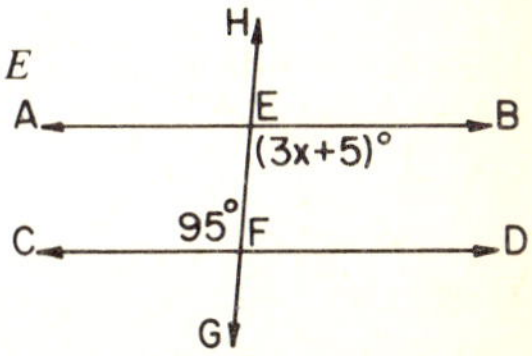

7. A student received test scores of 89, 86, 81, 94, and 75. Find the mean score.

8. If a card is drawn at random from a standard deck of 52 cards, what is the probability that the card is a red queen?

9. The measure of each base angle of an isosceles triangle is 15°. Find the number of degrees in the measure of the vertex angle.

10. Solve the following system of equations for *x*: $4x - y = 10$
 $x + y = 5$

11. In the accompanying diagram of circle *O*, diameters *AB* and *CD* intersect at *O*. If the measure of arc *AC* is 40°, find the number of degrees in the measure of angle *COB*.

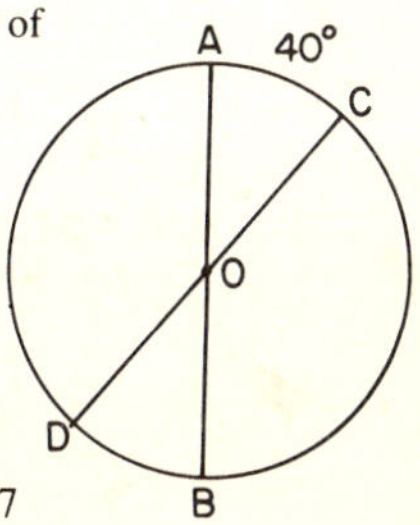

12. Factor: $x^2 + 5x - 6$

13. Solve for *x*: $0.4x - 1.7 = 0.7$

14. Solve for *x*: $\frac{x}{3} - 2 = 10$

15. In the accompanying diagram, triangle *ABC* is similar to triangle *XYZ*, with $\angle A \cong \angle X$, $\angle B \cong \angle Y$, and $\angle C \cong \angle Z$. If $AB = 35$, $XY = 7$, $BC = 10$, and $YZ = 2$, how many times larger is side *AC* than side *XZ*?

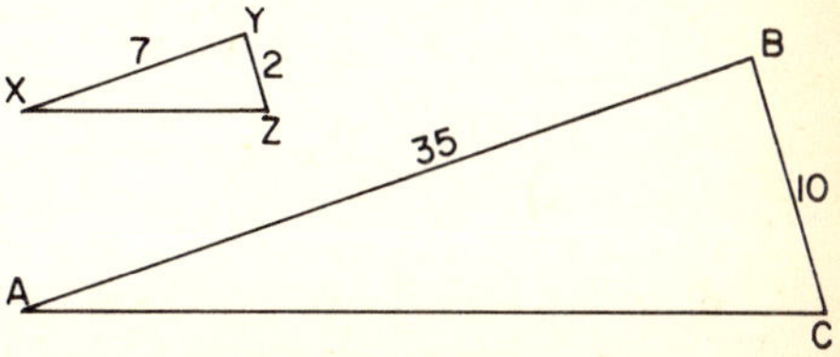

16. In the accompanying diagram, $\triangle ABC$ is a right triangle with the right angle at C. If $AB = 10$ and $BC = 6$, find AC.

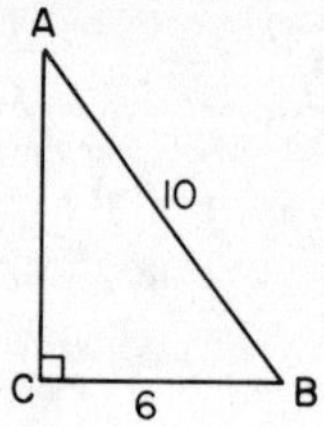

17. Write, in symbolic form, using p and q, the converse of $\sim p \rightarrow \sim q$.

18. What is the slope of the graph of the equation $y = -x + 1$?

19. From $3x^2 + x - 2$ subtract $x^2 - 2x + 3$.

20. Find the positive root of the equation $x^2 - 25 = 0$.

21. Two complementary angles are in the ratio of 8:1. Find the number of degrees in the measure of the smaller angle.

22. In the accompanying diagram, $\angle ACD$ is an exterior angle of triangle ABC. If $m\angle A = 40$ and $m\angle B = 60$, find $m\angle ACD$.

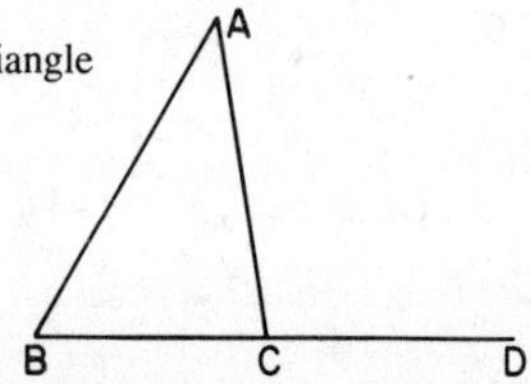

Directions (23–35): For *each* question chosen, write on the separate answer sheet the *numeral* preceding the word or expression that best completes the statement or answers the question.

23. The statement, "x is odd and x is greater than 2," is true when x equals
(1) 1 (2) 2 (3) 3 (4) 4

24. Let s represent "You score at least 65," and let p represent "You pass the exam." Which is the symbolic representation of the statement, "If you do not score at least 65, then you do not pass the exam"?
(1) $\sim s \wedge \sim p$ (2) $\sim s \rightarrow \sim p$ (3) $s \rightarrow p$ (4) $\sim p \rightarrow \sim s$

25. If n represents an odd integer, which represents the next larger consecutive odd integer?
(1) $n - 1$ (2) $2n$ (3) $n + 2$ (4) $n + 1$

26. For the set of data, 9, 9, 10, 11, 16, which statement is true?
(1) mean > median (2) mean < mode (3) median < mode (4) mean = mode

27. What are the coordinates of the y-intercept for the equation $y + 3x = 6$?
(1) (0,6) (2) (0, −6) (3) (0,3) (4) (0, −3)

28. The value of 6! is
(1) 6 (2) 30 (3) $\frac{1}{6}$ (4) 720

29. The circumference of a circle is 20π. What is the area of the circle?
(1) 10π (2) 20π (3) 100π (4) 400π

30. Given four geometric figures: a square, a rectangle, a trapezoid, and a circle. If one figure is selected at random, what is the probability that the figure has four right angles?
(1) $\frac{1}{4}$ (2) $\frac{1}{2}$ (3) $\frac{3}{4}$ (4) 0

31. Which are factors of $15y^2 - 5y$?
(1) $5y - 1$ and $3y + 5$ (3) $5y$ and $3y$
(2) $5y$ and $3y - 1$ (4) $5y - y$ and $3y + 5$

32. A triangle has a base of 12 centimeters and an area of 24 square centimeters. What is the height of the triangle?
(1) 6 cm (2) 2 cm (3) 3 cm (4) 4 cm

33. Which inequality is equivalent to $2x - 1 > 5$? (1) $x > 6$ (2) $x > 2$ (3) $x < 3$ (4) $x > 3$

34. The sum of $6\sqrt{6}$ and $\sqrt{54}$ equals (1) $3\sqrt{6}$ (2) $6\sqrt{60}$ (3) $9\sqrt{6}$ (4) $15\sqrt{6}$

35. In the truth table below, which is a correct heading for column III?
(1) $p \rightarrow q$ (2) $p \leftrightarrow q$ (3) $p \wedge \sim q$ (4) $\sim p \vee q$

Column I	Column II	Column III
p	q	?
T	T	T
T	F	F
F	T	F
F	F	T

Part II

Answer four questions from this part. Show all work unless otherwise directed.

36. Solve the following system of equations graphically and check: $x + y = 1$, $2x - y = 8$ [8,2]

37. Find the three largest consecutive integers whose sum is less than 86. [*Only an algebraic solution will be accepted.*] [5,5]

38. The frequency histogram below shows the distribution of scores on a math test.

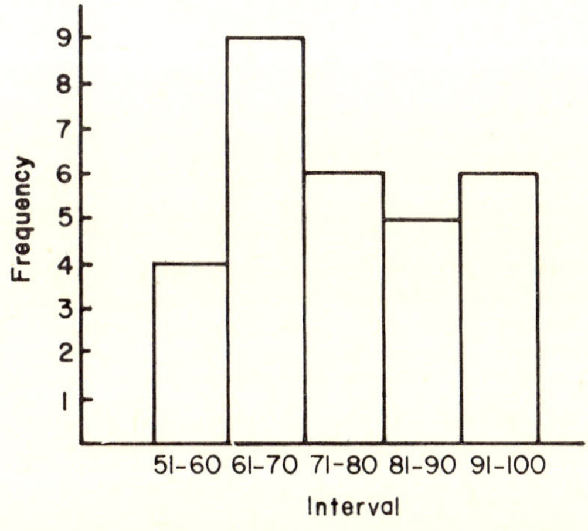

a. *On your answer paper*, copy and complete the table. [2]
b. How many students took the math test? [2]
c. How many students scored above 80? [2]
d. Using the table completed in part *a*, draw a cumulative frequency histogram. [4]

Scores	Frequency	Cumulative Frequency
51-60		
61-70		
71-80		
81-90		
91-100		

39. One positive number is 5 more than another. The sum of their squares is 53. Find both numbers. [*Only an algebraic solution will be accepted.*] [5,5]

40. In the accompanying diagram, *ABCD* is a rectangle and *AGE* is an isosceles triangle with $AG = EG$, $\overline{GF} \perp \overline{AD}$, *E* is the midpoint of $\overline{AD}$, $AF = FE$, $AB = 8$, and $AD = 24$.

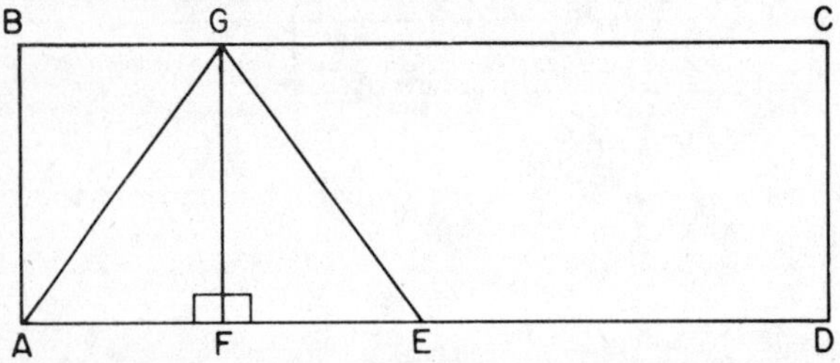

a. What is the length of $\overline{AE}$? [2]
b. What is the length of $\overline{AF}$? [2]
c. What is the length of $\overline{AG}$? [2]
d. What is the area of triangle *AGE*? [2]
e. What is the area of trapezoid *ADCG*? [2]

41. The letters *A, E, N, T* are written on four individual cards and placed in a container. Each has an equal likelihood of being drawn. One card is drawn from the container, the letter noted, and then the card is returned to the container. A second card is drawn and the letter noted.
a. Make a tree diagram or list the sample space showing all possible outcomes after both drawings. [4]
b. Find the probability that: (1) both letters drawn are the same [2]
(2) the first letter drawn is *A* and the second letter is *T* [2]
(3) for both drawings the letter *N* does not appear [2]

42. *a.* *On your answer paper*, copy and complete the truth table for the statement
$[(p \rightarrow q) \wedge \sim p] \rightarrow \sim q$. [8]

p	q	$p \rightarrow q$	$\sim p$	$\sim q$	$(p \rightarrow q) \wedge \sim p$	$[(p \rightarrow q) \wedge \sim p] \rightarrow \sim q$
T	T					
T	F					
F	T					
F	F					

b. Using your results from part *a*, is $[(p \rightarrow q) \wedge \sim p] \rightarrow \sim q$ a tautology? [1]
c. Justify the answer you gave in part *b*. [1]

INDEX